SLIDING FILAMENT MECHANISM IN MUSCLE CONTRACTION

Fifty Years of Research

ADVANCES IN EXPERIMENTAL MEDICINE AND BIOLOGY

Recent Volumes in this Series

Volume 557
BRAIN REPAIR
Edited by M. Bähr

Volume 558
DEFECTS OF SECRETION IN CYSTIC FIBROSIS
Edited by Carsten Schultz

Volume 559
CELL VOLUME AND SIGNALING
Edited by Peter K. Lauf and Norma C. Adragna

Volume 560
MECHANISMS OF LYMPHOCYTE ACTIVATION AND IMMUNE REGULATION X: Innate Immunity
Edited by Sudhir Gupta, William Paul, and Ralph Steinman

Volume 561
CHEMISTRY AND SAFETY OF ACRYLAMIDE IN FOOD
Edited by Mendel Friedman and Don Mottram

Volume 562
CHOLINERGIC MECHANISMS
Edited by José Gonzalez-Ros

Volume 563
UDPDATES IN PATHOLOGY
Edited by David C. Chhieng and Gene P. Siegal

Volume 564
GLYCOBIOLOGY AND MEDICINE
Edited by John S. Axford

Volume 565
SLIDING FILAMENT MECHANISM IN MUSCLE CONTRACTION: Fifty Years of Research
Edited by Haruo Sugi

SLIDING FILAMENT MECHANISM IN MUSCLE CONTRACTION

Fifty Years of Research

Edited by

Haruo Sugi

Teiko University
Tokyo, Japan

Springer

Proceedings of the 2004 Tokyo Muscle Symposium, "Sliding Filament Mechanism in Muscle Contraction: Fifty Years of Research," held in Tokyo, Japan, March 7–10, 2004.

ISSN: 0065 2598
ISBN-10: 0-387-24989-3 (Hardbound)
ISBN-13: 978-0387-24989-6 Printed on acid-free paper

Printed in the United States of America (EB)

9 8 7 6 5 4 3 2 1

springeronline.com

PREFACE

Fifty years have passed since the monumental discovery that muscle contraction results from relative sliding between the thick filaments, consisting mainly of myosin, and the thin filaments, consisting mainly of actin (A.F. Huxley and Niedergerke, Nature 173: 971-973, 1954; H.E. Huxley and Hanson, Nature 173: 973-976, 1954). Until the early 1970's, considerable progress have been achieved in the research field of muscle contraction. For example, A.F. Huxley and his coworkers put forward a contraction model, in which the myofilament sliding is caused by alternate formation and breaking of cross-links between the cross-bridges on the thick filament and the sites on the thin filament, while biochemical studies on actomyosin ATPase reactions indicated that, in solution, actin and myosin also repeat attachment-detachment cycles. Thus, when a Cold Spring Harbor Symposium on the Mechanism of Muscle Contraction was held in 1972, most participants felt that the molecular mechanism of muscle contraction would soon be clarified, at least in principle.

Contrary to the above "optimistic" expectation, however, we can not yet give a clear answer to the question, "what makes the filaments slide?" Of course, muscle investigators have continuously made every effort to progress their respective research work towards the understanding of muscle contraction mechanisms. Based on the idea that muscle contraction is best studies by using specimens in which the three-dimensional myofilamant-lattice structure is preserved, the author has organized six symposia on muscle contraction, each followed by publication of the Proceedings:

1. Cross-bridge Mechanism in Muscle Contraction, 665pp., University of Tokyo Press, Tokyo (1979)
2. Contractile Mechanism in Muscle, 921pp., Plenum Press, New York and London (1984)*
3. Molecular Mechanism of Muscle Contraction, 742pp., Plenum Press, New York and London (1986)
4. Mechanism of Myofilament Sliding in Muscle Contraction, 866pp., Plenum Press, New York and London (1993)
5. Mechanisms of Work Production and Work Absoprtion in Muscle, 663pp., Plenum Press, New York and London (1998)
6. Molecular and Cellular Aspects of Muscle Contraction, 701pp., Kluwer Academic/Plenum Publishers, New York, Boston, Dordrecht, London, Moscow

*Symposium held as US-Japan Cooperative Seminar organized by H. Sugi and G.H. Pollack.

Except for the first proceedings, the second to the sixth proceedings are published as series of Advances in Experimental Medicine and Biology (No. 170, 226, 332, 453 and 538).

In each publication, I always intended to include records of burning discussions among participants during the meeting, in order that the proceedings could be made as informative and attractive as possible. Fortunately, these proceedings have been widely regarded as milestones showing the progress of research work achieved at the time of each meeting.

This volume presents the proceedings of the symposium, which the author organized in celebration of the fiftieth anniversary of the discovery of the myofilament sliding in muscle contraction. In this connection, it was my great pleasure that I could have Professor Hugh E. Huxley, who was mainly responsible for the monumental discovery, in this meeting. At the beginning of this volume, he describes his exciting experiences during the course of his monumental discovery. The readers of this volume would realize a number of mysteries in this research field still remain despite the extensive studies accumulated in recent fifty years. Indeed, "Nature is much wiser than human being." I heartily hope that this volume would stimulate young investigators to start challenging the molecular mechanism of muscle contraction.

Finally, the author would like to express his thanks to Drs. J.A. Rall, J.W. Squire and S. Winegrad for their generous help in editing the discussion records, and to Mrs. Y. Suzuki for typing manuscripts.

Haruo Sugi

ACKNOWLEDGEMENT

The editor would like to express sincere thanks to the Uehara Memorial Foundation for Life Science and the Inoue Foundation for Science for generous financial support, which made this meeting possible.

The editor also owes a debt of gratitude to Dr. Kazutaka Kobayashi, Ms. Yuka Suzuki and Ms. Ibuki Shirakawa at Teikyo University for their enormous efforts in preparing the discussion records in this volume, and Drs. Teizo Tsuchiya, Takenori Yamada, Shigeru Chaen, Seiryo Sugiura and Kaoru Katoh for their help in carrying out this meeting.

CONTENTS

I. SLIDING FILAMENT MECHANISM AT THE MOLECULAR LEVEL

II. MECHANISM OF MUSCLE CONTRACTION

III. MUSCLE ENERGETICS

IV. REGULATORY MECHANISM AND E-C COUPLING

V. CARDIAC AND SMOOTH MUSCLE

VI. OTHER ASPECTS

I. SLIDING FILAMENT MECHANISM AT THE MOLECULAR LEVEL

EARLY DEVELOPMENTS IN MUSCLE RESEARCH AND THE ROLE OF NEW STRUCTURAL TECHNOLOGIES

H.E. Huxley[1]

1. INTRODUCTION

A few years after the end of World War II, in a small Medical Research Unit located, somewhat anomalously, in the Cavendish Laboratory (a physics laboratory) in Cambridge, England, a number of scientific discoveries were made which had far-reaching effects on the subsequent course of biological research. The best known of these was of course the proposal of the double helical base-paired structure of DNA, by Watson and Crick, in 1953. But in that same year Max Perutz, the head of the Unit, discovered how to determine the phases of X-ray reflections from protein crystals and thereby how to solve the atomic structure of protein molecules, which he and John Kendrew proceeded to do for haemoglobin and myoglobin during the next few years. This paved the way for the tens of thousands of different detailed protein structures which have now been determined, and, together with the basic knowledge of life processes that flowed from understanding of how DNA functions, has revolutionized biology and medicine in the last half century.

A little earlier than these momentous discoveries, during the years from 1949 through 1952, some other new discoveries were made in the same small unit, ones of a more specialized nature, but ones which did begin to set a new direction for work on the nature of muscle contraction. These discoveries flowed from the application of the same concepts as the two very dramatic ones mentioned above.

These concepts were extremely simple, almost simple-minded ones. The first was that it was very important to know what was the detailed structure of biological molecules and tissues, in the belief that information at the molecular and atomic level would be the essential key to understanding how all these biological processes worked, and that anything else was mere speculation. The second followed directly from this, and it was that methods had to be developed and applied in biology to carry out such structural work, and that a laboratory where this was done had to equip itself to carry out that development on the premises.

The professor of physics who headed the Cavendish Laboratory at that time was Sir Lawrence Bragg, who had invented X-ray diffraction analysis (and received the Nobel Prize for it at the age of 24), and it was he who had supported Perutz for many

[1] H.E. Huxley, Rosenstiel Center (MS-029), Brandeis University, 415 South St., Waltham, MA 02454-9110 USA

years, and who was instrumental in setting up the MRC Unit, run by Perutz and Kendrew, where I was a research student, and where Crick and Watson solved the structure of DNA. The Cavendish Laboratory had a very strong experimental tradition, exemplified in the Pt II Physics Practical Glass, where we used gold-leaf electroscopes to measure radioactive decay and α-particle ranges, learned to blow our own glassware and make cathode ray tubes, and repeat Millikan's oil drop determination of electronic charge. There was a large well-equipped mechanical workshop in the Cavendish, and a small student's workshop which our group used, with plenty of odd pieces of metal sheet and tubing, and a mechanic to help us use drills and lathes. And Max always stayed close to the bench!

So when the MRC Laboratory of Molecular Biology was set upon its own a few years later, an essential feature was the provision made for technological development, with a magnificently equipped and staffed mechanical workshop, a large electronic workshop, and smaller workshops elsewhere in the lab. This recognized that the structural work would require the highest level of technical support. The success of the laboratory testified to the strength of that approach.

2. EARLY MUSCLE WORK

For my own part, fascinated by experimental atomic physics, very conversant with the importance of knowledge of atomic structure in understanding so many of the properties of matter, it was a natural and easy transition to accept that detailed structural information was essential to even begin to understand biological processes. Vague theories were no good. And it was clear that new techniques were now waiting to be exploited.

X-ray diffraction was hardly new, but its use in biology was at that time quite limited. However, some years earlier Bernal had discovered that detailed diffraction patterns could be obtained from protein crystals if -- and only if – they were kept in a fully 'native' environment, hydrated, and in their mother-liquor. This was the foundation of protein crystallography. Another pioneer, Astbury, had looked at the wide-angle X-ray diagram of muscle, and found that there was no discernible change in it during contraction, indicating that the basic polypeptide chain configurations remained the same. So when I first started thinking about how muscles might contract, I got the idea that there must be larger structural units, protein molecules or assemblies of protein molecules, still way beyond the resolution of the light microscope, which interacted with each other and re-arranged themselves in some way so as to cause the muscle to shorten. To see this type of structure, which I thought might show structural repeats upward from fifty to a few hundred angstroms in size, I would need a low-angle X-ray camera, since the reflections would lie within 1° or less of the direct beam, and that might be why they had not been noticed previously.

Also, such cameras had to employ very narrow slits to collimate the X-ray beam, and usually needed to be quite long, to allow the pattern to spread out sufficiently. This meant that the total X-ray flux tended to be small, and the flux per unit area at the detector – film – was even smaller. Furthermore, I thought it was imperative to examine muscle in its native hydrated state, rather than dried down into a more concentrated state, as had been done in some previous work. So all this meant that I would only get extremely weak patterns and have impossibly long exposure times – even some protein crystals were then needing up to a month -- unless I could drastically increase the camera speed.

At that time, in the year 1949, there was a lot of interest at the laboratory in increasing the intensity available from X-ray tubes. One way to do this was to use a rapidly rotating anode, since the basic limitation was the rate at which heat could be dissipated from the incident electron beam, to avoid melting the copper target. This could be partly overcome by continuously presenting to the beam a fresh copper surface which had had time to cool during the cycle of rotation since its last exposure. Such machines were not easy to construct, since they operated in vacuums, with rotating seals, usually at 40,000 volts potential difference between anode and cathode, needed relatively large currents, and had to have water cooling. The ones which had been built previously were too unreliable for routine use. So a good electronic engineer was hired to construct a usable device, but it became clear that it would not be ready soon enough to help my thesis work.

Another way of increasing the available x-ray flux per unit area emitted by the target was to use a much smaller focal spot for the electron beam. Cooling of the irradiated area takes place laterally around the periphery of the spot, as well as vertically, into the depth of the target. So cooling is much more efficient for smaller or narrower spots, and the permissible flux per unit area in fact increases approximatly linearly as the inverse of the diameter of the focus. Ehrenberg and Spear, in Bernal's lab, in London, had built such a microfocus tube, in part for other reasons, and I was fortunate to obtain a prototype through my supervisor, John Kendrew, who had been a wartime colleague of Bernal (and who had been drawn into protein crystallography by him). This tube operated with a 50μ spot, which would be effectively forshortened to 5μ using a shallow viewing angle (5°). So I had an extremely bright, very narrow source, ideal for low angle diffraction, and was able to construct a camera with correspondingly narrow slits and only a few centimeters specimen to film distance which still gave me order-to-order resolution of several thousand angstroms, and a first order resolution well over 500 Å with relatively high (then!) recording speed. The patterns, recorded on film, needed to be viewed through a low-power microscope.

Looking for equatorial reflections in living muscle from filaments whose presence was indicated by very early electron micrographs, I soon found a clear set of reflections coming from a hexagonal array of filaments, spaced out about 400Å apart (Huxley, 1951). Moreover, in muscles in rigor, a second set of filaments seemed to be present, arranged in a regular pattern symmetrically between each set of three of the original filaments. Since it was known then that the muscle proteins actin and myosin seemed to form some kind of complex in the absence of ATP (the condition in a rigor muscle), and to dissociate in its presence, I concluded – an inspired guess, I suppose – that myosin and actin must be present in separate, stable filaments in muscle, the myosin filaments forming a permanent regular hexagonal array, and the actin filaments becoming regularly positioned in the array when they became attached by cross-linkages to the myosin filaments, who centers would be about 250Å away from those of the actin filaments. I presumed that it was interaction through those cross-links that produced muscle contraction, but because I assumed that this double array extended continuously through each sarcomere of the muscle, I did not envisage a sliding filament mechanism at this time. Instead, I wondered if actin depolymerization might be involved, because, even then, such depolymerization was recognized as a possibly biologically important process.

Additionally, I found that there was a clear set of axial reflections (I only had a slit camera), with a period of about 415Å – only approximate because the patterns I could record were so small – and a very strong third order at around 140Å. Remarkably, when a living muscle was passively stretched, this periodicity remained constant! (Huxley, 1953a) So I speculated in my thesis that the periodicity must come from one of the sets of filaments which was not attached to the Z-lines, probably the actin filaments, using the myosin filaments to transmit the force generated by their depolymerization during contraction!

3. WORK AT M.I.T.

The next step was first to find out whether my double array model was correct and then look for further evidence of how such a system might work. The best way to do this seemed to be electron microscopy, just beginning to be used in biology in a few labs in the world, one of which was F.O. Schmitt's lab at the Massachusetts Institute of Technology, where Dick Bear and Cecil Hall also worked. I was lucky enough to get a nice two-year fellowship (Commonwealth Fund) to do this, and arrived there in September 1952. Alan Hodge was also there as a postdoctoral fellow, from Australia, and he taught me how to operate the electron microscope. Together with Dave Spiro, we designed a simple microtome for ultra-thin sectioning (Hodge et al, 1953), just coming into use then, which Alan and Dave used for their own projects and I used to look at cross-sections of muscle, for my double array. I was very thrilled to soon find I could see it. The thicker filaments formed the basic hexagonal array, as I had supposed, and presumably contained myosin, since that was the major protein species, and the thinner filaments were actin, located at the trigonal positions of the lattice, as expected from the X-ray patterns. This convinced me that the combination of the two techniques was a very powerful tool indeed, and this became one of the main themes of the Structural Studies Division which Aaron Klug and I later directed at the much enlarged MRC Laboratory in Cambridge.

By Christmas 1952 I was ready to move on with this work, and by great good fortune, Jean Hanson, from the MRC Laboratory at King's College London, then arrived at M.I.T., also to learn electron microscopy. One of the specialties of the King's Lab had been different types of light microscope, including the phase-contrast light microscope. Jean was a zoologist, and had already studied a range of muscle types, but she had obtained particularly striking phase-contrast light micrographs of separated myofibrils from vertebrate striated muscle (rabbit psoas), which showed the sarcomere band pattern extremely clearly. This was a revelation to me, since I had never seen the muscle band pattern in the light microscope before, though of course I had seen it in electron micrographs. Jean was equally excited to see my EM and X-ray results, and we immediately decided to join forces, and to work together using both phase and electron microscopy.

4. MYOSIN FILAMENTS IN THE A-BAND

At that time it was generally assumed that the characteristic high density of the A-bands in striated muscle was due to the presence of some additional component other than actin and myosin. Smooth muscles contracted perfectly well, if more slowly, without such striations, and since sarcomeres could shorten down to much below the A-band length, it was assumed that filaments of the actin-myosin complex must extend continuously from one Z-line to the next. So perhaps this extra A-substance just enabled muscles to shorten faster?

We were absolutely astounded, therefore, when, in the phase contrast light microscope, we saw that myosin-extracting solutions removed the A-substance, but left behind a 'ghost' fibril, with a band of density on either side of each Z-line, extending in towards where the boundaries of the H-zone had originally been. (H-zone is a less-dense region in the center of the A-band). Within a day we realized what the explanation must be, and were soon able to confirm this by electron microscopy. Myosin-extracting solutions removed the thick filaments, leaving behind the sets of thin filaments (attached to the Z-lines), which had previously partially overlapped the thick filaments. When I had seen the double array in an end-on view, I had been looking at sections through the overlap region of the sarcomere, and what I thought was poor preservation or sectioning were just the I-regions!

This was a remarkable finding, which we published in Nature that year (Hanson & Huxley, 1953), though it took a long time to come out. We were told by Schmitt that we should not contaminate a perfectly good experimental paper with any speculation about mechanisms until we had further evidence. However, I did manage to slip some phrases to the effect that my constant X-ray axial periodicity, plus the overlapping arrays, suggested a sliding mechanism, into my write-up of the electron microscope cross-sectioning results (Huxley 1953b).

By January of 1954 we had good phase microscope data showing that the A-bands remained essentially constant in length during contraction of isolated myofibrils in ATP, and that the arrays of thin filaments, also of constant length, were drawn further and further into the A-bands as contraction proceeded. And so the sliding filament model was published in Nature in May of 1954, in two papers side by side, one from us, and one from A.F. Huxley and Niedergerke, who had been making similar observations on intact muscle fibers in the interference light microscope (Huxley & Hanson, 1954; A.F. Huxley and Niedergerke, 1954).

Subsequent development of the work took place in England again, and continued to be very dependent on new technical developments.

5. DEVELOPMENTS IN ELECTRON MICROSCOPY

Some of these were fairly straightforward. Thin sectioning for electron microscopy was still a very new technique, and I only slowly realized that most sections were over a thousand angstroms in thickness, and relatively lightly stained, so that, in longitudinal sections of muscle, several layers of filaments were superposed, obscuring the filament arrangement. By various incremental improvements of the microtome and of specimen preparations, I managed to reduce the minimum section thickness to 100-150Å, and to have sufficiently intense metal staining for single filaments to show up with good contrast. Remarkably, the double hexagonal lattice was often preserved with great regularity, so that single filament layers could be seen with thick and thin filaments lying side by side in the expected arrangement, within the thickness of the section. This began to persuade people that the overlapping filament structure really did exist. But the sliding filament mechanism still took many years of work to gain more converts.

I was disappointed that the staining methods that I had developed for thin sections showed little internal detail of the structure of the actin and myosin filaments, although it was possible to see crossbridges very clearly. However, I had accidentally discovered the so-called 'negative staining' technique in some work I was doing on tobacco mosaic virus, as a sideline (Huxley, 1957b), and I improved it further by using uranyl acetate as a negative stain, in work on another virus, and on ribosomes (Huxley & Zubay, 1960a & 1960b). The technique could only be used on small isolated objects, which were submerged or outlined in the stain, but it occurred to me that muscle might fragment easily if mechanically blended in a relaxing medium, and that turned out to be the case. The product was a nice suspension of separated and often unbroken myosin and actin filaments, which showed up excellently in negative stain (Huxley, 1963). This enabled me to recognize that both types of filament were constructed with a defined structural polarity, appropriate for the directions in which relative sliding forces needed to be developed in an overlapping filament system. This was further strong evidence in support of such a mechanism, and it also prompted me to suggest that similar directed movements might be involved in other forms of cell motility. It also enabled Jean Hanson and Jack Lowy (1963) to elucidate the helical structure of F-actin.

6. X-RAY DEVELOPMENTS IN THE 1960'S

The major developments came next from the X-ray field. I was now back in the new MRC laboratory in Cambridge and had access again to rotating anode X-ray tubes, several of which were operating routinely there. Ken Holmes and Bill Longley had also moved to the laboratory from Birkbeck College, in London, and had made a big improvement to such a tube by grafting on the cathode from a Boudoin X-ray tube, which provided a much smaller focus and hence light brilliance, very appropriate for the focusing quartz monochromator they were using in studies on Tobacco Mosaic Virus. Ken and I thought it would be interesting to combine this with a focusing mirror I had previously been using in conjunction with a commercial version of the Ehrenberg-Spear tube. In the course of playing around with the rotating anode-mirror-monochromator set up, I discovered that it was possible to use the entire input aperture of the monochromator, rather than a narrow collimated region (as was conventional), without excessive background scattering, with an enormous increase in total X-ray intensity and all the advantages of a monochromatic beam.

This made it possible for the first time to record the low angle meridional and layer-line patterns from live, contracting frog muscles, to show that the axial spacings from both myosin and actin remained essentially constant between rest and contraction (Huxley, Brown & Holmes, 1965), and to show, amongst other things, that the myosin layer line reflections from the helical arrangement of crossbridges around the myosin filaments, became very much fainter during contraction, showing that the crossbridiges must move when they developed the sliding force between the filaments (Huxley and Brown, 1967).

Holmes and I were able to continue to improve the technology because of the excellent workshop facilities and technicians in the lab. In a successful effort to achieve higher X-ray output, we constructed a large diameter (about 20 ins) rotating anode X-ray set, which eventually developed into the commercial Elliott 'Big Wheel'. But we realized that this was essentially the end of the line as far as increases X-ray intensity from these types of sources was concerned, since we had reached the limit of what the mechanical strength and the melting point of copper would allow, and other metals would give X-rays with a much less suitable wavelength, and present much bigger fabrication problems. Nevertheless, we still needed greatly increases X-ray fluxes, since there was potentially so much detailed information available in the diagrams from contracting muscle, and we needed to be able to record them with high time resolution to follow the changing patterns satisfactorily.

7. SYNCHROTRON RADIATION

Ken Holmes had always been thinking about more exotic X-ray sources, particularly for use in his work on insect flight muscle, and had become interested in the radiation emitted by electron synchrotrons, particle accelerators used by physicists to produce very high velocity electrons for collision experiments. Initially, it appeared that the available machines would not produce enough X-rays to be useful sources, but later, after he had moved to Heidelberg, and had learned from Gerd Rosenbaum about the characteristics of the DESY synchrotron in Hamburg, things appeared more hopeful, and he, Rosenbaum, and Witz carried out the crucial test experiment there in 1971 (Nature, Rosenbaum, Holmes, & Witz, 1971). This showed that a substantial gain was available over the best that could ever be obtained from a rotating anode X-ray tube, and that potentially enormous factors of improvement might be possible in the future from the electron storage rings being planned, in which much larger, and continuous, circulating currents of electrons and positrons would be present.

In any event, it took nearly ten years more before all the component of a satisfactory system were available. It took some years before a storage ring came on line, and was operating smoothly enough (not very smoothly!) for the muscle enthusiasts to begin to collect some really useful data. In the meantime, we gained a lot of essential experience with electronic data collection and analysis, and with the operation of remotely controlled cameras, at the synchrotron NINA, in Daresbury, U.K. (because of the high flux of radiation in the immediate environment of such a beamline, remote operation was mandatory). Once again, in the MRC lab, we were very fortunate that nearly all the equipment could be designed by excellent electronics and mechanical engineers and put together in our workshops (Haselgrove et al, 1977; Faruqi & Bond, 1980).

And then, in the early 1980s, working at the EMBL Outstation, in Hamburg (specially built to exploit synchrotron radiation), we were able to obtain good time-resolved patterns of the myosin and actin layer-lines with 5-10 msec time resolution, so that the time course of the changes in them during the onset of contraction could be compared with the time course of tension development. They were very closely related as we expected. More importantly, in measurements of the meridional 145Å reflection from the myosin crossbridge repeat, now with 1 msec time resolution, we found that a large drop in intensity was closely synchronized with a small quick length decrease, or increase, applied to a previously isometrically contracting muscle. This strongly indicated a tilting movement of actin-attached crossbridges, if they were approximately perpendicularly oriented in the isometric muscle, and provided the first direct demonstration of a direct relationship between crossbridge configuration and tension generation (Huxley et al 1981, 1983).

8. RECENT ADVANCES

Later development of new structural techniques in other laboratories have led to further remarkable extensions of our ability to obtain direct information about molecular motility processes. Perhaps the most striking of these has been the introduction of in vitro motility measurements in which force and movement can be measured by direct optical means on single myosin molecules interacting with single actin filaments, as exemplified, for instance, by Finer et al (1994), following the pioneering work of Kron and Spudich (1986) and Kishino and Yanagida (1988).

Another great step forward depended on extensive computerization of protein crystallographic analysis, to make the solution of very large structures possible, plus the use of synchrotron radiation and cryo techniques to collect data from very small and sensitive protein crystals. This made possible the solution of the high resolution structure of myosin subfragment one, the motor part of myosin, by Rayment et al. (1983a & 1983b), a very great help to our understanding of the details of the tilting movement of myosin crossbridges in the sliding filament mechanism.

More recently still, the latest generation of electron-positron storage rings provide extremely small X-ray sources with very high total flux. These make it possible to record X-ray patters from muscle at extremely high spatial resolution and at a time resolution of 1 msec or better, making use of imaging plates and improved CCD X-ray detectors (Linari et al, 2000; Huxley et al, 2000). Such patterns enable axial crossbridge movements to be measured with an accuracy of a few angstrom units.

9. CONCLUSION

Over the last fifty years or so, then, it has been very remarkable, and encouraging, to see how the well-directed scientific efforts of many people and many laboratories have been so successful in providing the tools for what were originally almost unimaginable opportunities to explore molecular reality.

REFERENCES

Faruqi, A.R. and Bond, C.C. (1980). Parallel readout multiwire proportional chambers for time-resolved X-ray diffraction experiments. Nucl. Instrum. Methods 179, 71-77.

Finer, J. T., Simmons, R. M., and Spudich, J. A. (1994). Single myosin molecule mechanics: piconewton forces and nanometre steps. Nature 368, 113-119.

Hanson, J., and Huxley, H. E. (1953). The structural basis of the cross-striation in muscle. Nature, Lond. 172, 530-532.

Hanson, J., and Lowy, J. (1963). The structure of F-actin and actin filaments isolated from muscle. J. Mol. Biol. 6, 46-60.

Haselgrove, J. C., Faruqi, A. R., Huxley, H. E., and Arndt, U. W. (1977). The Design and Use of a Camera for Low Angle X-ray Diffraction Experiments with Synchrotron Radiation. J. Physics 10, 1035-1044.

Huxley, A. F., and Niedergerke, R. (1954). Structural changes in muscle during contraction. Interference microscopy of living muscle fibres. Nature (London) 173, 971-973.

Huxley, H. E. (1951). Low-angle X-ray diffraction studies on muscle. Disc. Faraday Soc. 11, 148.

Huxley, H. E. (1953a). X-ray diffraction and the problem of muscle. Proc. Roy. Soc. B. 141, 58-63

Huxley, H. E. (1953). Electron-microscope studies of the organization of the filaments in striated muscle. Biochem. Biophys. Acta 12., 387-394.

Huxley, H. E., and Hanson, J. (1954). Changes in the cross-striations of muscle during contraction and stretch and their structural interpretation. Nature, Lond. 173, 973-976.

Huxley, H. E. (1957a). The double array of filaments in cross-striated muscle. J. Biophys. & Biochem. Ctyol. 3, 631-648.

Huxley, H. E., and Zubay, G. (1960). Electron microscope observations on the structure of microsomal particles from Escherichai Coli. J. Mol. Biol. 2, 10-18.

Huxley, H. E., and Zubay, G. (1960). The structure of the protein shell of turnip yellow mosaic virus. J. Mol. Biol. 2, 189-196.

Huxley, H. E. (1963). Electron microscope studies on the structure of natural and synthetic protein filaments from striated muscle. J. Mol. Biol. 7, 281-308.

Huxley, H. E., Brown, W., and Holmes, K. C. (1965). Constancy of axial spacings in frog sartorius muscle during contraction. Nature 206, 1358.

Huxley, H. E., and Brown, W. (1967). The low angle x-ray diagram of vertebrate striated muscle and its behaviour during contraction and rigor. J. Mol. Biol. 30, 383-434.

Huxley, H. E., Reconditi, M., Stewart, A., and Irving, T. (2000). Interference changes on the 14.5nm reflection during rapid length changes. Biophys. J. 78, 134A.

Kishino, A., and Yanagida, T. (1988). Force measurements by micromanipulation of a single actin filament by glass needles. Nature 334, 74-76.

Kron, S. J., and Spudich, J. A. (1986). Fluorescent actin filaments move on myosin fixed to a glass surface. Proc. Nat. Acad. Sci. 83, 6272-6276.

Linari M, Piazzesi G, Dobbie I, Koubassova N, Reconditi M, Narayanan T, Diat O, Irving M, Lombardi V. (2000) Interference fine structure and sarcomere length dependence of the axial x-ray pattern from active single muscle fibers. Proc Natl Acad Sci U S A. 97(13), 7226-31.

Rayment, I., Rypniewski, W., Schmidt-Base, K., Smith, R., Tomchick, D., Benning, M., Winkelmann, D., Wesenberg, G., and Holden, H. (1993). Three-dimensional structure of myosin subfragment-1: a molecular motor. Science 162, 50-58.

Rayment, I., Holden, H. M., Whittaker, M., Yohn, C. B., Lorenz, M., Holmes, K. C., and R.A, M. (1993). Structure of the actin-myosin complex and its implications for muscle contraction. Science 261, 58-65.

Rosenbaum, G., Holmes, K.C., and Witz, J., (1971). Synchrotron Radiation as a Source for X-ray Diffraction. (1971). Synchrotron Radiation as a Source for X-ray Diffraction. Nature 230, 434-437.

DISCUSSION

Gonzalez-Serratos: In your early experiments with Jean Hansen, you showed that, after extraction of myosin when only actin was left, the sarcomere decreased in length. How could they shorten without myosin?

Huxley: No. In the experiments I showed, myosin had been extracted after the contraction had taken place, so as to show more clearly the location of the I-segments in the shortened sarcomeres.

Pollack: In one of your EM slides showing two thin filaments between two thick filaments, we can see two kinds of bridges: between thick and thin and between thin and thin. We found the thin-thin connections even in the I-band (cf. Pollack, Muscles and Molecules, Ebner & Sons, Seattle, 1990). How do you interpret the thin-thin bridges that you see in the EM, and why do they also appear in the I-band?

Huxley: In the I-band, the thin-thin bridges may be some of the glycolytic enzymes which bind quite strongly to actin. In the A-band, they may also represent myosin crossbridges from myosin filaments above and below the actual plane of sectioning.

THE MOLECULAR BASIS OF CROSS-BRIDGE FUNCTION

Kenneth C. Holmes

1. INTRODUCTION

Our understanding of the physiology of muscle depends critically on the resolution of the available anatomy. Early insight was provided by light microscopy. However, the first radical new insight was provided by electron microscopy. Ultimately, an understanding in physicochemical terms is only possible if the structures of the components in various physiological states are known at atomic resolution. Some of these have become known in the last 15 years and now allow us to describe how the hydrolysis of ATP by the component proteins actin and myosin leads to movement.

HE Huxley was able to show that the filaments in the sarcomere were organised on a hexagonal lattice [1, 2]. The seminal works of HE Huxley and Jean Hanson [3] and AF Huxley and Niedergerke [4] showed that the two sets of filaments in the sarcomere glide over each other without altering their length. Through the work of HE Huxley, Jean Hanson and W. Hasselbach it was discovered that the thick filaments contain myosin and the thin filaments contain actin. The myosin molecule consists of two heavy chains and four light chains. A soluble proteolytic fragment of myosin, heavy mero-myosin that contains the 2 globular "heads" of myosin carries the ATP-ase activity [5], the rest of the molecule forming a long double α-helical coiled-coil involved in filament formation. The ATP-ase activity was later shown to reside in the "head" fragment (sub-fragment-1 or S1)[6-8] that constitutes the cross-bridge (see below).

Max Planck Institut für medizinische Forschung, 69210 Heidelberg, Germany

2. CROSS BRIDGE THEORIES

2.1 The swinging cross-bridge

After the discovery of the sliding filaments the question naturally arose; what made them slide? AF Huxley [9] argued that the source of the force must be independent elemental force generators since the force increases linearly with the degree of filament overlap. The myosin cross bridges were first visualised by electron microscopy [10, 11] and subsequently shown both to be the site of the ATP-ase and also to be the motor elements producing force and movement between the filaments. Two conformations of the cross-bridge could be detected in insect flight muscle by low angle X-ray scattering and electron microscopy [12]. The cross bridge attaches to the actin filament at about 45° in rigor and at right angles (90°) in the presence of ATP.

On the basis of the known structural data and their kinetic analysis Lymn and Taylor proposed the cross-bridge cycle [13, 14] (Fig. 1). The cross bridge was thought work by a kind of rowing action. Initially it would bind to actin in a 90° conformation, go over to an angled (45°) conformation and then release the products of hydrolysis. The rebinding of Mg-ATP rapidly dissociates the actin-myosin; myosin then hydrolyzes ATP and forms a stable myosin-products complex; actin recombines with this complex and dissociates the products thereby forming the initial actin-myosin complex that isomerises to the rigor complex. Force is generated during the last step.

The actual rowing movement could be measured by physiological experiments on contracting muscle and was shown to be about 100Å[15]. Since the cross-bridge was an elongated structure, such a distance could be accommodated by a rotating or swinging cross-bridge model (Fig 1a). Studies of the cross-bridge movement were undertaken by time resolved studies of contracting frog muscle using low angle x-ray fibre diffraction [16, 17]. These results are fully consistent with the swinging cross-bridge theory.

2.2 Swinging lever arm

Although the swinging cross bridge hypothesis of muscle contraction had become the textbook norm by the time of the Cold Spring Harbor Conference on Muscle in 1972 it proved remarkably difficult to catch a bridge *in flagranti delicto* (Cooke [18]).

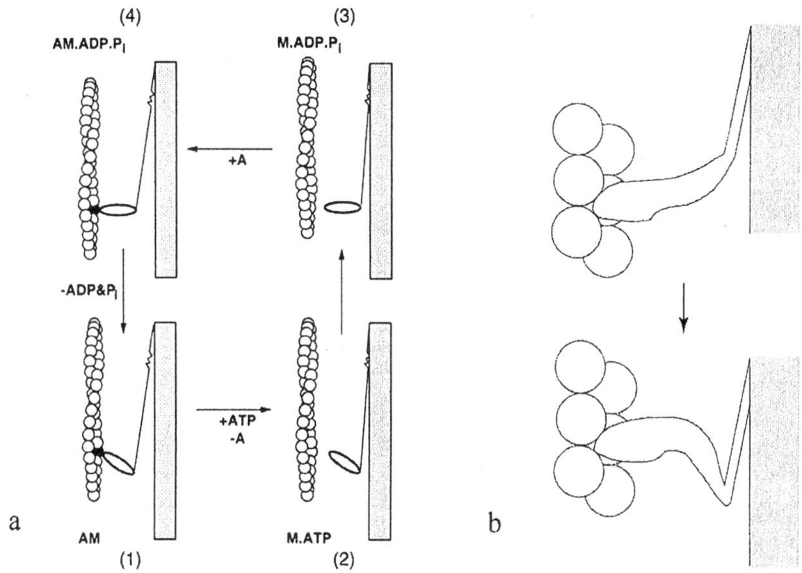

Figure 1. a The cross-bridge cycle from Lymn and Taylor [14]. The actin filament is represented as a double helix of spheres. The myosin cross-bridge is connected to the myosin thick filament (shaded). The cross-bridge movement is depicted as a hinge on the surface of actin. b. The swinging lever arm as proposed by Cooke [18]. The actin myosin interface has a constant geometry; the movement takes place in a lever arm distal to the actin filament.

Nevertheless, the swinging cross-bridge hypothesis provides by far the best framework for correlating and explaining the large muscle literature. The hypothesis has been modified over the years into a *swinging lever arm hypothesis* in which the bulk of the cross bridge is envisaged to bind to actin with a more or less fixed geometry and only the distal (C-terminal) part of the myosin molecule moves (Fig. 1b). A swinging lever arm explains why substantial changes in the cross bridge orientation were difficult to detect: only a small fraction of the cross-bridge mass moves. Furthermore, it became clear that the proportion of cross bridges in a muscle fibre taking part in a contraction was at any one time only a small fraction of the total, making the registration of active cross bridge movement doubly difficult.

The atomic structures of actin and myosin provided new impetus. The crystal structure of the myosin subfragment 1 [19] showed the myosin cross-bridge to have an extended C-terminal tail which looked like a lever arm and, moreover, a lever arm which was in the correct orientation and position to function as a lever arm [20]. In the last years large numbers of independent experiments provide results, which are in excellent accord with the idea that the C-terminal tail functions as a lever arm and indeed provide evidence that it can move. Purified myosin cross bridges (S1) can be attached to a substrate and used to transport actin filaments in vitro in the presence of ATP. A study by Spudich *et al* [21] showed that the speed of actin transport in motility assays was proportional to the length of the lever arm. This experiment has now been repeated a number of times on various myosins. Single molecule measurements come to similar conclusions. The bulk of the published data agree very well with the swinging lever arm model.

In addition, new crystal structures showed that the myosin cross bridge can exist in two orientations corresponding to the two end states of the power stroke.

3. PROTEIN CRYSTALLOGRAPHY

3.1 Structure of Actin

Actin (thin filament) fibres are helical polymers of g-actin (globular-actin) [22]. The structure of g-actin was first solved by protein crystallography as a 1:1 complex with the enzyme DNase I [23]. Orientated gels of f-actin yield X-ray fibre diagrams to about 6Å resolution. Holmes *et al* determined the orientation of the g-actin monomer that best accounted for the f-actin fibre diagram and thus arrive at first approximation to the atomic model of the actin filament[24]. Since a conformational change is involved in going from g- to f-actin, the g-actin structure has to be deformed in some way to fit the f-actin diffraction pattern. The various attempts at generating the f-actin structure that have been published have chosen various methods of defining these free parameters and have ended up with related, but different, solutions to the problem [24-26]. The method of Lorenz *et a l*[26] produces a very good fit to the fibre diffraction pattern but with a large number of free parameters and at the cost of poor stereochemistry. A new attempt to solve this problem has been made that is based on the sub-domain structure of actin and a small number of degrees of freedom [27]. The g-actin structure from Otterbein et al [28] was used as the starting structure.

3.2 Structure of Myosin

The cross-bridges comprise a part of the myosin molecule, namely subfragment-1 of heavy meromyosin (S1). X-ray crystallography [19] shows the chicken skeletal S1 to be tadpole-like in form (fig. 2) , with an elongated head, containing a 7-stranded β-sheet and numerous associated α-helices forming a deep cleft. The cleft separates two parts of the molecule, which are referred to as the upper 50K, and lower 50K domains. The C-terminal tail, sometimes called the "neck", which also provides the connection to the thick filament, forms an extended α-helix that binds two calmodulin-like light chains. The ATP binding site consists of a "P-loop" motive flanked by switch 1 and switch 2 elements, as are found in the G-proteins.

By fitting the atomic structures of f-actin and S1 into three dimensional cryo-electron microscope reconstructions one arrives at an atomic model of the actin myosin complex [20] (fig. 3). In particular, this model establishes the spatial orientation of the S1 myosin fragment in the active complex. One finds that the cleft in myosin extends from the ATP binding site to the actin binding site and that the opening and closing of this cleft is very likely to provide the communication between the ATP site and the actin binding site. The actin-binding site spans the upper and lower 50K domains. Furthermore, the very extended C-terminal α-helical neck of S1 is ideally placed to be a lever arm. The lever arm joins onto the bulk of the molecule via a small compact "converter domain" [29] which lies just distal to a broken α-helix containing two reactive thiol groups known as SH1 and SH2. Numerous experiments point to the putative "hinge" for the lever arm being in the SH1-SH2 region of the molecule (see [30] for review). The converter domain is the socket that carried the C-terminal helix (lever arm).

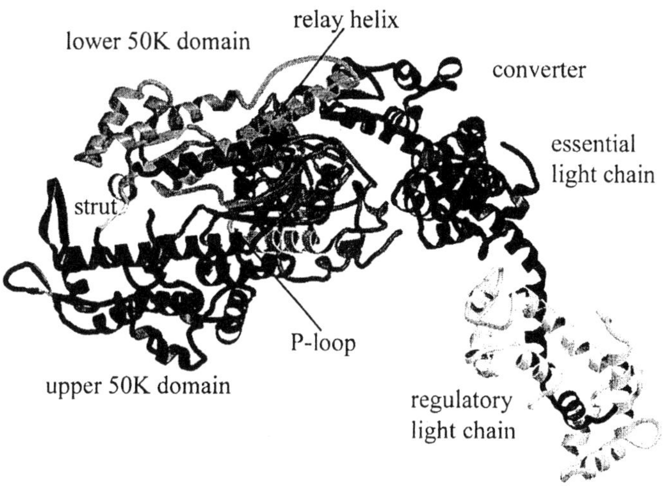

Figure 2 The structure of the myosin cross-bridge as a ribbon diagram[19]. The N-terminus is shown green and the nucleotide binding P-loop is shown yellow. The strut (yellow) connects the upper and lower 50K domains. Note the cleft separating the upper and lower 50k domains. The lower 50K domain is the primary actin-binding site. The upper and lower 50K domains are also connected by a disordered loop (not shown). The SH1 SH3 region lies underneath the right part of the relay helix. The C-terminal long helix (dark blue) carries two calmodulin-like light chains and joins on to the thick filament

3.3 Two main conformations of myosin have been discovered

According to the Lymn-Taylor scheme (fig. 1) the myosin cross bridge would be expected to have two discernible conformations: (1) when it first attaches to actin with the products of hydrolysis still bound with the lever at the beginning of the working stroke; and (2) at the end of the working stroke when the phosphate and ADP are released. This sequence is often referred to as the "power stroke". The end state is referred to as "rigor", since it is the state muscle enters on ATP depletion. It is also called "strong" because it binds to actin quite tightly. The initial state is called the "weak binding state" because of its low affinity for actin (see [31]), although it may be necessary to specify the pre-power stroke bound state as stereospecific weak binding.. We might anticipate that these two states of the myosin cross bridge might exist independently from actin and indeed protein crystallography shows this to be the case. The cross-bridge exists in two main conformations, the lever arm undergoes a 60° rotation between these two states, which have been identified as the beginning and end of the power stroke. The lever arm rotation is coupled with changes in the active site (the movement of the switch 2 element from closed to open) and to product release.

Figure 3 The pre (upper) and post (lower) power stroke states of the myosin cross bridge, as they would appear when attached to actin. The actin filament is shown on the right. These two states are referred to as CLOSED (upper) and OPEN (lower) because in the CLOSED state the switch 2 element has move in to close off the active site. This state is the ATP-ase. Note the large movement of the distal lever arm, which undergoes a 60° rotation between these two states,

3.3.1The OPEN state (post power stroke)

The first chicken S1 structure was solved without bound nucleotide. The chicken S1 crystal structure fits well into the electron micrograph reconstructions of the strong actin-myosin nucleotide-free interaction (decorated actin). Therefore the crystal structure of chicken S1 would appear to represent the end of the power stroke in a near rigor state. The switch 2 element lies away from the nucleotide binding pocket. Hence we refer to this state as the OPEN state. It has been found in the presence of a large variety of ATP analogues.

3.3.2 The CLOSED state (pre-power stroke)

Rayment *et al* have extensively studied a crystalline fragment of the *dictyostelium* myosin II cross-bridge which has been truncated after residue 761 (equivalent to 781 in chicken skeletal sequence). The truncation eliminates the lever arm and the associated light chains but retains the converter domain. The crystal structures of the 761 construct have been determined with a number of ATP analogs, particularly ADP.BeF_3 [32] and ADP.vanadate [33]. ADP.vanadate complexes are used as analogs of the transition state or possibly of the ADP.Pi state. While the ADP.BeF_3 state looks similar to rigor, the ADP.vanadate structure shows dramatic changes in shape of the S1 structure. There is a closing of the γ-phosphate binding pocket by moving switch 2 element about 5Å, This induces large movements in the C-terminal region of the molecule. The converter domain rotates through 60°. This new state has been called "CLOSED" (because the nucleotide pocket is closed) or "transition state" because it seems to be produced by transition state analogues of ATP.

Dominguez *et al*[34] have solved the structures of chicken smooth-muscle myosin truncated at 791 (smooth muscle sequence) or at 820 (expressed in insect cells using the baculovirus vector). The shorter construct stops at the end of the converter domain and the longer construct encompasses the essential light-chain binding site that is, the first half of the lever arm. The structures of both constructs have been solved as complexes with ADP.vanadate and ADP.BeF_3. Both structures show the myosin cross-bridge in the CLOSED form with the converter domain in the rotated position very similar to that obtained in the two *Dictyostelium* constructs. The authors refer to this state as the "pre-power stroke state". There are no substantial differences between ADP.AlF_4 complexes and ADP.BeF_3 complexes showing that the nature of the ligand (either a transition state analog or an ATP analog) does not control the protein conformation very closely. Since the smooth-muscle crystals display extensive non-crystallographic symmetry, CLOSED has now been obtained 16 times in a large variety of different environments: there is little chance that it arises as an artefact of crystal packing.

The two states are depicted in fig. 3 as they would appear if they were full length cross bridges attached to actin. (The missing lever arm has been added in). The lever arm moves some 12nm along the actin helix axis between the two states.

The movement of the lever arm and the status of the nucleotide-binding pocket are tightly coupled: pocket closed, lever up (beginning); pocket open, lever down (end). Only the CLOSED form is an ATPase. This is an essential control to make the cross-bridge cycle efficient.

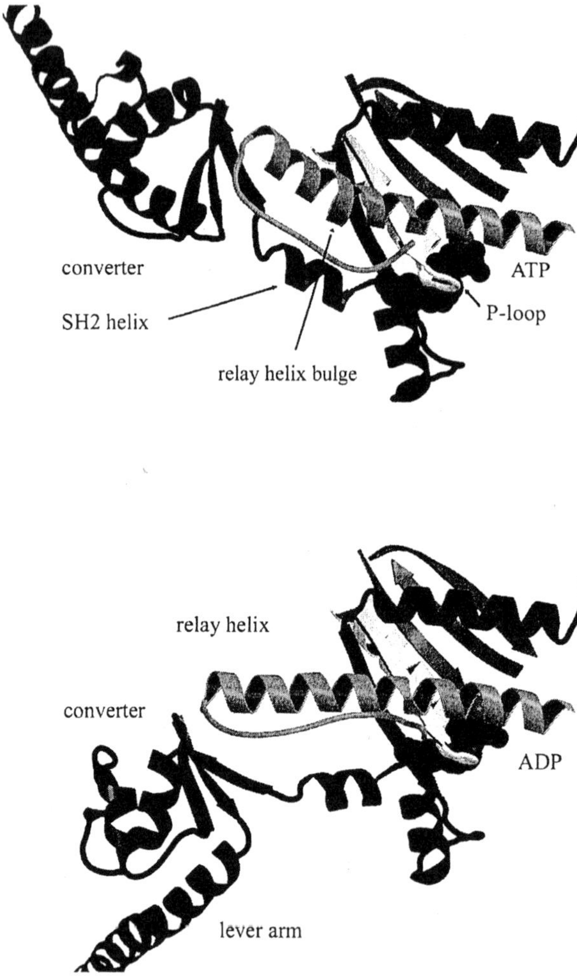

Figure 4 (Upper) Pre-power stroke and (lower) post-power stroke (near rigor) states. The removal of the bulge in the relay helix drives the rotation of the converter domain. The breaking of the relay helix appears to result from the helix being pressed against the β-sheet when the switch 2 element moves so as to hydrogen bond with the γ-phosphate. The colour coding is the same as in Fig. 2.

3.3.3 the relay helix bulge is the quintessence of muscle contraction

In the CLOSED state the upper/lower 50K domains rotate a few degrees with respect to each other so as to bring the invariant gly457 (466 in chicken skeletal myosin) amide group into H-bond contact with the g-phosphate - a movement of 5Å. At the same time the long relay helix, (residues 475-507 in *dictyosteleum*) bends and breaks to form a helix bulge (i.e. skips one hydrogen bond), which causes a rotation of the distal end of the helix and a rotation of the attached converter domain (711-781) by 60°. The relay helix breaks because it is forced into close contact with the β-sheet by the moving in of the switch 2 element. The fulcrum for the rotation of the converter domain is provided by the mutual rotation of the distal part of the SH1-SH2 helix around the distal part of the switch 2 helix. The relaxation of this bulge in the relay helix is the essential event that drives the lever arm down. It can come about through the moving out of switch 2 from the actin site on release of the g-phosphate. It can also come about by a twisting of the backbone β-sheet of the cross bridge. This second possibility has been shown by two recent structures of the myosin cross-bridge[35, 36].

4. STRCTURAL EVIDENCE FOR MOVEMENT OF THE LEVER ARM IN THE ACTIN-MYOSIN COMPLEX

Whereas the structure of "decorated actin" an actin filament with a myosin cross-bridge bound to each actin in the rigor state has been extensively studied [37-40] corresponding studies in the presence of ATP are difficult since the binding of ATP leads to rapid dissociation of the cross-bridges from actin. Time-resolved electron micrograph studies in fact show no bulk change of the cross bridge orientation on binding ATP before dissociation takes place [41] whereby a reorientation of the lever arm would not have been detected at the resolution attainable. High-resolution electron micrographs of actin decorated with smooth muscle myosin, however, show a 30-35° rotation of the lever arm on binding ADP[42, 43]. Although the main movement of the lever arm would be expected to be associated with phosphate release since this is a step associated with a large change in free energy, some fraction of the movement could arise from ADP binding and release. Moreover, this movement should be recoverable on adding ADP to actomyosin, which indeed it is. Although the effect has only been found in smooth muscle myosins this experiment was important in providing the first direct demonstration of a nucleotide-induced lever-arm swing.

5. MYOSIN V, A PROCESSIVE MYOSIN

Myosin II and myosin V have rather similar structures except that the lever arm on myosin V is long enough to binds 6 calmodulin light chains rather than 2 (myosin II cross bridges are about 16 nm long, myosin V cross bridges about 31 nm long). Whereas the two heads of myosin II act independently the two heads of myosin V bind conjointly to the actin filament to produce a processive linear motion along actin by a hand over hand mechanism. The myosin V lever arms are long enough to allow two-headed binding that spans the actin helical repeat (approximately 36 nm). Electron microscopy has shown that while working, myosin V spans the actin helical repeat [44]. The heads are mostly 13 actin subunits apart. Typically the lead head appears curved. The leading head may correspond to the beginning of the working stroke of the motor. Besides providing a graphical demonstration of the swinging lever arm myosin V offers the possibility of studying the elusive top-of-power-stroke state.

7. WEAK AND STRONG BINDING

The actin binding site straddles a cleft between two sub-domains, the upper and lower 50K domains.. Kinetic analysis shows that the initial binding of myosin to actin is "weak" and that weak binding isomerises to "strong"[31]. Strong binding drives the power stroke and affects ATP and ADP affinity. All myosin II crystal structures whether OPEN or CLOSED have an open 50K domain cleft and appear to be in the weak actin binding form. Since myosin V binds more tightly to actin it there is a chance that the myosin cross-bridge alone might be found in the strong binding form. X-ray crystallography has recently given the structure of the myosin V cross-bridge and indeed it appears to be in the strong binding form[35]. Moreover, the interaction between the myosin cross-bridge and actin can be studied by cryo-electron microscopy[40]. Together these studies explain the reciprocal linkage between actin binding and nucleotide binding: on strong binding to actin the 50k domain cleft closes leading to a movement of the switch 1 element out of the active site; the rebinding of ATP closes the active site by pulling in switch 1 thereby opening the cleft and weakening actin binding.

8. CONCLUSION

Three of the four states anticipated by the Lymn-Taylor cross-bridge cycle are now known in atomic detail. The "unbound" states, which are more easily available to protein crystallographic analysis, have yielded most information. The connections between ATP binding and the conformation of the myosin cross-bridge in solution are well understood. Besides being an ATPase the myosin "head" has two essential functions: a shape change induced by product release that drives contraction; a large change of affinity for actin induced by binding ATP. X-ray crystallography, in conjunction with electron microscopy has recently yielded an explanation of both these phenomena in molecular terms. The detailed structures of the actin bound states are gradually becoming available by combining crystal structures analysis with high-resolution electron microscopy. However, the structure of the ephemeral strong binding at the beginning of the power stroke can at present only be inferred.

DISCUSSION

Ishiwata: The structural change of the motor domain of S1 obtained with and without Pi (inorganic phosphate) in the presence of ADP is very impressive. But, I think that the ADP-bound structure is that of the simple complex of ADP and S1. If this structural change is responsible for the power stroke, it means that the force is generated by the dissociation of Pi from the S1-ADP complex. I suppose that the structure of S1-ADP complex obtained after the Pi release step in the ATP hydrolysis-cycle is different from that of the S1-ADP complex.

Holmes: I wish I could answer your question, but I am a mere crystallographist so that I only construct a model based on the crystallographic structures available so far.

Huxley: Dr. Brenner has suggested that initial tension development by cross-bridges takes place by a different process from that involved in the sliding movement. The closure of the creft in the 50K subunit of S1 could be the source of this initial force.

MOLECULAR SYNCHRONIZATION IN ACTOMYOSIN MOTORS - FROM SINGLE MOLECULE TO MUSCLE FIBER VIA NANOMUSCLE

Shin'ichi Ishiwata*#, Yuta Shimamoto*, Daisuke Sasaki* and Madoka Suzuki*

1. INTRODUCTION

What mysteries exist in the sliding filament mechanism? No one doubts that the thick (myosin) and thin (actin) filaments "slide" along side each other during the shortening of muscles. In the sliding filament mechanism, it is widely considered that the head part of myosin molecules interacts with actin to form a cross-bridge and its conformational change occurs in association with ATP hydrolysis: that is, the cross-bridge works as a lever arm. Here, it is assumed that a conformational change in actin is not required, implying that the actin filaments are a rigid body, so that a thin (actin) filament is sometimes called a "track" for molecular motors. Myosin molecules of muscle (Myosin II) are considered to be typical molecular motors. Thus, in the sliding filament mechanism, the cross-bridge is considered to work as an independent force generator; in other words, the force developed in each track is the sum of the forces developed by each cross-bridge.

What are the questions that remain to be answered in the sliding filament mechanism? We would like to raise the following questions to be answered in the future: 1) Are cross-bridges an independent force generator under all conditions, not only under the conditions of full activation but also under intermediate conditions? 2) Is the

*Department of Physics, School of Science and Engineering, and #Advanced Research Institute for Science and Engineering, Waseda University, 3-4-1 Okubo, Shinjuku-ku, Tokyo 169-8555, Japan

track really a rigid body (only a basis of the movement of molecular motors)?

The contractile system of striated muscle, i.e., skeletal and cardiac, consists of a hierarchy, i.e., from the single molecular level to a myofilament and a sarcomere composed of an assembly of actin and myosin molecules and several accessory proteins, and to a myofibril and a muscle fiber. An in vitro motility assay system demonstrated that single actin filaments can slide on myosin molecules randomly adhered to the two-dimensional glass surface, and tension can be developed on a single filament level, implying that the organized structure of myofilament lattice is not indispensable to the tension development.

Now, we raise the following question from a different point of view: Are all of the contractile properties of fibers (myofibrils) solely attributable to the characteristics of single molecular systems? In practice, the spontaneous oscillatory contraction of sarcomeres (length oscillation of the saw-tooth waveform of each sarcomere and the traveling wave of the lengthening phase to the adjacent sarcomeres) observed in single myofibrils under the conditions intermediate between relaxation and contraction (named SPOC) seems to be characteristic of the organized contractile system. We consider that myosin-II motors, which function as a stochastic molecular machine on a single molecular level, are synchronized not only with the external mechanical impulse but also with the force generated by other motors when assembled into a sarcomere and a myofibril (a series connection of sarcomeres). This may be a typical example showing that molecular motors do not necessarily function as independent force generators but *inter-molecular synchronization* is essential for the higher-ordered function in the organized system.

We examined whether the SPOC phenomenon occurs in the "nanomuscle" system that we have recently developed (as reported in the last Symposium supported by the Fujihara Foundation of Science; Suzuki et al., 2003): this newly devised motile system is composed of a single A-band and a single actin filament of which the B-end is trapped with optical tweezers through an attached plastic bead. The preliminary results showed that, under some intermediate condition for activation, an actin filament moved in a saw-tooth waveform composed of a slow shortening phase and a rapid yielding phase, but a periodic oscillation like that of a sarcomere oscillation in SPOC was not observed. This suggests that regulatory proteins and/or higher-ordered structure are necessary for regular SPOC to occur. It is to be noted that the series connection of sarcomeres is indispensable for the SPOC wave to occur. The elastic framework present in the sarcomere and/or between sarcomeres may partly be involved in such a collective mode of motion.

Molecular motors are classified into processive and non-processive motors. The processive motor works as a single molecule and can walk along the track, e.g., myosin V along an actin filament and kinesin along a microtubule. On the other hand, the non-processive motor cannot walk alone but can function only as an assembly. A typical example of this kind of motor is a muscle motor, i.e., myosin II. We discuss that, in processive motors such as kinesin and myosin V, the *intra-molecular synchronization* between the two bound heads is essential for the processivity. That is, the asymmetry (the distinction between the leading and the trailing heads) occurs through the internal force that exists between the heads bound to the track, such that the binding affinity of nucleotides becomes different for the two bound heads, which brings about the dissociation of the trailing head, and directional walking becomes possible.

Now, we focus on the "inter- and intra-molecular synchronization" occurring in the function of molecular motors at various levels of hierarchy.

2. MATERIALS AND METHODS

2.1. Proteins.

Actin and myosin II were prepared from rabbit skeletal white muscle according to a standard procedure (Kondo and Ishiwata, 1976). F-actin (FA) was labeled with rhodamine-phalloidin (Rh-Ph) and the bead-tailed FA, i.e., FA with the barbed (B-) end attached to a polystyrene bead (1 _m in diam.) through gelsolin, was prepared as previously reported (Suzuki et al., 1996; Nishizaka et al., 2000). Myosin V was prepared from chick brain (Ali et al., 2002); gelsolin from bovine plasma (Funatsu et al., 1990; Nishizaka et al., 1995 & 2000), and kinesin and tubulin from bovine and porcine brains, respectively (Kawaguchi and Ishiwata, 2001; Uemura et al., 2002; Kawaguchi et al., 2003).

2.2. Myofibrils and Muscle Fibers.

Myofibrils were prepared by homogenizing rabbit psoas glycerinated muscle fibers as described previously (Ishiwata et al., 1993; Yasuda et al., 1996). Glycerinated muscle fibers (a single fiber for skeletal muscle and a small bundle of cardiac muscle) were prepared by incubating in glycerol (50%) solution and then treating with Triton X-100. "Nanomuscle" was prepared by the treatment of myofibril with gelsolin (Suzuki et al., 2003; Suzuki and Ishiwata, 2004). All experiments were performed at room temperature.

2.3. Optical Microscopy.

The microscopy system and the method used for carrying out the image analysis were basically the same as those reported previously (Nishizaka et al., 2000). The fluorescence images of an F-actin and a bead were visualized using an ICCD camera (ICCD-350F; Video Scope International, Washington D.C.). A Nd:YAG laser (T10-V-106C; 2.5 W, Spectra-Physics Lasers, Inc., Mountain View, CA) was used as the optical tweezers.

3. RESULTS AND DICUSSION

The contractile system of muscle constructs the hierarchical structure, as schematically illustrated in Fig. 1: that is, from a single molecular system composed of a single myosin molecule and a single actin filament, to a protein assemblage such as thick and thin filaments, and to the sarcomere composed of the myofilament lattice structure and the myofibrils composed of a series connection of sarcomeres. Each hierarchy has contractile properties characteristic of its own. That is, each myosin motor functions as a stochastic nano-machine. Besides, myosin II is a "non-processive" motor that does not walk along the thin filament by itself. This is in contrast with a "processive" motor such as myosin V walking along an actin filament towards its B-end and kinesin walking along a microtubule towards its plus-end. Thus, we call myosin II a running (jumping) motor (Kinosita et al., 1998). Myosin II motors exhibit a physiological function only when they are assembled into the thick filaments. In the case of striated muscle,

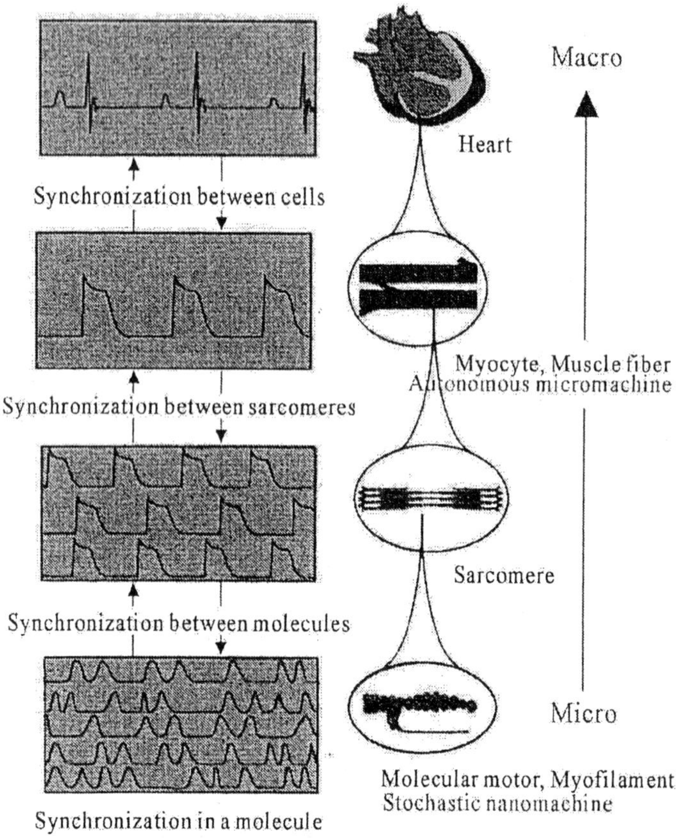

Figure 1. A schematic illustration showing the hierarchical structure of the contractile system of cardiac muscle. Molecular motors in striated muscle, i.e., myosin II, function as stochastic nano-machines. Myosin II molecules cannot work alone but can exhibit a physiological function only as an assembly, where intermolecular synchronization plays an essential role.

the thick filaments need to be incorporated further into a sarcomere and a myofibril that is composed of a series of connections of sarcomeres and then a muscle fiber that is composed of a parallel connection of myofibrils. The contractile properties are organized stepwise as the system is elevated to the upper level of hierarchy, organized from a random distribution to an ordered assembly.

We have demonstrated that myosin V is a left-handed spiral motor, which walks along an actin filament step by step with a step size of approximately 36 nm, a little shorter than a half pitch of a right-handed long helix of actin filaments (Ali et al., 2002). The walking mechanism is called the hand-over-hand mechanism. The processive motor can walk directionally in spite of the fact that the two heads of each molecular motor are identical. It is to be noted, however, that the directionality is created by the asymmetrical structure of actin filaments and microtubules, in which protomers (i.e., actin or tubulin molecules) are oriented in one direction and form a helical polymer. Thus, the two identical heads of molecular motors are attached to the track with the same orientation. Here, there still remains a question: how the motors can step uni-directionally. If the rate of detachment of one head were identical to the other, the probability of a step forward would be equal to that of a step backward. Therefore, what is the mechanism leading to the directional movement for the processive motors?

In this respect, many data have been accumulated suggesting that the internal load imposed between the two bound heads, which creates an asymmetrical distortion of the bound heads, may play a key role in determining the directionality of the stepping movement.

We found recently that the apparent binding affinity of ADP to kinesin depends on the loading direction (Uemura and Ishiwata, 2003). There is a way to distinguish the binding mode of protein motors with the "track" filaments. That is, we can determine whether there is a strong binding or a weak binding, and single-headed binding or double-headed binding, through measurements of the unbinding force under each nucleotide condition, i.e., nucleotide-free (strong-binding), ADP-bound (weak-binding) and AMP-PNP-bound (strong-binding) states (Kawaguchi and Ishiwata, 2001; Uemura et al., 2002; Kawaguchi et al., 2003). We confirmed that, on adding ADP stepwise to the nucleotide-free kinesin-microtubule complex, the proportion of the weak-binding state increased according to the simple hyperbolic relationship with the MgADP concentrations (Uemura and Ishiwata, 2003). Interestingly, we found that the apparent binding affinity of ADP depends on the loading direction: the MgADP concentration at which the proportion of the weak-binding state becomes equal to that of the strong-binding state was about 90 μM for backward loading, whereas it was about 10 μM for forward loading, indicating that ADP has a tendency to bind to the kinesin head in which the load is applied in the forward direction. This implies that, if we assume that an internal load exists between the two bound heads, the ADP prefers to bind to the trailing head, so that the probability of detachment of the trailing head increases. This can produce the directionality of the movement for the processive motors.

Now, let us move on to a discussion of the *in vitro* motility assay system for the acto-myosin II complex (Kinosita et al., 1991; Kato et al., 1999). First, we show that the cross-bridges produce the torque on actin filaments. When the front part of the actin filaments is fixed to a glass surface in an in vitro assay system, the myosin II molecules that adhered to the glass surface push the rear part of the filament, so that the middle part of the filament is buckled. Because the left-handed super-coil was always formed after the buckling, we concluded that the rear part of the filament slides like a right-handed screw, implying that the myosin II motors produce the right-handed torque (Nishizaka et al., 1993; Sase et al., 1997).

In in vitro motility assay systems, some curious phenomena have been observed. For example, a ring of actin filaments is formed because the pointed end of FA is sometimes linked to the barbed end during circular motion (Tanaka et al., 1992). The phalloidin, a stabilizer for the actin-actin bond, functions as glue. Even a twisted-ring for which the structure is analogous to a super-coil of DNA is formed when both ends of the twisted filament are linked. Besides, the filament sliding occurs even in such rings accompanied by the twisting motion. Thus, it is to be stressed that the in vitro motility assay system is artificial in some respects because the sliding of the actin filaments occurs on the 2D substrate like a glass surface on which myosin II molecules are randomly adhered.

To bridge the gap between the in vitro motility assay system and the fiber system, we have been trying to develop a new experimental system named "bio-nanomuscle" (Suzuki et al., 2003; Suzuki and Ishiwata, 2004). The bio-nanomuscle is composed of a single A-band and a bead-tailed actin filament that is trapped with optical tweezers. The single A-band was prepared by the treatment of a myofibril with gelsolin, an actin filament severing protein, to selectively remove the thin filaments. The bead-tailed actin filament trapped with optical tweezers was approached to the edge of the A-band. As soon as the free end of the actin filament touched the end plane or the outer surface of the A-band, the filament began to be pulled into the thick filament lattice and showed a large fluctuation in tension development that is balanced by the trapping force of the optical tweezers. We confirmed that the time average of the developed tension is proportional to the degree of overlap between the thick and the actin filaments. We also found that the average force developed per unit length of an actin filament is smaller when the interaction occurs outside the A-band than when the interaction occurs inside. This result is attributable to the difference in the number of thick filaments interacting with a single FA between the inside and the outside of the A-band.

Finally, let us describe the auto-oscillatory properties characteristic of each single myofibril, i.e., the phenomenon named SPOC (Spontaneous Oscillatory Contraction; Okamura and Ishiwata, 1988; Ishiwata and Yasuda, 1993). Under the solvent conditions intermediate between relaxation and contraction, the contractile system shows SPOC: The length of each sarcomere spontaneously oscillates with a saw-tooth waveform composed of a quick stretching phase and a slow shortening phase, and this oscillation (the stretching phase) propagates to the adjacent sarcomeres (more strictly speaking, to the adjacent half sarcomeres) as a traveling wave. When the myofibrils are bundled, several traveling waves appear here and there within the same bundle of myofibrils. It appears that there is a particular sarcomere (a "trigger" sarcomere), at which the SPOC wave is born, and the SPOC wave travels from the trigger sarcomere in both directions. Once the SPOC wave is born, the same pattern of SPOC wave continues for a while. But, the sarcomere from which the SPOC wave appears to begin is not necessarily fixed in the myofibril. The trigger sarcomere is exchangeable for others if the SPOC is once reset by washing with the relaxing solution, implying that the trigger sarcomere is not a particular damaged one but is interchangeable.

We have several results indicating that the elastic properties of connectin/titin (Fukuda et al., 2001) are *not* the major components required for the SPOC to occur. Firstly, the SPOC occurs even at short sarcomere lengths (shorter than 2.4 μm) where the resting tension is not developed (the connectin/titin is considered to be slacked). Secondly, when the sarcomere was stretched to longer than 2.4 μm and up to 2.8-2.9 μm, the SPOC period was slightly elongated but the major feature of the SPOC did not change. Only when the sarcomere length was elongated longer than about 3.0 μm did the SPOC stop, suggesting that the passive tension due to the elongation of the connectin/titin becomes comparable to or overcomes the active tension developed by cross-bridges.

Even when the total length of myofibrils is maintained constant (isometric condition), the length of every sarcomere oscillates but is forced to be out of phase (not synchronized) with each other. When the total length of a myofibril is long (the myofibril is composed of many, several tens of sarcomeres), the SPOC wave occurs here and there (locally) under the condition that the total length of the myofibril is maintained constant (for movies, see the following homepage of our laboratory: http://www.phys.waseda.ac.jp/bio/ishiwata). The movie shows that the sarcomere oscillation occurs without a change in the length of the A-band; only the lengths of both the I-band and the H-zone change. One can visualize the filament sliding repeatedly on the same single myofibril during SPOC.

We can measure the tension oscillation developed under the SPOC conditions not only for the muscle fibers but also for the single myofibrils. The difficulty in examining the SPOC in muscle fibers is that the tension oscillation tends to occur only transiently, although the sarcomere length oscillation continues within the fiber. This is probably because the SPOC solvent condition is not maintained uniformly within a fiber, so that the sarcomere oscillation tends to become out of phase. On the other hand, in myofibrils, the SPOC (both oscillations in the sarcomere length and the tension) continues for a while, even for an hour, if the SPOC solution is continuously flowed. So, using the microscopic technique of holding both ends of a myofibril with a pair of glass micro-needles, we can study the dynamics of SPOC under various conditions (Anazawa et al., 1992; Yasuda et al., 1996). This situation is schematically illustrated in Fig. 2.

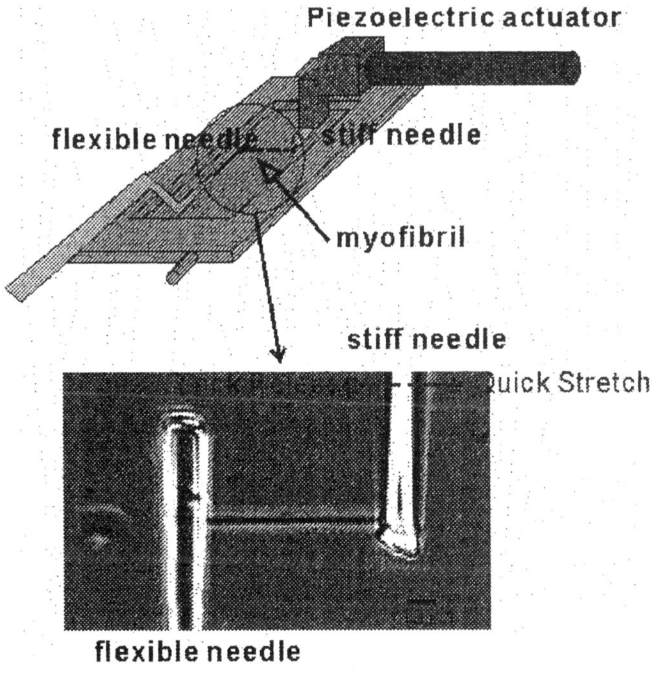

Figure 2. A schematic illustration of how to measure the tension development of myofibrils under optical microscopy. The total length of an oscillating myofibril is measured from the distance between the two glass micro-needles. The developed tension can be estimated from the deflection of a flexible needle from its equilibrium position. Video movies showing the SPOC of a myofibril are available from our homepage (http://www.phys.waseda.ac.jp/bio/ishiwata).

When one end of the myofibrils is fixed to a flexible micro-needle, both the tension and the sarcomere lengths oscillate under SPOC conditions (this boundary condition is named an "auxotonic" condition). When the myofibrils are composed of about 20-30 sarcomeres, it is easy to observe the SPOC wave traveling from one end to the other of the myofibril (*metachronal* SPOC). We found that the metachronal SPOC wave traveling over the whole length of the myofibril is easily to be induced upon stretching the myofibril to some extent under the condition that the average sarcomere length is less than about 3 μm.

Under an isotonic condition, which is achieved using a feedback control of the position of the rigid needle to keep the position of the flexible needle fixed (tension is maintained constant), we found that the traveling velocity of the SPOC wave tended to become very fast, such that the oscillation of every sarcomere tended to be synchronized to each other (the oscillation tends to be in phase, i.e., *synchronous* SPOC; Yasuda et al., 1996). This was the first evidence showing that the SPOC pattern can be mechanically controlled, implying that mechano-chemical coupling should exist in the molecular mechanism of SPOC.

What happens when the external mechanical impulse (by stepwise stretching or shortening of the myofibrils) is suddenly applied during the SPOC? We found that quick yielding occurs on all of the oscillating sarcomeres in response to the external mechanical impulse if the magnitude and the rate of the impulse are over some critical values.

The interesting finding is that a strong instability against the external mechanical impulse exists for the ADP contraction that is induced by the activation with the addition of ADP under relaxation conditions (Shimizu et al., 1992). Under ADP-contraction conditions, both slow shortening of the sarcomeres (about one tenth of the shortening velocity under usual activation conditions) and slow tension development (up to 70-80% of active isometric tension) occur, but no oscillations occur. Under ADP-contraction conditions, we can demonstrate that the yielding of every sarcomere occurs instantaneously upon the application of a mechanical impulse, so that the whole length of the myofibril is quickly elongated and then slowly shortens up to the original length where the isometric tension is balanced with the external load. If the mechanical impulse is applied repetitively, the myofibril can respond to each impulse, meaning that the oscillation of the myofibril is controlled by the external mechanical oscillation (Shimamoto et al., 2004 & manuscript in preparation).

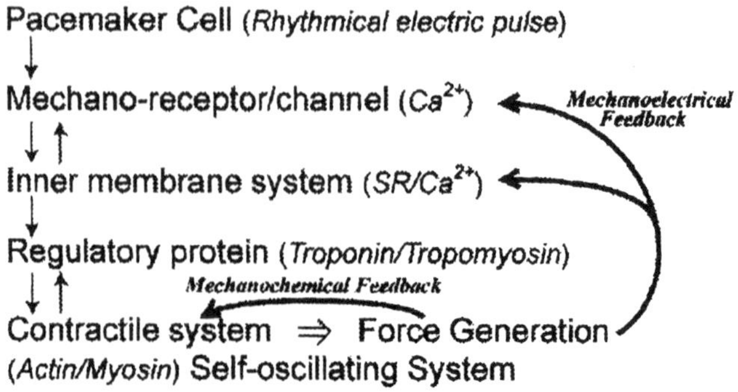

Figure 3. Hypothesis on the feedback regulation and molecular synchronization in cardiac muscle.

SPOC also occurs in the contractile system of cardiac muscle (Fabiato and Fabiato, 1978; Linke et al., 1993; Fukuda et al., 1996). It is interesting to note that SPOC more readily occurs in cardiac muscle than in skeletal muscle. That is, the state diagram showing the regions in which contraction, relaxation or SPOC occur was constructed for various concentrations of MgADP, Pi, free Ca^{2+} (pCa) and pH at several fixed concentrations of MgATP (Ishiwata and Yasuda, 1993; Ishiwata et al., 1993 & 1998; Fukuda and Ishiwata, 1999). This state diagram clearly showed that the SPOC region for cardiac muscle is much larger than that for skeletal muscle.

It is widely known that the free Ca^{2+} concentration in cardiac cells does not increase up to a fully activated level; that is, the activation is at most in the intermediate level, which just suits the condition for Ca-SPOC (the SPOC that occurs at a pCa of around 6 in the presence of MgATP: Fabiato and Fabiato, 1978; Linke et al., 1993; Fukuda et al., 1996). We should stress here that the contractile system of muscle itself is a self-oscillating system, which can respond to external forces. The state, contraction or relaxation, of the contractile system of muscle is controlled by the level of free Ca^{2+} concentration that is regulated by electric impulses (Ebashi and Endo, 1968), implying that the contractile system follows only the external signal. However, it is interesting to speculate that the flexible control of a heart beat due to an electric pulse (originally produced by pacemaker cells) is possible because the contractile system intrinsically possesses auto-oscillatory properties that can adapt to environmental changes. This situation is schematically illustrated in Fig. 3, in which we assume that not only the "mechano-electrical feedback" but also the "mechano-chemical feedback" plays a role in the regulation of the heart beat mechanism.

In summary, the contractile system of striated (skeletal and cardiac) muscle exhibits properties characteristic of each level of hierarchy, i.e., single-molecular level, myofibril and muscle fiber and so on. We have described several properties observed at each level of hierarchy, focusing on the synchronization that occurs in between the two heads within a single motor protein (intra-molecular synchronization) and in between myosin II motors within each sarcomere (inter-molecular synchronization) and in between

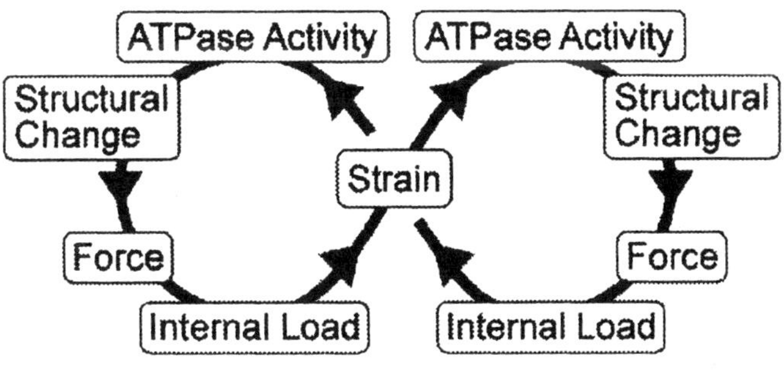

Figure 4. A feedback loop that exists in the intra- and inter-molecular synchronization of the mechano-chemical coupling mechanism of molecular motors.

sarcomeres within each myofibril. The feedback loop that exists in the coupling between the enzymatic activity and the mechanical event seems to be essential for the molecular mechanism of biological motors (Fig. 4).

This research was partly supported by Grants-in-Aid for Specially Promoted Research, for the Bio-venture Project and for The 21st Century COE Program (Physics of Self-organization Systems) at Waseda University from the Ministry of Education, Sports, Culture, Science and Technology of Japan.

4. REFERENCES

Ali, Y. M., Uemura, S., Adachi, K., Itoh, H., Kinosita, Jr., K., and Ishiwata, S., 2002, Myosin V is a left-handed spiral motor on the right-handed actin helix, *Nature Struct. Biol.* **9**: 464-467.

Anazawa, T., Yasuda, K., and Ishiwata, S., 1992, Spontaneous oscillation of tension and sarcomere length in skeletal myofibrils. Microscopic measurement and analysis, *Biophys. J.* **61**: 1099-1108.

Ebashi, S., and Endo, M., 1968, Calcium ions and muscle contraction, *Prog. Biophys. Mol. Biol.* **18**: 123-183.

Fabiato, A., and Fabiato, F., 1978, Myofilament-generated tension oscillations during partial calcium activation and activation dependence of the sarcomere length-tension relation of skinned cardiac cells, *J. Gen. Physiol.* **72**: 667-699.

Fukuda, N., Fujita, H., Fujita, T., and Ishiwata, S., 1996, Spontaneous tension oscillation in skinned bovine cardiac muscle, *Pflugers Arch.* **433**: 1-8.

Fukuda, N., and Ishiwata, S., 1999, Effects of pH on spontaneous tension oscillation in skinned bovine cardiac muscle, *Pflugers Arch.* **438**: 125-132.

Fukuda, N., Sasaki, D., Ishiwata, S., and Kurihara, S., 2001, Length dependence of tension generation in rat skinned cardiac muscle. Role of titin in the Frank-Starling mechanism of the heart, *Circulation.* **104**: 1639-1645.

Funatsu, T., Higuchi, H., and Ishiwata, S., 1990, Elastic filaments in skeletal muscle revealed by selective removal of thin filaments with plasma gelsolin, *J. Cell Biol.* **110**: 53-62.

Ishiwata, S., Anazawa, T., Fujita, T., Fukuda, N., Shimizu, H., and Yasuda, K., 1993, Spontaneous tension oscillation (SPOC) of muscle fibers and myofibrils. Minimum requirements for SPOC, *In* Mechanism of myofilament sliding in muscle contraction. (ed. by H. Sugi & G. H. Pollack) pp. 545-556. (Plenum Press)

Ishiwata, S., Funatsu, T., and Fujita, H. 1998, Contractile properties of thin (actin) filament-reconstituted muscle fibers, *In* Mechanism of work production and work absorption in muscle. (ed. by H. Sugi & G. H. Pollack) pp. 319-329. (Plenum Press)

Ishiwata, S., and Yasuda, K., 1993, Mechano-chemical coupling in spontaneous oscillatory contraction of muscle, *Phase Transitions.* **45**: 105-136.

Kato, H., Nishizaka, T., Iga, T., Kinosita, Jr., K., and Ishiwata, S., 1999, Imaging of thermal activation of actomyosin motors, *Proc. Natl. Acad. Sci. U.S.A.* **96**: 9602-9606.

Kawaguchi, K., and Ishiwata, S., 2001, Nucleotide-dependent single- to double-headed binding of kinesin *Science*, **291**: 667-669.

Kawaguchi, K., Uemura, S., and Ishiwata, S., 2003, Equilibrium and transition mechanics between single- and double-headed binding of kinesin revealed by single-molecule mechanics, *Biophys. J.* **84**: 1103-1113.

Kinosita, Jr., K., Yasuda, R., Noji, H., Ishiwata, S., and Yoshida, M., 1998, F1-ATPase: A rotary motor made of a single molecule, *Cell.* **93**: 21-24.

Kinosita, Jr., K., Itoh, H., Ishiwata, S., Hirano, K., Nishizaka, T., and Hayakawa, T., 1991, Dual-view microscopy with a single camera: Real-time imaging of molecular orientations and calcium, *J. Cell Biol.* **115**: 67-73.

Kondo, H., and Ishiwata, S., 1976, Uni-directional growth of F-actin, *J. Biochem.* **79**: 159-171.

Linke, W.A., Bartoo, M. L., and Pollack, G. H., 1993, Spontaneous sarcomere oscillations at intermediate activation levels in single isolated cardiac myofibrils, *Circ. Res.* **73**: 724-734.

Nishizaka, T., Yagi, T., Tanaka, Y., and Ishiwata, S., 1993, Right-handed rotation of an actin filament in an in vitro motile system, *Nature.* **361**: 269-271.

Nishizaka, T., Miyata, H., Yoshikawa, H., Ishiwata, S., and Kinosita, Jr., K., 1995, Mechanical properties of a single protein motor of muscle studied by optical tweezers, *Nature* **377**: 251-254.

Nishizaka, T., Seo, R., Tadakuma, H., Kinosita, Jr., K., and Ishiwata, S., 2000, Characterization of single

actomyosin rigor bonds: Load dependence of lifetime and mechanical properties, *Biophys. J.* **79**: 962-974.
Okamura, N., and Ishiwata, S., 1988, Spontaneous oscillatory contraction of sarcomeres in skeletal myofibrils, *J. Muscle Res. Cell Motil.* **9**: 111-119.
Sase, I., Miyata, H., Ishiwata, S., and Kinosita, Jr., K., 1997, Axial rotation of sliding actin filaments revealed by single-fluorophore Imaging, *Proc. Natl. Acad. Sci. U.S.A.* **94**: 5646-5650.
Shimamoto, Y., Maejima, H., Suzuki, M., Sasaki, D., Yasuda, K., and Ishiwata, S., 2004, Stability of the spontaneous oscillatory contraction (SPOC) in single myofibrils studied by mechanical response and fluorescence imaging, *Biophys. J.* **86**: 564a (Abstract).
Shimizu, H., Fujita, T., and Ishiwata, S., 1992, Regulation of tension development by MgADP and Pi without Ca^{2+}. Role in spontaneous tension oscillation of skeletal muscle, *Biophys. J.* **61**: 1087-1096.
Suzuki, M., Fujita, H., and Ishiwata, S., 2003, Bio-nanomuscle project: Contractile properties of single actin filaments in an A-band motility assay system, *In* Molecular and cellular aspects of muscle contraction. (ed. by H. Sugi), Kluwer Acad./Plenum Press. pp. 103-110.
Suzuki, M., and Ishiwata, S., 2004, Contractile systems of muscles, *In* Reflexive polymers and hydrogels: Understanding and designing fast-responsive polymeric systems. (ed. by N. Yui, R. Mrsny and K. Park), CRC Press LLC. pp. 33-48.
Suzuki, N., Miyata, H., Ishiwata, S., and Kinosita, Jr., K., 1996, Preparation of bead-tailed actin filaments: Estimation of the torque produced by the sliding force in an in vitro motility assay, *Biophys. J.* **70**: 401-408.
Tanaka, Y., Ishijima, A., and Ishiwata, S., 1992, Super helix formation of actin filaments in an in vitro motile system, *Biochim. Biophys. Acta.* **1159**: 94-98.
Uemura, S., and Ishiwata, S., 2003, Loading direction regulates the affinity of ADP for kinesin, *Nature Struct. Biol.* **10**: 308-311.
Uemura, S., Kawaguchi, K., Yajima, J., Edamatsu, M., Toyoshima, Y. Y., and Ishiwata, S., 2002, Kinesin-microtubule binding depends on both nucleotide state and loading direction, *Proc. Natl. Acad. Sci. U.S.A.* **99**: 5977-5981.
Yasuda, K., Shindo, Y., and Ishiwata, S., 1996, Synchronous behavior of spontaneous oscillations of sarcomeres in skeletal myofibrils under isotonic conditions, *Biophys. J.* **70**: 1823-1829.

DISCUSSION

Pollack: You suggested that the SPOC may be related to pacemaking. Are you suggesting that the oscillations you observe are associated with corresponding electrical oscillations?

Ishiwata: I think that the SPOC is not coupled with electrical oscillations, but originates from the contractile system. I have an idea that the role of pacemaker cells is to trigger the auto-oscillatory properties inherent to the contractile system.

Gonzalez-Serratos: Dr Inesi and I published a paper showing, in permeabilized isolated heart cells, the same type of traveling waves that you presented. At that time, we proposed that this traveling contraction waves were caused by Ca^{2+} uptake and release by the sarcoplasmic reticulum, resulting in a Ca^{2+} gradient. In your preparation, there is no sarcoplasmic reticulum and you attributed the traveling wave only to contractile protein interaction. How would you reconcile both results and propositions.

Ishiwata: To examine the SPOC, we are using detergent-treated myofibrils and muscle fibers, so that Ca^{2+} uptake and release by the SR do not take place. We speculate that the mechanochemical coupling (force-dependent modulation of enzymatic activity) propagates to adjacent sarcomeres. Unfortunately, it is at present difficult to prove the validity of this speculation.

Cecchi: 1) Is it possible from the effect of BDM on the dwell time to estimate the effect on force? 2) Can you estimate the actin filament compliance with your nanomuscle model?

Ishiwata: 1) It is possible to examine the effect of BDM on force development of myosin V. We found that the size of substeps was maintained constant in the presence of BDM under various loads, suggesting that BDM does not affect the amount of force development. BDM only prolongs the dwell time.
2) We have not yet examined it.

DISTRIBUTION OF CROSSBRIDGE STATES IN CONTRACTING MUSCLE

Hugh E. Huxley, Massimo Reconditi, Alexander Stewart, and Tom Irving[1]

1. INTRODUCTION

The X-ray diffraction pattern from surviving frog muscle continues to provide significant new information about crossbridge behaviour and to highlight problems still unsolved. Recent studies of interference effects on the meridional reflections from the myosin filaments have revealed important new details about the configuration of the myosin heads and their changes during rapid mechanical transients (Linari et al, 2000; Lombardi et al, 2000; Huxley et al, 2000; Piazzesi et al, 2001; Irving et al, 2002; Piazzesi et al, 2002; Reconditi et al, 2003; and Huxley et al, 2001, 2002, 2003). These interference effects arise because all the thick filaments have exactly the same construction, so that the centers of scattering mass of the axial arrays of myosin heads in the two halves of each filament have exactly the same axial separation in each filament. So each one gives the same interference fringes on the reflection that arise from the underlying axial repeat of the crossbridges in each half filament. The position of the interference fringes on the first myosin meridional reflection (M3) at 14.56 nm in contracting muscle provides an extremely sensitive measure of the changes in the axial position of the center of scattering mass of actin-attached myosin heads, able to detect movements of one or two angstroms.

2. THE X-RAY EVIDENCE

When a sudden small length decrease is allowed to take place in a previously isometrically contracting muscle, a relatively large shift takes place in the position of the interference fringes relative to the envelope of the M3 reflection. If the myosin heads in each half-filament occupy a length, say, of 0.7 μm, and if the H-zone gap between the two arrays is 0.2 μm, (total length of filament is about 1.6 μm), then the envelope of the M3 reflection from each half filament will have a total width of $2/0.7\ \mu m^{-1} = 2.86\ \mu m^{-1}$ The interference distance between the two arrays (distance between outermost myosin head in one half-filament to innermost head of the other half-filament) will be $0.9\ \mu m^{-1}$.

[1] Hugh E. Huxley, Rosenstiel Center, Brandeis University, Waltham, MA 02454 USA. Massimo Reconditi, Department of Physiological Sciences, University of Firenze, 50134 Firenze, Italy. Alexander Stewart, Rosenstiel Center, Brandeis University, Waltham, MA 02454 USA.. Tom Irving, Department of BCPS, Illinois Institute of Technology, BioCAT, Argonne, IL 60439 USA.

Hence there will always be at least two fringes crossing the central maximum of the envelope of the M3 reflection (and all the other myosin meridional reflections), and the ratio of the heights of the resultant sampled peaks will vary as the fringes shift. Since the M3 reflection (14.56 nm) occurs near to the sixty-second order of the interference fringe system, a 2 nm decrease in the interference distance (originally 0.9 μm ≡ 900 nm), (which would be produced by a 1 nm movement of the heads in the two half-sarcomeres towards each other), will result in a fringe shift approximately equal to $2x62/900$ of the fringe separation, or about $1/7$ of a fringe. The observed ratio of the intensities of the two main fringes on the M3 reflection from isometrically contracting muscle is approximately 0.8 (higher angle fringe is the weaker one). A 1 nm movement of the center of scattering mass of the heads in each half-filament towards each other would change that ratio to approximately 0.47, so that a few angstroms of such a shift should be readily measurable.

In practice, we analyze the changes in terms of the sliding filament, tilting lever model, with the catalytic subunit of myosin S1 attached to the actin filament, and axial actin movement relative to the myosin filament backbone is produced by a change in the angle of the lever arm connecting the catalytic subunit to the S_1S_2 junction. In this model, an inward movement of the catalytic subunit by 1 nm, pulling the actin filament with it, changes the ratio by a smaller amount than above, down to .551. This corresponds to a shift in the center of mass approximately only 0.7 nm, since the movement of the center of mass of the lever arm is only about one half of that of the catalytic subunit.

3. COMPARISON WITH MODEL PREDICTIONS

When these predictions were compared with the ratio changes actually observed, it was found that the ratio only changed initially by about half the expected amount, and that instead of progressively decreasing to almost zero, as would be expected when the outer fringe reached the first zero of the envelop of the diffraction peak, the ratio saturated at a value of about 0.25 – 0.30. This would be readily explained by the presence of some fixed component, perhaps unattached partners of the attached heads, with approximately the same scattering contribution as the attached heads, in the isometric muscle. So at first, the total center of mass movement is reduced by a factor of two. Then, as the lever arm tilts further, the axial profiles of the catalytic subunit and of the lever arm become increasingly misaligned, so that the scattering contribution of the moving head is progressively reduced, while that of the fixed component stays constant. Thus the movement of the center of effective scattering mass of the head and the fixed component combined slows down and eventually stops, and then reverses, as the lever arm angle continued to tilt. These effects can all be computed, and give good agreement with experimental observation (Huxley et al, 2001; Piazzesi et al, 2002).

However, the identify of the additional fixed, diffracting component has not been firmly established. One possibility is that the 'second' unattached head of these myosin molecules whose 'first' head is attached to actin is stabilized in some way by its partner, with its lever arm oriented near to the position where catalytic subunit and lever arm are maximally aligned in projection onto the filament axis. The attached head, in isometric muscle, has its lever arm tilted some five to ten degrees or more beyond this position, i.e. further away from the rigor position (assumed to correspond to the end of the working stroke). This arrangement will give good modelling of the initial increase in intensity with smaller releases (as the two heads become aligned) and gives a good match to the experimental observations both of ratio and total intensity variation with extent of release.

4. IDENTITY OF THE FIXED COMPONENT

Although the fixed component could in theory be these 'second heads', there is no over-riding reason why that has to be the case, and there are a number of reasons why the situation may be a lot more complicated.

1) The position of the interference fringes on the M6 reflection (at 7.28 nm) in isometric contraction is quite different from what would be the case if this reflection was largely the second order diffraction from the myosin heads giving the M3 reflection. Instead, the reflection must mainly come from some other structure, very likely in the thick filament backbone. The same is true of the M9 and M15 reflections. If the backbone gives reflections at both these submultiples of the 14.56 nm spacing, it is extremely likely that it would also give a contribution at the 14.56 nm fundamental spacing, and contribute to the fixed component there.

2) If the fixed component of the M3 reflection is produced just by the 'second heads', then, for their contribution to be approximately equal to that of the attached heads in isometric contraction, the dispersion of their lever arms about the S_1S_2 junction must just happen to be the same as the dispersion of the lever arms of the attached heads. This seems an unlikely coincidence.

3) Polarized fluorescence studies of lever arm orientations in contracting single fibers indicate that only a small fraction (10 – 20%) of all the myosin heads is involved in tension development at any one time (Hopkins et al, 1998, 2000). There is therefore a very large population, 80-90%, of unattached heads. All these heads are presumably carrying ATP or its split products, awaiting attachment to actin so that the products can dissociate and the head go through its tension-generating conformation change. When they do attach, they need to be suitably placed to develop force efficiently. This will not be the case if the S2 regions are crumpled in any way, or if they are significantly misaligned to the fiber axis. That is, the S2 regions must be so configured that they can transmit force and movement, without any lost motion, from the tilting lever arm of the now attached head.

Thus it seems very likely that the unattached heads have their S2 regions straight and parallel to the fiber axis for most of the time. Thus the only ways for the myosin heads to be disordered is by rotation of the catalytic subunit about the lever arm, which is likely to be limited, and by rotation of the lever arm about the S_1S_2 hinge, which may be limited only by steric constraints in the fiber. The intensity of diffraction from such heads can be computed.

5. CONTIBUTION FROM UNATTACHED HEADS

If this rotational dispersion, say of ± 90°, in two direction, is uniform, one's initial supposition would be that the X-ray scattering mass of the head would be spread out over such a large volume that it would make a negligible contribution to the 14.56 nm axial periodicity. However, longer consideration, and direct computation, shows that this is not the case. When the heads are oriented so that the lever arm and the catalytic subunit are aligned, i.e. so that their combined axial projection – which is what determines the strength of the meridional reflections – is as sharp as possible, they will contribute a much larger peak in the axial density project than they do in other orientation. Thus if we have a uniform distribution of orientations, the total axial density distribution will still show a significant peak, repeating axially at 14.56 nm. This is confirmed by actual computation of the projected density (Fig 1).

Such a structure will, therefore, still contribute to the 14.56 nm meridional reflection. This will be, obviously, considerably weaker than that from an optimally oriented head, but if there are 5 or 10 times as many detached heads as attached ones, then the total intensity contribution that they give is 25 to 100 times as great as a a single

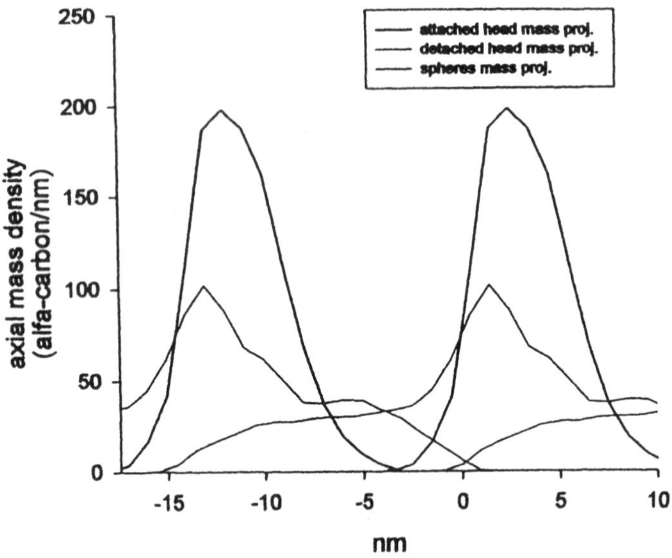

Figure 1: Computed axial profiles of myosin head density; black curve, all heads attached, with ±23° lever arm dispersion. Red curve, all heads detached, with ±80° dispersion

such head, and becomes comparable with that from an attached head. This is confirmed by computing the relative intensity of diffraction from two such arrays, with the attached heads having a lever arm angle dispersion of about ± 23°, as deduced from the behaviour of the M6 reflection. (This reduces its intensity contribution to about 56% of that given by zero dispersion). Table I shows the relative values for equal numbers of attached and detached heads; with large lever arm dispersions of the detached heads. Clearly, when these relative intensities (detached/attached) are multiplied by a factor between 25 and 100, the contribution of the detached heads can be seen to be quite considerable.

A possible difficulty with this argument is the observation that the M3 reflection is extremely weak in rapidly shortening muscle, where nearly all the heads are thought to be detached at any given time. However, such rapid cycling may lead to increased disorder of the S2 regions (e.g., by compression due to late detachment at the end of the working stroke).

6. DISTIBUTION OF LEVER ARM ANGLES

The behaviour of the M6 reflection shows that the major part of it must come from some fixed component, rather than being the second order term generated by the axial repeat of the attached crossbridges, or that of unattached myosin heads. In either of these latter cases, the predicted position of the interference fringes is completely wrong (Huxley et al, 2002, 2003).. Also, the increase in M6 intensity during quick releases – only 40-60% - is much lower than would be expected if the attached myosin heads contributed a large part of the intensity of the reflection, and the small change in the ratio of the intensities of the two interference peaks in the M6 reflection during release is also inconsistent with such a supposition.

Table 1. Diffraction from attached heads (mean lever arm angle 50° from rigor) and unattached heads (mean lever arm angle 45° from rigor)

	Intensity from equal numbers of heads	Amplitude from equal numbers of heads	Number of unattached heads to give same amplitude (and intensity) as one attached head	% of heads attached for contributions from attached and unattached heads to be equal
Attached heads, 23% dispersion	100	10	(1)	N/A
Unattached heads, 60° dispersion	7.06	2.65	3.77	20.9%
Unattached heads, 70° dispersion	4.41	2.10	4.76	17.3%
Unattached heads, 80° dispersion	3.98	1.99	5.05	16.5%
Unattached heads, 90° dispersion	4.05	2.01	4.97	16.7%

For the contribution from the crossbridges to be small, there must be a substantial dispersion of their axial positions about their average repeat of 14.56 nm in isometric contraction. It seems unlikely that this arises from an inherent irregularity in this underlying backbone repeat of the thick filaments, since reflections from that repeat extend out to high angles (beyond 10Å) even in contracting muscle. Elasticity in the S2 component can also be ruled out, since the known compliance of the crossbridges – of the order of 2.5 nm/T_0 – plus the dispersion of myosin backbone repeats due to the tension gradient along the thick filaments is quite insufficient to decrease the contribution to M6 relative to M3 sufficiently.

This leaves dispersion of lever arm angles as the remaining possible source of the dispersion of axial positions of the centers of scattering mass of the crossbridges. The shape of the relationship between M6 intensity increase and extent of release shows that a uniform dispersion of about ± 23° needs to be present, and this value also gives good agreement with the observed total intensity of the reflection, and with the small change in peak ratios (Huxley, in preparation).

Evidence about average lever arm angle in isometric contraction

Another experimental parameter which has to be watched in these modellings is the increase in total intensity of the M3 reflection during small releases, amounting usually to 10-15%. This can be accounted for in general terms if the isometric myosin heads are attached to actin when their lever arms are tilted, away from rigor (presumed to be the end of the working stroke), slightly beyond the position at which the axial projections of the lever arm and the catalytic subunit of myosin are maximally aligned, and where they generate maximum intensity in the M3 reflection (Irving et al, 1995). The same general explanation holds true when a second, fixed, diffracting structure needs to be invoked to explain the behavior of the interference fringes on the M3 reflection, and the presence of substantial dispersion simply means that the sharpness of the maximum on the intensity curve is decreased, and the mean isometric angle of the lever arms has to be somewhat further out from the maximum intensity position.

In practice, we find that a mean isometric angle 63° away from the Rayment rigor position gives very satisfactory agreement with the observed intensity increase. Thus the range of positions extends from about 40° away from rigor up to 83° away.

7. IMPLICATIONS OF ANGULAR DISTRIBUTION

This distribution of cross-bridge angles is very interesting. During quick releases, the behavior of the interference fringes on M3 shows that the lever arms must tilt over until their mean angle is close to the rigor position, and a shift of the mean angle in the same direction is also shown during steady shortening. But in the isometric state, there are apparently very few, or no, crossbridges in that angular range, so that some of the ones that were there as the isometric contraction developed, and internal shortening was taking place, must have detached well before the end of the working stroke, as movement stopped. So the muscle will only be able to do work with greatest efficiency when the shortening velocity is such that most myosin heads complete their stroke before detaching. (It seems less likely that a prematurely detached crossbridge could store its energy in some way and reattach).

The myosin heads near the top of the distribution, about 83° away from the rigor end of stroke position, must move their catalytic subunits over 12 nm axially to reach that position. That happens to be the maximum release (per half sarcomere) at which any rapid tension recovery occurs in quick release experiments on single fibers (Huxley & Simmons, 1971), and has been interpreted by them as the maximum possible movement that an attached head can undergo. (The 1-2 msec needed for tension recovery being insufficient for new heads to attach). This strongly suggests that such heads may be ones involved in tension redevelopment after a quick release, when new tension appears, with a 1-2 msec time delay, considerably beyond the point (about 5 nm release) at which all the original tension is discharged.

In the original model by A.F. Huxley and Simmons, the crossbridge's active stroke was postulated as occurring in a number of steps, possibly tilting through a succession of pairs of attachment points to actin. However, in the case of striated muscle myosin, there has been no structural (e.g. crystallographic) evidence so far that the working stroke is likely to take place in anything other than a single stroke. However, suppose that a myosin head could attach to actin initially without developing active tension, over a range of lever arm position, and then, when subsequent sliding of the actin filament (as a result of tension developed by other attached myosin heads) tilted its lever arm into a critical range, suppose that it could be activated with a 1-2 msec time constant to generate tension. Then such a process might correspond to the necessary two steps for attached heads.

8, LENGTH OF THE WORKING STROKE

There has also been a disturbing discrepancy between the length of the working stroke determined in single molecule experiments, when values around 5 nm are usually found, and the evidence from quick release experiments which strongly suggest a total working stroke of around 10-12 nm. The length of the active working stroke in the model we are suggesting would need to be at least 6-7 nm, if the activated heads were ones near the bottom end of the distribution of lever arms, some 43° away from the rigor position, and thus still significantly longer than in the single molecule experiments. However, one needs to remember that in those experiments the S_1 heads are either directly attached to a substrate, in which case the normal movement of the lever arms may be partially inhibited, or else attached via some type of tether. In the latter case the movement of the end of the lever arm will in general not be aligned with orientation of the tether between its attachment point to the substrate and to the S1 molecule, so that the effective displacement will be reduced by a cosine term which averages out over all possible orientations to a factor of 3.142/2. So the approximately 5 nm displacements measured may represent molecular displacements of 7-8 nm, within the range suggested by the present structural observations.

It might be expected that there would be an elastic component within each myosin head which could store the full movement of the working stroke, when the critical chemical change took place in the molecule. But in practice, a quick release of only 5 nm will reduce single fiber tension to zero, and at least nearly half of that will be taken up by filament elasticity. Thus the compliance of the assembly of attached crossbridges on a given actin filament must be constrained to be much smaller than the crossbridge active stroke. In the arrangement we have proposed, there would otherwise be no reason why crossbridge movement of 7-8 nm should not immediately take place upon a quick release, driven by the activated crossbridges in the isometric contraction, unless the movement is constrained by the other strongly attached heads.

These speculations have become somewhat extended here, but they may serve to illustrate the questions that still have to be answered in order to fully understand this mechanism.

REFERENCES

Hopkins, S. C., Sabido-David, C., Corrie, J. E. T., Irving, M., and Goldman, Y. E. (1998). Fluorescent polarization transients from rhodamine isomers on the myosin regulatory light chain on skeletal muscle fibers. Biophysical J. 74, 3093-3110.

Hopkins, S.C., Sabido-David, C., van der Heide, U., Ferguson, R.E., Brandmeir, B.O., Dale, R.E., Kendrick-Jones, J., Corrie, J.E.T., Trentham, D.R., Irving, M., & Goldman, Y.E. (2000). Orientation Changes of the Myosin Light Chain Domain during Filament Sliding in Active and Rigor Muscle. J. Mol. Biol 318, 1275-1291.

Huxley, A. F., and Simmons, R. M. (1971). Proposed mechanism of force generation in striated muscle. Nature 233, 533-538.

Huxley, H.E., Reconditi, M., Stewart, A., and Irving, T. (2000). "Interference changes on the 14.5 nm reflection during rapid length changes." Biophys. J. 78, 134A.

Huxley, H.E., Reconditi, M., Stewart, A., Irving, T., Fischetti, R. (2001). Use of X-ray interferometry to study crossbridge behavior during rapid mechanical transients. Biophys. J. 80, 266A.

Huxley, H.E., Reconditi, M., Stewart, A., and Irving, T. (2002). "Crossbridge and backbone contributions to interference effects on meridional X-ray reflections." Biophys. J. 82, 5A.

Huxley, H.E., Reconditi, M., Stewart, A., and Irving, T. (2003). "What the higher order meridional reflections tell us." Biophys. J. 84, 139A.

Irving, M., Allen, T. S. C., Sabido-David, C., Craik, J. S., Brandmeier, B., Kendrick-Jones, J., Corrie, J. E. T., Trentham, D. R., and Goldman, Y. E. (1995). Tilting of the light chain region of myosin during step length changes and active force generation in skeletal muscle. Nature 375, 688-691.

Irving, M., Piazzesi, G., Linari, M., Lucii, Y., Nrayanan, T., Boesecke, P., Stewart, A., Fischetti, R., Irving, T., Piazzesi, G., and Lombardi, V. (2002). "X-ray interference measurements of myosin head motions during isotonic shortenig of skeletal muscle fibers." Biophys. J. 82, 371A.

Linari, M., Piazzesi, G., Dobbie, I., Koubassova, N., Reconditi, M., Narayanan, T., Diat, O., Irving, M, and Lombardi, V. (2000). Interference fine structure and sarcomere length dependence of the axial X-ray pattern from active single muscle fibers. Proc. Natl. Acad. Sci. 97, 7226-7231.

Linari, M., Piazzesi, G., Dobbie, I., Koubassova, N., Reconditi, M., Narayanan, T., Diat, O., Irving, M, and Lombardi, V. (2000). Interference fine structure and sarcomere length dependence of the axial X-ray pattern from active single muscle fibers. Proc. Natl. Acad. Sci. 97, 7226-7231.

Lombardi, V., Piazzesi,G., Linari, M., Vannicelli-Casoni, M.E., Lucii, L., Boesecke, P., Narayahan, T., & Irving, M. (2000). X-ray interference studies of the working stroke in single muscle fibres. Biophys. J. 78, 134A.

Piazzesi, G., Reconditi, M., Linari, M., Lucii, L., Sun, Y-B., Nagayanan, T., Boesecke, P., Lombardi, V., and Irving, M. (2002). Mechanism of force generation by myosin heads in skeletal muscle. Nature 415, 659-662.

DISCUSSION

Brenner: ~45% of heads responding to movement, does this mean that ~45% of heads are attached to actin?

Huxley: In the model I described, ~45% of the heads which contribute to the diffraction pattern respond to the movement, and ~55% remain fixed. But in this model, there could be a large population of other heads which are too dispersed axially to contribute. Also, some other fixed structure with the myosin periodicity could make up some or all of the fixed component.

Cecchi: 1) Can you distinguish peak intensity changes due to S1 movement towards the M line from those due to S1 tilting? 2) If I understood well, the mean position of S1 lever arm during isometric contraction is about 1 nm away from the perpendicular. Is that right?

Huxley: 1) All the change in total intensity is due to the tilting of S1 (which spreads out its axial profile). In the model, the lever arm tilting is of course directly coupled to movement of the actin-attached catalytic subunit towards the M-line. 2) In our experiments, the isometric position appears to be about 1-2nm beyond the position, which gives the maximum intensity after correction for filament compliance.

X-RAY DIFFRACTION STUDIES OF STRIATED MUSCLES

John M. Squire, Carlo Knupp, Manfred Roessle, Hind A. AL-Khayat, Thomas C. Irving, Felicity Eakins, Ngai-Shing Mok, Jeffrey J. Harford and Michael K. Reedy

1. INTRODUCTION

Following the pioneering work of Hugh Huxley, Gerald Elliott and their respective collaborators (e.g. Huxley and Brown, 1967; Elliott *et al.*, 1967), X-ray diffraction studies of muscle have provided a wealth of detailed information about muscle ultrastructure and about the mechanisms of muscle activation and force-production. However, until recently, most of the X-ray diffraction data, apart from the spacings of the peaks which can be accurately determined, have been analysed in a qualitative or semi-quantitative way enabling suggestions to be made about mechanism, but not providing rigorous proof. Arguments have been presented elsewhere about why the X-ray diffraction technique is potentially one of the very few experimental approaches which can provide unambiguous information about the crossbridge mechanism in muscle (Squire *et al.*, 1997; 2003c,d,e). Now, with the advent of very fast computers, with knowledge of the structures of various muscle proteins (particularly actin and the myosin head; Kabsch *et al.*, 1990; Rayment *et al.*, 1993a respectively) from protein crystallography, and with improved analytical protocols (Squire *et al.*, 2003a,b), muscle diffraction patterns can be analysed in a rigorous way to give conclusions which are much more soundly based than was ever possible previously. Some of the advances that have been made in rigorously fitting low-angle X-ray diffraction patterns from the highly-ordered insect flight and bony fish muscles have been described elsewhere (Hudson *et al.*, 1997; AL-Khayat *et al.*, 2003;

John Squire, Hind AL-Khayat, Carlo Knupp, Jeffrey Harford, Liam Hudson, Ngai-Shing Mok, Felicity Eakins, Biological Structure and Function Section, Biomedical Sciences Division, Faculty of Medicine, Imperial College London, Exhibition Road, London SW7 2AZ, UK. Carlo Knupp now at School of Optometry and Vision Sciences, University of Cardiff, Cardiff CF10 3NB. Manfred Roessle at ESRF, rue Jules Horowitz, Grenoble, France. Tom Irving, BioCAT, Dept. Biological, Chemical and Physical Sciences, Illinois Institute of Technology, Chicago, IL 60616, USA. Michael Reedy, Dept of Cell Biology, Duke University, Durham, NC 27710, USA.

Squire *et al.*, 1998; 2003a,b,c,d). Building on that work, this review now addresses three main questions: [1] Do time-resolved low-angle X-ray diffraction patterns from bony fish muscle provide useful information about contractile events in active muscle?, [2] Do X-ray patterns from bony fish muscle provide any information about the distribution in the sarcomere of proteins such as titin, C-protein (MyBP-C), troponin and nebulin which are essential for proper muscle function? and [3] Do the low-angle X-ray diffraction patterns from insect flight muscle provide information about the highly-developed stretch-activation mechanism in that muscle? Some of the answers to these questions are unequivocal. Others are just beginning to emerge, but illustrate fascinating possibilities about the molecular events involved in the sliding filament mechanism.

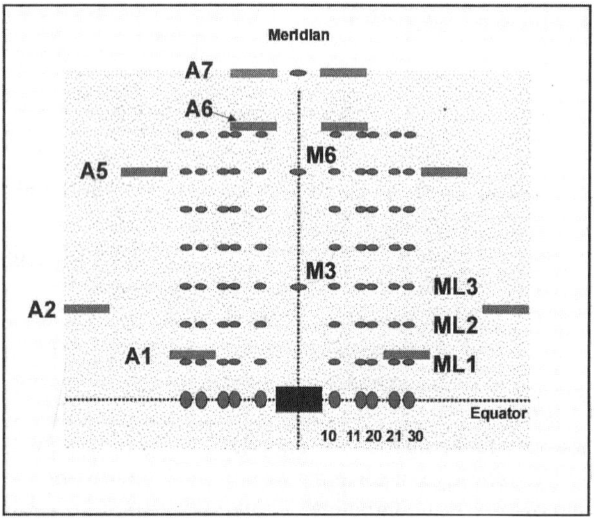

Figure 1. Simulation of the top half of the low-angle X-ray diffraction pattern from bony fish muscle. (Harford and Squire, 1986). Layer-lines ML1 to ML3 are orders of a 3 x 143 Å repeat. The meridional reflection on the third layer-line is the well-known M3 reflection at 143 Å. Labelling on the left shows some of the actin layer-lines based on a repeat of about 365 Å. A1 is the first actin layer-line at about 365 Å, A2 is the second actin layer-line at 365/2 Å, known to be responsive to movements in tropomyosin. A6 and A7 are the layer-lines at 59 Å and 51 Å respectively. The whole myosin pattern is crossed by vertical row-lines which index like the equatorial reflections as the 10, 11, 20, 21 and 30 etc reflections from the hexagonal A-band lattice.

2. BASIC FEATURES OF BONY FISH MUSCLE DIFFRACTION PATTERNS

The main features of fish muscle low-angle X-ray diffraction patterns are summarised in Figure 1. Only the top half of the pattern is shown; the pattern can be mirrored in the horizontal line (the equator) through the position of the undiffracted X-ray beam. The vertical line through this point, a line parallel to the long axis of the muscle, is called the meridian. The pattern consists of two main sets of layer-lines, one from the myosin filaments and with an axial repeat of 429 Å, and one from actin filaments with an axial repeat of about 365 Å. Some of the myosin layer-lines are labelled ML1 (429 Å) to ML3 (143 Å) in Figure 1. The third layer-line at a spacing of 143 Å has a strong meridional component labelled M3, which corresponds to the equivalent reflection from frog muscle studied extensively by Irving, Lombardi and their colleagues (Irving *et al.*, 1992; Lombardi *et al.*, 1995; Piazzesi *et al.*, 2002; Reconditi *et al.*, 2003) to probe the myosin head mechanism in active muscle. The myosin layer-lines are crossed by vertical row-lines which pass through the equatorial reflections and are labelled in Figure 1 as 10, 11, 20, 21 and 30. The equator provides information about the muscle when viewed down the fibre axis, and the equatorials and row-lines relate to the hexagonal lattice of myosin and actin filaments in the muscle A-band (as discussed later: see Figure 5(b)).

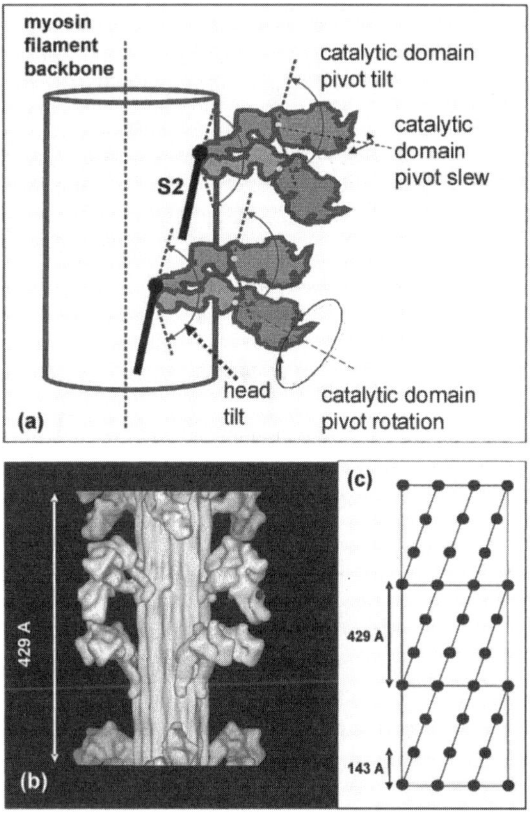

Figure 2. (a) Description of the myosin head positions on the myosin filament surface in terms of some key parameters. The position of an individual head in space can be defined by parameters such as the head tilt (illustrated), and the head slew about an axis parallel to the filament axis and head rotation about its own long axis (not illustrated). With the neck position thus determined, the head shape can then be defined by the catalytic domain pivot tilt, pivot slew and pivot rotation assuming that the catalytic (motor) domain and the neck domain (lever arm) both act as more or less rigid bodies at the resolution being considered. These parameters can change the head shape, among other things, from the rigor shape (Rayment *et al.*, 1993) to a 'pre-powerstroke' shape (M.ADP.Pi; Xu *et al*, 1999; Dominguez *et al.*, 1998). (b) The crossbridge array in bony fish muscle determined by Hudson *et al.* (1997). (c) The radial net showing the 3-stranded helix of myosin head pairs which approximately describes the myosin filament array in vertebrate striated muscle thick filaments.

The myosin layer-lines in diffraction patterns from resting muscle show first of all that the myosin head pairs of the myosin molecules that form the myosin filaments lie roughly on a helical net as in Figure 2(c) with the heads lying approximately on three helical strands each of pitch 3 x 429 Å and with 9 head pairs spaced at about 143 Å axial intervals in one turn of each strand. The fact that there are three strands reduces the axial repeat to just 429 Å (Squire, 1972). The myosin head shape is known from the work of Rayment *et al.* (1993a), and other crystallographic studies show that the position of the neck region (lever arm) of the myosin head on the motor (catalytic) domain can vary depending on the nucleotide bound, with analogues such as ADP.VO_4 (smooth muscle myosin; Dominguez *et al.*, 1998) and ADP.AlF_4 (molluscan myosin; Houdusse *et al.*, 2000) giving a shape which is supposed to mimic the pre-powerstroke state ([A]M.ADP.Pi). This means that the head shape and position can be parameterised as in Figure 2(a) and that a parameter search can be carried out to find the optimal head positions and shapes to account for the intensities on the observed myosin layer-lines. This procedure can now be carried out on properly stripped intensity data (e.g. with CCP13 software; Squire *et al.*, 2003; www.ccp13.ac.uk) using a simulated annealing approach with the program MOVIE (Hudson, 1996; AL-Khayat *et al.* 2004). It has been applied successfully to bony fish muscle (Hudson *et al.*, 1997; Squire *et al.*, 1998; see Fig. 2(b)) and to insect flight muscle (AL-Khayat *et al.*, 2003; see later discussion).

3. TIME-RESOLVED X-RAY DIFFRACTION FROM BONY FISH MUSCLE

An important probe of myosin head behaviour in active muscle, pioneered by H.E.Huxley (e.g. Huxley and Brown, 1967; Huxley *et al.*, 1980; 1982) is time-resolved X-ray diffraction. Since the molecular contributors to particular parts of the X-ray pattern can sometimes be separated (Fig. 1), these parts can have different time-courses and they thus have the potential to illustrate the sequence of molecular events that takes place during muscle activation and force-generation (see Harford and Squire, 1992; 1996; Martin-Fernandez *et al.*, 1994; Squire, 1998; 2000; Squire *et al.*, 2003c). Time-resolved diffraction studies can now be carried out relatively easily because of the development of the RAPID X-ray detector (Lewis *et al*, 1997). This detector can be read out in much less than 1 millisecond permitting very fast time-resolution (e.g. 0.1 to 1 ms) if the diffraction peaks being studied are strong enough. We have used the RAPID detector on beamline 16.1 at the Daresbury Synchrotron to record changes on a millisecond time-scale in the low-angle X-ray pattern from contracting bony fish mucle. The muscles were treated as described by Harford and Squire (1992). Some preliminary results from this study are shown in Figure 3. The full analysis will be presented elsewhere (Mok *et al.*, ms in preparation). The plots show the rise of tension at the start of tetanic contractions of plaice fin muscle and the corresponding intensity or peak position changes that occur. These have all been normalised to show the %age change from the initial value in patterns from resting muscle to the final values at the tension plateau. In summary, the equatorial 11 reflection changes very fast (Fig. 3(a)), with a time to 50% change of about 20 ms, it overshoots slightly, and then steadies to a high plateau value. We have shown this previously (Harford and Squire, 1992), along

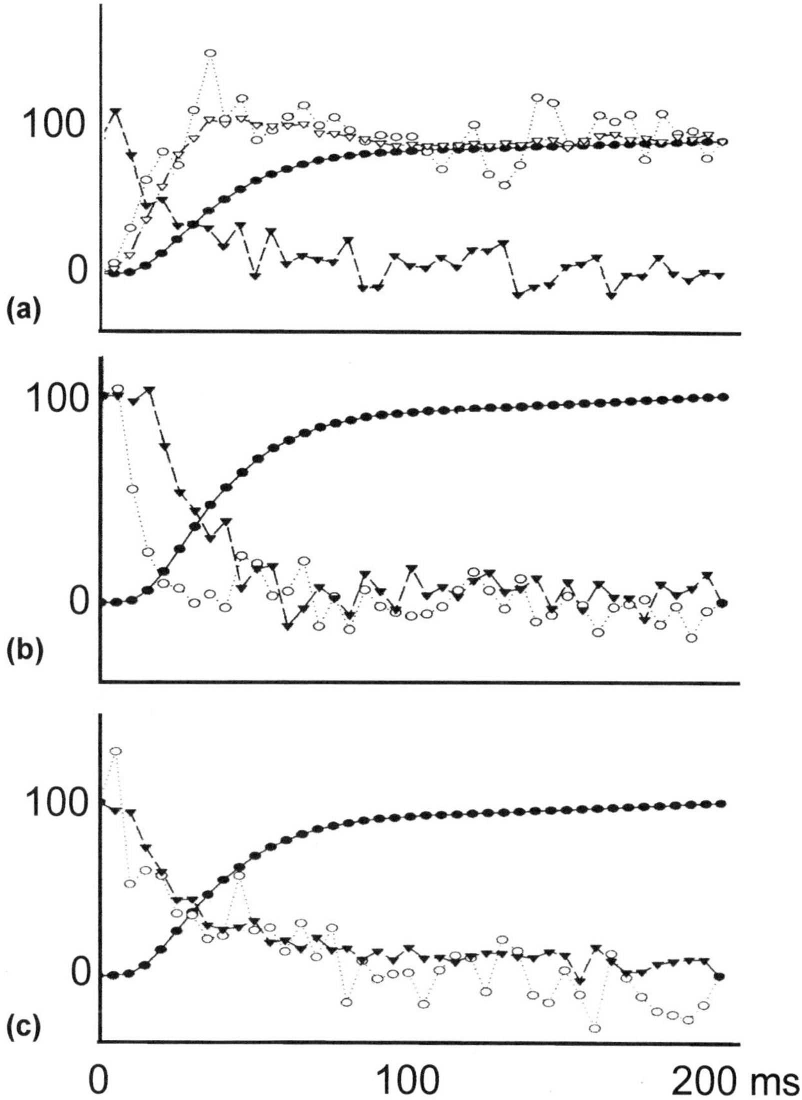

Figure 3. Preliminary analysis of the tension and intensity timecourses from active bony fish muscle (plaice) X-ray diffraction patterns recorded using the RAPID detector on beamline 16.1 at the CLRC Daresbury Synchrotron Radiation Source. In each panel the solid circular symbols show the tension timecourse. This and the illustrated intensity changes have been normalised in terms of %age change between the initial relaxed state and the tension plateau value. Following activation the maximum tension is reached in about 70 to 80 ms with a time to 50% change of about 35 ms. In (a) the open triangles show the intensity increase of the 11 equatorial reflection. It changes much more quickly than tension and overshoots slightly before reaching a steady value at the tension plateau. The open circles show the change of the 2nd actin layer-line (A2 in Figure 1) which occurs very much in step with the change in the 11 reflection. The solid triangles show the change of the inner part of the actin 1st layer-line (A1). (b) Open circles show the change in the M3 intensity and filled triangles show the %age change in spacing of this reflection. (c) Open circles show the change of the 3rd myosin layer-line (ML3) off the meridian and filled triangles shown the change in intensity at the 11 and 20 row-line positions on the myosin 1st layer-line (ML1).

with the fact that the 10 intensity changes much more slowly and is virtually in step with tension with 50% change occurring after about 35 ms.

Fig. 3(a) also shows that the 2nd actin layer-line (A2 in Fig. 1), thought to be associated with the tropomyosin shift during muscle activation (Kress *et al*., 1986; Squire and Morris, 1998) changes at about the same speed as the 11 equatorial reflection. Figure 3(b) shows that the drop in the intensity of the M3 meridional reflection also occurs very quickly. 50% change occurs at about 15 ms. However the change in spacing of the M3 (it changes from about 143.2 Å in resting patterns to over 144 Å in active patterns), is much slower; like the tension, the 50% change occurs at close to 35 ms. Figure 3(c) shows that the off-meridional myosin layer-lines, ML1 and ML3, change relatively slowly compared to the M3 meridional peak, with a 50% change after about 25 ms. Finally, the closed triangles in Figure 3(a) show the change at the inner end of the 1st actin layer line (A1). Of significance here is the fact that there is an intensity drop, that the 50% change occurs at around 20 ms, but that the change then seems to slow down. More detailed fitting of these time-courses will enable all of these changes to be properly evaluated. However, even at this stage, all of the observations fit into a plausible scenario.

The story that can come from this is as follows. Upon electrical stimulation of the plaice muscle, Ca^{2+} ions are released and bind to troponin C, thereby switching on the thin filament. This involves a movement of tropomyosin (Squire and Morris, 1998; Vibert *et al*., 1997; Craig and Lehman, 2002) as evidenced by the change in intensity of the 2nd actin layer-line (A2). It is envisaged that, even in resting muscle, myosin heads in the resting structure illustrated in Figure 2(b) occasionally pop out of their regular positions (as in any equilibrium structure) and can momentarily come near actin. If the thin filament is not switched on, nothing happens and the head pops back again. However, after the stimulus and the resulting activation of the thin filament, such a head can bind in the putative weak-binding, first attached state (Harford and Squire, 1992; Holmes, 1997) which causes the increase in the 11 equatorial intensity and the drop in intensity of the M3 meridional reflection. After a short time, some heads then move to the strong binding (rigor-like) states, thus changing the 10 equatorial reflection, which is sensitive to strong but not weak states, and producing tension. So there will be a series of changes which are early and related to thin filament activation and the binding of heads in the weak-binding state, and then later events associated with the transition to strong-binding and force-production.

Two apparent anomalies remain, the drop in intensity of part of the 1st actin layer-line (Fig. 3(a)) and the slow change of the myosin layer-lines (Fig. 3(c)). If myosin heads are labelling actin to produce tension, one might expect the actin layer-lines to increase in intensity due to the added mass. Indeed this occurs in parts of the pattern. However, the movement of tropomyosin involved in thin filament activation, which increases the 2nd actin layer-line, will itself cause a <u>decrease</u> in the 1st actin layer-line. We might therefore expect a reduction in part of the A1 in step with the A2 increase, but this may start to slow down later when there are plenty of strong-binding heads stereospecifically-labelling actin. The myosin layer-lines are also interesting. It was shown by Squire and Harford (1988) and Harford and Squire (1992), and will be further documented elsewhere (Squire *et al*., 2004b), that weak-binding heads on actin give a set of layer-lines with a 429 Å repeat, like the myosin pattern. The drop in intensity of these layer-lines due to loss of the resting myosin head array (Fig. 2(b)) is therefore partially compensated by the generation of new myosin-like layer-lines from the weak-binding heads on actin. The net effect is that the time-course of changes in these layer-lines is slowed down.

4. NEW HIGH DEFINITION LOW-ANGLE X-RAY DIFFRACTION PATTERNS FROM BONY FISH MUSCLE - A FEW SURPRISES

Following on from the last Sections, where much of the low-angle X-ray diffraction pattern from bony fish muscle has been explained and the observed time-courses can be accounted for in terms of plausible myosin head movements, we now discuss some totally new observations which will be reported and analysed fully elsewhere (Squire *et al.*, 2004a).

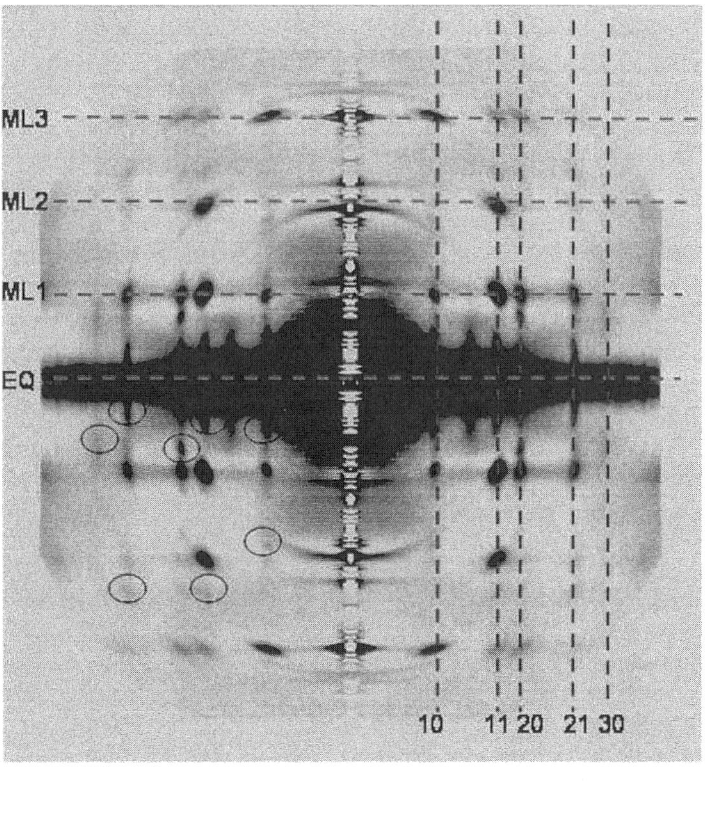

Figure 4. Low-angle X-ray diffraction pattern from resting plaice fin muscle recorded on CCD detector with a camera length of 10m on line ID02 at the ESRF in Grenoble, France. The horizontal dashed lines show the myosin layer-lines labelled ML1, ML2 and ML3 in Figure 1 and the equator (EQ) and the vertical dashed lines show the 10, 11, 20, 21 and 30 row-lines. Some of the previously unobserved reflections are ringed in the lower left quadrant. Data from Squire *et al.* (2004a).

Figure 4 shows a low-angle X-ray diffraction pattern from resting plaice muscle recorded on a high-resolution CCD detector with a 10m camera length on the ID02 beamline at the ESRF Synchrotron in Grenoble. The horizontal dotted lines show the positions of the equator and first three myosin layer-lines (ML1, ML2 and ML3; cf. Fig. 1) and the vertical dotted lines show the row-lines through the 10, 11, 20, 21 and 30 equatorial reflections. However, what is very apparent is that there are very many intensity peaks, some highlighted by rings in Figure 4, which do not fit on the myosin layer-lines. Nor do they fit on the expected actin layer-line positions.

What we have done is to try to account for these new reflections in terms of plausible arrangements of the 'extra' proteins titin, C-protein (MyBP-C), troponin and nebulin. We have previously proposed that in resting vertebrate striated muscle the N-terminal ends of MyBP-C may be attached to actin, thereby being in a labelling pattern that is partially defined by the actin geometry (Squire *et al.*, 2003a). It turns out that this arrangement, which was originally developed to help to explain some apparently anomalous features of the meridian of the X-ray patterns from these muscles, can also explain some of the new reflections. Others seem to require to be explained by something with the axial periodicity of troponin. Although troponin pairs are spaced axially at 385 Å intervals on the thin filament, in vertebrate striated muscles they lie on a slowly twisting helix. The structure gives meridional reflections at orders of 385 Å (apart from any interference effects) and off-meridional reflections at axial positions roughly halfway between the meridionals. They also give very closely spaced layer-lines just above and just below the meridional peaks. However, troponin and C-protein do not explain all of the new features in Figure 4. We have also sought explanations for these and other reflections in terms of possible arrangements of titin and nebulin, but the required models are perhaps less plausible than the troponin and C-protein stories. There is, therefore, much in the pattern that still requires explanation. However, one feature that does come out, if the troponin story is correct, is that, since the new reflections all lie on the same row-lines as the other myosin reflections and these row-lines are associated with the hexagonal A-band lattice (Fig. 5(b); Harford *et al.*, 1994), the implication is that the troponins, hence the thin filaments themselves, are also well-ordered in the fish muscle A-band. Even in vertebrate muscle, the troponin molecules may lie in positions with specific structural relationships, yet to be determined, to the myosin heads.

5. ACTIN TARGET AREA LABELLING IN FISH AND INSECT MUSCLES

A specific relationship between filaments in the A-band is a key and well-known feature of asynchronous insect flight muscle. Figure 5 illustrates the effects that lattice geometry and filament axial repeats can have on the nature of the interactions between myosin and actin as the filament overlap changes. As illustrated in Figure 5(a), one can imagine a myosin head on a particular origin (C) on the myosin filament surface searching round in space under Brownian motion until it finds a suitable binding site on actin at point B. However, there will be a limit to the range over which the head can search, here defined by the azimuthal reach $\pm\Delta\theta$ and the axial reach $\pm\Delta Z$ of the head, and there will be preferred azimuthal positions for the actin monomer as defined by the angles ϕ and $\Delta\phi$ which define actin target area positions and size as discussed previously by Squire and

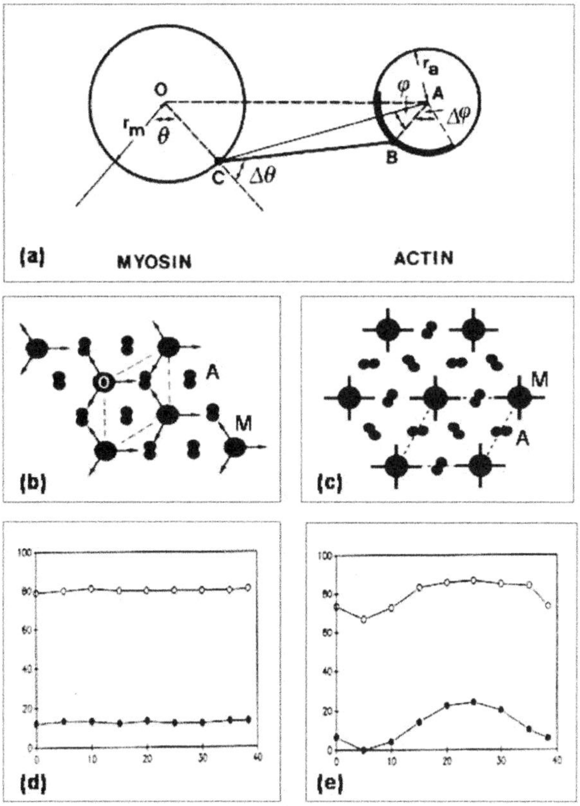

Figure 5. (a) Geometry of interaction of myosin heads from a site C on the myosin filament (left) to an actin site (B) on an adjacent actin filament (right) (after Squire and Harford, 1988; Squire, 1992). For a given head start position (C; defined by the angle θ), the interaction geometry can be defined in terms of an azimuthal search limit (Δθ) and a corresponding axial search limit (ΔZ; not illustrated) of the head, combined with the position of the actin site relative to the line CA joining the myosin head origin position to the actin filament axis. This is defined by the angles ϕ and Δϕ. The precise implications of the geometry in (a) depend very much on the lattice type that is involved. (b) The filament lattice arrangement (the simple lattice; see Luther *et al.*, 1981) in bony fish muscle illustrating that the 3-fold symmetric myosin filaments (M) all have the same orientations as do all the actin filaments (A). (c) As for (b) but for the 4-fold symmetric myosin filament array in insect (*Lethocerus*) flight muscle. The myosin filaments (M) all have the same orientations at least locally in the *Lethocerus* A-band (although there may be different orientations elsewhere in the same myofibril giving a domain structure; Freundlich and Squire, 1983). However, the actin filaments (A) show systematic rotations. (d,e) The effects of applying the search parameters in (a) to the two lattices in (b) and (c) and then changing the axial positions of the actin and myosin filaments as would occur with changes in sarcomere length. Because of actin target area labelling (see Squire, 1972; Squire and Harford, 1988; Squire, 1992), the ease of attachment of a head to an actin site depends on the azimuth of the actin site. In (e), because the actin filament array matches the myosin head geometry (Wray, 1979), for a given set of head search parameters (results for two sets are shown in (d) and (e)), as the sarcomere length changes the probability of attachment also changes. Because the common periodicity in this case is 385 Å, the pattern in (e) repeats over this distance. Wray (1979) suggested that this lattice matching and mis-matching as the sarcomere length changes might be associated with the stretch-activation properties of asynchronous insect flight muscle. This possibility is supported by the response of vertebrate muscle in (d) where the level of attachment does not change significantly as the filament overlap changes. [After Squire, 1992]

Harford (1988) and Squire (1972). The effects of changing these search and target parameters for different filament lattice geometries can be explored as in Squire (1992) using the program MusLABEL, which will shortly be made available on the CCP13 website (www.ccp13.ac.uk; Squire and Knupp, 2004). Figure 5(b) and (c) show the filament lattices in bony fish muscle and insect (*Lethocerus*) flight muscle respectively. In (b) for fish muscle the myosin filaments all have the same orientations around their long axes as do the actin filaments. In (c), for insect muscle, the myosin filaments are identically oriented, at least over substantial domains (Freundlich and Squire, 1983), but the actin filaments have systematic rotations between them. Also in insect muscle the myosin and actin axial repeats are closely related, being 3 x 385 Å and 385 Å respectively, whereas in fish muscle they are quite different; 429 Å and 365 Å. Using the MusLABEL program one can set the parameters to get a small number of heads attached or a large number and then, with these parameters fixed, simply change the relative axial positions of the myosin and actin filament arrays. The effects are quite different in the two muscle types.

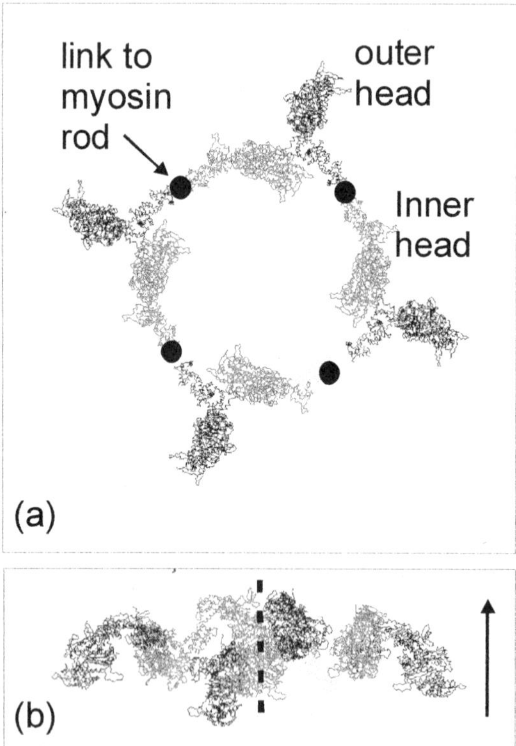

Figure 6. The structure of a single myosin head 'crown' in relaxed insect (*Lethocerus*) flight muscle viewed down the myosin filament axis from the M-band (a) and in nearly side-on view (b). The arrow in (b) is towards the Z-band. The dots in (a) show the origin positions of the head pairs where they would link back to the myosin rod. The structure has an outer head apparently supported by an inner head lying around the myosin filament backbone and from a different myosin molecule. The two heads from one molecule form an S-shaped structure in this view. In (b) the outer catalytic domain of the right-hand head is circled. It is almost in the correct configuration to attach directly to actin. Data from the analysis of AL-Khayat *et al.* (2003).

In vertebrate muscle the number of heads attached is largely independent of the axial shift of the filaments, whereas in insect muscle there is a periodic fluctuation in the number attached as the filaments slide (Fig. 5(e)), just as observed in the physiological stretch-activation response reported by Abbott and Cage (1984), as accounted for by Wray (1979) in terms of match and mis-match of the thick and thin filament lattices, and as specifically modelled by Squire (1992).

6. MYOSIN HEAD ORGANISATION IN RELAXED INSECT FLIGHT MUSCLE AND THE STRETCH-ACTIVATION MECHANISM

The simulated annealing method described above for bony fish muscle (Fig. 2; Hudson *et al.*, 1997) has also been applied to resting insect (*Lethocerus*) flight muscle diffraction data (AL-Khayat *et al.*, 2003; Squire *et al.*, 2003e). In this case the head array appears to be purely helical with a 145 Å spacing between head levels (crowns) and the result of the analysis for one crown of heads is shown in Figure. 6. Here the filaments are 4-stranded so the crown has 4-fold rotational symmetry (Morris *et al.*, 1991). The organisation of the heads is quite different from that proposed for fish muscle. Firstly, with regard to head shape, the modelled structure is very close to the pre-powerstroke shapes, mimicking (A)M.ADP.Pi, as observed by Dominguez *et al.* (1998) and Houdusse *et al.* (2000) using protein crystallography. Secondlly, with regard to head position, one head from each myosin molecule (the inner head) lies around the filament surface and appears to provide support for another head from an adjacent myosin molecule which projects out from the filament surface (the outer head). This shape and configuration has the effect of putting the catalytic domain of the outer heads in a position that is close to that needed for the motor domain to bind to actin.

Some of the possible implications of this arrangement are analysed in Figure 7. Here it is imagined that one is a viewer situated on the axis of the myosin filament (i.e. at position O in Figure 5(a)) and looking outwards along the direction of an outer head in Figure 6 assumed to point directly towards an actin filament (at position A in Fig. 5(a)). It is also assumed that the actin filament is presenting an actin target area to the myosin head, so that there is an actin monomer suitably positioned for the myosin head to attach to. Figure 7 illustrates two possible scenarios, (a) and (b), concerning what might happen. In (a), the myosin head on the left is in the resting position of the outer heads illustrated in Figure 6, but we are here looking more or less out along the neck region or lever arm towards the catalytic domain. Assuming, to start with, that the head is not already interacting in any way with the actin filament, the middle image in (a) shows what the head would have to do, without changing its shape, to make an attachment to actin with the motor domain in the known rigor (strong binding) conformation. The head would have to tilt down slightly and to rotate by about 60° to get the catalytic domain in the right place. This would then presumably be mimicking the initially attached, weak-binding, pre-powerstroke state. The change in head state to strong-binding would then be associated with a change in shape of the attached head from the pre-powerstroke shape to the rigor shape (Rayment *et al.*, 1993b), as illustrated in the right hand picture in Figure 7(a). The net result would be a 100 Å movement between the actin-attached end of the head and the outer part of the lever arm attached to myosin S2.

The second scenario in Figure 7(b) is rather different. It has recently been shown that asynchronous insect flight muscle has two kinds of troponin C (TnC), one of which is associated with normal Ca^{2+}-activation and the other solely with stretch-activation

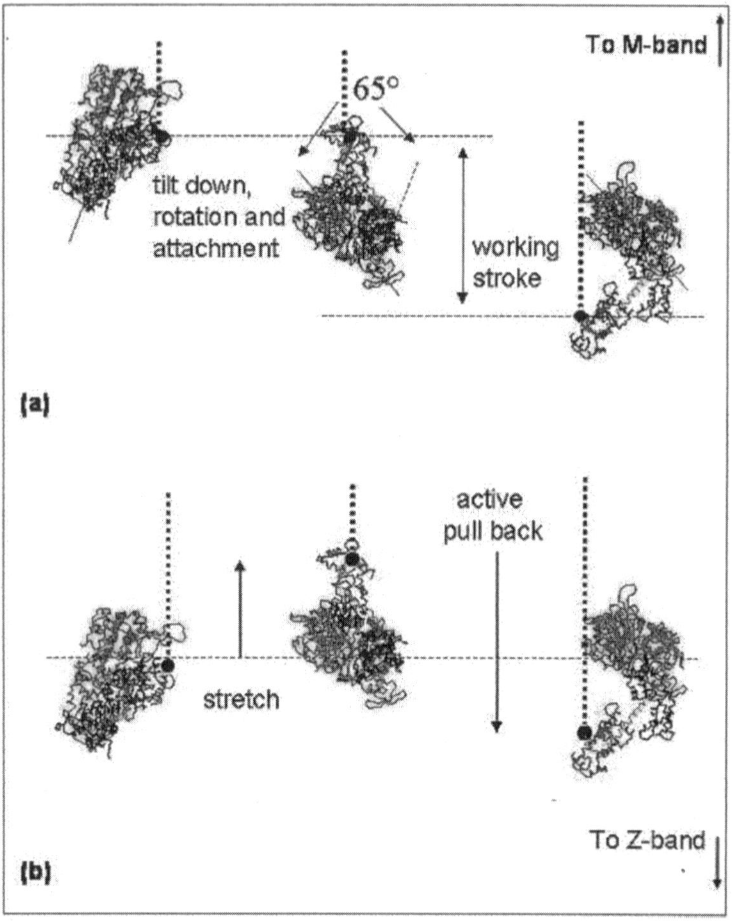

Figure 7. Illustration of two different scenarios for the attachment of myosin heads to actin in insect (*Lethocerus*) flight muscle. In all cases the right hand, circled, head in Figure 6 is being viewed from the middle of the myosin filament looking directly towards an adjacent actin filament (not shown) which is presumed to be in a position where it presents an actin monomer in an appropriate orientation for this particular myosin head to attach. In both (a) and (b) the left figures have the myosin head in exactly the configuration that was modelled by AL-Khayat *et al.* (2003) for the outer heads in relaxed insect flight muscle. The central figure in each has these heads with exactly the same shape as illustrated on the left, but with the head tilted and rotated to bring the motor domain into exactly the configuration to attach to actin with the catalytic domain in the rigor position on actin. This motor domain position is unchanged in the right hand figures, but the neck or lever arm has now moved on the motor domain to generate the rigor myosin head shape. The axial movement of the outer end of the neck where it links to myosin S2 (illustrated by the blob on the end of the vertical dashed line) between the middle and right hand images is about 100 Å. As detailed in the text, the scenario in (a) might represent normal Ca^{2+}-activated contraction, whereas (b) might represent stretch-activation.

(Agianian *et al.*, 2004). In this muscle, if enough Ca^{2+} is supplied, the muscle produces fairly normal sustained contractions. However, if the muscle is given a certain minimum level of Ca^{2+} and is subjected to a small stretch, the tension increases rapidly after a short delay, thus allowing the muscle to carry out oscillatory work (Pringle, 1978; Tregear, 1983). The mechanical responses of this muscle as a function of Ca^{2+} level have recently been studied by Linari *et al.* (2004) who have shown that the higher the initial Ca^{2+} level, causing an increased baseline steady tension, the less added stretch-activated tension occurs, so that the two processes sum to give a constant combined tension.

The contractile machinery in insect flight muscle is also unusual in that, as well as two types of TnC, there is an unusual troponin T (TnT) that has a negatively-charged extension at the C-terminus which is absent in the vertebrate TnT. There is also an unusual inhibitory component (TnH) which is like the vertebrate TnI, but with an extra Pro-Ala-rich sequence fused to its C-terminus (Bullard *et al.*, 1988). There is also evidence, at least in Drosophila, for an extension on the myosin regulatory light chain which can interact with actin (Moore *et al.*, 2000). These features may suggest that even in resting muscle, or at least with a small amount of Ca^{2+} present, the outer myosin heads may be interacting in some way with the thin filament complex, but without the head going through its force-generating cycle. A small amount of stretch could then make the attached head swing slightly, as in Figure 7(b) middle, so that (i) an appropriate contact with actin can be made, so that (ii) the head can perhaps have an effect locally on the position of the tropomyosin strand adjacent to its attachment site on actin thus locally activating the filament, and so that (iii) the head becomes free to go through its force generating steps to end up in the rigor configuration. The net effect of the stretch is that the head pulls back. With the systematic matching of the myosin and actin filament arrays in the insect A-band (Fig. 5(e)), such stretch-activation events would take place at multiple sites where the heads and actin line up. Change of the filament overlap would first take all of the heads away from their optimal interaction sites for stretch-activation and then bring them back into register after 385 Å, as discussed earlier (Fig. 5(e)). In summary, many different results on the structure of insect flight muscle may be coming together to start to explain the all-important stretch-activation feature of the muscle that enables the insects to fly.

7. CONCLUSION

In this short review a number of recent X-ray diffraction results on the highly ordered striated muscles in insects and in bony fish have been briefly described. What is clear is that this technique applied to muscles which are amenable to rigorous analysis, taken together with related data from other sources (e.g. protein crystallography, biochemistry, mechanics, computer modelling) can provide not only the best descriptions yet available on the myosin head organisations on different myosin filaments in the relaxed state, but can also show the sequence of molecular events that occurs in the contractile cycle, and may also help to explain such phenomena as stretch-activation. X-ray diffraction is clearly an enormously powerful tool in studies of muscle. It has already provided a wealth of detail on muscle ultrastructure; it is providing ever more fascinating insights into the molecular events involved in the 50-year old sliding filament mechanism, and there remains a great deal more potential that is as yet untapped.

ACKNOWLEDGEMENTS

We are indebted to Bruce Baumann for assistance in collecting fibre X-ray diffraction patterns from IFM at the Argonne/ APS/ BioCAT beamline. We acknowledge specific support to JMS for the insect work from a UK BBSRC project grant (# 28/S10891) and for the fish work from the Wellcome Trust (#061729). We acknowledge support of the ESRF for work on ID-02, experiment number SC-888, particularly the help of Drs. T. Narayan, P. Boesecke and P. Panine. MKR was supported by NIH AR-14317. CCP13 software was developed as part of UK BBSRC/ EPSRC funded projects (e.g. # 28/B10368 and 28/B15281; www.ccp13.ac.uk). Use of the Advanced Photon Source was supported by the U.S. Department of Energy, Basic Energy Sciences, Office of Energy Research, under Contract No. W-31-109-ENG-38. BioCAT is a U.S. National Institutes of Health-supported Research Centre RR08630.

REFERENCES:

Agianian, B., Krzic, U., Qiu, F., Linke, W.A., Leonard, K. and Bullard, B., 2004, A troponin switch that regulates muscle contraction by stretch instead of calcium. *EMBO J.*, **23**, 772-779.

AL-Khayat, H.A., Hudson, L., Reedy, M.K., Irving, T.C. and Squire, J.M., 2003, Myosin head configuration in relaxed insect flight muscle: X-ray modelled resting crossbridges in a pre-powerstroke state are poised for actin binding. *Biophys. J.* **85**, 1063-1079.

AL-Khayat, H.A., Hudson, L., Reedy, M.K., Irving, T.C. and Squire, J.M., 2004, Modelling oriented macromolecular assemblies from low-angle X-ray fibre diffraction data with the program MOVIE: Insect Flight Muscle as example. *Fibre Diffraction Review* **12**, 50-60.

Craig, R. and Lehman, W., 2002, The ultrastructural basis of actin filament regulation. *Results Probl. Cell Differ.* **36**, 149-169.

Dominguez, R., Freyzon, Y., Trybus, K. M. and Cohen, C., 1998, Crystal structure of a vertebrate smooth muscle myosin motor domain and its complex with the essential light chain: visualization of the pre- power stroke state. *Cell* **94**, 559-571.

Elliott, G.F., Lowy, J. and Millman, B.M., 1967, Low-angle X-ray diffraction studies of living striated muscle during contraction. *J. Molec. Biol.* **25**, 31-45.

Freundlich, A. and Squire, J.M. "Three-dimensional structure of the Insect (*Lethocerus*) flight muscle M-band" J. Mol. Biol. **169**, 439-453 (1983).

Harford, J.J. and Squire, J.M., 1986, The 'crystalline' myosin crossbridge array in relaxed bony fish muscles, *Biophys. J.* **50**, 145-155.

Harford, J. J. and Squire, J. M., 1992, Evidence for structurally different attached states of myosin cross-bridges on Actin during contraction of fish muscle. *Biophys. J. 63*, 387-396.

Harford, J.J. and Squire, J. M., 1997, Time-resolved diffraction studies of muscle using synchrotron radiation. *Rep. Prog. Phys.* **60**, 1723-1787.

Harford, J.J., Luther, P.K. and Squire, J.M., 1994, Equatorial A-band and I-Band X-ray Diffraction from Relaxed and Active Fish Muscle: Further details of Myosin Crossbridge Behaviour. *J. Molec. Biol.* **239**, 500-512.

Holmes, K. C., 1996, Muscle proteins-their actions and interactions. *Curr. Opin. Struct. Biol.* **6**, 781-789.

Houdusse, A., Szent-Gyorgyi, A. G., and Cohen, C., 2000, Three conformational states of scallop myosin S1. *Proc. Natl. Acad. Sci. U S A* **97**, 11238-11243.

Hudson, L., 1996, PhD Thesis, University of London.

Hudson, L., Harford, J. J., Denny, R. C., and Squire, J. M., 1997, Myosin head configuration in relaxed fish muscle: resting state myosin heads must swing axially by up to 150 A or turn upside down to reach rigor. *J. Molec. Biol.* **273**, 440-455.

Huxley, H.E. and Brown, W., 1967, The low-angle X-ray diagram of vertebrate striated muscle and its behaviour during contraction and rigor. *J. Molec. Biol.* **30**, 383-434.

Huxley, H.E., Faruqi, A.R., Bordas, J., Koch, M. and Milch, J.R., 1980, The use of synchrotron radiation in time-resolved X-ray diffraction studies of myosin layer-line reflections during muscle contraction. *Nature* **284**, 140-143.

Huxley, H.E., Faruqi, A.R., Kress, M., Bordas, J. and Koch, M.H.J., 1983, Changes in the X-ray reflections from contracting muscle during rapid mechanical transients and their structural implications. *J. Molec. Biol.* **169**, 469-506.

Irving, M., Piazzesi, G., Lucii, L., Sun, Y. B., Harford, J. J., Dobbie, I. M., Ferenczi, M. A., Reconditi, M., and Lombardi, V., 2000, Conformation of the myosin motor during force generation in skeletal muscle. *Nat. Struct. Biol.* 7, 482-485.

Kabsch, W., Mannherz, H.G., Suck, D., Pai, E.F., and Holmes, K.C., 1990, Atomic structure of the actin: DNase I complex. *Nature* **347**, 37-44

Kress, M., Huxley, H.E., Faruqi, A.R. and hendrix, J., 1986, Structural changes during activation of frog muscle studied by time-resolved X-ray diffraction. *J. Molec. Biol.* **188**, 325-342.

Lewis, R.A., Helsby, W.I., Jones, A.O., Hall, C.J., Parker, B., Sheldon, J., Clifford, P., Hillen, M., Sumner, I., Fore, N.S., Jones, R.W.M. and Roberts, K., 1997, The "RAPID" high rate large area X-ray detector system. *Nucl. Inst.Methods Phys. Res. A,* **392**, 32-41.

Linari, M., Reedy, M.K., Reedy, M.C., Lombardi, V. and Piazzesi, G., 2004, Ca-Activation and Stretch-Activation in Insect Flight Muscle. *Biophys. J.* **87**, 1101-1111.

Lombardi, V., Piazzesi, G., Ferenczi, M.A., Thirlwell, H., Dobbie, I. and Irving, M., 1995, Elastic distortion of myosin heads and repriming of the working stroke in muscle. *Nature* **374**, 553-555.

Luther, P. K., and Squire, J. M., 1980, Three-dimensional structure of the vertebrate muscle A-band. II. The myosin filament superlattice. *J. Molec. Biol.* **141**, 409-439.

Luther, P.K., Munro, P.M.G. and Squire, J.M., 1981, Three-dimensional structure of the vertebrate muscle A-band III: M-region structure and myosin filament symmetry. *J. Molec. Biol.* **151**, 703-730.

Luther, P. K., Squire, J. M., and Forey, P. L., 1996, Evolution of myosin filament arrangements in vertebrate skeletal muscle. *J. Morphol.* **229**, 325-335.

Martin-Fernandez, M.L., Bordas, J., Diakun, G., Harries, J., Lowy, J., Mant, G.R., Svennson, A. and Towns-Andrews, E, 1994, Time-resolved X-ray diffraction studies of myosin head movements in live frog sartorius muscle during isometric and isotonic contractions. *J. Mus. Res. Cell Motil.* **15**, 319-348.

Moore, J.R., Dickinson, M.H., Vigoreaux, J.O. and Maughan, D.W., 2000, The effect of removing the N-terminal extension of the *Drosophila* myosin regulatory light chain upon flight ability and the contractile dynamics of indirect flight muscle. *Biophys. J.* **78**, 1431-1440.

Morris, E. P., Squire, J. M., and Fuller, G. W., 1991, The 4-stranded helical arrangement of myosin heads on insect (*Lethocerus*) flight muscle thick filaments. *J. Struct. Biol.* **107**, 237-249.

Piazzesi , G., Reconditi, M., Linari, M., Lucii, L., Sun, Y.B., Narayan, T., Boesecke, P., Lombardi, V. and Irving, M., 2002, Mechanism of force generation by myosin heads in skeletal muscle. *Nature* **415**, 659-662.

Pringle, J.W.S., 1978, The Croonian Lecture, 1977: Stretch-activation of muscle: function and mechanism. *Proc. Roy. Soc. Lond. B.* **201**, 107-130.

Rayment, I., Rypniewsky, W. R., Schmidt-Bäse, K., Smith, R., Tomchick, D. R., Benning, M. M., Winkelmann, D. A., Wesenberg, G., and Holden, H. M., 1993a, Three-dimensional structure of myosin subfragment-1: a molecular motor. *Science* **261**, 50-58.

Rayment, I., Holden, H. M., Whittaker, M., Yohn, C. B., Lorenz, M., Holmes, K. C., and Milligan, R. A., 1993b, Structure of the actin-myosin complex and its implications for muscle contraction. *Science* **261**, 58-65.

Reconditi, M., Koubassova, N., Linari, M., Dobbie, I., Narayan, T. Diat, O., Piazzesi, G. Lombardi, V. and Irving, M., 2003, The conformation of myosin head domains in rigor muscle determined by X-ray interference. *Biophys. J.* **85**, 1098-1110.

Squire, J.M., 1972, General model of myosin filament structure II: myosin filaments and crossbridge interactions in vertebrate striated and insect flight muscles" *J. Molec. Biol.* **72**, 125-138.

Squire, J.M., 1992, Muscle filament lattices and stretch-activation: The match/ mismatch model reassessed., *J. Musc. Res. Cell Motil.* **13**, 183-189.

Squire, J.M., 1997, Architecture and function in the muscle sarcomere. *Curr. Opin. Struct. Biol.* 7, 247-257.

Squire, J. M., 1998, Time-resolved X-ray diffraction. In *Current Methods In Muscle Physiology*, H. Sugi, ed. (Oxford, Oxford Univ. Press), pp. 241-285.

Squire, J. M., 2000, Fibre and Muscle Diffraction. In *Structure and Dynamics of Biomolecules*, E. Fanchon, E. Geissler, L.-L. Hodeau, J.-R. Regnard, and P. Timmins, eds. (Oxford, UK, Oxford Univ. Press), pp. 272-301.

Squire, J.M. and Harford, J.J., 1988, Actin filament organisation and myosin head labelling patterns in vertebrate skeletal muscles in the rigor and weak-binding states. *J. Mus. Res. Cell Motil.* **9**, 344-358

Squire, J.M. and Morris, E.P., 1998, A new look at thin filament regulation in vertebrate skeletal muscle. *FASEB J.* **12**, 761-771

Squire, J.M. and Knupp, C., 2004, Simulation of Muscle Diffraction Patterns using the MusLABEL Program. *J. Mus. Res. Cell Motil.* – in press. [Also see: www.ccp13.ac.uk]

Squire, J. M., Cantino, M., Chew, M., Denny, R., Harford, J., Hudson, L., and Luther, P., 1998, Myosin rod-packing schemes in vertebrate muscle thick filaments. *J. Struct. Biol.* **122**, 128-138.

Squire, J.M., Luther, P.K. and Knupp, C., 2003a, Structural evidence for the interaction of C-protein (MyBP-C) with actin and sequence identification of a possible actin-binding domain. *J. Molec. Biol.* **331**, 713-724.

Squire, J.M., Roessle, M and Knupp, C., 2004a, Setting the scene for muscle motor action: New X-ray Diffraction analysis on C-protein (MyBP-C), Titin, Nebulin and Troponin in the A-band. J. Molec. Biol. In press.

Squire, J.M., Harford, J.J., AL-Khayat, H.A., Roessle, M. and Knupp, C., 2004b, Steric constraints on muscle motor action: Evidence for Target Area Labelling in Rigor Bony Fish Muscle. (in preparation).

Squire, J.M., AL-Khayat, H.A., Arnott, A., Crawshaw, J., Denny, R., Diakun, G., Dover, S.D., Forsyth, V.T., He, A., Knupp, C., Mant, G., Rajkumar, G., Rodman, M.J., Shotton, M. and Windle, A.H., 2003b, New CCP13 software and the strategy behind further developments: Stripping and modelling of fibre diffraction data. *Fibre Diffraction Review* **11**, 7-19.

Squire, J.M., Knupp, C., AL-Khayat, H.A. and Harford, J.J., 2003c, Millisecond time-resolved low-angle X-ray diffraction: a powerful, high-sensitivity technique for modelling real-time movements in biological macromolecular assemblies. *Fibre Diffraction Review* **11**, 28-35.

Squire, J.M., AL-Khayat, H.A., Harford, J.J., Hudson, L., Irving, T.C., Knupp, C., Mok, N-S. and Reedy, M.K., 2003d, Myosin filament structure and myosin crossbridge dynamics in fish and insect muscles. In 'Molecular and cellular aspects of muscle contraction' (Ed. H. Sugi). Advances in experimental medicine and biology. Volume 538, pp. 251-266. Kluwer/ Plenum.

Squire, J.M., AL-Khayat, H.A., Harford, J.J., Hudson, L., Irving, T., Knupp, C. and Reedy. M.K., 2003e, Modelling muscle motor conformations using Low-angle X-ray diffraction. *Proc. Bio-Nanotechnol.* **150**, 103-110.

Tregear, R.T., 1983, Physiology of insect flight muscle. In Peachey LD, Adrian, RH, Geiger SR (eds). Skeletal Muscle. Bethesda, MD: American Physiological Society. pp. 487-506.

Vibert, P.J., Craig, R. and Lehman, W., 1997, Steric-model for activation of muscle thin filaments. *J. Molec. Biol.* **266**, 8-14.

Wray, J.S., 1979, Filament geometry and the activation of insect flight muscle. *Nature* **280**, 325-326.

Xu, S., Gu, J., Rhodes, T., Belknap, B., Rosenbaum, G., Offer, G., White, H., and Yu, L. C., 1999, The M.ADP.P(i) state is required for helical order in the thick filaments of skeletal muscle. *Biophys. J.* 77, 2665-2676.

CONFORMATIONAL CHANGE AND REGULATION OF MYOSIN MOLECULES

Mitsuo Ikebe*, Xiang-dong Li, Katsuhide Mabuchi#, and Reiko Ikebe

1. DEVELOPMENT OF MYOSIN BIOCHEMISTRY

It has been a long time since myosin was first named by Wilhelm Kuhne in the mid 19th century as a component of skeletal muscle extract precipitated at low salt. In 1939, Engelhartd found that myosin is an ATPase. A critical finding was made by Albert Szent-Gyorgyi in 1942 who found that myosin interacts with actin and produces superprecipitation with the addition of ATP. The finding opened a door for the study of muscle contraction and actomyosin biology. The enzymatic function of myosin has been studied by biochemical means. One of the most important findings of myosin enzymatic function is the discovery of Pi-burst by Tonomura et al. (Tonomura et al., 1962). They found that the timecourse of the Pi release of myosin ATPase shows initial rapid phase followed by a steady state Pi release. It was thought that the initial Pi-burst reflectd the formation of phosphorylated intermediate, but it was realized later that this is due to the formation of myosin/ADP/P intermediate. The kinetic model of actomyosin ATPase subsequently developed (Lymn and Taylor, 1971; Stein et al., 1981; Siemankowaki et al., 1985), and it is now known that each kinetic step is closely correlated with the mechanical cycle of cross-bridge movement. A critical feature of myosin in addition to the ATPase activity and the actin binding activity is the filament formation. In the 1960's, a number of studies were made to determine the molecular structure of myosin using various physico-chemical techniques. The molecular shape of skeletal myosin was visualized by electron microscopy and it was found that myosin has two globular heads connected with a long tail (Slayter and Lowey, 1967) that is critical for the thick filament formation. While the filament formation is a key feature of a myosin molecule, an unusual myosin like protein was first discovered from an amoeba. Pollard and Korn (Pollard and

Department of Physiology, University of Massachusetts Medical School, Worcester, MA 01655, and #Boston Biomedical Research Institute, Watertown, MA 02472
*To whom correspondence should be addressed: Department of Physiology, University of Massachusetts Medical School, 55 Lake Avenue North, Worcester, MA 01655
Fax: 508-856-4600; E-mail: Mitsuo.Ikebe@umassmed.edu.

Korn, 1971) isolated a myosin like protein from Acanthamoeba that has ATPase activity and the actin binding ability, but no elongated tail domain of myosin and resembles myosin S1, the head portion of muscle myosin. This unique myosin was named myosin I. During the last decade, a number of myosin like proteins were found and it is now known that myosin constitutes a superfamily (Fig. 1). The myosin superfamily is currently organized into at least 18 classes based upon phylogenetic sequence comparisons of the motor domain (Reck-Peterson et al., 2000; Mermall et al., 1998; Sellers, 2000; Berg et al., 2001). Among them,

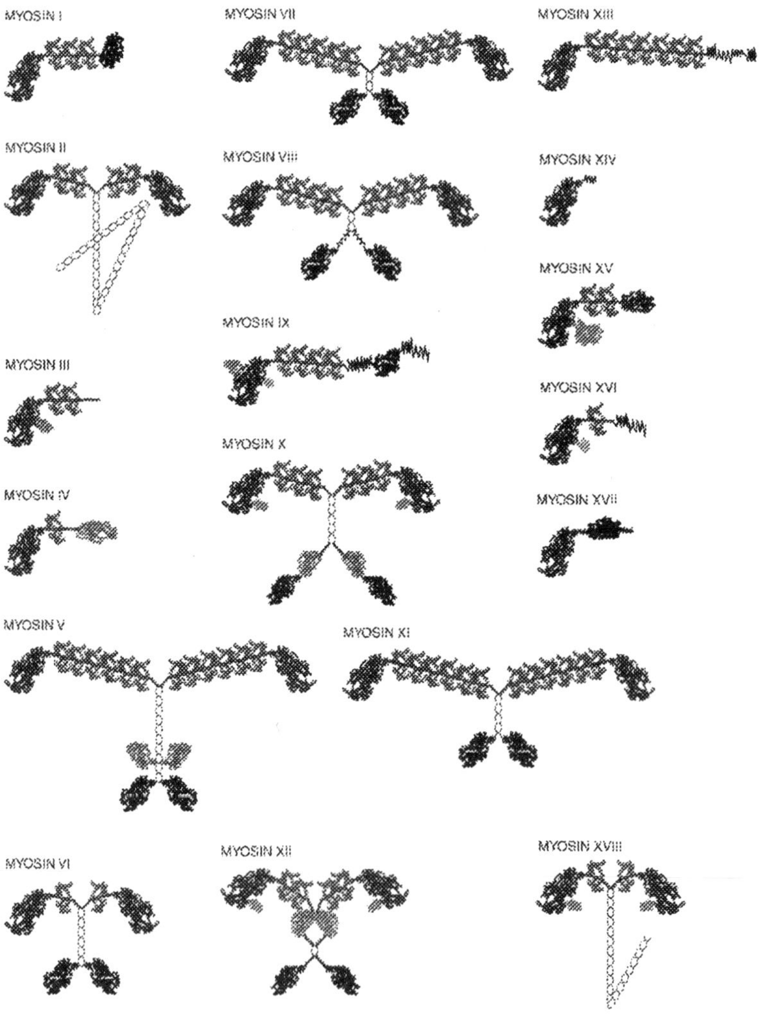

Figure 1. Schematic drawing of myosin superfamily members.

only the class II myosin (conventional myosin) can form thick filament, and it is thought that the class-specific tail domain plays a role in binding to the target molecules, thus determining the cellular localization and function of the each member of myosin.

2. REGULATION OF ACTOMYOSIN

The regulation of actomyosin was first studied with skeletal muscle. Ebashi and Ebashi (Ebashi and Ebashi, 1964) found regulatory components in skeletal actomyosin preparation and this was named troponin. A number of studies have established that troponin-tropomyosin complex, the thin filament liked regulatory components, is responsible for the Ca^{2+} dependent regulation of striated muscle actomyosin. On the other hand, a myosin-linked regulation mechanism was first observed in Molluscan muscle actomyosin (Kendrick-Jones et al., 1970). It was found that a class of light chain, called regulatory light chain (RLC), dissociates from myosin by EDTA, and this abolishes the Ca^{2+} dependent regulation of actomyosin, thus making actomyosin constitutively active. Initially, the Ca^{2+} binding site was thought to be RLC, but subsequently it was realized that the Ca^{2+} binding site is in the other light chain, essential light chain (ELC) and RLC stabilizes Ca^{2+} binding to ELC (Xie et al., 1994). The regulation of smooth muscle actomyosin is also myosin linked, but the mode of regulation is different from that of Molluascan actomyosin. It was found that smooth muscle myosin is phosphorylated in the presence, but not in the absence, of Ca^{2+}, and the phosphorylation is required for the activation of actomyosin (Sobieszek, 1977 ; Gorecka et al., 1976; Ikebe et al., 1977). Once myosin is phosphorylated, Ca^{2+} is no longer required for the activation. Furthermore, it was found that the phosphorylation changes the conformation of smooth muscle myosin. It was originally found in Watanabe's laboratory that smooth muscle myosin showed the different sedimentation velocity in high and low ionic strength. The sedimentation velocity was 6S at high salt and 10S at low salt, and it was thought that the 10S

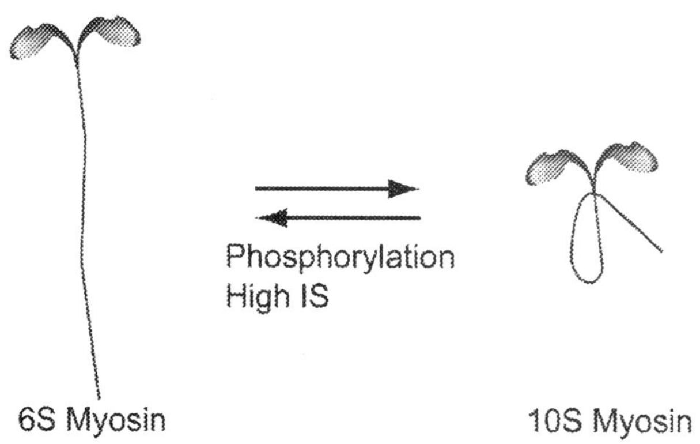

Figure 2. Schematic representation of 10 S and 6 S conformation of smooth muscle myosin II.

component is the dimer of myosin (Suzuki et al., 1978). Subsequently, it was found that this is not the case and the tail of myosin bent back towards the head-neck junction in 10S myosin (Onishi and Wakabayashi, 1982; Trybus et al., 1982; Craig et al., 1983) (Fig. 2). Importantly, the phosphorylation of RLC favors the formation of 6S conformation (Onishi et al., 1983; Ikebe et al., 1983; Trybus and Lowey, 1984). Quite interestingly, we found that myosin Va undergoes similar conformational change that is accompanied by the change in the ATPase activity. The following section describes this finding in detail.

3. MYOSIN V UNDERGOES CONFORMATION TRANSITION COUPLED WITH ITS MOTOR ACTIVITY

Myosin V is an unconventional myosin that processively moves along an actin filament. Myosin V has two heads, each of which consists of a motor domain and an expanded neck domain that contains six IQ motifs that bind calmodulins or light chains. Myosin V was isolated from chicken brain and found that the actin activated ATPase activity is significantly activated by Ca^{2+} (Cheney et al., 1993). However, the recombinant truncated myosin V, having the head domain and the coiled-coil domain expressed in Sf9 cells, showed no activation of the ATPase activity in Ca^{2+} (Trybus et al., 1999; Homma et al., 2000). The reason why tissue isolated myosin Va and expressed myosin Va fragments show different properties is unknown. An obvious possibility is that the tail domain influences the Ca^{2+} dependent regulation. Alternatively, the difference may be attributed to the differences in the post translational modifications of recombinant and tissue isolated myosin Va or the subunit compositions and it is reported that tissue isolated myosin Va contains dynein light chain (Espindola et al., 2000). We succeeded in functionally expressing recombinant full-length myosin Va and studied Ca^{2+} dependent regulation of myosin Va by examining the expressed full-length myosin Va and various truncated variants.

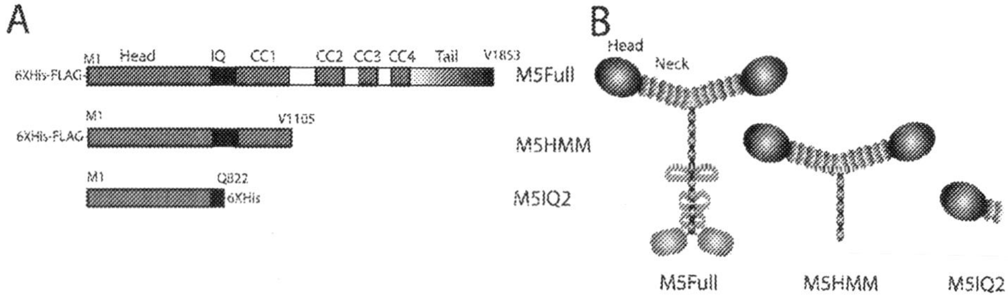

Figure 3. Schematics of Myosin Va constructs. (A) Schematic primary structure of myosin Va constructs expressed in this study. CC represents coiled-coil sequence. IQ represents the calmodulin light chain binding domain. His-Tag and FLAG-tag were added at N-terminus of M5Full and M5HMM, while His-Tag was added at C-terminus of M5IQ2. (B) Schematic structure of myosin Va constructs based on its amino acid sequence. The chain represents coiled-coil structure.

3.1. Expression of recombinant full-length myosin Va and its fragments.

We succeeded in functionally expressing a full-length mouse myosin Va (M5Full). Myosin Va heavy chain was co-expressed with calmodulin light chain and purified as described in MATERIALS AND METHODS (Fig. 3). The expressed M5Full was composed of heavy chain with apparent molecular weight of 190 kDa and calmodulin light chain. Myosin Va HMM (M5HMM) containing the entire head, IQ domain and coiled-coil domain, and M5IQ2 containing entire head and 2 IQ domains were also isolated. These constructs were used for the experiments described in this study.

3.2. ATPase activity of recombinant Myosin Va and its fragments.

Fig. 4A shows the actin-activated ATPase activity of M5Full in the presence and absence of free Ca^{2+}. The actin-activated ATPase activity of M5Full showed strong Ca^{2+} dependence and the activity was increased for approximately 9 folds in the presence of 100 μM free Ca^{2+} (Fig. 4A and Table I). The result of M5Full is similar to that of tissue-isolated myosin Va (Cheney et al., 1993). On the other hand, Ca^{2+} did not significantly change the activity of M5HMM in the presence of 12 μM calmodulin (Fig. 4B). In the absence of exogenous calmodulin, actin-activated ATPase activity of M5HMM was decreased at high Ca^{2+} as was reported previously (Homma et al., 2000), and it is thought that the decrease in the ATPase activity is due to the dissociation of calmodulin light chain. The ATPase activity of M5IQ2 showed reverse Ca^{2+} dependence and the activity was slightly higher in EGTA than in Ca^{2+} (Fig. 4C). Both truncated constructs have high ATPase activity in EGTA condition, which is similar to that of M5Full in Ca^{2+} condition. The results suggest that the Ca^{2+} dependent

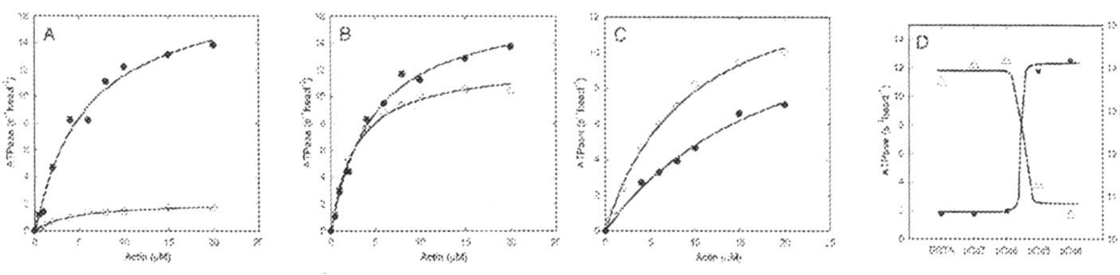

Figure 4. MgATPase activity of myosin Va constructs. (A-C). The actin dependence of ATPase activity of M5Full (A), M5HMM (B), and M5IQ2 (C) in EGTA (open circle) and pCa4 condition (closed circle). The solid lines in (A-C) are fit to the Michaelis-Menten equation. The fitting data are summarized in Table I. (D) Effects of Ca^{2+} on the ATPase activity (closed circle) and sedimentation coefficient (open triangular) of M5Full. The ATPase activity was measured at 25°C in a solution of 20 mM MOPS (pH7.0), 0.1 M KCl, 1 mM $MgCl_2$, 1 mM DTT, 0.25 mg/ml BSA, 12 μM calmodulin, 20 μM actin, 0.5 mM ATP, 2.5 mM PEP, 20 U/ml pyruvate kinase. EGTA (1 mM) was added for EGTA condition, whereas 1 mM $CaCl_2$ and various concentrations of EGTA were added for pCa7-pCa4 conditions.

Table 1. V_{max} and K_{actin} of Actin-activated ATPase activity of Myosin Va

	EGTA condition			Ca^{2+} condition		
	V_0 ($s^{-1}head^{-1}$)	V_{max} ($s^{-1}head^{-1}$)	K_{actin} (μM)	V_0 ($s^{-1}head^{-1}$)	V_{max} ($s^{-1}head^{-1}$)	K_{actin} (μM)
M5Full	0.05	2.15	4.38	0.06	18.52	6.02
M5HMM	0.09	12.48	2.75	0.09	17.30	4.84
M5IQ2	0.07	15.07	9.21	0.07	14.45	19.87

Assay conditions were as described in the legend to Figure 1. Basal activity (V_o) was deducted. Curves are the least squares fits of the data points based upon the equation: $V = (V_{max} * [actin]) / (K_{actin} + [actin])$.

regulation observed for M5Full is due to the inhibition in EGTA rather than the activation in Ca^{2+}. Figure 4D shows free Ca^{2+} concentration dependence of the actin activated ATPase activity of M5Full. The activity was increased at higher than pCa5 suggesting that Ca^{2+} binding to calmodulin light chain is responsible for the activation. The result suggests that the tail domain of myosin Va is responsible for Ca^{2+} dependent regulation.

3.3. Ca^{2+} dependent conformational change of Myosin Va.

We examined the conformational changes of myosin Va constructs. First, we analyzed the sedimentation coefficient ($S_{20,w}$) of myosin Va in EGTA and Ca^{2+} conditions by velocity sedimentation analysis. The apparent $S_{20,w}$ of myosin Va decreased significantly from 13.9 S in EGTA to 11.3 S in pCa5 condition (Fig 5; Table 2). Remarkably, the stimulation of actin-activated ATPase activity of M5Full by Ca^{2+} is accompanied with the decrease of $S_{20,w}$ (Figure 4D; Table 2).

M5HMM, containing partial coiled-coil domain, failed to show a decrease of $S_{20,w}$. Therefore, the results suggest that the observed change in the $S_{20,w}$ of M5Full is due to the large change in the conformation. It is predicted that M5Full forms a more compact conformation in the presence of low Ca^{2+} than in high Ca^{2+}. On the other hand, the $S_{20,w}$ of M5Full at high ionic strength was slightly increased from 9.4 in EGTA to 9.7 in pCa5 (Table 2). Both values were similar to that in Ca^{2+} at low ionic strength, suggesting that the formation of a compact structure is abolished at high ionic strength. In contrast to M5Full, the $S_{20,w}$ for M5HMM rather increased from 8.7 S in EGTA to 9.5 S in Ca^{2+} condition (Fig. 5; Table 2). Ca^{2+} induced the slight decrease of $S_{20,w}$ with M5IQ2 (Fig. 5; Table 2). While the change was small, the same results were obtained repeatedly, therefore, the change in $S_{20,w}$ of M5IQ2 may reflect the Ca^{2+} induced release of the bound calmodulin (Homma et al., 2000).

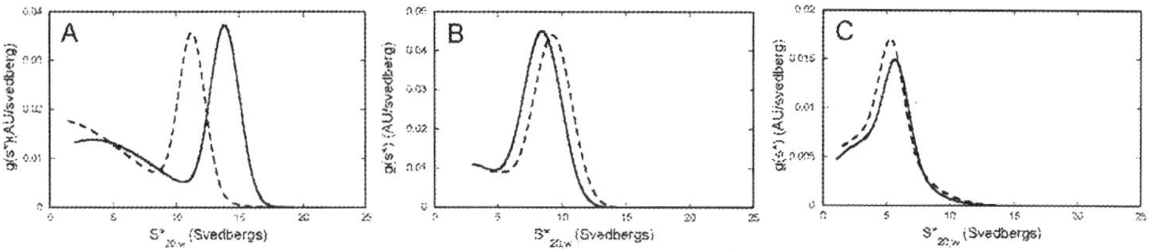

Figure 5. Apparent sedimentation coefficient distributions for myosin Va constructs. Sedimentation velocity was determined in solutions containing 20 mM MOPS (pH7.0), 0.2 M NaCl, 1 mM DTT, 1 mM EGTA (solid line). EGTA was replaced by EGTA-$CaCl_2$ buffer for pCa5 condition (broken line). The velocity runs were carried out at 42,000 rpm at 20°C for M5Full (A), M5HMM (B), and M5IQ2 (C). The x axis, $S^*_{20,w}$ is the apparent sedimentation coefficient.

Table 2. Sedimentation coefficient ($S_{20,w}$)of myosin Va constructs.

	EGTA	pCa7	pCa6	pCa5	pCa4
M5Full*	13.9	14.1	14.2	11.3	10.6
M5Full**	9.4	\	\	9.7	\
M5HMM*	8.7	\	\	9.5	\
M5IQ2*	5.8	\	\	5.4	\

* Analytical centrifugation was run in a solution of 20 mM MOPS (pH7.0), 0.2 M NaCl, 1 mM $MgCl_2$, 1 mM DTT and various concentration of EGTA and $CaCl_2$ as described in Materials and Methods. ** Same condition as * except 0.6 M NaCl was used.

To visualize the nature of the conformational change of M5Full, we examined the structure of M5Full by rotary shadowing of electron microscopy. Fig. 6 shows the representative images of M5Full in various conditions. At high ionic strength at low Ca^{2+}, M5Full showed an extended conformation that was similar to those images previously reported (Cheney et al., 1993). On the other hand, we found a folded shape of M5Full at low ionic strength in the presence of EGTA, in which the tail domain was folded back towards the head-neck region. In the Ca^{2+} condition, we predominantly found an extended conformation even at low ionic strength. These results are consistent with centrifugation analysis and show that 14 S M5Full represents a folded conformation, while 11 S M5Full represents an extended conformation. Furthermore, we found that the head of myosin Va appears large and globular with no obvious neck domain in high Ca^{2+} regardless of ionic strength. On the other hand, in low Ca^{2+} myosin Va showed a smaller globular head connected with long neck domain (Fig 6).

We found that myosin Va (M5Full) exhibits a Ca^{2+} dependent large change in the $S_{20,w}$ at physiological ionic strength. Based upon rotary shadowing of electron microscopy, it was found that M5Full forms a folded conformation at low Ca^{2+} and low ionic strength. In this conformation, myosin Va tail bent back to the head-neck region of the molecule. On the other hand, M5Full forms a more extended conformation at high Ca^{2+}. At high ionic strength, an extended conformation dominates over a folded conformation regardless of the Ca^{2+} concentration that is reflected by the decrease in $S_{20,w}$. Ca^{2+} concentration required for the shift in conformation is pCa6-pCa5 based upon the change in the $S_{20,w}$. This suggests that the conformational change is initiated by the binding of Ca^{2+} to calmodulin light chain that is associated at the neck of myosin Va. It should be mentioned that myosin Va showed several apparently different shapes in low Ca^{2+} at low ionic strength, although the molecules showed compact structures. This suggests that the attachment of the tail domain to the neck domain is not tight and there is enough flexibility to allow myosin Va molecules to take various conformations.

We also found that Ca^{2+} changes the head-neck conformation regardless of the ionic strength. At high Ca^{2+}, the characteristic long neck domain of myosin Va is not evident and two large globular heads are observed. We think that the globular motor domain of myosin Va is folded back to the neck domain at high Ca^{2+} to create apparent large globular head (Fig 6). At low ionic strength, the attachment of the motor domain to the neck domain in Ca^{2+} may prevent the interaction of the tail domain at the neck domain thus inhibiting the formation of a folded conformation. The Ca^{2+} induced increase in $S_{20,w}$ with M5Full in high salt (9.4S to 9.7S) and M5HMM (8.7S to 9.5 S) supports the model (Fig. 6B) in which the head folds back to neck in high Ca^{2+}.

Interestingly, the change in the conformation of M5Full was closely correlated with the change in the actin-activated ATPase activity. At low ionic strength, Ca^{2+} markedly increases the actin-activated ATPase activity and this is accompanied by the change in the conformation from a folded to an extended. On the other hand, there was no change in the ATPase activity by Ca^{2+} at high ionic strength, where the compact folded structure of myosin Va is not found. A similar relationship between the conformational change and ATPase activity of myosin has been known for vertebrate smooth muscle and non-muscle myosin II as described above (Onishi et al., 1983; Ikebe et al., 1983; Trybus et al., 1984).

A folded myosin II has a low ATPase activity while an extended myosin II shows significantly higher ATPase activity (Ikebe et al., 1983). The change in the conformation of myosin II is regulated by regulatory light chain phosphorylation (Onishi et al., 1983; Ikebe et al., 1983; Trybus et al., 1984) and it is thought that the phosphorylation of the light chain induces the conformational change at the neck region where the light chain associates and this stabilizes the association of the tail domain to form a folded conformation. For myosin Va, it is anticipated that Ca^{2+} binding to calmodulin light chain induces the conformational change of calmodulin at the neck domain, which destabilizes the association of the tail domain to the head-neck region of myosin Va.

Recently, a similar tail inhibition model was proposed for the regulation of kinesin, i.e., kinesin is in a folded conformation such that the kinesin globular tail domain interacts with and inhibits the kinesin motor activity (Verhey and Rapoport, 2001). Full-length kinesin undergoes a 9S to 6S conformational transition, i.e. compact to extended conformation, whereas C-terminal domain truncated kinesin constitutively in extended form (Stock et al., 1999). Correspondingly, the ATPase activity of full-length kinesin is activated by cargo binding, whereas C-terminal domain truncated kinesin is constitutively active. Furthermore, it was found that the expressed C-terminal globular domain inhibits the ATPase activity of C-terminal domain truncated kinesin (Coy et al., 1999). The present study suggests that there is a similarity in the regulatory mechanism between kinesin and myosin Va.

A question is whether the myosin V motor function is regulated via a conformational transition found in this study in cells. A critical issue is whether the myosin driven cargo movement is regulated by the change in the cytosolic Ca^{2+}, and it requires further study to clarify this question. Another issue is whether the myosin V targeting molecules affect the motor activity of myosin V since the myosin V binding partners are bound to the C-terminal globular domain of myosin V. It is plausible that the binding of the tail domain to partner molecules may release the inhibition of the ATPase, thus regulating myosin Va motor function. It requires further study to see whether cargo binding at the tail domain can activate myosin Va motor function.

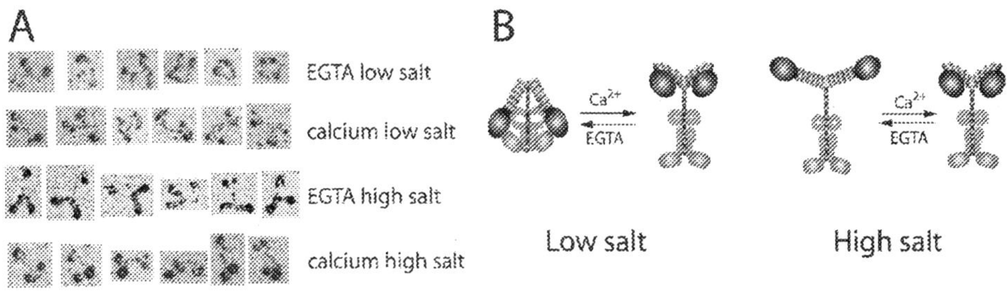

Figure 6. Electron micrographs M5Full and model for the regulation of M5Full.
A. Electron microscopic images of M5Full in various conditions. M5Full samples diluted to about 4 nM by dilution buffer were absorbed onto a freshly cleaved mica surface for 30 s. Unbound proteins were rinsed away, and then the specimen was stabilized by brief exposure to uranyl acetate. Dilution buffers were as follows: EGTA-low salt, 1 mM $MgCl_2$, 30% glycerol, 1 mM EGTA, 50 mM ammonia acetate (pH7.0); Ca^{2+}-low salt, 1 mM $MgCl_2$, 30% glycerol, 0.1 mM $CaCl_2$, 50 mM ammonia acetate (pH7.0); EGTA-high salt, 1 mM $MgCl_2$, 30% glycerol, 1 mM EGTA, 600 mM ammonia acetate (pH7.0); Ca^{2+}-high salt, 1 mM $MgCl_2$, 30% glycerol, 0.1 mM $CaCl_2$, 600 mM ammonia acetate (pH7.0). B. Schematic model of M5Full conformations. The motor domain is folded back to associate the neck domain in high Ca^{2+}. At low salt condition, the tail globular domain interacts with the head-neck region to produce a compact structure in low Ca^{2+}. This is inhibited in high Ca^{2+} because the binding of the motor domain to the neck domain interferes the association between the tail and the head-neck of myosin Va.

4. MATERIALS AND METHODS

4.1. Construction of Myosin Va expression vector.

A baculovirus transfer vector for mouse myosin Va (M5Full) in pFastBac (Invitrogen, CA) was produced as follows. A unique Spel site was created at nucleotide 3316 of DHM5 (Homma et al., 2000). A cDNA fragment 3316-5602 franked by Speland Kpn1 sites were introduced to DHM5 to produce a full-length myosin Va construct. The nucleotides at the created Spel site were changed to resume the authentic sequence. An N-terminal tag (MSYYH HHHHH DYKDD DDKNI PTTEN LYFQG AMGIR NSKAY) containing a sequence of hexa-histidine-Tag and FLAG-tag was added to the N-terminus of M5Full. For M5HMM, a stop codon was introduced at the nucleotide 3316. Vector containing nucleotide 1-2468 of myosin Va (M5IQ2) was prepared as described previously (Homma et al., 2000).

4.2. Expression and purification of Myosin Va.

To express recombinant M5Full, Sf9 cells (about 1X10[9]) were co-infected with two separate viruses expressing the myosin Va heavy chain and calmodulin, respectively, and the recombinant myosin Va was purified as described (Homma et al., 2000). All concentrations of Myosin Va in this paper are referred to the concentration of head.

ATPase assay: Since myosin Va ATPase activity is significantly inhibited by the product ADP, we use an ATP regeneration system to measure its activity. The MgATPase activity was measured at 25°C in a solution containing 10 - 100 nM Myosin Va, 20 mM MOPS (pH7.0), 1 mM $MgCl_2$, 0.25 mg/ml BSA, 1 mM DTT, 2.5 mM PEP, 20 U/ml pyruvate kinase, 12 μM calmodulin, 100 mM KCl, 0.5 mM ATP, 1 mM EGTA-$CaCl_2$ buffer system, and various concentrations of actin.

4.3. Analytical Ultracentrifugation.

The myosin Va sample was dialyzed against a buffer of 20 mM MOPS (pH7.0), 0.2 M NaCl, 0.1 mM EGTA, 1 mM DTT. Just before running, $MgCl_2$ and EGTA concentration was adjusted to 1 mM for EGTA condition. For Ca^{2+} condition, $MgCl_2$ and $CaCl_2$ were adjusted to 1 mM with various EGTA concentrations.

Sedimentation velocity was run at 42,000 rpm at 20°C for 2 hours in a Beckman Optima XL-I analytical ultracentrifuge. Sedimentation boundaries were analyzed using a time-derived sedimentation velocity program, i.e. DcDt (Stafford, 1992) or DcDt plus (Philo, 2000). The density and viscosity calculated with Sednterp for a solution containing 0.2 M NaCl and 1 mM $MgCl_2$ were 1.00658 g/ml and 1.0213 cp, respectively. Those values for a solution of 0.6 M NaCl and 1 mM $MgCl_2$ were 1.0228 g/ml and 1.0590 cp, respectively. The values of the partial specific volumes of M5Full (0.7323), M5HMM (0.7296), and M5IQ2 (0.7328) were calculated from the amino acid composition of heavy chain and calmodulin.

4.4. Electron microscopy.

M5Full samples diluted to about 4 nM were absorbed onto a freshly cleaved mica surface. Unbound proteins were rinsed away, and then the specimen was stabilized by brief exposure to uranyl acetate. The specimen was visualized by the rotary shadowing technique according to Mabuchi (Mabuchi, 1991) with an electron microscope (Philips 300 electron microscope) at 60 kV.

5. REFERENCES

Berg, J. S., Powell, B. C., and Cheney, R. E. (2001) A millennial myosin census, *Mol. Biol. Cell* **12**, 780-794.

Cheney, R.E., O'Shea, M.K., Heuser, J.E., Coelho, M.V., Wolenski, J.S., Espreafico, E.M., Forscher, P., Larson, R.E., and Mooseker, M.S. (1993) Brain myosin-V is a two-headed unconventional myosin with motor activity, *Cell*, **75**, 13-23.

Coy, D. L., Hancock, W. O., Wagenbach, M., and Howard, J. (1999) Kinesin's tail domain is an inhibitory regulator of the motor domain, *Nat. Cell Biol.* **1**, 288-292.

Craig, R., Smith, R., and Kendrick-Jones, J. (1983) Light-chain phosphorylation controls the conformation of vertebrate non-muscle and smooth muscle myosin molecules, *Nature*, **302**, 436-439.

Ebashi, S., and Ebashi, F. (1964) A NEW PROTEIN COMPONENT PARTICIPATING IN THE SUPERPRECIPITATION OF MYOSIN B, *J. Biochem.*, **55**, 604-613.

Ebashi, S., and Ebashi, F. (1964) A NEW PROTEIN FACTOR PROMOTING CONTRACTION OF ACTOMYOSIN*Nature*, **203**, 645-646.

Espindola, F. S., Suter, D. M., Partata, L. B., Cao, T., Wolenski, J. S., Cheney, R. E., King, S. M., and Mooseker, M. S. (2000) The light chain composition of chicken brain myosin-Va: calmodulin, myosin-II essential light chains, and 8-kDa dynein light chain/PIN, *Cell Motility & the Cytoskeleton* **47**, 269-281.

Gorecka, A., Aksoy, M.O., and Hartshorne, D.J. (1976) The effect of phosphorylation of gizzard myosin on actin activation, *Biochem. Biophys. Res. Commun.*, **76**, 325-331.

Homma, K., Saito, J., Ikebe, R., and Ikebe, M. (2000) Ca(2+)-dependent regulation of the motor activity of myosin V, *J. Biol. Chem.*, 275, 34766-34771.

Ikebe, M., Onishi, H., and Watanabe, S. (1977) Phosphorylation and dephosphorylation of a light chain of the chicken gizzard myosin molecule, *J. Biochem.*, **82**, 219-302.

Ikebe, M., Hinkins, S., and Hartshorne, D.J. (1983) Correlation of enzymatic properties and conformation of smooth muscle myosin, *Biochemistry*, **22**, 4580-4587.

Kendrick-Jones, J., Lehman, W., and Sent-Gyorgyi, A.G. (1970) Regulation in molluscan muscles, *J. Mol. Biol.*, **54**, 313-326.

Lymn, R.W., and Taylor, E.W. (1971) Mechanism of adenosine triphosphate hydrolysis by actomyosin, *Biochemistry*, **10**, 4617-4624.

Mabuchi, K. (1991) Heavy-meromyosin-decorated actin filaments: a simple method to preserve actin filaments for rotary shadowing, *J. Struct. Biol.*, **107**, 22-28.

Mermall, V., Post, P. L., and Mooseker, M. S. (1998) Unconventional myosins in cell movement, membrane traffic, and signal transduction, *Science* **279**, 527-533.

Onishi, H., and Wakabayashi, T. (1982) Electron microscopic studies of myosin molecules from chicken gizzard muscle I: the formation of the intramolecular loop in the myosin tail, *J. Biochem.*, **92**, 871-879.

Onishi, H., Wakabayashi, T., Kamata, T., and Watanabe, S. (1983) Electron microscopic studies of myosin molecules from chicken gizzard muscle II: The effect of thiophosphorylation of the 20K-dalton light chain on the ATP-induced change in the conformation of myosin monomers, *J. Biochem.*, **94**, 1147-1154.

Philo, J. S. (2000) A method for directly fitting the time derivative of sedimentation velocity data and an alternative algorithm for calculating sedimentation coefficient distribution functions, *Anal. Biochem.* **279**, 151-163.

Pollard, T.D., and Korn, E.D. (1971) Filaments of Amoeba proteus. II. Binding of heavy meromyosin by thin filaments in motile cytoplasmic extracts, *J. Biol. Chem.*, **248**, 4682-4690.

Reck-Peterson, S. L., Provance, D. W., Jr., Mooseker, M. S., and Mercer, J. A. (2000) Class V myosins, *Biochim. Biophys. Acta* **1496**, 36-51.

Sellers, J. R. (2000) Myosins: a diverse superfamily, *Biochim. Biophys. Acta* **1496**, 3-22.

Siemankowaki, R.F., Wiseman, M.O., and White, H.D. (1985) ADP dissociation from actomyosin subfragment 1 is sufficiently slow to limit the unloaded shortening velocity in vertebrate muscle, *Proc. Natl. Acad. Sci. U.S.A.*, **82**, 658-662.

Slayter, H.S., and Lowey, S. (1967) Substructure of the myosin molecule as visualized by electron microscopy, *Proc. Natl. Acad. Sci. U.S.A.*, **58**, 1611-1618.

Sobieszek, A. (1977) Ca-linked phosphorylation of a light chain of vertebrate smooth-muscle myosin, *Eur. J. Biochem.*, **73**, 477-483.

Stafford, W. F., 3rd (1992) Boundary analysis in sedimentation transport experiments: a procedure for obtaining sedimentation coefficient distributions using the time derivative of the concentration profile, *Anal. Biochem.*, **203**, 295-301.

Stein, L.A., Chock, P.B., and Eisenberg, E. (1981) Mechanism of the actomyosin ATPase: effect of actin on the ATP hydrolysis step, *Proc. Natl. Acad. Sci. U.S.A.*, **78**, 1346-1350.

Stock, M. F., Guerrero, J., Cobb, B., Eggers, C. T., Huang, T. G., Li, X., and Hackney, D. D. (1999) Formation of the compact confomer of kinesin requires a COOH-terminal heavy chain domain and inhibits microtubule-stimulated ATPase activity, *J. Biol. Chem.*, **274**, 14617-14623.

Suzuki, H., Onishi, H., Takahashi, K., and Watanabe, S. (1978) Structure and function of chicken gizzard myosin, *J. Biochem.*, **84**, 1529-1542.

Tonomura, Y., Kitagawa, S., and Yoshimura, J. (1962) The initial phase of myosin A-adenosinetriphosphatase and the possible phosphorylation of myosin A, *J. Biol. Chem.*, **237**, 3660-3666.

Trybus, K.M., Huiatt, T.W., and Lowey, S. (1982) A bent monomeric conformation of myosin from smooth muscle, *Pro.Natl. Acad. Sci. U.S.A.*, **79**, 6151-6155.

Trybus, K.M., and Lowey, S. (1984) Conformational states of smooth muscle myosin. Effects of light chain phosphorylation and ionic strength, *J. Biol. Chem.*, **259**, 8564-8571.

Trybus, K.M., Krementsova, E., Freyzon, Y.(1999) Kinetic characterization of a monomeric unconventional myosin V construct, *J. Biol. Chem.*, **274**, 27448-27456.

Verhey, K. J., and Rapoport, T. A. (2001) Kinesin carries the signal, *Trends in Biochemical Sciences* **26**, 545-550.

Xie, X., Harrison, D.N., Schlichiting, J., Sweet, R.M., Kalabokis, V.N., Szent-Gyorgyi, A.G., and Cohen, C. (1994) Structure of the regulatory domain of scallop myosin at 2.8 A resolution, *Nature*, **368**, 306-312.

DISCUSSION

Pfittzer: How do you think about the stabilization of smooth muscle myosin filament?

Ikebe: Dephosphorylation induces smooth muscle myosin filament formation in vitro. Whether or not myosin filament formation in intact smooth muscle is regulated is unclear, but it is believed that majority of myosin forms filament in vivo, even in a resting state. Spacio-temporal change in dephosphorylated myosin in smooth muscle after stimulation is under investigation.

II. MECHANISM OF MUSCLE CONTRACTION

DRIVING FILAMENT SLIDING:

Weak binding cross-bridge states, strong binding cross-bridge states, and the power stroke

Bernhard Brenner*[#], Enke Mählmann*, Thomas Mattei*, and Theresia Kraft*

1. INTRODUCTION

The terms weak binding cross-bridge states and strong binding cross-bridge states by now are widely used in the literature on muscle contraction and cell motility, although the physiological significance of the actin attachment of cross-bridges in weak binding states is still questioned quite frequently. In addition, the term 'weak binding cross-bridge state' is often used when binding to actin with low affinity is described without clear distinction whether the binding occurs to the activated or inactive actin filament. Furthermore, the role of weak binding cross-bridge states (weak binding cross-bridge conformation) in relation to the power stroke is still disputed.

In this paper we therefore review our work on (i) characterizing weak and strong binding states of the actomyosin cross-bridge, (ii) possible physiological relevance of attachment of cross-bridges in weak binding states to the actin filament, and on (iii) the relation of weak and strong binding cross-bridge states to the power stroke of the actomyosin cross-bridge. All this work was performed with the structurally intact contractile system of skinned fibers of the M. psoas of the rabbit to preserve the natural steric relations within the contractile apparatus. By this approach we can exclude interactions with actin, specifically of cross-bridges in weak binding states, which are the result of lost steric relations and thus the loss of possible steric hindrance of myosin head attachment by regulatory proteins as is possible in solution studies on isolated proteins. Initial characterization of the weak binding states of the myosin head had been performed

*Dept. Molecular and Cell Physiology, Medical School Hannover, D-30625 Hannover, Germany. [#] To whom correspondence should be addressed.

under relaxing conditions to avoid active cross-bridge cycling in the presence of ATP, i.e., to avoid generation of a mixture of weak and strong binding states of the myosin head. In later studies, binding of weak binding states not only to inactive but also to activated actin filaments was also studied by using suitable ATP-analogs or, more recently, by using analogs of the ADP.P_i states.

2. WHAT DEFINES WEAK BINDING STATE VS. STRONG BINDING STATE?

Weak binding vs. strong binding is a property of the myosin head that only depends on the state of the bound nucleotide or, more specifically, on the state of the γ-phosphate in the nucleotide binding pocket (cf. Fig. 1). Myosin heads with no γ-phosphate, or no analog of the γ-phosphate, are in strong binding states (strong binding conformation). Myosin heads with ATP (or ATP-analogs like ATPγS or AMPPNP) or with ADP and inorganic phosphate or phosphate analogs in the nucleotide binding pocket are considered weak binding states. The existence of the MDP_i^{II}/$AMDP_i^{II}$ pair had been postulated in earlier biochemical work (e.g., Stein et al., 1979, 1984), and on the basis of studies of release of phosphate from caged phosphate into skinned muscle fibers (Dantzig et al., 1992; Millar and Homsher, 1990). It remained, however, unclear whether it is a typical strong binding state of the actomyosin complex or a weak binding state or even a state with new properties. Recent work on phosphate analogs, reviewed below, allows to characterize and classify this intermediate.

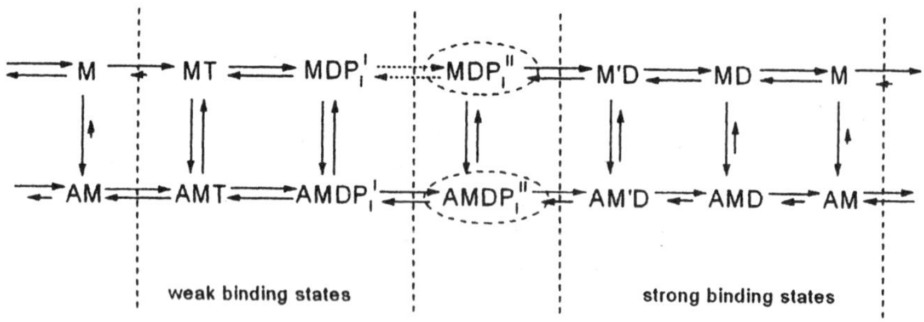

Figure 1. Kinetic Scheme of the actomyosin ATPase with the most important biochemical intermediates. M = myosin, A = actin, T = ATP, D = ADP, P_i = inorganic phosphate. Relative length of arrows qualitatively indicates equilibrium constants of the associated reaction step.

In summary, weak vs. strong binding is a property of the myosin head, and this property is independent on (i) whether the myosin head is bound to actin or not and (ii) whether the actin filament is in an activated form or not.

3. PROPERTIES THAT DISTINGUISH WEAK BINDING STATES OF THE MYOSIN HEAD FROM STRONG BINDING STATES

3.1 Actin affinity

Actin affinity is the property that originally led to the terms 'weak binding state' vs. 'strong binding state' on the basis of biochemical solution studies (e.g., White and Taylor, 1976; Stein et al., 1979).

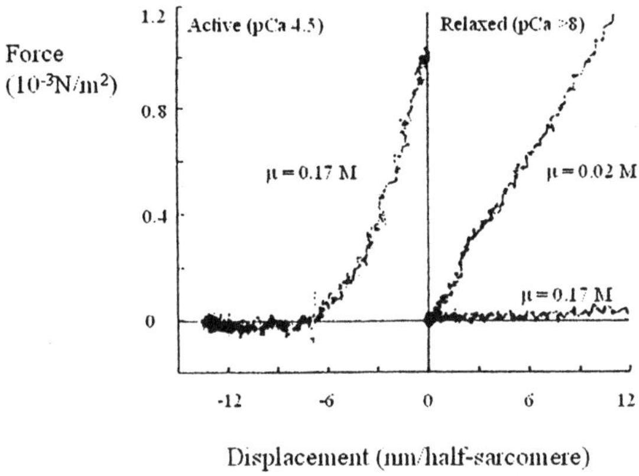

Figure 2. Fiber stiffness of a relaxed skinned fiber of rabbit psoas muscle at low (μ = 0.02M) and high (μ = 0.17M) ionic strength vs. active fiber stiffness during isometric contraction (modified from Brenner et al., 1982).

At a given ionic strength, actin affinity of weak binding states is much lower than actin affinity of strong binding states of the myosin head. In fact, actin affinity in weak binding states is so low that a large fraction of myosin heads in weak binding states is found attached to actin filaments only at low ionic strengths, e.g. at 0.02-0.05M. Thus, attachment of weak binding states to actin was initially detected and characterized by comparison of fiber stiffness at ionic strengths of 0.02M vs. 0.17M (cf. Fig. 2). The difference in stiffness of a relaxed fiber at low vs. high ionic strength was attributed to cross-bridge attachment as this component was found to scale with overlap between actin and myosin filaments (Brenner et al., 1982; Fig. 3). The stiffness seen at high ionic strength was attributed mainly to parallel elastic and viscoelastic components within the skinned fibers. Further support for cross-bridge origin of the stiffness increase seen with lowering ionic strength came later with the observation that the actin binding fragment of the protein caldesmon selectively inhibited this component of fiber stiffness (Brenner et al., 1991; Kraft et al., 1999).

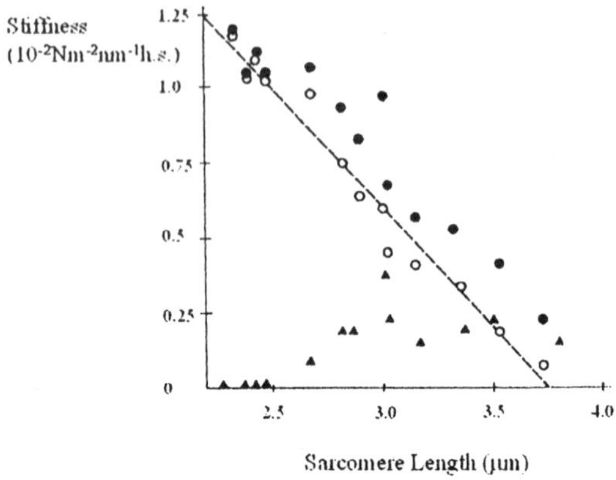

Figure 3. Fiber stiffness at different sarcomere lengths. Filled circles, stiffness at low ionic strength (μ = 0.02M); filled triangles, fiber stiffness at high ionic strength (μ = 0.17M); open circles, difference in fiber stiffness between low and high ionic strengths. Long dashed line, linear least squares fit to data represented by open circles; h.s., half sarcomere (modified from Brenner et al., 1982).

3.1.1. *Fraction of myosin heads in weak binding states that is attached to actin under relaxing conditions at near-physiological ionic strength and temperature*

The low actin affinity observed for weak binding states in relaxed muscle fibers raised the question of what fraction of time a cross-bridge in a weak binding state (or weak binding conformation) spends attached to an actin filament in a muscle fiber or, related to a large ensemble of myosin heads, which fraction of myosin heads in weak binding states is, at any one moment, attached to an actin filament, including physiological ionic strengths and temperature conditions. This question was addressed by making use of the already mentioned fact that an actin binding fragment of caldesmon can selectively inhibit actin attachment of myosin heads in weak binding states (Brenner et al., 1991). Thus, measuring the component of relaxed fiber stiffness that could be titrated away by addition of the actin binding caldesmon fragment we estimated the fraction of weak binding cross-bridges attached to actin under different experimental conditions. This includes higher temperatures as well as higher ionic strengths where stiffness of relaxed fibers is small. For such conditions we had initially assumed that all of the observed fibers stiffness is due to parallel elastic and viscoelastic components (Brenner et al., 1982). However, using this actin binding caldesmon fragment we found that even at physiological ionic strengths part of the observed fibers stiffness can be titrated away by the actin binding caldesmon fragment (Kraft et al., 1995). For physiological ionic strength (μ = 0.17M) and temperature of 20^{0}C we estimated that about 5% of myosin heads in weak binding states are attached to actin under relaxing conditions, i.e., that under these nearly physiological conditions a myosin head in a weak binding conformation spends about 5% of total time attached to actin (Kraft et al., 1995).

3.2. Kinetics of dissociation from actin

A second property that is distinctly different for weak and strong binding states of the myosin head are the kinetics of interaction of myosin heads with actin, specifically the kinetics for dissociation from actin.

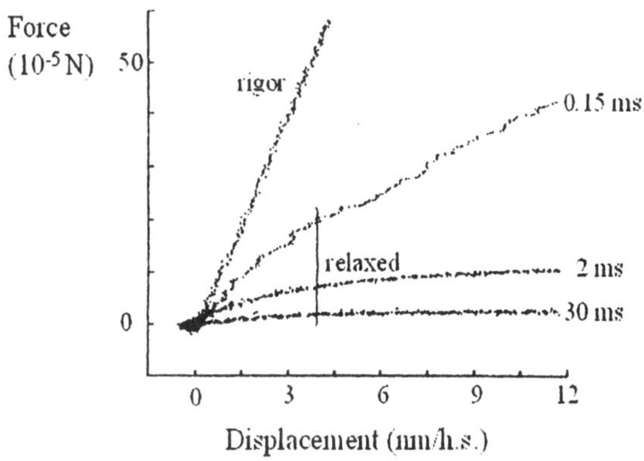

Figure 4. Original plots of force vs. length change recorded during ramp-shaped stretches of different speed imposed to one end of fibers. Number next to each trace recorded under relaxing conditions is time required for stretch by 6 nm/half sarcomere. Note that in rigor the plot of force vs. imposed length change is very insensitive to speed of stretch (c.f. Fig. 5); h.s., half sarcomere. Observed fiber stiffness was defined as the chord stiffness (force increase/imposed length change) when imposed length change had reached 2nm/half sarcomere. Modified from Brenner et al. (1982)

Characterization of the rate constant of myosin head dissociation from actin (k^-) was based on the observation that observed fiber stiffness under essentially all experimental conditions depends on the speed of the length change that is applied to one end of a fiber in order to measure fiber stiffness (Figs. 4, 5). We had shown by model simulations that the observed time course of force response during imposed stretches of different speed (Fig. 4) and the resulting dependence of apparent fiber stiffness on speed of the applied length change (stiffness-speed-profile; Fig. 5) allows to derive the rate constant (probability) of dissociation (k^-) from actin (Brenner, 1990). As a good approximation, the value of k^- can be taken as the value of speed of stretch ((nm/h.s.)/s) where stiffness has fallen to 1/e of its maximum level (Schoenberg 1985). Thus, a shift of the stiffness-speed-profile along the abscissa allows to estimate changes in k^-.

From the stiffness-speed profiles shown in Fig. 5 cross-bridges in relaxed muscle (MgATP, low Ca^{++}) dissociate with a probability of some $10^4 s^{-1}$ while in the presence of pyrophosphate dissociation occurs with some 1-100s^{-1}, in rigor with rate constants << 1/100 s^{-1}. Note that to account for the very wide speed dependence of fiber stiffness (much more than 2-3 orders of magnitude) a range of rate constants is required. This is not surprising since due to the mismatch between actin and myosin periodicities in fibers, attached cross-bridges are strained to different extent. Different strain, however, is expected to result in different rate constants for interaction with actin, e.g. strain-dependence of the rate constant of dissociation (Hill, 1974).

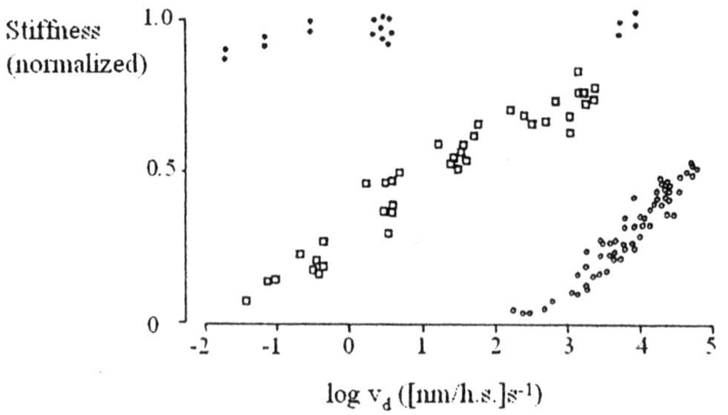

Figure 5. Observed fiber stiffness plotted vs. log of speed of stretch. Use of logarithm of speed of stretch to allow for changes in speed of stretch over several (seven) orders of magnitude. Open circles, relaxing conditions at μ = 0.02M; open squares, in the presence of 4mM Mg-pyrophosphate; filled circles, rigor conditions. (modified from Brenner et al., 1986)

3.3. Affinity for binding to activated vs. inactive actin filaments

A third characteristic difference between myosin heads in a weak vs. strong binding state (weak binding vs. strong binding conformation) is the dependence of actin affinity on interactions of the myosin head with the tropomyosin controlled sites that become accessible for binding when thin filaments are activated, e.g. at high Ca^{++}-concentrations (e.g., Greene and Eisenberg, 1980).

To address this question required an approach that allowed us to study binding of cross-bridges in weak binding states to the activated actin filament while, at the same time, active cross-bridge cycling is prevented. Otherwise, active cross-bridge cycling would result in occupancy of strong binding states by some cross-bridges, i.e., generation of a mixture of weak and strong binding states that makes definite characterization of weak binding states difficult. One approach is to use non-hydrolyzable, or slowly hydrolyzed nucleotide analogs such that strong binding states are not accessible even when thin filaments are activated. As a first example, we used the ATP-analog ATPγS that is only slowly hydrolyzed (Bagshaw et al., 1973; Goody and Hofmann,1980). To examine effects on actin affinity of weak binding cross-bridge states by (additional) interactions with the tropomyosin controlled sites, we compared attachment of myosin heads to actin in the presence of ATPγS, both at high and low Ca^{++}-concentrations by using equatorial intensities of X-ray diffraction patterns. This is based on the observation that attachment of cross-bridges to actin, both in weak and strong binding states, affects the intensity ratio, $I_{1,1}/I_{1,0}$, of the two innermost equatorial reflections, the 1,0 and 1,1 reflections of X-ray diffraction patterns (cf. Brenner et al., 1984, 1985). Fig. 6 shows that with increasing ionic strength, as fewer and fewer cross-bridges in weak binding states are attached to actin, the $I_{1,1}/I_{1,0}$ ratio continuously decreases, both in the absence and presence of Ca^{++}. In the presence of Ca^{++} the $I_{1,1}/I_{1,0}$ ratio is only slightly higher than in the absence of Ca^{++}. For example, at an ionic strength of 0.17M the increase in the $I_{1,1}/I_{1,0}$ ratio when Ca^{++} is raised is less than the increase in $I_{1,1}/I_{1,0}$ resulting from reduction in ionic strength from 0.017M to 0.12M in the absence of Ca^{++}. Reduction of ionic strength from 0.17M to 0.12M is expected to increase actin affinity of myosin heads about 2-fold (Greene et al., 1983). Thus, in the presence of Ca^{++}, actin affinity of weak binding myosin heads is at most 2-fold larger than in the absence of Ca^{++} (cf. Kraft et al. 1992).

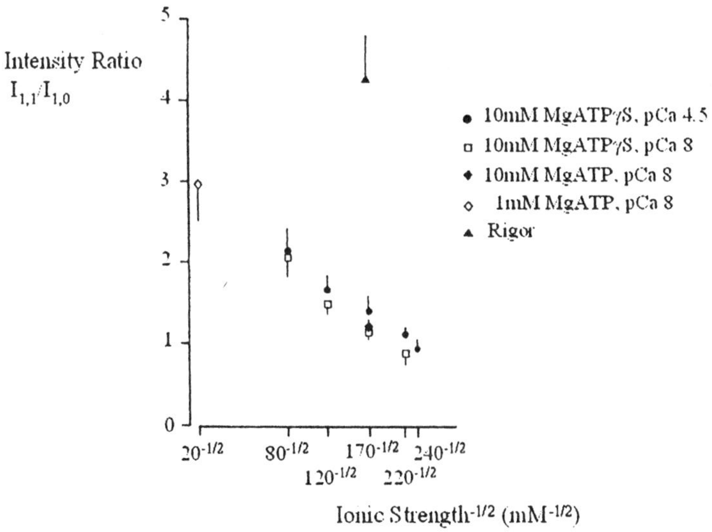

Figure 6. Intensity ratio $I_{1,1}/I_{1,0}$ of X-ray diffraction patterns recorded in the presence of ATPγS at different ionic strengths, both at high and low Ca^{++}. Modified from Kraft et al. (1992)

In conclusion, for weak binding cross-bridge states activation of thin filaments by Ca^{++} has only a rather small (≤ 2 fold) effect on actin affinity. This is quite different from strong binding cross-bridge states, e.g. in the presence of pyrophosphate, where actin binding is strongly affected by Ca^{++}-activation of the thin filaments (Brenner et al., 1986b), for solution studies cf. (Greene and Eisenberg, 1980).

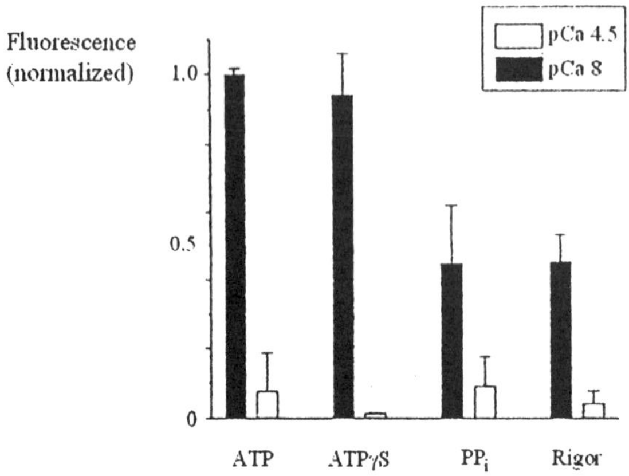

Figure 7. NBD-fluorescence with different nucleotides/nucleotide analogs at high and low Ca^{++}. Note that with ATP and ATPγS fluorescence at low [Ca^{++}] is high while in the presence of pyrophosphate (PP_i) fluorescence is quenched as it is in rigor or when Ca^{++} is raised. Quench of NBD-fluorescence is thus an indicator for activation of thin filaments which is induced by either increased [Ca^{++}] or by rigor cross-bridges or cross-bridges in presence of PP_i, i.e., by actin attachment of strong binding states of the myosin head (modified from Brenner et al., 1999).

3.4. Effect of myosin binding to actin on the state of the thin filament

A fourth characteristic property that allows to distinguish weak binding states of the myosin head from strong binding states is the ability to interfere with the state of the regulatory proteins. This results in activation of the thin filament when strong binding cross-bridges bind to regulated actin filaments at low [Ca^{++}]. This has been demonstrated by the effect of myosin head binding on the fluorescence of NBD-labeled troponin (cf. Brenner et al., 1999; Brenner and Chalovich, 1999). In contrast, attachment of weak binding states of the myosin head to actin filaments does not interfere with regulatory protein function, i.e., does not activate the thin filaments at any ionic strength studied, as is evidenced by the lack of any effect of ionic strength on the fluorescence of NBD-labeled troponin under relaxing conditions (Fig. 8). In conclusion, attachment to actin of cross-bridges in weak binding states does not interfere with regulatory protein function, i.e., does not activate thin filaments, while attachment of cross-bridges in strong binding states activates the thin filament. Note, this property is independent on ionic strength.

This also answers a frequently asked question whether a strong binding cross-bridge state is changed to a weak binding state of the myosin head by raising ionic strength and thus weakening actin affinity or, whether a weak binding state by lowering ionic strength can be transformed into a strong binding state of the myosin head. Since the ability of a strong binding cross-bridge state to activate the thin filament still remains at high ionic strength while the inability of a weak binding cross-bridge state to change the state of activation of the thin filament remains unchanged at low ionic strength, the properties 'weak binding' and 'strong binding', i.e., ability to activate thin filaments or not, do not simply depend on the magnitude of actin affinity but is rather a property that is independent of ionic strength but rather determined by the state of the nucleotide, specifically of the γ-phosphate.

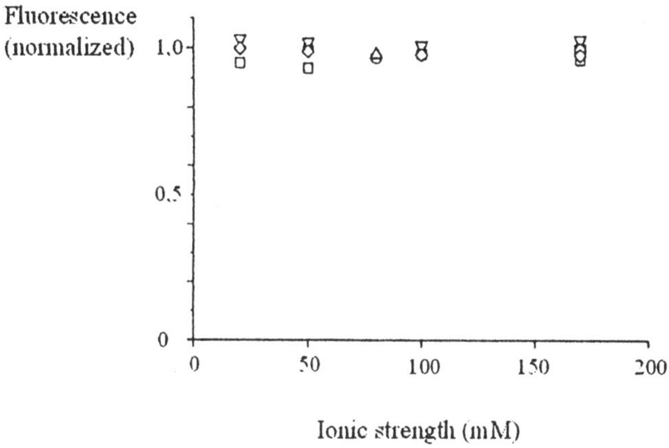

Figure 8. Fluorescence of NBD-labeled troponin when attachment of cross-bridges in weak binding states is increased by lowering ionic strength under relaxing conditions (MgATP, pCa 8). While actin attachment of cross-bridges in weak binding states increases with lowering ionic strength, fluorescence is not quenched as is expected if cross-bridges in weak binding states could interfere with regulatory proteins resulting in (partial) activation of thin filaments (modified from Brenner et al., 1999).

4. THE MODE OF CROSS-BRIDGE ATTACHMENT TO ACTIN: SPECIFIC VS. NON-SPECIFIC BINDING AND STEREOSPECIFIC VS. NON-STEREO-SPECIFIC BINDING

Another frequently raised question concerns the mode of actin attachment of weak and strong binding conformations of the myosin head, specifically, whether attachment of weak binding cross-bridge states to actin is specific or not, i.e., whether it goes to specific sites on actin. We had addressed this question by using the actin binding fragment of caldesmon. Although this is a highly elongated molecule, actin binding of cross-bridges in weak binding states, e.g., relaxing conditions (MgATP, pCa 8) or in the presence of MgATPγS (pCa 8 or pCa 4.5) could be inhibited (Brenner et al., 1991; Kraft et al., 1995). This suggested that attachment of weak binding cross-bridges involves specific sites on actin. From the pronounced ionic strength dependence and the fact that attachment occurs under relaxing conditions it is thought that attachment of cross-bridges in weak binding states involves the electrostatic sites on the actin monomers that were shown not to be controlled by the tropomyosin (e.g. Lehman et al., 1994; 1995). Modeling of 2D-Xray patterns suggested that the attachment of weak binding myosin heads under relaxing conditions mainly involves the N-terminus of actin (Gu et al., 2002). Altogether, actin binding of myosin heads in weak binding states involves specific sites, i.e., is site-specific.

Table 1. Summary of properties of weak binding states vs. strong binding states of the myosin head

Distinct differences between weak and strong binding states of the myosin head

	weak binding state	strong binding state
actin affinity (at comparable ionic strength)	low	high
actin binding kinetics (rate constant for dissociation)	very fast	much slower
effect of activation of thin filament on actin affinity	small (≤ 2 fold)	large (> 50-100 fold)
ability to activate thin filament	no effect	activates thin filaments even in absence of Ca^{++}

A separate question, however, is whether the actin binding of cross-bridges in weak binding states is also stereospecific, as it is for strong binding cross-bridge states, i.e., that at least part of the attached myosin heads, e.g. the catalytic domain, have a defined orientation relative to the actin monomers. As a result, stereospecific attachment labels the helical features of the actin filament and thus is expected to enhance actin based layer lines in 2D-X-ray diffraction patterns. Such enhancement of actin based layer lines is characteristic for binding in rigor or in the presence of $MgPP_i$ or during isometric contraction. Associated with enhancement of the actin based layer lines is weakening or loss of the myosin based layer lines. We therefore tested the question about stereospecific vs. non-stereospecific binding of myosin heads in weak binding states by recording 2D-X-ray diffraction patterns under relaxing conditions at high and low (0.05M) ionic strengths (Xu et al., 1997; Kraft et al., 1999) where in the latter condition many more of the myosin heads in the weak binding states are attached to actin. No enhancement of actin based layer lines was found nor did we see much change in the myosin based layer lines. Lack of enhancement of actin based layer lines suggests that while the actin attachment of cross-bridges in weak binding states goes to specific sites, it is non-stereospecific, i.e., the attached heads do not have a fixed orientation relative to the actin monomers to which the heads are attached and thus do not have a fixed orientation relative to the actin helix.

Next we asked the question whether actin attachment of cross-bridges in weak binding states might become stereospecific when the actin filament is activated by Ca^{++} (cf. Holmes, 1995) or when the actin filament has changed to its closed form by lowering ionic strength below 0.05M, e.g. to 0.03M (Head et al., 1995; Geeves and Conibear, 1995). Lowering ionic strength to 0.03 M under relaxing conditions (MgATP, pCa 8) did not enhance actin based layer lines nor were the myosin based layer lines much weakened (Kraft et al., 1999). In the presence of MgATPγS as an ATP-analog to avoid turnover at high calcium, a small enhancement of the 5.9nm actin based layer line was found when Ca^{++} was raised to pCa 4.5. This small enhancement was somewhat larger than the enhancement of the 5.9nm actin based layer line when fibers were first stretched beyond filament overlap, i.e., when the observed enhancement can not be due to cross-bridge attachment but instead is presumably due to changes in the thin filaments by Ca^{++}-activation. The small additional enhancement on the 5.9nm actin based layer line in full overlap might indicate a small actin labeling effect, e.g. due to some restriction of the otherwise highly variable angles of attachment when myosin heads in weak binding states also interact with the tropomyosin controlled sites at high Ca^{++} (Kraft et al., 1999).

Thus, when the actin filament is activated by Ca^{++} but not by lowering ionic strength to 0.03M, a slight restriction of the non-stereospecific binding of myosin heads in weak binding states is observed.
This is very different from cross-bridges in strong binding states where attachment is stereospecific, i.e., the attached myosin heads have a quite fixed orientation relative to the actin monomers and thus to the actin helix. This results in quite prominent enhancement of actin based layer lines when such cross-bridges attach to actin, e.g. in rigor (Kraft et al., 1999, 2002), in the presence of ADP or during isometric contraction (Kraft et al., 2002).

In summary, myosin heads in weak binding states bind to specific sites on actin but non-stereospecifically when the actin filament is inactive but with some restriction of non-stereospecific binding when actin filaments are activated by Ca^{++}.

4. INTERACTION OF MYOSIN HEADS IN WEAK BINDING STATES WITH THE TROPOMYOSIN CONTROLLED BINDING SITES ON ACTIN

Since there is only a small effect of activation of thin filaments on the intensity ratio of the two innermost equatorial reflections, $I_{1,1}/I_{1,0}$, of X-ray diffraction patterns (Fig. 6) the question may be raised whether myosin heads in weak binding states (weak binding conformation) are at all able to interact with the tropomyosin controlled binding sites on actin that become accessible at high Ca^{++}-concentrations. That myosin heads in weak binding states indeed interact with these sites at high Ca^{++}-concentrations is suggested by the slight enhancement of the 5.9nm actin based layer line. Interactions with the tropomyosin controlled sites, however, are much more evident when fiber stiffness is plotted vs. speed of applied stretch for fibers incubated with MgATPγS both at low and high Ca^{++}-concentrations (Fig. 9). As a result of interactions with the tropomyosin controlled binding sites on actin at high Ca^{++}, the stiffness speed relation at high Ca^{++} is shifted by more than 2 orders of magnitude to slower stretch velocities. This indicates more than 100-fold slower probability of dissociation from actin at high Ca^{++} compared with low Ca^{++}-concentrations (Kraft et al., 1992).

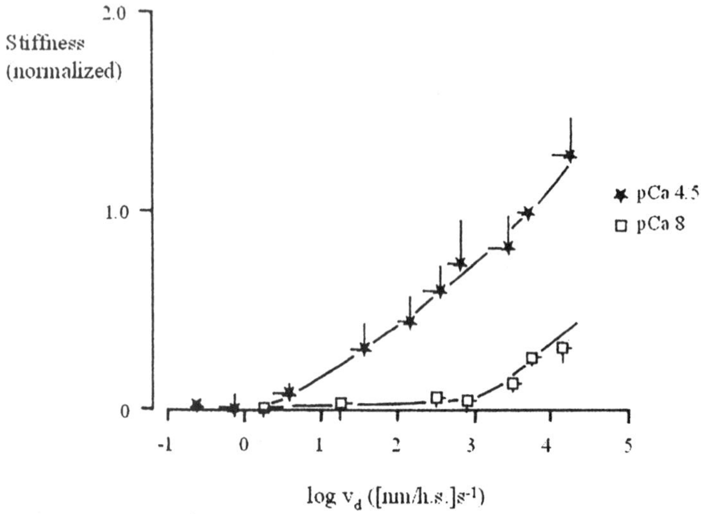

Figure 9. Stiffness speed relations obtained in the presence of 10 mM MgATPγS at low and high Ca^{++}-concentrations. Ionic strength 0.075M, temperature –3°C. Note that 10mM MgATPγS and low temperature were needed to keep myosin heads saturated with MgATPγS at high Ca^{++}- concentrations (cf. Fig. 10); modified from Kraft et al., 1992.

Somewhat surprisingly, however, we found that to keep myosin heads fully saturated with MgATPγS also at the high Ca^{++}-concentrations we had to (i) increase the concentration of MgATPγS to 10-20mM and (ii) lower temperature had to –3°C, the lowest possible temperature without freezing of the solution. In contrast, in the absence of Ca^{++} about 50μM of MgATPγS were sufficient to keep myosin heads saturated at otherwise identical ionic and temperature conditions (Fig. 10).

5. PHYSIOLOGICAL SIGNIFICANCE OF ACTIN ATTACHMENT OF MYOSIN HEADS IN WEAK BINDING STATES

All these data imply that although the myosin heads in weak binding states can make additional interactions with tropomyosin controlled sites at high Ca^{++}-concentrations, actin affinity is increased at most 2-fold, while at the same time the nucleotide affinity of the weak binding myosin heads when interacting with these additional sites is lowered about 1000-fold. We have interpreted this (Brenner et al., 1998, 1999) as an indication that a myosin head in a weak binding state becomes distorted when also interacting with the tropomyosin controlled sites, thus (i) the increase in actin affinity is almost absent despite the additional interactions between myosin head

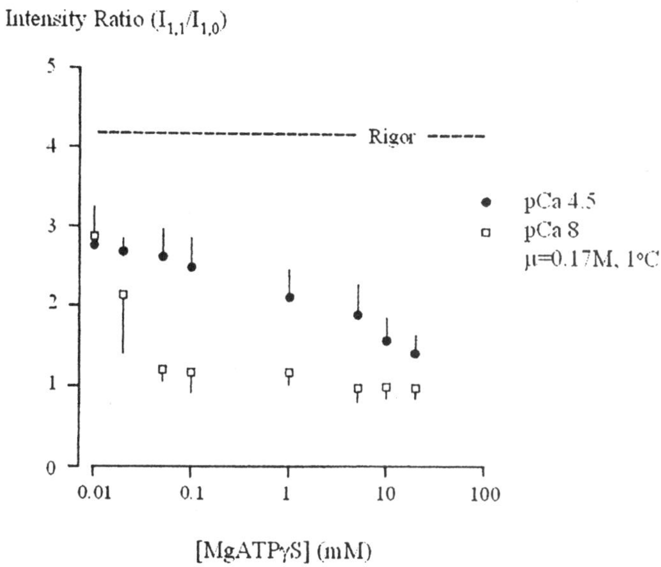

Figure 10. Intensity ratio $I_{1,1}/I_{1,0}$ of X-ray diffraction patterns recorded in the presence of different concentrations of ATPγS both at high and low Ca^{++}-concentrations. Note that in the presence of Ca^{++} the $I_{1,1}/I_{1,0}$-ratio as a measure of nucleotide-free cross-bridges decreases up to the highest nucleotide concentrations, just about approaching the value in the absence of Ca^{++} at 20mM of ATPγS, while in the absence of Ca^{++} a constant $I_{1,1}/I_{1,0}$-ratio is reached already at 50μM ATPγS (modified from Kraft et al. 1992).

and actin in the presence of Ca^{++}, and (ii) the nucleotide affinity is very much weakened. This behavior was also found for MgGTP and MgAMPPNP (Frisbie, et al., 1997), i.e., for molecules with a covalently bound γ-P_i. Moreover, we also found a weakening of AlF_4 trapped in the active site during active contraction forming an MgADP.AlF_4 complex (Chase et al., 1993), i.e., AlF_4 could be untrapped simply by activating the thin filaments by high Ca^{++} but in the presence of ATPγS which prevents active cross-bridge cycling and force generation that had previously been postulated to be required to untrap AlF_4 from the active site (Chase et al., 1993). Since we have identified the myosin head with the MgADP.AlF_4 complex formed during active cross-bridge cycling to be an analog of the weak binding conformation, the weakening of the AlF_4 simply by binding of the myosin head to the activated thin filament implies that interaction with the tropomyosin controlled sites on actin has a profound effect on the weak binding conformation of the myosin head. This results in weakening of (i) the nucleotide/nucleotide analog when the γ-phosphate (or γ-phosphate analog) is still covalently bound to the ADP moiety or (ii) of the AlF_4 trapped in the active site during active turnover. If the weakening effect seen with the AlF_4 also applies for P_i in the corresponding MgADP.P_i-complex, attachment of the weak binding states of the myosin head to the activated thin filament destabilizes the γ-phosphate, i.e., initiates phosphate release. In other words, attachment of weak binding myosin heads to activated actin and its profound effects on the weak binding conformation of the myosin head appears to be responsible for the actin activation of the myosin ATPase as actin activation is the result of faster release of inorganic phosphate from the active site in the presence of activated actin.

In summary, our data suggest that interaction of myosin heads in weak binding states with the activated thin filament (tropomyosin controlled sites) has a profound effect on the weak binding conformation of the myosin head leading to destabilization of the γ-P_i in the active site, i.e., speeding up P_i-release and thus resulting in actin activation of the myosin ATPase.

6. DISTORTION OF THE WEAK BINDING CONFORMATION OF THE MYOSIN HEAD UPON ATTACHMENT TO ACTIVATED THIN FILAMENTS DOES NOT RESULT IN A GLOBAL STRUCTURAL CHANGE.

As already pointed out, 2D-Xray diffraction patterns recorded when myosin heads in their weak binding conformation (weak binding states) attach to actin both at high and low Ca^{++} show no large change in the myosin based layer lines and only a slight enhancement of the 5.9nm actin based layer line, much less than seen with stereospecific attachment of myosin heads in strong binding states (Kraft et al., 1999). This implies that although generating some distortion of the catalytic domain of the attached myosin head in the weak binding conformation (weakening of nucleotide affinity, untrapping of AlF_4, almost no enhancement of actin affinity despite additional interactions with actin), interaction with the tropomyosin controlled sites does not induce a global structural change in the weak binding conformation of the myosin head domain. One might speculate that the distortion in the weak binding conformation of the myosin head generated by the interactions with the tropomyosin controlled sites may be a kind of 'forced' closure of the cleft between upper and lower 50kDa domain with destabilization of the γ-phosphate.

In contrast, binding of myosin heads in strong binding states to the activated thin filament results in very little difference in nucleotide affinity but in large increase in actin affinity (Brenner et al., 1986b) consistent with little if any distortion of the myosin head domain upon additional binding to the tropomyosin controlled sites. This might be accounted for by the cleft between upper and lower 50kDa domain being easily closed upon interaction with the tropomyosin controlled sites at high Ca^{++} when the γ-phosphate or γ-phosphate analogue is released or at least displaced (strong binding conformation of myosin heads) while such cleft closure meets large resistance when the γ-phosphate or γ-phosphate analogue is still strongly coordinated in the nucleotide binding pocket (weak binding conformation of myosin head).

Aside from the detailed structural properties, myosin heads in a weak vs. strong binding conformation apparently have quite different conformation when attached to activated thin filaments. This implies that upon binding to activated thin filaments, myosin heads in these two conformations will pass through different intermediates, i.e., will not even transiently pass through identical intermediates as postulated by the concept of Geeves et al. (1984; McKillop and Geeves, 1993).

7. RELATION OF WEAK AND STRONG BINDING STATES OF THE MYOSIN HEAD TO THE POWER STROKE

Based on the effects of release of phosphate from caged phosphate on force generation under isometric conditions it had previously been postulated that phosphate is released in a two step process. It was further postulated that the second M.ADP.P_i-intermediate contributes to fiber stiffness and possibly to active tension (Dantzig et al., 1992). During steady state force generation this state, however, is at most only a transient intermediate with little occupancy. Thus, it had so far not been possible to clearly characterize and classify this intermediate in terms of weak vs. strong binding properties.

In a recent attempt we succeeded to generate an analog state of this second M.ADP.P_i-intermediate (cf. Fig. 1) by trapping AlF_4 in the active sites of myosin heads that were bound to actin in the presence of MgADP (Mattei et al., 2004; Kraft et al., 2004). In contrast to the analog state in which AlF_4 had been trapped in the active site during active cross-bridge cycling (Chase et al., 1993) and which shows all properties of weak binding states of the myosin head summarized above, cross-bridges with the ADP.AlF_4-complex trapped in the active site by the second approach show all the features characteristic for strong binding states of the myosin head. Most characteristically, this second ADP.AlF_4 analogue state shows (i) a much larger actin affinity when binding to the activated thin filament (high Ca^{++}) compared to binding to the inactive thin filament (low Ca^{++}), (ii) much slower dissociation from activated thin filaments, and (iii) stereospecific binding to the activated thin filament as evidenced by enhanced actin based layer lines. Thus, assuming analogy for the second M.ADP.P_i-intermediate, these data suggest that the second ADP.P_i-analog is the first strong binding state of the myosin head. It has neither intermediate nor new properties other than those already characterized for the weak vs. strong binding conformation of the myosin head.

Very interestingly, trapping essentially all myosin heads in this second ADP.AlF_4 state results in an 2D-Xray pattern very reminiscent of the pattern observed under isometric conditions for which we had provided evidence that a large fraction of myosin heads are in a strong binding conformation (Kraft et al., 2002). The similarity of these 2D-patterns supports this point and is also consistent with our conclusion that during isometric force generation a large fraction of cross-bridges occupy states early in the power stroke (Brenner et al., 1995).

Acknowledgements:

This work was supported by grants from the Deutscl : Forschungsgemeinschaft to BB (BR 849/1-4) and TK (KR 1187/18-1). The authoı also like to thank Drs. J.M. Chalovich, E. Eisenberg, M. Schoenberg, L.C. Yu for tl many years of collaboration the earlier part of the work reviewed in this paper.

REFERENCES

Bagshaw CR, Eccleston JF, Trentham DR, and Yates DW. Transient kinetic studies of the Mg^{++}-dependent ATPase of myosin and its proteolytic fragments. *Cold Spring Harb Symp Quant Biol* 37: 127-135, 1973.

Brenner B, Schoenberg M, Chalovich JM, Greene LC, and Eisenberg E. Evidence for cross-bridge attachment in relaxed muscle at low ionic strength. *Proc Natl Acad Sci USA* 79: 7288-7291, 1982.

Brenner B, Yu LC, and Podolsky RJ. X-ray diffraction evidence for cross-bridge formation in relaxed muscle fibers at various ionic strengths. *Biophys J* 46: 299-306, 1984.

Brenner B, and Yu LC. Equatorial X-ray diffraction from single skinned rabbit psoas fibers at various degrees of activation. Changes in intensities and lattice spacing. *Biophys J* 48:829-834, 1985.

Brenner B, Chalovich JM, Greene LC, Eisenberg E, and Schoenberg M. Stiffness of skinned rabbit psoas fibers in MgATP and MgPP$_i$ solutions. *Biophys J* 50: 685-691,1986.

Brenner B, Yu LC, Greene LC, Eisenberg E, and Schoenberg M. Ca^{2+}-sensitive cross-bridge dissociation in the presence of MgPP$_i$ in skinned rabbit psoas fibers. *Biophys J* 50: 1101-1108, 1986(b).

Brenner B. Muscle mechanics and biochemical kinetics, in: *Molecular Mechanism of Muscular Contraction*, John Squire, ed., Macmillan Press Ltd., London, pp. 77-149, 1990.

Brenner B, Yu LC, and Chalovich JM. Parallel inhibition of active force and relaxed fiber stiffness in skeletal muscle by caldesmon. Implications for the pathway to force generation. *Proc Natl Acad Sci USA* 88: 5739-5743, 1991.

Brenner B, Chalovich JM, and Yu LC. Distinct molecular processes associated with isometric force generation and with rapid tension recovery after quick release. *Biophys J* 68:106s-111s, 1995.

Brenner B, Kraft T, and Chalovich JM. Fluorescence of NBD-labeled troponin-I as a probe for the kinetics of thin filament activation. *Adv Exp Med Mol Biol* 453:177-185, 1998.

Brenner B, and Chalovich JM. Kinetics of thin filament activation probed by fluorescence of N-((2-Iodoacetoxy)ethyl)-N-methyl)amino-7-nitrobenz-2-oxa-1,3-diazole-labelled troponin I incorporated in skinned fibers of rabbit psoas muscle. Implications for regulation of muscle contraction. *Biophys J* 77:2692-2708, 1999.

Brenner B, Kraft T, Yu LC, and Chalovich JM. Thin filament activation probed by fluorescence of N-((2-Iodoacetoxy)ethyl)-N-methyl)amino-7-nitrobenz-2-oxa-1,3-diazole-labelled troponin I incorporated in skinned fibers of rabbit psoas muscle. *Biophys J* 77:2677-2691, 1999.

Chase PB, Martyn DA, Kushmerick MJ, and Gordon AM. Effects of inorganic phosphate analogues on stiffness and unloaded shortening of skinned muscle fibres from rabbit. *J Physiol Lond* 460:231-46, 1993.

Dantzig JA, Goldman YE, Millar NC, Lacktis J, and Homsher E. Reversal of the cross-bridge force-generating transition by photogeneration of phosphate in rabbit psoas muscle fibres. *J Physiol Lond* 451:247-278, 1992.

Frisbie SM, Chalovich JM, Brenner B, and Yu LC. Modulation of cross-bridge affinity for MgGTP by Ca^{++} in skinned fibers of rabbit psoas muscle. *Biophys J* 72:2255-2261, 1997.

Geeves MA, Goody RS, and Gutfreund H. Kinetics of acto.S1 interaction. *J Muscle Res Cell Motil* 5:351-361, 1984.

Geeves MA, and Conibear PB. The role of three state docking of myosin S1 with actin in force generation. *Biophys J* 68:194s-201s, 1995.

Goody RS, and Hofmann F. Stereochemical aspects of the interaction of myosin and actomyosin with nucleotides *J Muscle Res Cell Motil* 1:101-115, 1980.

Greene LE, and Eisenberg E. Cooperative binding of myosin subfragment-1 to the actin-troponin-tropomyosin complex. *Proc Natl Acad Sci USA*: 77:2616-2620, 1980.

Greene LE, Sellers JR, Eisenberg E, and Adelstein RS. Binding of gizzard smooth muscle myosin subfragment-one to actin in the presence and absence of ATP. *Biochemistry* 22:530-535, 1983.

Gu J, Xu S, and Yu LC. A model of cross-bridge attachment to actin in the A*M*ATP state based on x-ray diffraction from permeabilized rabbit psoas muscle. Biophys J 82:2123-33, 2002.

Head JG, Ritchie MD, and Geeves MA. Characterization of the equilibrium between blocked and closed states of muscle thin filaments. *Eur J Biochem* 227:694-699, 1995.

Holmes KC. The actomyosin interaction and its control by tropomyosin. *Biophys J* 68:2s-5s, 1995.

Kraft T, Yu LC, Kuhn HJ, and Brenner B. Effect of Ca^{++} on weak cross-bridge interaction with actin in the presence of the nucleotide analog ATPγS. *Proc Natl Acad Sci USA* 89: 11362-11366, 1992.

Kraft T, Chalovich JM, Yu LC, and Brenner B. Parallel inhibition of active force and relaxed fiber stiffness by caldesmon fragments at physiological temperature and ionic strength conditions. Additional evidence that weak cross-bridge binding to actin is an essential intermediate for force generation. Biophys J 68: 2404-2418, 1995.

Kraft T, Xu S, Brenner B, and Yu LC. The effect of thin filament activation on the attachment of weak binding cross-bridges: A 2D-Xray-diffraction study on single muscle fibers. *Biophys J* 76:1494-1513,1999.

Kraft T, Mattei T, Radocaj A, Piep B, Nocula Ch, Furch M, and Brenner B. Structural features of cross-bridges in isometrically contracting skeletal muscle *Biophys J* 82:2536-2547, 2002.

Kraft T, Mählmann E, Mattei T, and Brenner B. Effects of myosin binding to actin on structural and functional properties of the myosin head domain. *Biophys J* 86:266a, 2004

Lehman W, Craig R, and Vibert P. Ca^{2+}-induced tropomyosin movement in *Limulus* thin filaments revealed by three-dimensional reconstruction. *Nature* 368:65-67, 1994.

Lehman W, Vibert P, Uman P, and Craig R. Steric-blocking by tropomyosin visualized in relaxed vertebrate muscle thin filaments. *J Mol Biol* 251:191-196, 1995.

Mattei T, Mählmann E, Piep B, Kraft T, and Brenner B. MgADP.AlF$_4$-binding to cross-bridges under rigor conditions: mechanical and structural properties of the resulting cross-bridge state. *Biophys J* 86:266a, 2004.

McKillop DFA, and Geeves MA. Regulation of the interaction between actin and myosin subfragment 1: evidence for three states of the thin filament. *Biophys J* 65:693-701, 1993.

Millar NC, and Homsher E. The effect of phosphate and calcium on force generation in glycerinated rabbit skeletal muscle fibers. A steady state and transient kinetic study. *J Biol Chem* 265:20234-20240, 1990.

Schoenberg M. Equilibrium muscle cross-bridge behavior. Theoretical considerations. *Biophys J* 48:467-475, 1985.

Stein LA, Schwartz RP, Chock PB, and Eisenberg E. Mechanism of actomyosin adenosine triphosphatase. Evidence that adenosine 5'-triphosphate hydrolysis can occur without dissociation of the actomyosin complex. *Biochemistry* 18;3895-3909, 1979.

Stein LE, Chock PB, and Eisenberg E. The rate-limiting step in the actomyosin adenosintriphosphatase cycle. *Biochemistry* 23:1555-1563, 1984.

White HD, and Taylor EW. Energetics and mechanism of actomyosin adenosine triphosphatase. *Biochemistry* 15:5818-5826, 1976.

Xu S, Malinchik S, Gilroy D, KraftT, Brenner B, and Yu LC. X-ray diffraction studies of cross-bridges weakly bound to actin in relaxed skinned fibers of rabbit psoas muscle. *Biophys J* 72:2292-2303, 1997.

MYSTERIES ABOUT AMPLITUDE AND EFFICIENCY OF CROSS-BRIDGE POWERSTROKE

Haruo Sugi[1], Shigeru Chaen[1,2] and Ibuki Shirakawa[1]

1. THE HUXLEY CONTRACTION MODEL

Since the monumental discovery of the sliding filament mechanism in the hexagonal lattice of thick and thin filaments by H. E. Huxley and his colleagues (Huxley and Hanson, 1954), the contraction model of A. F. Huxley (1957) has long been central in the field of muscle mechanics. The main framework of this model is as follows:

1. Muscle contraction results from cyclic attachment and detachment between the myosin heads, i.e. the cross-bridges, and the sites on the thin filament. A cross-bridge splits one ATP molecule in each attachment-detachment cycle.

2. As shown in Fig. 1A, a cross-bridge (M), connected to the thick filament with springs (S_1 and S_2), is fluctuating around its equilibrium position (0) due to thermal agitation. When it attaches to an active site (A) on the thin filament. it exerts either positive or negative force (F) proportional to its distance (x) from 0 by pulling S_1 or S_2 ($F = 0$ when $x = 0$, and $F = kx$ when $x > 0$ or $x < 0$, k is the stiffness of the spring.

3. The rate constants for attachment of M to, and its detachment from, A (f and g, respectively) are functions of x as shown in Fig. 1B. M attaches to A if M and A are within the range of x where f has finite values ($0 \leq x \leq h$) to exert positive force. If the A-M link moves across 0 to the negative x region, it breaks fairly rapidly because f is zero and g has a large value.

In Fig. 1C, the steady-state distribution of A-M links (expressed by fractional numbers $0 \leq n \leq 1$) is shown against x at various steady shortening velocities V (expressed relative to V_{max}). In the isometric condition ($P = P_0$, $V = 0$), A-M links show square-shaped distribution only in the positive x region where f has finite values ($0 \leq x \leq h$, Fig. $1C_1$). As the steady shortening velocity increases, the number of A-M links

[1]Department of Physiology, School of Medicine, Teikyo University, Itabashi-ku, Tokyo 173-0003, Japan; [2]Department of Physics, Faculty of Liberal Arts and Science, Nihon University, Setagaya-ku, Tokyo, Japan.

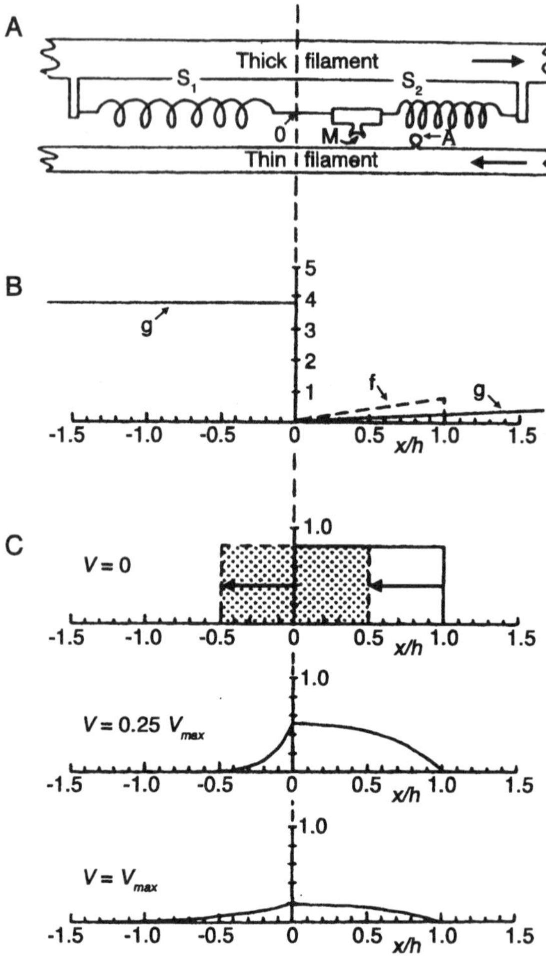

Figure 1. A. F. Huxley's contraction model (Huxley 1957). (A) Diagram showing the myosin head *M* connected to the thick filament with springs S_1 and S_2, and the active site *A* on the thin filament. Direction of relative myofilament sliding is indicated by arrows. (B) Dependence of *f* and *g* on the distance *x* of *M* from the equilibrium position 0. *M* can only attach to *A* to form an *A-M* link at $0 \leq x \leq h$, where *f* has a finite value. (C_1 – C_3) Diagram showing distribution of *A-M* links with respect to *x* at three different steady-shortening velocities: in C_1 the broken line and dotted area show a rapid shift of *A-M* link distribution in response to a quick length decrease of an isometrically tetanized muscle fiver to drop the force to zero; C_2 and C_3 show changes in *A-M* link distribution with increasing steady velocity of shortening.

exerting positive forces decreases as a result of a moderate value of *f* ; the proportion of M that slides past A without formation of A-M link increases as the velocity of filament sliding increases. On the other hand, A-M links brought into the negative *x* region tend to exist over a fairly large distance as the sliding velocity increases (Fig. 1C_2). Finally, the sliding velocity reaches a value, under which the positive and negative forces exerted by A-M links at both sides of 0 are equal (Fig. 1C_3). This velocity corresponds to the maximum shortening velocity (V_{max}) under zero external load.

In this model, the myofilaments are assumed to be rigid, and the series elastic component SEC therefore originates only from S_1 and S_2 connecting M to the thick filament. During the plateau of isometric tetanus, all A-M links generate positive forces ≥ 0 by pulling S_1. The quick decrease in force in response to quick release occurs due to elastic recoil of stretched S_1, resulting in a rapid shift of the A-M link distribution into the negative x region. If the shift amounts to $h / 2$, the positive and negative forces on S_1 and S_2 become equal immediately after the release, and the force just drops from P_0 to zero (dotted area in Fig. $1C_1$). Since the amount of quick release required to drop P_0 to zero is about 1 % of muscle length (i.e. about 10 nm in each 1 _ m long half-sarcomere), the value of h is ~ 20 nm.

The Huxley contraction model can well explain the force-velocity and the force-energy relations in skeletal muscle (A. V. Hill, 1938). Its simplicity and beauty have attracted attention of investigators in the field of muscle mechanics, and a number of experiments have hitherto been made and the results obtained have been interpreted within the framework of this model. In this article, we aim at considering some selected topics in the field of muscle mechanics, which still remain mysterious and should be clarified in future.

2. TRANSIENT PHENOMENA FOLLOWING STEP CHANGES IN LOAD AND IN LENGTH

An effective means to obtain insight into the mechanism producing force and motion in muscle is to apply step changes in load to an isometrically contracting muscle fiber to examine the resulting isotonic shortening. Using this technique, Civan and Podolsky (1966) found that, following step decreases in load from P_0 to $P < P_0$, the fiber showed oscillatory changes in shortening velocity, i.e. isotonic velocity transients, before a steady isotonic shortening determined by the new load P was reached (Fig. 2A). The isotonic condition is suitable for simulating the transient phenomena by determining the A-M link distribution for each consecutive time segment and then shift the A-M link distribution along the force (x) axis to maintain the isotonic condition. In the Huxley contraction model, the square-shaped A-M link distribution in the isometric condition shifts to the left immediately after a step load decrease. During the subsequent time segment, the A-M links increase in the positive x region, while they decrease in the negative x region. To maintain the isotonic condition, therefore, the new A-M link distribution should be shifted to the left. The above process is repeated until the steady A-M link distribution associated with steady isotonic shortening is reached.

Podolsky and his colleagues (Podolsky, Nolan and Zavaler, 1969) showed that the oscillatory isotonic velocity transients could only be simulated by the rate constants and the cross-bridge force as functions of x as shown in Fig. 2C. The Podolsky model differs from the Huxley contraction model in that (1) the rate function for making A-M links f is large so that every M passing through the positive x region forms A-M link, (2) the rate constant for breaking A-M link g in the negative x region is very small near the equilibrium position 0, and suddenly increases for x larger than a critical value. They also made additional assumptions that (1) the series elasticity originates not only from the A-M link but also from the thin filament in the I-band, and (2) breaking of A-M links are associated with ATP splitting only in the small g region, while A-M links brought into the large g region are broken by being pushed by other A-M links exerting positive

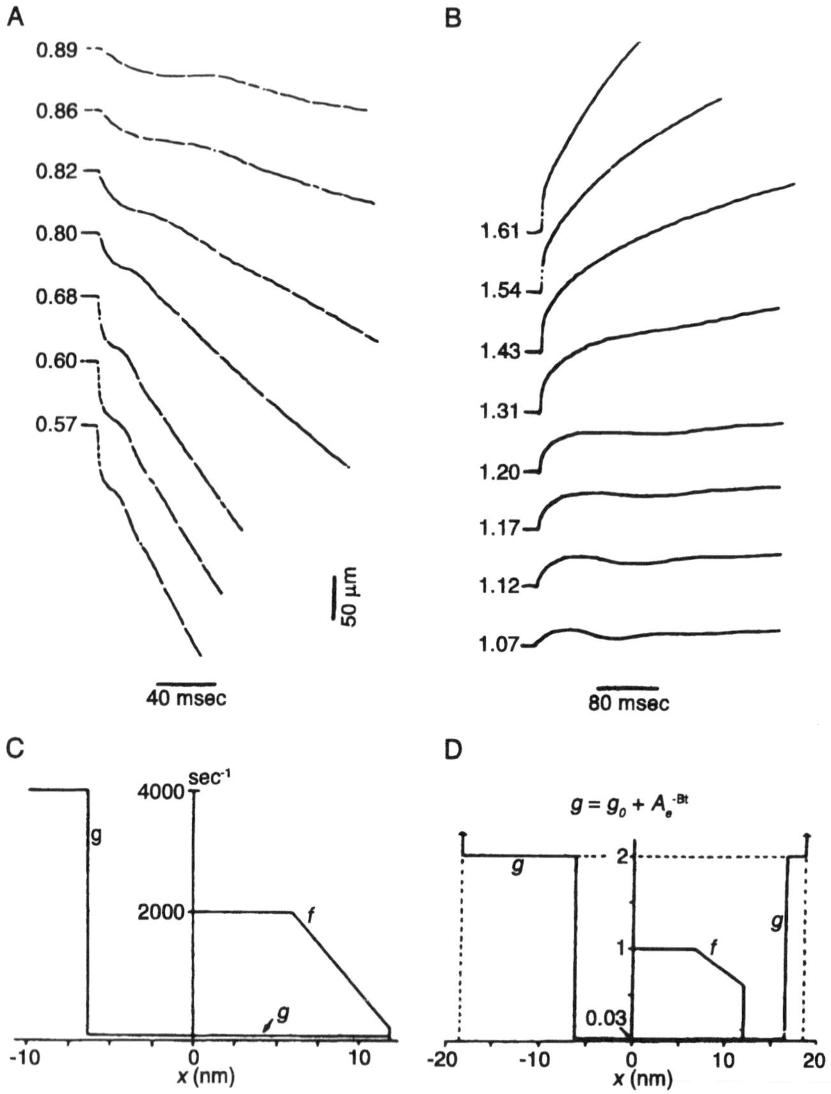

Figure 2. Isotonic velocity transients and the contraction models to stimulate the transients. (A) Isotonic velocity transients following quick decreases in force (= load) from P_0 to $P < P_0$. Note oscillatory changes in shortening velocity before steady velocities are reached. (B) Isotonic velocity transients following quick increases in force from P_0 to P . P_0. Note oscillatory changes in fiber length in response to small increases in force. In both A and B, the values of P relative to P_0 are shown alongside each record (Sugi and Tsuchiya 1981). (C) Podolsky-Nolan model for calculating the velocity transients following quick decreases in force. Note large values of f and a small value of g between f and the major part of g in the negative x region (Podolsky and Nolan 1973). (D) Sugi-Tsuchiya model for calculating isotonic velocity transients following both quick decreases and quick increases in force. Note an additional gap between f and the major part of g in the positive x region together with the assumption that a quick increase in force produces an instantaneous increase in g followed by exponential recovery at every value of x (Sugi and Tsuchiya 1981).

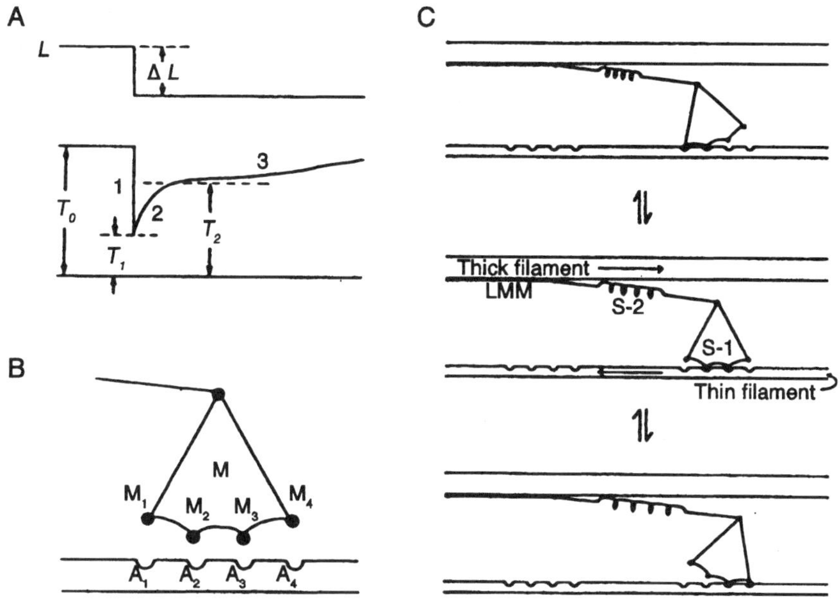

Figure 3. Huxlley-Simmons contraction model (A. F. Huxley and Simmons 1971). (A) Diagram shoing the components of the isometric force transients following a quick release. (B) Diagram showing the myosin S-1 head (*M*) with four binding sites (M_1-M_4) and part of the thin filament with four binding sitses (A_1-A_4). (C) Mechanism of force generation by rotation of the myosin head on the thin filament by pulling the spring (S-2) between the myosin head and the thick filament (A.F. Huxley 1974).

force. The Podolsky model described above can also account for the force-velocity and the force-energy relations in skeletal muscle, when some additional assumptions are made.

Later, Sugi and Tsuchiya (Sugi and Tsutiya, 1981) studied the isotonic velocity transients following step increases in load from P_0 to $P > P_0$ with the results shown in Fig. 2B. The most characteristic feature was that the transients following a small load increase showed a remarkable length oscillation consisting of alternate lengthening and shortening of the fiber. They could simulate the results using the contraction model show in Fig. 3D with the additional assumption that the value of g increases everywhere immediately after a step load increase and then returns to the initial value with an exponential time course.

Although studies on the isotonic velocity transients provided considerable insights into the mechanism producing force and motion in muscle, it has not been possible to decide which or which part of these models are correct. Next, we will consider other kinds of approaches to the understanding of muscle contraction mechanism.

3. ISOMETRIC FORCE TRANSIENTS

Another effective means to obtain information about the mechanism of muscle contraction is to apply step changes in length to an isometrically contracting muscle fiber.

In frog muscle fibers, however, the sarcomere spacings tend to be shorter and less extensible near the fiber ends than at the middle region where the sarcomere spacings are reasonably uniform. Consequently, the applied length changes does not take place uniformly along the fiber length. To overcome this difficulty, Gordon et al. (1966 *a*, *b*) developed ingenious spot-follower method, with which fiber length was controlled at the middle part between the two position markers on the fiber. With this method, they could establish a well-known sarcomere length-force relation, which has been generally taken as evidence that the cross-bridges are independent force generators.

Huxley and Simmons (1971) also used the spot-follower method to study the isometric force transients in response to step decreases in fiber length with results and interpretations illustrated in Fig. 3. They interpreted the initial force drop coincident with the length step (from T_0 to T_1, Fig. 3A) as being due to the cross-bridge elasticity, while they interpreted the initial quick force recovery (from T_1 to T_2) as being due to rotation of the cross-bridges that remain attached to the thin filament either before and after the length step. Huxley and Simmons proposed a model diagrammatically shown in Fig. 3B, C. Here, the cross-bridge (M) and the spring connecting it to the myosin filament (S-2) are drawn more realistically than Huxley's 1957 model. Both M and the thin filament are assumed to have more than one combining site (for example M_1 to M_4 and A_1 to A_4, Fig. 3B). M first attaches to A by forming links M_1A_1 and M_2A_2, while the spring is at its slack length (Fig. 3C_1). Then, according to the assumption of the potential energy diagram along they attachment sites A_1 to A_4, M changes its attachment sites from M_1A_1-M_2A_2 to M_2A_2-M_3A_3 and then to M_3A_3-M_4A_4. As a result, M rotates along the thin filament to pull the spring S-2, and the resulting force in S-2 shows up as isometric force if the thick and thin filaments are not allowed to slide past each other, or it causes the relative filament sliding. The ATP splitting is assumed to be coupled with the detachment of M from A. A quick release applied to an isometrically tetanized fiber causes a decrease in force in the spring, resulting in a quick force drop from T_0 to T_1 (Fig. 3A). This is followed by the increased probability that the attached M rotates on the thin filament to pull back the spring, resulting in the quick force recovery (phase 2). The subsequent slow force recovery is due to turnover of A-M links.

The main difference between the Huxley-Simmons model and the Huxley 1957 model is that, in the former, there are more than one A-M links, including the initial nonforce-generating state, while in the latter the A-M link has only one state and generates force as soon as A-M link is formed. The presence of the nonforce-generating A-M link is in accord with the phenomenon that both the stiffness increase and the transfer of mass from the thick to the thin filament as evidenced by the change in equatorial X-ray diffraction pattern (for a review. see Huxley and Faruqi 1983) occur ahead of the force development (see, Fig. 4A). The Huxley-Simmons model has been useful in relating the hypothetical A-M links to the enzyme kinetics of actomyosin ATPase in solution. The amount of extension of the spring S-2, when the force in a myosin head increases to the maximum value T_0, has been carefully estimated by A. F. Huxley and his coworkers (Ford et al. 1977) to be less than 4 nm.

The only experimental evidence for the cross-bridge origin of the T_1 and T_2 curves was that these curves were scaled according to the number of attached cross-bridges. If the steady isometric force P_0 was altered by changing the amount of filament overlap, the curves relating the level of T_1 and T_2 to the shortening step (T_1 and T_2 curves) were scaled according to the amount of P_0. Both T_1 and T_2 curves were not, however, obtained up to the level of zero force, since the feedback loop in the spot-follower did not

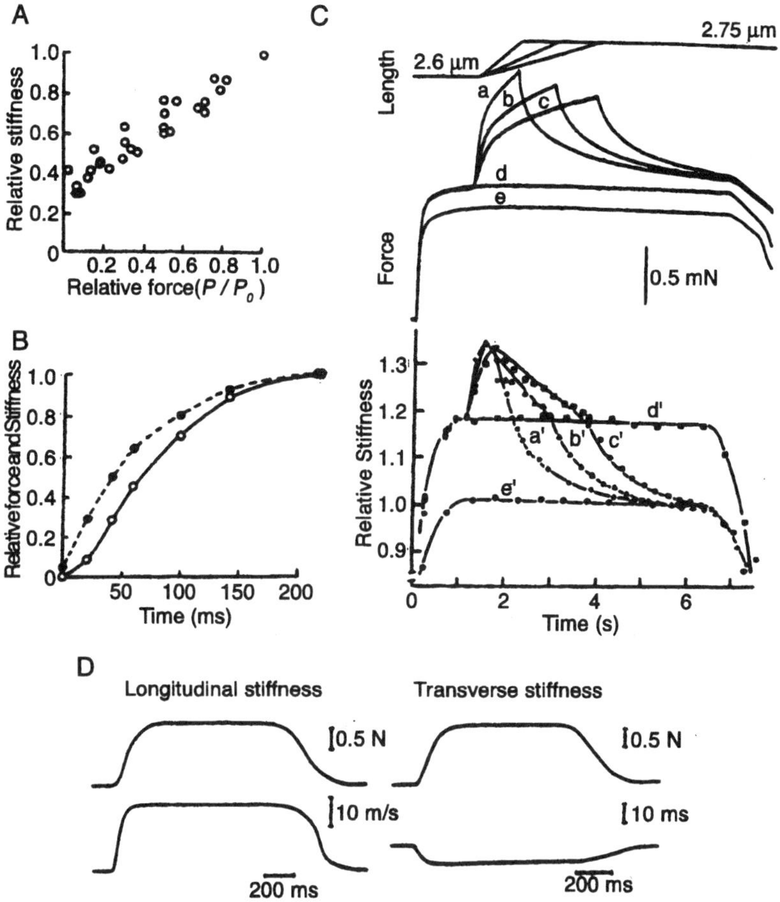

Figure 4. Muscle and muscle fibers stiffness changes in various experimental conditions. (A) Dependence of muscle fiber stiffness on the amount of load (= force) during steady isotonic shortening (Julian and Sollins 1975). (B) Increase in muscle fiber stiffness (*filled circles*) and force (*open circles*) during rise of isometric teatnic force (Cecchi et al. 1984). (C) Force and stiffness changes of a tetanized fiber in response to slow stretches. The fiber was first tetanized isometrically at 2.6 μm sarcomere length, and then slowly stretched to 2.75 μm sarcomere length with three different velocities. *Upper, middle* and *lower traces* show time course of changes in the fiber length, the force and the stiffness, respectively. Traces a, b and c in the force records correspond to traces a', b' and c' in the stiffness trace, respectively. Traces d and e are records of ordinary isometric forces at sarcomere length of 2.6 μm to 2.75 μm, respectively. Note that the stiffness increases abruptly on application of stretch, then decrease linearly during stretch, and eventually decreases to a steady level equal to that of ordinary isometric tetanus at a sarcomere length of 2.75 μm after completion of stretch, whereas the isometric force attained after completion of stretch is appreciably higher than the ordinary isometric force at a sarcomere length of 2.6 μm despite the decrease in the amount of myofilament overlap (Sugi and Tsuchiya 1988). (D) Changes in the force (*upper traces*) and the stiffness (*lower traces*) of a muscle during isometric tetanus . The stiffness was measured as propagation velocity of ultrasonic waves (3-7 MHz) between the two piezoelectric transducers. Upward and downward deflections of the stiffness records represent increase and decrease of stiffness, respectively. Note that the transverse stiffness decrease during contraction (Hatta et al. 1988).

work well at low force range, where the slope of length-step versus force relation was not steep. Therefore, the evidence for the validity of the model was incomplete. Meanwhile, Huxley and Simmons (1971) used very rapid length steps and examined the resulting force transients on a very fast time base, so that the force in the fiber was not uniform in every fiber cross-section; in other words, it was impossible in principle to actually record and control "segmental force" in the length-controlled segment. In fact, Sugi and his colleagues have demonstrated with various methods, including ultrahigh-speed cinematography and ultrahigh sensitivity sarcomere length measurement, that there are considerable nonuniformities in both length and force during the period of quick force recovery (Sugi and Kobayashi, 1983), indicating that the quick force recovery does not reflect fundamental molecular events responsible for force and motion in muscle. Now, it is generally agreed that the series elastic component originates not only from the cross-bridges but also from the thin filament (for a review, see Sugi and Tsuchiya 1998), while no concrete evidence has been obtained for the cross-bridge rotation during muscle contraction (for reviews, see Sugi, 1992, 1993; Sugi and Tsuchiya, 1998).

4.MUSCLE FIBER STIFFNESS

In the Huxley contraction model, in which myofilaments are assumed to be rigid, small length perturbations (~ 0.1 % of L_0, kHz range) applied to a tetanized muscle fiber would be taken up by the elastic link connecting each cross-bridge to the thick filament in both positive and negative x regions (Fig. 1). Thus the resulting force changes serve as a measure of total number of attached cross-bridges, if the compliance of recording system are reasonably reduced. Julian and Sollins (1975) measured muscle fiber stiffness by applying such small length perturbations during steady isotonic shortening, and showed that the stiffness decreased with decreasing isotonic load (= force), approaching a minimum value (about 20 % of the maximum value at the plateau of isometric tetanus) as the load tended to zero (Fig. 4A). This result is in fairly good agreement with the Huxley contraction model with respect to the total number of attached cross-bridges at various steady shortening velocities (Fig. 1B, C). On the other hand, the fiber stiffness rises from zero to the maximum value during isometric force development, reflecting an increase in the number of attached cross-bridges. Cecchi *et al.* (1982, 1986) have shown that muscle fiber stiffness rises ahead of force by more than 15 ms (Fig. 4B) during isometric force development (Fig. 4B). This can be taken to indicate that, as predicted by the Huxley-Simmons model, cross-bridge first attaches to the thin filament, and then undergoes structural changes to produce force after a definite delay.

Sugi and Tsuchiya (1988) have demonstrated a marked dissociation between the stiffness and the isometric force when an isometrically tetanized fiber is subjected to a slow stretch. As illustrated in Fig. 4C, the stiffness first rises to a steady level as the fiber develops a steady isometric force on tetanic stimulation. On application of a slow stretch at the plateau of the preceding isometric tetanus, the stiffness shows an abrupt rise (probably due to synchronized extension of attached cross-bridges) which is then followed by a gradual decrease to a new steady level appreciably below the level of the preceding tetanus, reflecting a decrease in the amount of myofilament overlap caused by the stretch. The marked decrease of the stiffness starts when the stretch is still going on and the force is rising, and the final stiffness level attained after the end of stretch is definitely below the initial stiffness level, although the isometric force attained after the end of stretch is appreciably higher than the normal isometric force at the stretched fiber length. These results indicate that cross-bridge elasticity may differ depending on their physiological states.

The stiffness measurements described so far are concerned with relative values expressed as the magnitude of force changes caused by small constant length changes in kHz range. Sugi and his coworkers (Tamura *et al.*, 1982; Hatta *et al.*, 1988) developed a method, with which muscle stiffness was measured with ultrasonic waves in MHz region, with negligibly small perturbations to the muscle. The stiffness measurement is based on the equation,

$$V_f = (C_f/\rho)^{1/2},$$

where V_f is the phase velocity of ultrasonic waves, C_f is the muscle stiffness, and ρ is the density of muscle; this is essentially the measurement of propagation velocity of ultrasonic waves in muscle. The advantage of this method is that the stiffness can be measured not only along the muscle length as with other stiffness measurements (longitudinal stiffness), but also across the muscle transverse section (transverse stiffness). On tetanic stimulation, the longitudinal stiffness rises ahead of force by more than 20 ms during isometric force development, while it lags behind the force during relaxation of isometric force (Fig. 4D, left). On the other hand, the transverse stiffness decreases during isometric tetanus (Fig. 4D, right). This interesting phenomenon, possibly reflecting change in the state of water between the myofibrils as ATP hydrolysis goes on, should further be studied using single fibers. On the other hand, both the longitudinal and transverse stiffness remained unchanged if a muscle in rigor state was stretched by 1 - 2 %, so that the rigor force increased twofold. If the ultrasonic waves are assumed to shake water molecules, the state of water may change dynamically during active force generation but not in rigor state.

Evidence has been accumulating that, muscle fiber stiffness measured with step length changes over a wide range of velocities extending above the kHz range, exhibit feature completely different from those obtained with kHz range perturbations (Bagni et al, 1992, 1995; Stehle and Brenner, 2000). This is not surprising, since force generation in muscle is a dynamic phenomenon involving changes in small molecules, macromolecules, myofilaments and higher structures; the above hierarchy in muscle structure responsible for contraction would exhibit different features according to frequency of externally applied perturbations.

5. THE DISTANCE OF CROSS-BRIDGE POWERSTROKE

The distance cross-bridge powerstroke *d* is defined as the distance, over which a cross-bridge attached to the thin filament exerts positive force to produce force and motion utilizing energy derived from ATP hydrolysis. On the basis of Huxley-Simmons model, *d* corresponds with the distance of myofilament sliding achieved by the cross-bridge rotation (Fig. 3C). Since the myosin head is pear-shaped with the larger diameter of ~ 20 nm, it gives a structural constraint that *d* is ~ 10 nm and never more than 20 nm. The development of *in vitro* motility assay system, in which fluorescently labeled actin filaments are made to slide on myosin fixed to a glass surface (Kron and Spudich, 1986), has made it possible to roughly estimate the value of *d* by measuring the amount of ATP utilized and the distance of actin filament sliding on myosin, assuming d in each cross-bridge is coupled with hydrolysis of one ATP molecule. Due to lack of accuracy in measuring the amount of ATP utilized, however, the estimated values of cross-bridge powerstroke ranged from ~ 8 nm to > 200 nm (Toyoshima et al., 1990; Harada et al., 1990), and no definite conclusion has been reached with this kind of approach.

On the other hand, a number of motor proteins other than muscle myosin, which is now called myosin II, have been found, and their motor functions have been compared using the in vitro motility assay systems. Based on both density of motor proteins on a glass surface and the minimum length of actin filaments (or microtubules) to move on myosin without being detached, the concept of "duty ratio" , i.e. the fraction of time in which a motor protein is attached to its filament, has proved to be useful in understanding functional diversity of motor proteins (for a review, see Howard, 1997).

The duty ratio of skeletal muscle myosin is regarded to be small, being 0.01 - 0.1 (Uyeda et al., 1990; Leibler and Huse, 1993). In terms of the duty ratio P, the value of d, coupled with hydrolysis of one ATP molecule, can be expressed as:

$$P = (d \bullet \mathrm{A}) / \mathrm{V}_{\mathrm{max}} \quad (1),$$

where A is ATPase rate per cross-bridge and V_{max} is the maximum velocity of actin-myosin sliding under zero external load. If we put 20 s^{-1} cross-bridge^{-1} (Toyoshima et al., 1987) for A and ~ 3 mms^{-1} (Oiwa et al., 1990) for V_{max} in equation 1, a value of d of 4 - 16 nm, consistent with the Huxley-Simmons model (1971), can be obtained. This of course does not necessarily prove the validity of equation 1 in a muscle fiber shortening with V_{max}; in the Huxley contraction model, in which a huge number of cross-bridges are interacting with the thin filaments, the cross-bridges, once attached to the thin filament in the positive x region (Fig. 1A), is brought into the negative x region before being detached from the thin filament. Based on the Huxley model, Higuchi and Goldman (1991) attempted to determine the interaction distance of cross-bridges Di, extending over both positive x and negative x regions. The equation they used is:

$$\mathrm{Di} = \mathrm{Ds\ Sa}\ [\mathrm{M_0}] / [\mathrm{ATP_u}] \quad (2),$$

where Ds is the total distance of myofilament sliding, Sa is the average muscle fiber stiffness during shortening (expressed relative to rigor stiffness), $[M_0]$ is the cross-bridge concentration in the fiber (154 μM), and $[ATP_u]$ is the ATP concentration utilized for fiber shortening. They used the technique of laser-flash photolysis of caged ATP to induce shortening of a demembranated muscle fiber. The value of Di obtained was ~ 28 nm. This experiment includes many uncertainties in the values to be put in equation 2, especially S_a, and gives no direct information about the value of d.

To determine the distance of cross-bridge powerstroke without measuring muscle fiber stiffness, which appears to be an incomplete measure of the number of working cross-bridges, Yamada et al. (1993) used the techniques of laser-flash photolysis of caged ATP and high-speed video recording of the whole demembranated muscle fiber. Under nearly isometric conditions in which the cross-bridges might only exist in the positive x region according to the Huxley contraction model (Fig. 1), they found that the minimum uniform ATP-induced fiber shortening was ~ 10nm per half sarcomere, a value which can be regarded as d within the framework of Huxley contraction model (Fig. 5). Using an in vitro motility assay system, in which a myosin-coated microneedle was made to slide along actin filaments by iontophoretically applied ATP (Oiwa et al., 1993), the unitary actin-myosin sliding (analogous to d) was also found to be ~ 10 nm.

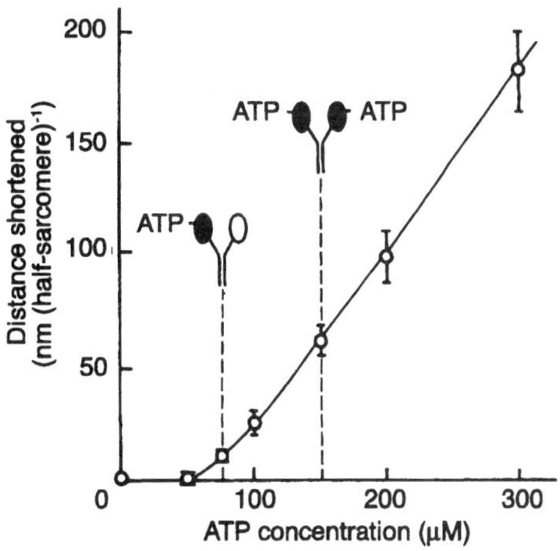

Figure 5. Relation between the concentration of released ATP and the velocity and distance of muscle fiber shortening. The initial shortening velocity and the distance shortening are plotted against the ATP concentration in A and B, respectively. Vertical dashed lines and diagrams of myosin molecules show the ATP concentrations equal to, and a half of, the total myosin head concentration within the fiber (150 μM) (Yamada et al., 1993).

On the other hand, the most straightforward way to measure the distance of cross-bridge powerstroke is to observe living muscle thick filaments under an electron microscopy (magnification, x 10,000) using a gas environmental chamber, with the filaments can be kept alive by circulating water vapor (Fubani et al.), and record ATP-induced movement of cross-bridges on the thick filament. By putting position markers on the cross-bridges on the cross-bridges extending from the synthetic thick filaments (myosin-paramyosin core complex), Sugi et al. (1997) have succeeded in recording the cross-bridge movement induced by iontophoretically applied ATP. Their results are summarized as follows (Fig. 6):

1. The mean position of each cross-bridge does not change with time despite its thermal motion.

2. In response to ATP, the cross-bridges move along the fiber length. The mean distance of the ATP-induced cross-bridge movement is ~ 30 nm, a value larger than the structural constraint of cross-bridge powerstroke (Fig. 6).

3. The cross -bridges does not move in response to ADP.

4. As the experimental system does not contain actin filaments, the above results indicate most of the cross-bridge motion can take place with our their attachment to actin.

At present, the observed cross-bridge movement is likely to be a recovery stroke preceding the powerstroke. The distance of cross-bridge motion of ~ 30 nm can only be accounted for by assuming shortening of the subfragment-2 region (length, ~ 43 nm), connecting the myosin head to the thick filament backbone (Harrington et al., 1988; Sugi et al., 1992). We are now preparing the experiments, in which synthetic bipolar myosin filaments are made to slide past actin filaments, in the hope to clarify mysteries about the cross-bridge powerstroke.

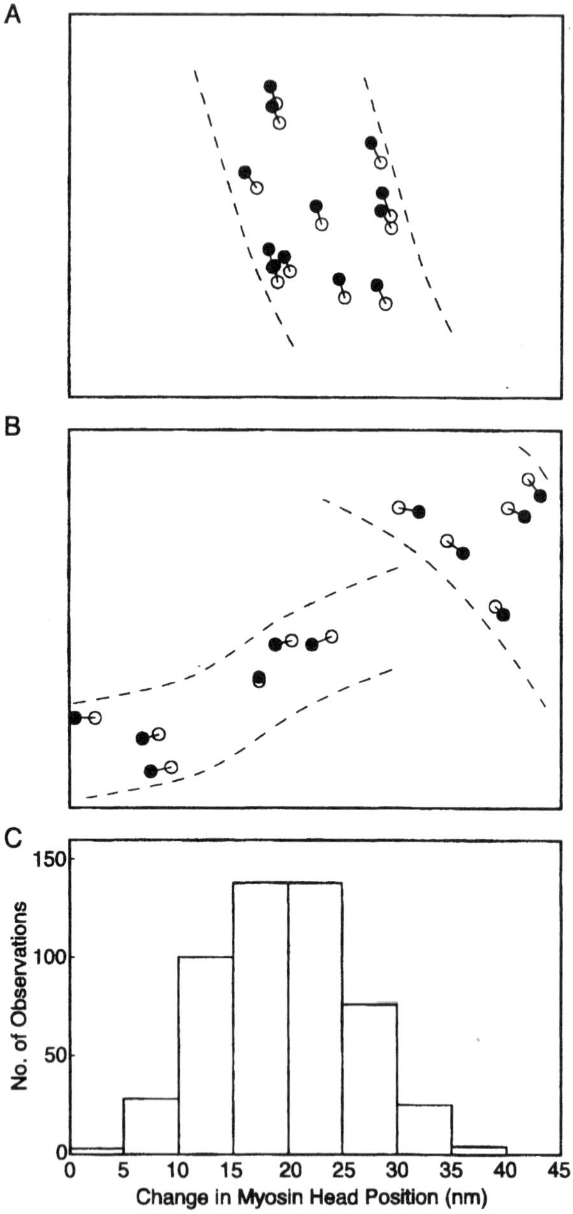

Figure 6. Movement of the individual myosin heads on the filaments in response to ATP application. (A and B) Examples showing changes in the center of mass position of the same particles after ATP application. Filled and open circles (diameter, 15 nm) are also drawn around the center of mass positions of the same particles in the records before and those after ATP application, respectively (Sugi et al., 1997).

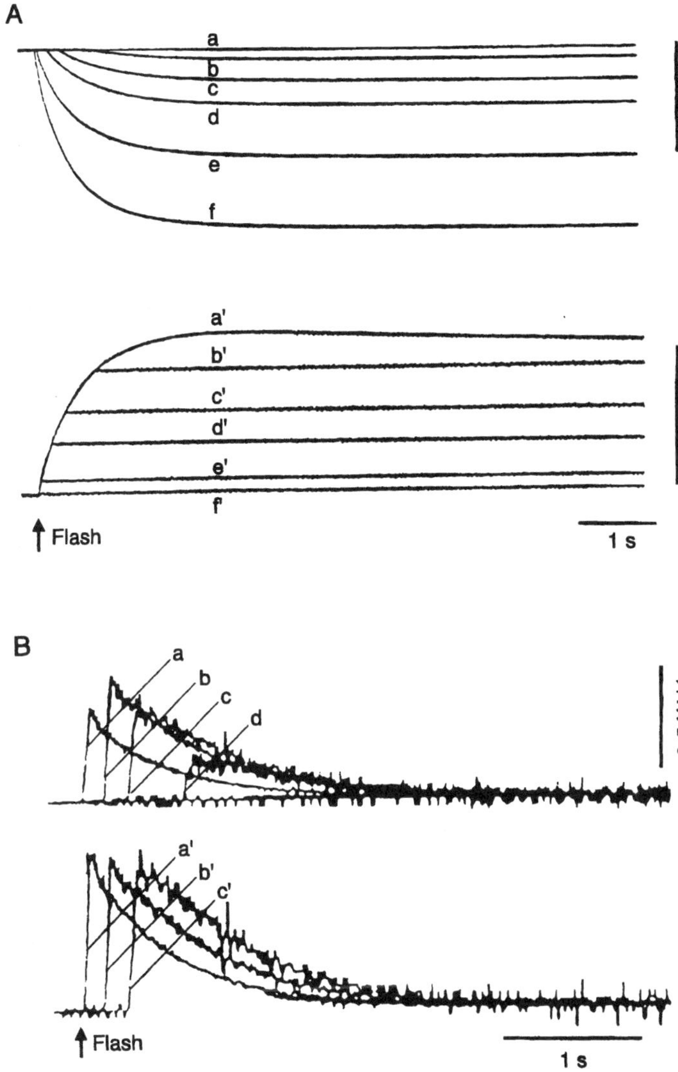

Figure 7. (A) Laser flash-induced mechanical response. Upper and lower traces show fiber length and force changes, respectively. Length recordings a, b, c, d, e and f correspond to force recordings a', b', c', d', e' and f', respectively. The loads were P_0 (isometric contraction *a, a'*), 0.78 P_0 (*b, b'*), 0.53 P_0 (*c, c'*), 0.35 P_0 (*d, d'*), 0.09 P_0 (*e, e'*), and 0 (unloaded shortening, *f, f'*), respectively. (B) Power output during flash-induced fiber shortening. Upper traces: Power output recordings under four different afterloads. Lower traces: Power output records normalized relative to the peak values attained. The load was 0.09 P_0 (*a, a'*), 0.35 P_0 (*b, b'*), 0.53 P_0 (*c, c'*) and 0.78 P_0 (*d*). The recordings were obtained from the experiment shown in Fig. 1 (Sugi et al., 2003).

Apart from the experiments described above, an idea of "biased Brownian ratchet mechanism" has been proposed to account for some in vitro motility experiments, suggesting *d* values well beyond the structural constraints of the cross-bridge (Vale and Oosawa, 1990). I wonder how this idea harmonizes with the vast physiological results accumulated in the past. In my feeling, the criteria for "single molecule events", obtained from experiments, in which a single myosin molecule on a latex bead is made to interact with a single actin filament, are not strict among the motility assay people, and may be associated with a number of pitfalls, considering the bead diameter ~ 10^6 times larger than the size of a myosin head.

6. THE MECHANICAL EFFICIENCY OF CROSS-BRIDGE POWERSTROKE

The mechanical efficiency, with which chemical energy derived from ATP hydrolysis is converted into mechanical work, in demembranated muscle fibers has now been estimated by measuring the amount of ATP utilized for work production, using fluorescence of a phosphate-binding protein (He et al., 1999) or NADH (Reggiani et. al., 1997; Sun et al., 2001). During myofilament sliding, however, the cross-bridges not only attach to the thin filament in the positive force region to perform their powerstroke-producing positive forces, but also produce negative forces in the negative force region before being detached from the thin filament (Fig. 1). On this basis, the overall mechanical efficiency of muscle fibers may be much smaller than that of individual cross-bridges during their powerstroke, since positive forces are always opposed by negative forces, due to asynchronous cross-bridge activity. To accurately estimate the mechanical efficiency of individual cross-bridges when they perform their powerstroke, it is necessary to perform experiments under condition in which the cross-bridges start their powerstroke synchronously.

For this purpose, we used an experimental condition, in which each cross-bridge (= myosin head) in the demembranated muscle fiber could hydrolyze ATP only once during a single mechanical response (Sugi et al., 1998, 2003). This condition is achieved by soaking the fibers in a solution containing ATP at a concentration almost equal to the total myosin head concentration in the fiber, exposing them in air to blot solution around the fibers, and then rapidly activating them to contract by using laser flash photolysis of caged calcium. Immediately before activation, the number of ATP molecules in the fiber is made almost equal to the number of myosin heads, so that all the myosin heads are in the state M•ADP•Pi, in which ATP is already hydrolyzed by the myosin head (M) but its products are still bound to M (Bagshaw, 1984).

When the fibers are maximally activated by photo-released Ca^{2+}, they first develop an isometric force equal to the afterload P, start shortening isotonically, and then eventually stop shortening as the fibers enter into rigor state after complete exhaustion of ATP (Fig. 7A). As shown in Fig. 7B, the power output in the flash-activated fibers reached a peak at the early phase of fiber shortening, and then decreased with time. The higher the initial peak, the larger the area under the power output trace, i.e. the amount of work done by fiber shortening. The power output records were almost identical when normalized with respect to their peak values, except for the load close to P_0. The distance of fiber shortening, when the power output reached a maximum did not exceed 10 nm per half sarcomere. This suggests that, at the beginning of fiber shortening, the cross-bridges, in the form of M•ADP•Pi, start their powerstroke almost synchronously, while sensing the amount of load to determine their energy output.

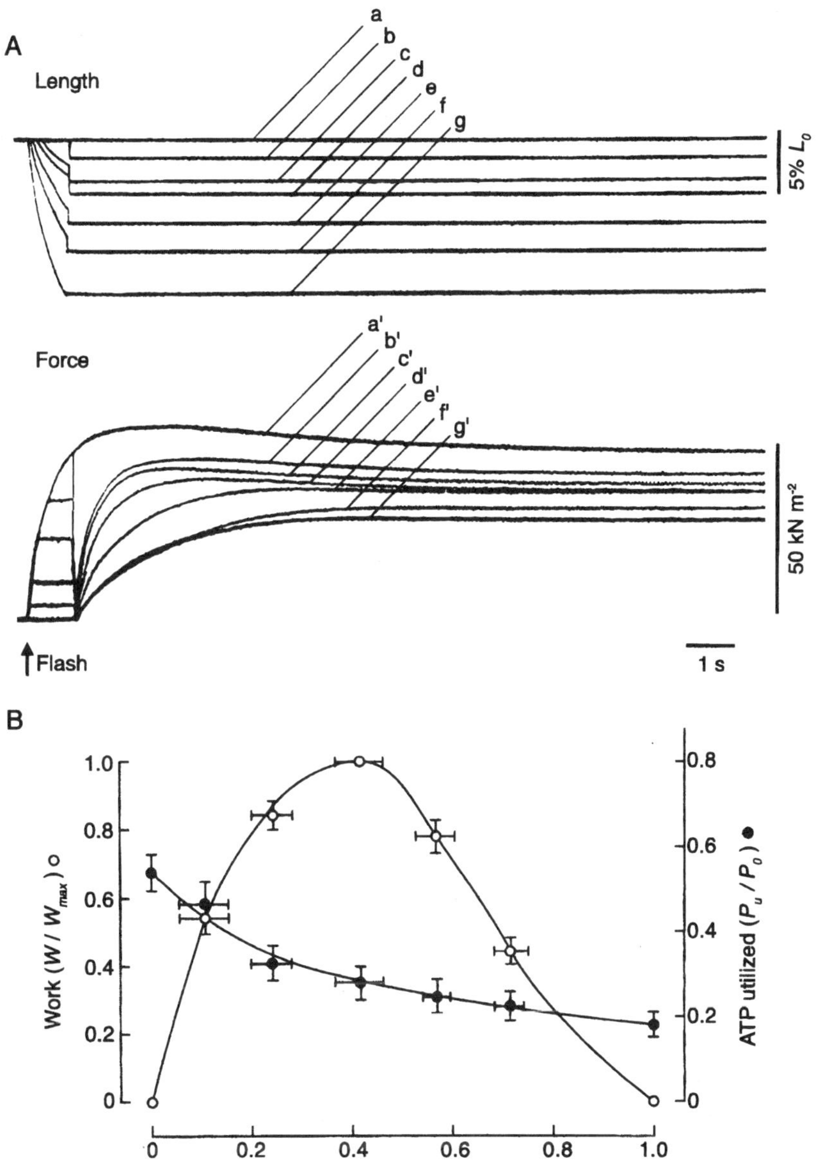

Figure 8. (A) Fiber length and force changes of a preparation that was first made to shorten isotonically under five different afterloads, and then subjected to quick releases at 1s after activation to drop the force to zero. After each release, the preparation redeveloped isometric force at the decrease fiber length. Length recordings *a* – *g*, correspond to force recordings *a'* – *g'*, respectively. The load was P_0 (isometric contraction *b, b'*), 0.63 P_0 (*c, c'*), 0.41 P_0 (*d, d'*), 0.20 P_0 (*e, e'*), 0.09 P_0 (*f, f'*), and 0 (unloaded shortening, *g, g'*). Recordings *a, a'* were obtained during isometric contraction without quick release. (B) Dependence of the amount of ATP utilized for mechanical response (*P*u, filled circles) and the amount of work done (*W*, open circles) on the isotonic load (*P*) at 1 s after flash activation. Values were means ± S.E.M. obtained from eight different data sets (Sugi et al., 2003).

During isometric contraction, in-phase stiffness, i.e. the magnitude of force changes in response to sinusoidal length changes, increased approximately in parallel with isometric force, while quadrature stiffness, i.e. the 90° out-of phase stiffness component, reached a maximum at approximately 0.3s after activation, and stayed almost unchanged for the following 3-4 s. This indicates that there were no appreciable changes in the number of force-generating cross-bridges during isometric force development preceding fiber shortening, since quadrature stiffness is taken as a measure of the fraction of active cross-bridges (Goldman et al., 1984). Furthermore, under conditions identical to the present experiments, no appreciable increase of internal resistance against fiber shortening takes place at least for the first 1-2 s after activation (Sugi et al., 1998). It may therefore be safe to conclude that, at least for 1-2 s after activation, the cross-bridges may not readily from rigor links after releasing Pi and ADP, irrespective of whether the fiber is shortening or kept isometric.

Fig. 8A shows a typical experiment in which the fibers were activated to contract isometrically or isotonically under five different afterloads for 1s, and then subjected to a quick release to drop the force to zero, whereon the fiber length was clamped and the fibers developed after a quick release (P_r), i.e. a measure of the amount of ATP remaining in the fiber at 1 s after activation, was maximum when $P = P_0$ (isometric contraction and minimum when $P = 0$ (unloaded shortening). Similar results were obtained from 7 different preparations examined. The amount of ATP utilized at 1 s after flash activation ($P_u = P_0 - P_r$) was therefore maximum during unloaded shortening (P = 0), and minimum during isometric contraction ($P = P_0$).

On the other hand, the possibility that cross-bridges forming rigor links with the thin filaments may produce rigor force to contribute to the isometric force development after a quick release can be precluded by the extremely slow development of rigor force in demembranated muscle fibers (Kobayashi et al., 1998).

Fig. 8B shows the dependence of the amount of work done (W, expressed relative to the maximum value, W_{max}) and the amount of ATP utilized for the whole mechanical response (P_u, expressed relative to P_0) on the isotonic load (P). The data points were obtained from 8 different datasets. The value of P_u at P = 0 was approximately 3 times larger than that at $P = P_0$.

The amount of ATP utilized for the whole mechanical response (P_u) is the sum of the amount of ATP utilized for the preceding isometric force development (P_i) and that utilized for subsequent isotonic shortening (P_s). The value of P_i as a function of isotonic load were obtained by applying a quick release to isometrically contracting fibers at various times after activation and measuring the amount of force development after each quick release. Thus, the value of P_s could be obtained by subtracting the value of P_i for a given isometric force equal to the isotonic load from P_u for the whole mechanical response. The mechanical efficiency of individual cross-bridges (E), averaged over the period of work production, can be estimated as $E = W / (P_u - P_i) = W / P_s$. The dependence of E (expressed relative to the maximum value, E_{max} on the isotonic load is shown schematically in Fig. 9 together with W, P_u, P_i and P_s. The value E versus P relationship was bell-shaped, with a broad peak at $0.5 - 0.6\ P_0$.

Although the mechanical efficiency of individual cross-bridges is obtained as relative values in the present study, we made a conservative estimation of its absolute value as follows. The average fiber cross-sectional area of 8 preparations, from which the data shown in Fig. 4 were obtained, was $6.1 \pm 0.1 \times 10^{-5}$ cm^2, while the fiber length was $\leq 2.5 - 3$ mm. To avoid underestimation of fiber volume leading to overestimation

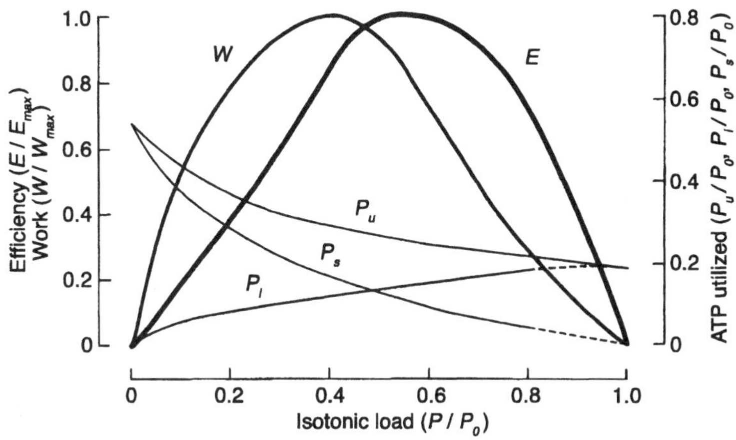

Figure 9. Dependence of the mechanical efficiency of individual cross-bridges (*E*) on the amount of isotonic load (*P*) obtained from the results shown in Figs 4 and 5. Values are scaled to adjust the value at 1.0 P / P_0 to that in Fig. 4. The amount of work done (*W*), ATP utilized for preceding isometric force development (*Pi*), and ATP utilized for isotonic shortening (*Ps*) are also shown as function of load (Sugi et al., 2003).

of the efficiency, we use the maximum fiber length 3 mm to obtain mean fiber volume of 1.8 x 10^{-5} cm^3. Assuming a cross-bridge concentration of 200 _ moll^{-1} (higher than the widely used values of 145 or 150 μmol^{-1}), the amount of M•ADP•Pi immediately before flash activation is estimated to be 3.6 x 10^{-6} μmol (200 x 1.8 x 10^{-5} x 10^{-3}) = 3.6 x 10^{-12} mol. In Fig. 7, the value of E is maximum at P = 0.53 P_0 , and the corresponding value of P_s is 0.13 P_0, where corresponds to the initial amount of M•ADP•Pi of 3.6 x 10^{-12} mol. The number of ATP molecule utilized for work production is calculated to be. 2.8 x 10^{11} (3.6 x 10^{-12} x 0.13 x 6 x 10^{23}). Assuming the energy released by ATP hydrolysis of 50 kJmol^{-1} (Bagshaw, 1984; Oiwa et al., 1991), the energy available from one ATP molecule is 8.3 x_10^{-20} J (50 x_10^3 / 6 x_10^{23}). The energy released from ATP molecules during work production is 2.3 x_10^{-8} J (2.8 x_10^{11} x_8.3 x_10^{-20}). In Fig. 6, the amount of work done at 0.53 *P*o is 1.6 x_10^{-8} J. The maximum mechanical efficiency of individual cross-bridges is therefore estimated to be (1.6 x_10^{-8})/ (2.3 x_10^{-8}) = 0.7. Since the above estimation is conservative, the actual maximum mechanical efficiency of individual cross-bridges is suggested to be 0.8 - 0.9, being close to unity.

These results may constitute evidence that the maximum mechanical efficiency of individual cross-bridges may be very high, probably close to unity (0.8 – 0.9). In this connection, it is of interest to note that it has also been suggested that the mechanical efficiency of the ATP-dependent rotary motion of F_0-F_1 ATPase at the mitochondrial membrane is close to unity (kinoshita et al., 2000).

7. REFERENCES

Bagni, M. A., Cecchi, G., Colomo, F. and Garzella P., 1992, Are weakly binding bridges present in resting intact muscle fibers?, *Biophys. J.* **63**: 1412-1415.

Bagni, M. A., Cecchi, G., Colomo, F. and Garzella P., 1995, Absence of mechanical evidence for attached weakly binding cross-bridges in frog relaxed muscle fibers, *J. Physiol.* **482**: 391-400.

Bagshaw, C. R., 1994, *Muscle Contraction*, Chapman & Hall, London.

Cecchi, G., Griffiths, P. J., and Taylor, S., 1982, Muscular contraction: kinetics of cross-bridge attachment studied by high-frequency stiffness measurements. *Science.* **217**: 70-72.

Cecchi, G., Griffiths, P. J., and Taylor, S., 1984, The kinetics of cross-bridge attachment and detachment studied by high frequency stiffness measurements, in: *Contractile Mechanisms in Muscle*, H. Sugi, G. H. Pollack eds., Plenum, New York, pp. 641-655.

Cecchi, G., Griffiths, P. J., and Taylor, S., 1986, Stiffness and force in activated frog skeletal muscle fibers, *Biophys. J.* **49**: 437-451.

Civan, M. M., and Podolsky, R. J., 1966, Contractile kinetics of striated muscle fibres following quick changes in load, *J. Physiol.* **184**: 511-534.

Ford, L. E., Huxley, A. F., and Simmons, R.M., 1971, Tension responses to sudden length change in stimulated frog muscle fibres near slack length, *J. Physiol.* **269**: 441-515.

Goldman, Y. E., Hibberd, M. G., and Trentham, D. R., 1984, Relaxation of rabbit psoas muscle fibers from rigor by photochemical generation of adenosine-5'-triphosphate, *J. Physiol.* **354**: 577-604.

Gordon, A. M., Huxley, A. F., and Julian, F. J., 1966, Tension development in highly stretched vertebrate muscle fibres, *J. Physiol.* **184**: 143-169.

Gordon, A. M., Huxley, A. F., and Julian, F. J., 1966, The variation in isometric tension with sarcomere length in vertebrate muscle fibres, *J. Physiol.* **184**: 170-192.

Harada, Y., Sakurada, K., Aoki, T., Tohmas, D. D., and Yanagida, T., 1990, Mechanochemical coupling in actomyosin energy transduction studied by in vitro movement assay, *J. Mol. Biol.* **216**: 49-68.

Harrington, W. F., Ueno, H., and Davis, J. S., 1988, Helix-coil melting in rigor and activated cross-bridges of skeletal muscle, in: *Molecular Mechanism of Muscle Contraction*, H. Sugi, G. H. Pollack eds., Plenum, New York, pp. 307-318.

Hatta, I., Sugi, H., and Tamura, Y., 1988, Stiffness changes in frog skeletal muscle during contraction recorded using ultrasonic waves, *J. Physiol.* **403**: 193-209.

He, Z-H., Chillingworth, R. K., Brune, M., Corrie, J. E. T., Webb, M. R., and Ferenczi, M. A., 1999, The efficiency of contraction in rabbit skeletal muscle fibers, determined from the rate of release of inorganic phosphate, *J. Physiol.* **517**: 839-854.

Higuchi, H., and Goldman, Y. E., 1991, Sliding distance between actin and myosin filaments per ATP molecule hydrolysed in skinned muscle fibres, *Nature.* **352**: 352-354.

Hill, A. V., 1938, The heat of shortening and the dynamic constants of muscle, *Proc. Roy. Soc. Lond.* **B126**: 136-195

Howard, J., 1997, Molecular motors: structural adaptations to cellular functions, *Nature.* **389**: 561-567.

Huxley, A. F., 1957, Muscle structure and theories of contraction, *Prog. Biophys. Biophys. Chem.* **7**: 255-318.

Huxley, A. F., 1974, Muscular contraction, *J. Physiol.* **243**: 1-43.

Huxley, A. F., and Simmons, R.M., 1971, Proposed mechanism of force generation in striated muscle, *Nature.* **233**: 533-538.

Huxley, H. E., and Faruqi, A. R., 1983, Time-resolved X-ray diffraction studies on vertebrate striated muscle, *Annl Rev. Biophys. Bioeng.* **12**: 381-417.

Huxley, H. E., and Hanson, J., 1954, Changes in the cross-striations of muscle during contraction and stretch and their structural interpretation, *Nature.* **173**: 973-976.

Julian, F. J., and Sollins, M. R., 1975, Variation of muscle stiffness with force at increasing speeds of shortening, *J. Gen. Physiol.* **66**: 282-302.

Kinosita, K. Jr., Yasuda, R., Noji, H., and Adachi, K., 2000, A rotary molecular motor that can work at near 100% efficiency, *Phil. Trans. R. Soc. B.* **355**: 473-489.

Kobayashi, T., Kosuge, S., Karr, T., and Sugi, H., 1998, Evidence for bidirectional functional communication between myosin subfragments 1 and 2 in skeletal muscle fibers, *Biochem. Biophys. Res. Commun.* **246**: 539-542.

Kron, S. J., and Spudich, J. A., 1986, Fluorescent actin filament move on myosin fixed to a glass surface, *Proc. Natl. Acad. Sci. USA.* **83**: 6272-6276.

Oiwa, K., Chaen. S., and Sugi, H., 1991, Measurment of work done by ATP-indiced sliding between rabbit muscle myosin and algal cell actin cables in vitro, *J. Physiol.* **437**: 715-763.

Oiwa, K., Kawakami, T., and Sugi, H., 1993, Unitary distance of actin-myosin sliding studied using an in vitro force-movement assay system combined with ATP iontophoresis, *J. Biochem.* **114**: 28-32.

Podolsky, R. J., Nolan, A. C., and Zavaler, S. A., 1969, Cross-bridge properties derived from muscle isotonic velocity transients, *Proc. Natl. Acad. Sci. USA.* **64**: 504-511.

Reggiani, C., Potma, E. J., Bottinelli, R., Canepari, M., Pellegrino, M. A., and Stienen, G. J. M., 1997, Chemo-mechanical energy transduction in relation to myosin isoform composition in skeletal muscle fibers of the rat, *J. Physiol.* **502**: 449-460.

Stehle, R., and Brenner B., 2000, Cross-bridge attachment during high-speed active shortening of skinned fibers of the rabbit psoas muscle: implications for cross-bridge action during maximum velocity of filament shortening, *Biophys. J.* **78**: 1458-1473.

Sugi, H., 1992, Molecular mechanisms of actin-myosin interaction in muscle contraction, in: *Muscle Contraction and Cell Motility*, H. Sugi, ed., Springer, Berlin and Heidelberg, pp. 131-171.

Sugi, H., 1993, Molecular mechanism of ATP dependent actin-myosin interaction in muscle contraction, *Jpn. J. Physiol.* **43**: 435-454.

Sugi, H., Akimoto T., Sutoh, K., Chaen, S., Oishi, N., and Suzuki, S., 1997, Dynamic electron microscopy of ATP-induced myosin head movement in living muscle thick filaments, *Proc. Natl. Acad. Sci. USA.* **94**: 4378-4382.

Sugi, H., Iwamoto, H., Akimoto T., and Kishi, H., 2003, High mechanical efficiency of the cross-bridge powerstroke in skeletal muscle, *J. Exp. Biol.* **206**: 1201-1206.

Sugi, H., Iwamoto, H., Akimoto T., and Ushitani, H., 1998, Evidence for the load-dependent mechanical efficiency of individual myosin heads in skeletal muscle fibers activated by laser flash photolysis of caged calcium in the presence of a limited amount of ATP, *Proc. Natl. Acad. Sci.* **95**: 2273-2278.

Sugi, H., and Kobayashi, T., 1983, Sarcomere length and tension changes in tetanized frog muscle fibers after quick stretches and releases, *Proc. Natl. Acad. Sci. USA.* **80**: 6422-6425.

Sugi, H., and Tsuchiya, T., 1981, Isotonic velocity transients in frog muscle fibres following quick changes in load, *J. Physiol.* **319**: 239-252.

Sugi, H., and Tsuchiya, T., 1988, Stiffness changes during enhancement and deficit of isometric force by slow length changes in frog skeletal muscle fibres, *J. Physiol.* **407**: 215-229.

Sugi, H., and Tsuchiya, T., 1998, Muscle mechanics I: intact single muscle fibres, in: *Current Methods in Muscle Physiology; Advantages, Problems and Limitations*, H. Sugi, ed.,Oxford University Press, Oxford, pp. 3-31.

Sun, Y-B., Hilber K., and Irving, M., 2001, Effect of active shortening on the rate of ATP utilization by rabbit psoas muscle fibers, *J. Physiol.* **531**: 781-791.

Tamura, Y., Hatta, I., Matsuda, T., Sugi, H., and Tsuchiya, T., 1982, Changes in muscle stiffness during contraction recorded using ultrasonic waves, *Nature.* **299**: 631-633.

Toyoshima, Y. Y., Kron, S. J., and Spudich, J. A., 1990, The myosin step size: measurement of the unit displacement per ATP hydrolyzed in an in vitro motility assay, *Proc. Natl. Acad. Sci. USA.* **87**: 7130-7134.

Uyeda, T. P. Q., Kron, S. J., and Spudich, J. A., 1990, Myosin step size estimation from slow sliding movement of actin over low densities of heavy meromyosin, *J. Mol. Biol.* **214**: 699-710.

Vale, R. D., and Osawa, F., 1990, Protein motors and Maxwell's demons: does mechanochemical transduction involve a thermal rachet, *Adv. Biophys.* **26**: 97-134.

Yamada, T., Abe, O., Kobayashi, T., and Sugi, H., 1993, Myofilament sliding per ATP molecule in rabbit muscle fibres studied using laser flash photolysis of caged ATP, *J. Physiol.* **466**: 229-243.

DISCUSSION

Maughan: Prof. Sugi, your elegant ultrasonic wave (MHz) experiments are very impressive. Does this technique prove movement of small molecules, including water? Do you in fact believe water is the major molecule probe?

Sugi: Yes, I've discussed the results with Andrew Huxley and he agrees with me.

Maughan: Thus, could the MHz signal measure crossbridge movement indirectly by its action of pushing water during its movement?

Sugi: I believe that, with ultrasonic waves, we are shaking water molecules in between the myofilaments. Water molecules are known to undergo dynamic changes during contraction. It should be noted that the MHz wave stiffness is insensitive to resting tension, indicating that the MHz waves do not detect changes in strain of myofilaments. In addition, the MHz stiffness in rigor fibers do not change by stretch despite a great increase in rigor force, again indicating that the MHz stiffness does not detect strain of cross-bridge.

Pollack: Your idea that the MHz stiffness changes reflect the state of water is consistent with the ^{1}H NMR studies of Yamada et al. (Yamada, Biochim, Biophys. Acta 1379: 224-232, 1998). Several people in the audience (Drs. Yamada and Ogata) have reported changes in water structure during the transition from rest to activity. Therefore it seems that your hypothesis may be correct: the difference of ultrasonic velocity may well be related to the state of water.

SARCOMERE DYNAMICS, STEPWISE SHORTENING AND THE NATURE OF CONTRACTION

Gerald H. Pollack, Felix A. Blyakhman, Xiumei Liu, and Ekaterina Nagornyak*

1. INTRODUCTION

This paper has a dual goal. First it outlines the methods that have evolved to track the time course of sarcomere length (SL) with increasingly high precision. Serious attempts at this began roughly at the time the sliding filament theory was introduced in the mid-1950s, and have progressed to the point where resolution has reached the nanometer level. Second, and within the context of these developments, it considers one of the more controversial aspects of these developments: stepwise shortening. Stepwise shortening was first observed a quarter century ago (Pollack et al., 1977), and has provided no shortage of controversy for the muscle-contraction field.

2. EARLY MEASUREMENTS

The first serious measurement of the time course of SL was carried out by Gordon et al., (1966). These investigators placed opaque markers on the surface of single isolated frog muscle fibers and measured the intervening distance as a function of time. Through feedback control it was possible to control the length of the segment, thereby maintaining invariance of segment length and permitting construction of the isometric length-tension relation.

Another, rather ingenious approach, was based on high-resolution optical diffraction (Haugen and Sten-Knudsen, 1976). In these experiments a diffraction order was split by a prism, one fraction focused on one photodiode, the remaining fraction onto another. SL changes as small as 0.2 A were reported, although the dynamic range of SL was extremely small. The method was used mainly for measurements of latency relaxation.

* Dept. of Bioengineering, Univ. of Washington, Seattle WA 98195

3. LATER MEASUREMENTS

True dynamic measurements based on optical diffraction were realized with the advent of linear photodiode arrays. Measurements using this approach were made in both cardiac and skeletal muscle (Pollack et al., 1977, 1979). The specimen was trans-illuminated with a variably compressed He-Ne laser beam. A single first order of the diffraction pattern was projected onto either a Schottky barrier photodiode, kymographic film, or, in most experiments, a linear photodiode array, from which the centroid of the first diffraction order could be located with high time resolution (Iwazumi and Pollack, 1979). In these early experiments contraction was auxotonic, i.e., sarcomere shortening occurred by virtue of abundant compliant tissue caused by clamping.

The results of such experiments showed that shortening occurred in steps (Pollack et al., 1977). That is, the shortening waveform was punctuated by a series of brief periods in which shortening was zero or almost zero, conferring a staircase-like pattern on the waveform. Records obtained simultaneously with the above-mentioned alternative sensors gave similar patterns, implying that the stepwise nature of the pattern did not arise out of some unsuspected artifact associated with one or the other sensors.

Nevertheless, criticism of these experimental results came from the observation that there was a lack of correlation between the steps seen in the two first orders (Rudel and Zite-Ferenczy, 1979). The issue was resolved by analyzing the striation pattern itself. This work proceed two ways: first by analyzing the time course of the striation banding pattern obtained by using high-speed cinemicrography (Delay et al., 1981); and, later by striation image analysis using an on-line, phase-locked loop method (Myers et al., 1982).

For the-high speed film analysis, single fibers of frog semitendinosus muscle were illuminated with a laser beam. The image of the striations was projected onto the film plane of a high-speed rotating prism camera operating at 4,000 frames per second. The optical diffraction pattern was also obtained during the same contraction for comparison. It was was projected through the vertical tube of the microscope onto a photodiode array, from which SL was obtained as described above.

Each film frame was analyzed using multiple scan microdensitometry to obtain mean striation spacing. Frame-by-frame analysis gave the time course of sarcomere shortening. Pauses and steps were apparent in both the cine analysis and the diffraction analysis. Although the steps obtained with the two methods were not identical, there was considerable overlap of waveforms. Possibly the differences arose because the sampled volume in the two cases was different. Nevertheless, the fact that steps of similar nature could be seen using both approaches implied that the approaches gave valid information on SL dynamics.

The time demands imposed by cine-film analysis prompted the development of an on-line approach. Here, the image of the striation pattern, obtained using an incoherent Xe-light source, was projected onto a photodiode array. The array was scanned across the striations every 256 μs, each scan giving a waveform whose frequency was equal to the spatial frequency of the striated image. This signal was then passed into a phase-locked loop circuit, which extracted the principal frequency and converted it to SL. The phase-locked loop method is a known means of extracting the frequency of a quasi-periodic signal, remaining effective even when the noise level is relatively high. The method was demonstrated to have excellent linearity, and could resolve SL changes as small as 1 nm. The results showed stepwise shortening patterns, which, at least qualitatively, resembled those obtained with the methods above (Jacobsen et al., 1983).

An alternative approach that did not depend on analysis of any features of the striation pattern was developed to measure SL dynamics. Follwing the pioneering work of Gordon et al. (1966) and subsequent work by Edman and Hoglund (1981), we devised a method to track the length of a fiber segment. The segment was demarcated by a pair of thin hair tips taken from the ear of a black cat. Segment length could be tracked optically (Granzier and Pollack, 1985). The results confirmed all of the optical methods. Stepwise shortening was seen regularly during both active shortening and unstimulated releases. The shortening patterns obtained with the segment-length method bore close resemblance to patterns obtained with the microscopic methods. In fact, when segment length and diffraction methods were employed on the same fiber during the same contraction, records were typically superimposable (Granzier, et al. 1990).

Several optical methods to track SL developed in other laboratories confirmed the presence of steps. Housmans (1984) plunged microelectrode tips into the belly of a papillary muscle and used an optical scanner to follow inter-marker spacing. Although some records showed smooth shortening, many showed shortening with "hesitations" interposed between steps. Repeatable stepwise behavior was particularly evident when the load was kept modest. In a more detailed study, Tameyasu et al (1985) used a high-speed video system to follow shortening in single isolated frog heart cells. Segments were demarcated by natural markers such as glycogen granules. Stepwise shortening was found regularly. The number of detectable steps was limited, as the records became progressively less distinct with shortening. Nevertheless, the stepping pattern was repeatable from contraction ot contraction as well as from specimen to specimen. Finally, the high-speed cine method of Sugi (1968) was applied to the question of steps. Early records had been obtained by activating a fiber locally by passing depolarizing currents near the fiber's edge. Because contraction was localized, the fiber did not translate appreciably during contraction and it was thereby possible to track a consistent set of striations. Altlhough some shortening patterns were smooth, many showed clear stepwise shortening (Toride and Sugi, 1989).

In sum, all of these early methods showed a line of consistency. Stepwise shortening was observed in many laboratories using a variety of methods. Perhaps because of the unexpected level of synchrony implied by the results, skepticism was appreciable, and various papers along the way criticized certain aspects the earliest results (Rudel and Zite-Ferenczy, 1979; Altringham et al., 1984; Huxley, 1984) — all of which inspired responses and additional controls (Pollack, 1984, 1986). To the knowledge of these authors, the latter results have not been criticized; nor have the confirmations in other laboratories.

The most recent experimental approaches, described next, provide evidence for steps at the near molecular level, which show strong quantitative consistency.

4. RECENT APPROACHES

4.1. Single myofibrillar sarcomeres

The first of two independent approaches probes the dynamics of single sarcomeres within a single myofibril. The single myofibril is advantageous for mechanical measurements. Sarcomeres are all in series, so they bear the same tension, which is measurable at the end. Consequently, tension and single-sarcomere dynamics can be unambiguously related. This is not true in larger-scale specimens because tension distribution among parallel myofibrils is unknown. The single myofibril is also attractive because it is the smallest unit that fully preserves muscle's characteristic structure; thus, with some care in preparation, large-scale relevance is likely. Yet, the scale is small enough that molecular-level information might be extractable.

The single myofibril apparatus consists of a piezoelectric motor with needle that attaches to one end of the specimen, and a force transducer that attaches to the other. The force transducer is made by nanofabrication (Fauver et al., 1997). The specimen is attached either by a few turns around the naturally adherent tip or by use of a quick-curing silicon-based glue. The specimen is viewed in an inverted microscope (Zeiss Axiovert). The image of the striations is then projected along a photodiode array, which gives an intensity trace corresponding to the banding structure, A-bands positive, I-bands negative. The centroid of each A-band is computed, and the distance between contiguous centroids is determined in order to yield SL. By tracking this distance over repeat scans, the time course of SL is measured.

Steps of shortening were apparent both in the relaxed myofibril (Yang et al., 1998) and in activated myofibrils (Blyakhman et al., 1999). Although resolution was only ~ 5 nm in these preliminary experiments, step size histograms showed peaks at 5.4 nm, 8.1 nm, and 10.8 nm. These values are multiples of 2.7 nm. Hence, the hypothesis was advanced that with improved resolution, a fundamental quantum value of 2.7 nm might be detected. This has been confirmed under a variety of circumstances with a variety of specimens following apparatus improvements.

The most effective improvement involved implementation of the so-called "minimum-risk" algorithm. The standard algorithm identified the centroid of each A-band; the distance between contiguous centroids then gave sarcomere length. In the new algorithm (Sokolov et al., 2003), centroids are likewise computed, but the centroids from one scan are compared to the centroids from the previous scan. Because the computation is differential, the signal-to-noise ratio is higher, by 3 – 5 times, depending on the level of noise. All previous results were reproduced with this algorithm, only signals were cleaner and histogram peaks narrower. An example of results obtained analyzing a computer-generated trace with Gaussian noise superimposed, comparing standard vs. newer algorithm, is shown in **Fig. 1.**

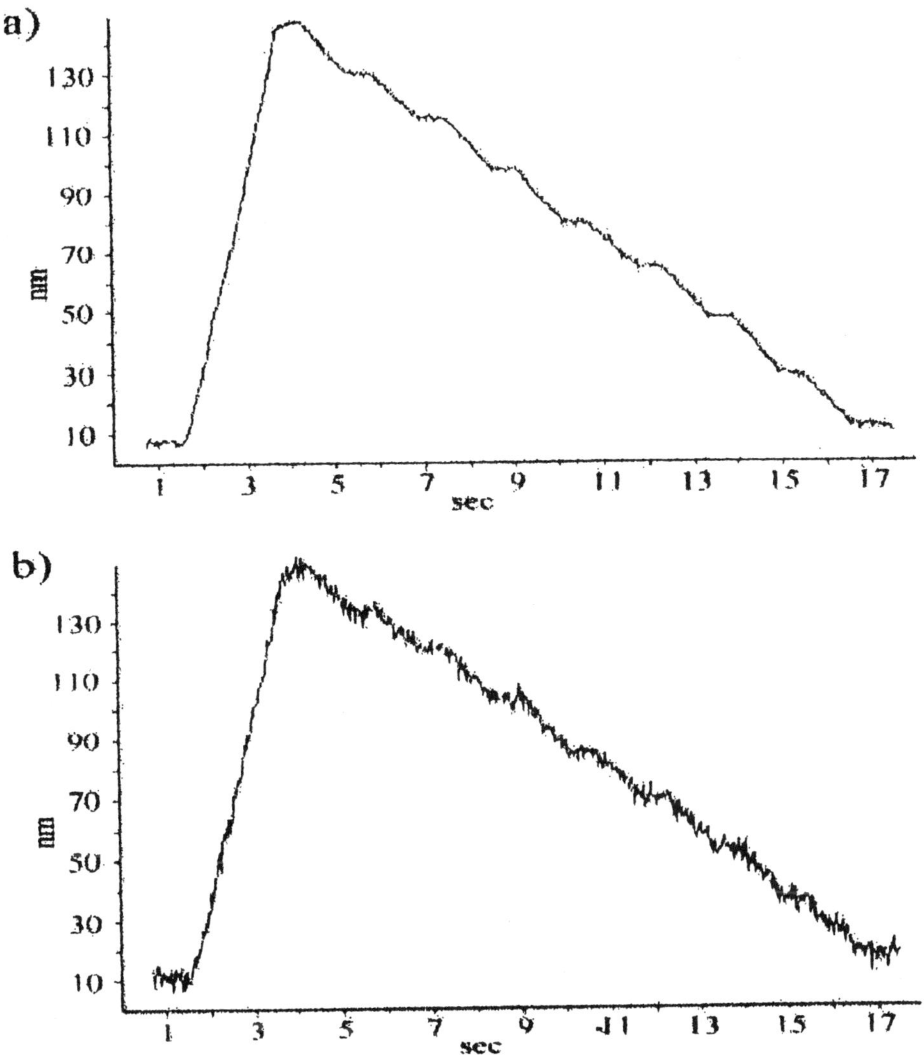

Figure 1. Analysis of sarcomere-length change using 'minimum-risk' algorithm (top), vs. standard centroid algorithm (bottom).

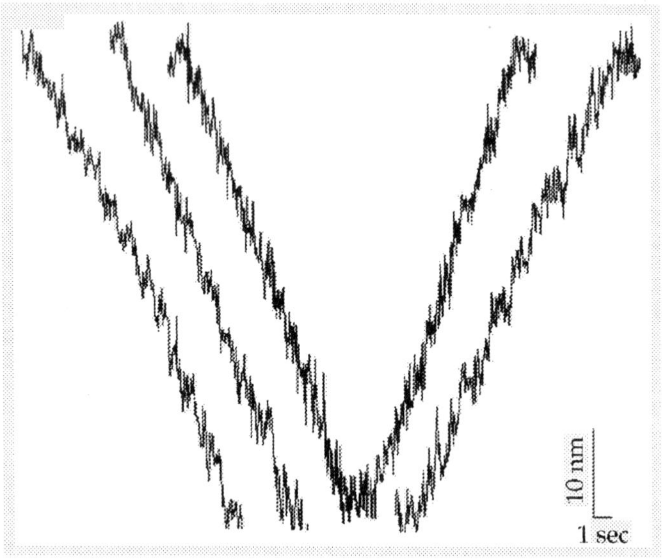

Figure 2. Records of SL change obtained from activated single sarcomeres of rabbit cardiac myofibrils. Third record shows entire trace; other are partial records. Taken from Yakovenko et al. (2002).

With this algorithm, steps became clearer. An example of stepwise shortening traces obtained during contraction is shown in **Fig. 2.**

Steps collected from many shortening traces were plotted as histograms. The histograms showed peaks that were indeed integer multiples of 2.7 nm, as anticipated. An example is shown in **Fig. 3**.

4.1.1. Control Experiments

Numerous experiments were carried out to test for artifact. With a consistent, signature-like step-size paradigm of n x 2.7 nm — 2.7 nm is equal to the linear repeat of actin monomers along the thin filament — it seemed particularly important to test whether this paradigm might have arisen as an artifact of some feature of the apparatus. The controls described below proved uniformly negative for artifact. All of this has been published or is in press (*cf.* Yang et al., 1998; Blyakhman et al., 1999, 2001; Yakovenko et al., 2002; Nagornyak et al., 2004); it is summarized below.

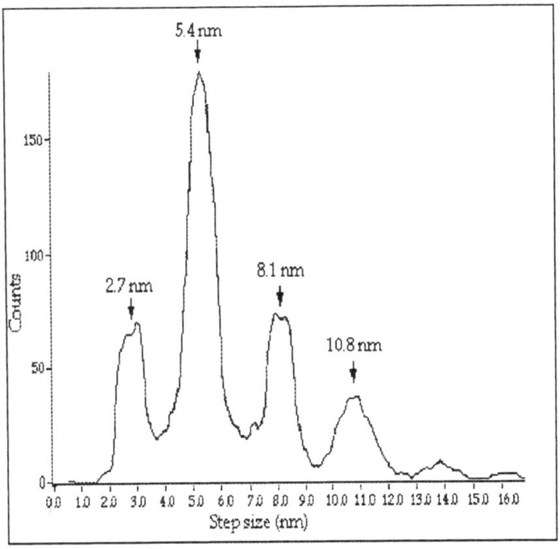

Figure 3. Histogram of step size obtained from records from single myofibrillar sarcomeres. From Yakovenko et al. (2002).

1. *Translating a sarcomere along the photodiode array did not generate steps.* We examined the detected position (A-band centroid) of the sarcomere nearest the motor, which should translate smoothly because the motor was programmed to move in a smooth ramp. No steps were found. Hence the photodiode array's discreteness did not generate the steps.
2. *Random noise superimposed on a ramp produced occasional step-like features, but their sizes were random.* Hence, the steps are not a product of noise.
3. *Different magnification preserved the 2.7-nm paradigm.* A 1.6x slider lens was placed in the microscope in order to change the magnification. Stepwise shortening traces were detected and analyzed. Step-size values remained n x 2.7 nm despite the change of magnification. Again, the photodiode array itself is not responsible for this step paradigm.
4. *Different signal-processing algorithms preserved the 2.7 nm paradigm.* Initially, computation of SL was achieved by identifying the centroids of contiguous A-bands and tracking the span between them. Computation of Z - Z centroids gave indistinguishable step-size paradigms. Later, the "minimum-risk algorithm was substituted. It too produced the 2.7-nm paradigm. Hence, the size paradigm did not arise out of unsuspected features of the signal-processing algorithm.
5. *Different analytical algorithms gave the 2.7-nm paradigm.* Initially, a cursor was used to identify beginning and endpoints of each pause; best-fit straight lines for each pause were then computed by an algorithm, and step size between successive

pauses were determined. We found that the 2.7-nm paradigm was preserved when the power-spectral analysis method (Svoboda et al., 1993) was used. That algorithm is fully automatic. Hence user bias could not have played a role.

6. *Steps seen in myofibrils in other than activated conditions were* not *2.7 nm*. When Ca-activated myofibrils were replaced by relaxed myofibrils, all other factors remaining identical, the quantal step size was different. A mean value of 2.3 nm was found. Although this histogram peak is close to 2.7 nm, the second and third orders of the respective histograms are widely separated and readily distinguishable. Hence, some unknown feature of the apparatus apparently does not generate a spurious step quantum of 2.7 nm.

Thus, all tests conducted to date have proved negative for artifact.

Following these tests for artifact, many physiological studies were carried out with this approach. Here we recap some of the more prominent single sarcomere results from the papers cited above: (1) A-A spans and A-Z spans gave similar step traces. This implies that steps are half-sarcomere events, and the two half-sarcomeres step asynchronously. Contiguous sarcomeres also showed asynchronous stepping. Such asynchrony is seemingly inevitable: because the imposed boundary condition at the end of the myofibril is a smooth ramp, the summed shortenings of all series sarcomeres (or half-sarcomeres) must be smooth. This would not be possible if all sarcomeres stepped synchronously. (2) Steps were found not only during active shortening but also as the activated myofibril was forcibly stretched. The lengthening-step paradigm was also n x 2.7 nm, although in lengthening specimens the lower order peaks were typically relatively larger than higher order ones. (3) The 2.7-nm quantum value was found with great consistency. It was seen in insect-flight muscle, rabbit cardiac muscle, and rabbit psoas muscle. It was independent of SL. Based on the centroid positions of higher order peaks, we could fairly accurately estimate the true quantum position to lie consistently between 2.69 nm and 2.71 nm. (4) Noting that histogram peaks in passive specimens were broader than those in activated specimens, we began to dissect possible dependencies. In rabbit psoas muscle, step size was found to vary inversely with SL. There was an approximately linear dependency, ranging from 2.3 nm for 1.5 μm < SL < 2.0 μm, down to 2.0 nm for 2.5 μm < SL < 3.5 μm. Active sarcomeres showed no SL dependence. (5) Activated specimens stretched to near overlap showed a combination of passive and active step-size paradigms.

Perhaps the main conclusion from the single myofibril experiments is that step size is consistently an integer multiple of 2.7 nm, whether the sarcomere is actively shortening or actively lengthening. This is true in cardiac, skeletal and insect-flight muscle.

4.2. Single filament pairs

Complementing the single sarcomere method described above is the method involving single actin and myosin filaments sliding over one another. The advent of "nanolevers" developed in this laboratory (Fauver et al., 1998) presaged the development of this approach. Nanolevers work in much the same way as larger levers: they deflect in response to a force, applied normal to their long axis, near its tip. Because

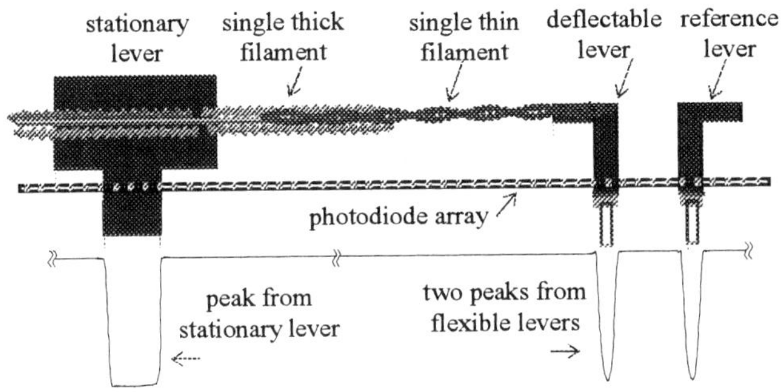

Figure 4. Measurement scheme for single filament dynamics. See text for description. Traces beneath diagram are actual intensity traces.

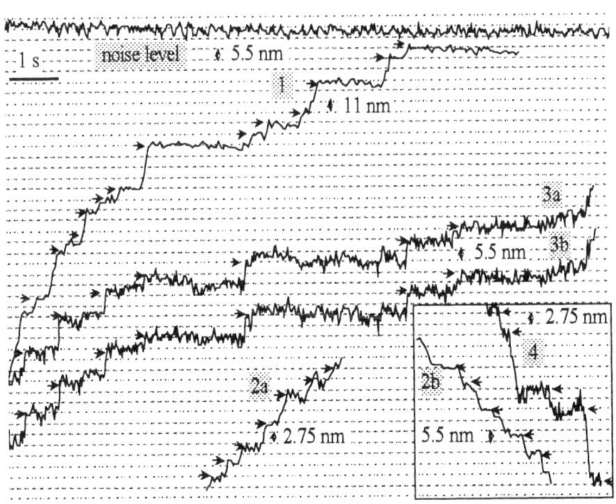

Figure 5. Records of actin filament translation over thick filament. All records show progressively increasing overlap except those in box, which show periods of occasional backwards sliding. Note that scales on records are different. From Liu and Pollack (2004)

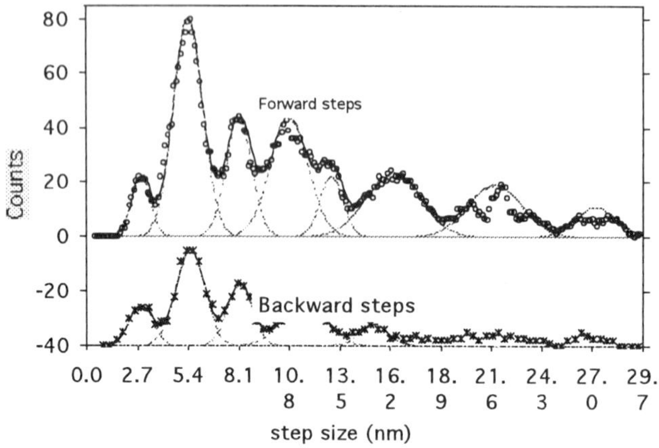

Figure 6. Histograms of step size for forward and backward sliding. Liu and Pollack (2004).

the nanolevers are sub-micron in cross-section, they can be used to measure piconewton forces or nanometer displacements. With a single filament placed between a nanolever and a motor, stress-strain relations can be measured. With the substitution of overlapping actin and myosin filaments, dynamics of sliding can and have been measured.

A schematic of the apparatus is shown in **Fig. 4**. A thick filament is attached to the stationary lever, and an actin filament is attached to a deflectable lever, which bends leftward as the filaments slide past one another. The image of the lever tips is projected onto a linear photodiode array. Actual traces are shown beneath the lever. Lever-tip position is determined from the centroids of the respective image peaks.

Representative traces of filament sliding are shown in **Fig. 5**. Sliding begins rapidly, and gradually slows down as the deflected lever provides a progressively increasing load. The traces show the near-final segments of the traces, where shortening is sufficiently slow relative to the time resolution of the apparatus to allow steps and pauses to be easily discerned. In some traces, when contractile force transiently diminishes below the load force, backwards sliding occurs. This is shown on the lower right. Step sizes were measured from individual traces and plotted as a histogram, which is shown in **Fig. 6**. For both forward and backward steps, the size paradigm was similar to that found in the myofibrils: steps of n x 2.7 nm. Steps were similar whether processing was done using an algorithm similar to that used for myofibrils, or one from NIH Image based on a polynomial fit. The results of these experiments are published (Liu and Pollack, 2004).

As a control to check for systematic artifact, spider silk was substituted for the filament pair. The silk strands were stretched and released by the nanolever apparatus. No steps at 2.7

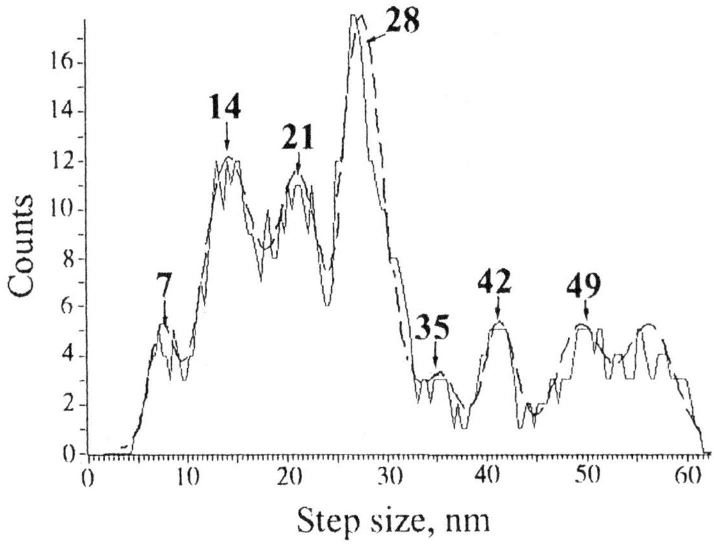

Figure 7. Step size distribution measured in spider silk. Horizontal axis in nanometers.

nm were observed. Instead, steps of *n* x 7 nm dominate the histogram **(Fig. 7)**. If additional experiments confirm this observation, it will be interesting to determine whether a correlation exists between structure and dynamics. [A recent paper using AFM to stretch spider silk reports 14 and 28 nm steps, the second and fourth order of what we have observed (Oroudjev et al., 2002)]. These preliminary results lend confidence that the 2.7 nm steps do not arise out of some unsuspected feature of the nanolever apparatus.

5. CONCLUSIONS

Recent experiments on single sarcomeres and single filament pairs confirm earlier measurements demonstrating steps. The earlier measurements were plagued with potential artifacts, stemming largely from the relatively large volume of sampled tissue. Those limitations are ameliorated in the small specimens studied more recently.

Even though the older methods had been fraught with uncertainties and resolution limitations, step size histograms showed "best-fit" steps that were integer multiples of 2.6 – 2.7 nm (Pollack, 1990, Fig 9.10). With the higher precision of the more recent methods, the value 2.7 nm is confirmed. It is confirmed not only in the single sarcomere, but also in the single filament pair. It seems fairly certain, then, that the essential event of muscle shortening (and lengthening) is a step whose size is an integer multiple of 2.7 nm.

How this signature-like result translates into a molecular mechanism is beyond the scope of this paper — except to say that 2.7 nm is equal to the linear repeat of actin monomers, projected onto the long axis of the thin filament. This result and its implications are discussed at length in a recent book by one of us (Pollack, 2001). The book considers not only the muscle-contraction mechanism, but also the mechanism of various intracellular processes, all based on a single generic process.

6. REFERENCES

Altringham, J. D., Bottinelli, R., and lacktis, J. W., 1984, Is stepwise shortening an artifact?, Natue **307**: 653-655.

Blyakhman, F., Shklyar, T., and Pollack, G.H., 1999, Quantal length changes in single contracting sarcomeres, *J. Muscle Res. Cell Motil,* **20**: 529-538.

Blyakhman, F., Tourovskaya, A. and Pollack, G. H., 2001, Quantal sacromere-length changes in relaxed single myofibrils, Biophys. J. **81**: 1093-1100.

Delay, M.J., Ishide, N., Jacobson, R.C., Pollack, G.H. and Tirosh, R., 1981, Stepwise sarcomere shortening: Analysis by high-speed cinemicrography, *Science* **213**:1523-1525.

Edman, K.A.P. and Hoglund, O., 1981, A technique for measuring length changes of individul segments of an isolated muscle fibre, *J. Physiol.* **317**:8-9.

Fauver, M., Dunaway, D., Lilienfield, D., Craighead, H., and Pollack, G.H., 1998, Microfabricated cantilevers for measurement of subcellular and molecular forces, *IEEE Trans. Biomed. Engr.* **45**: 891-898.

Gordon, A.M., Huxley, A.F. and Julian, F.J., 1966, Tension development in highly stretched vertebrate muscle fibers, *J. Physiol.* **184**:143-169.

Granzier, H.L.M. and Pollack, G.H., 1985, Stepwise shortening in unstimulated frog skeletal muscle fibers, *J. Physiol.* **362**:173-188.

Granzier, H.L.M., Mattiazzi, A. and Pollack, G.H., 1990, Sarcomere dynamics during isotonic velocity transients in single frog muscle fibers, *Am J. Physiol (Cell)* **259**: C266-278.

Haugen, P and Sten-Knudsen, O., 1976, Sarcomere lengthening and tension drop in the latent period of isolated frog skeletal muscle fibers, *J. Gen Physiol.* **68**: 247-265.

Housmans, P., 1984, *Discussion in Contractile Mechanisms in Muscle* (eds., Pollack, G.H. and Sugi, H.), Plenum Press, N.Y., pp. 782-784.

Huxley, A. F., 1984, Response to "Is stepwise sarcomere shortening an artifact?", *Nature* **309**:713-714.

Iwazumi, T. and Pollack, G.H., 1979, On-line measurement of sarcomere length from diffraction patterns in cardiac and skeletal muscle, *IEEE Trans Biomed. Eng.* **26**(2):86-93.

Liu, X. and Pollack, G. H., 2004, Stepwise sliding of single actin and myosin filaments, *Biophys. J.* **86**:353-358.

Myers, J., Tirosh, R., Jacobson, R.C. and Pollack, G.H., 1982, Phase locked loop measurement of sarcomere length with high time resolution, *IEEE Trans Biomed. Eng.* **29**(6):463-466.

Nagornyak, E., Blyakhman, F. and Pollack, G.H., 2004, Effect of sarcomere length on step size in relaxed psoas muscle, *J. Mus. Res. Cell Motil.* In press.

Oroudjev, E., Soares, J. Arcidiacono, S., Thompson, J. B., Fossey, A. A. and Hansma, H. G., 2002, Segmented nanofibers of spider dragline silk: atomic force microscopy and single-molecule force spectroscopy, *Proc. Natl. Acad. Sci.* **99**(suppl 2):6460-6465.

Pollack, G.H , Iwazumi, T., ter Keurs, H.E.D.J. and Shibata, E.F., 1977, Sarcomere shortening in striated muscle occurs in stepwise fashion, *Nature* **268**:757-759.

Pollack, G.H., Vassallo, D.V., Jacobson, R.C., Iwazumi, T., and Delay, M.J., 1979, Discrete nature of sarcomere shortening in striated muscle, In: *Cross-bridge Mechanism in Muscle Contraction,* H. Sugi and G.H. Pollack, ed., Univ. of Tokyo Press/Univ. Park Press, pp. 23-40.

Pollack, G. H., 1984, Response to "Is stepwise shortening an artifact?", *Nature* **309**:712-714.

Pollack, G. H., 1986, Quantal mechanisms in cardiac contraction, *Circ. Res.* **59**: 1-8.

Pollack, G. H., 1990, *Muscles and Molecules: Uncovering the Principles of Biological Motion*, Ebner and Sons, Seattle.

Pollack, G. H., 2001, *Cells, Gels and the Engines of Life,* Ebner and Sons, Seattle.

Rudel, R. and Zite-Ferenczy, F., 1979, Do laser diffraction studies on striated muscle indicate ste4pwise sarcomere shortening?, *Nature* **278**:573-575.

Sokolov, S., Grinko, A. A., Tourovskaia, A. V., Reitz, F. B., Blykhman, F. A. and Pollack, G. H., 2003, 'Minimum average risk' as a new peak detection algorithm to myofibrillar dynamics, *Comput. Meth and Prog. in Biomed.* **72**(1):21-26.

Sugi, H., 1968, Local activation of frog muscle fibers with linearly rising currents, *J. Physiol* **199**: 549-567.

Svoboda, K., Schmidt, C. F., Schnapp, B. J. and Block, S. M., 1993, Direct observation of kinesin stepping by optical trapping interferometry, *Nature* **365**:721-727.

Tameyasu, T., Toyoki, T. and Sugi, H., 1985, Non-steady motion in unloaded contractions of single frog cardiac cells, *Biophys J.* **48**:461-465.

Toride, M. and Sugi, H., 1989, Stepwise sarcomere shortening in locally activated frog skeletal muscle fibers, *Proc. Natl. Acad. Sci. USA* **72**:1729-1733.

Yakovenko, O., Blyakhman, F. and Pollack, G. H., 2002, Fundamental step size in single cardiac and skeletal sarcomeres, *Am J. Physiol (Cell)* **283**(9):C735-C743.

Yang, P., Tameyasu, T. and Pollack, G.H., 1998, Stepwise dynamics of connecting filaments measured in single myofibrillar sarcomeres, *Biophys. J.* **74**:1473-1483.

DISCUSSION

Gonzalez-Serratos: 1) I noticed that in all your records within a single ramp there was a large variation in the duration of steps. 2) Also in your records, these are clear steps that have a "negative" slope (up slope) instead of being no slope or down slope.

Pollack: Pause duration depends on the speed of the ramp imposed. With faster ramps, pauses are shorter, with slower ramps pauses are longer. Within a given record, pauses tend to be of variable duration – there appears to be a stochastic contribution to pause duration. Pause slopes also vary. Some pauses show a slightly upward slope, some slightly downward. On average slopes are close to zero. We assume that the non zero slopes are related to the presence of noise, because , on average, the slope is zero.

Cecchi: Referring to the model you presented, it seems to me that it could be right only if myosin head are assumed to be fixed in the same position.

Pollack: Yes. If cross-bridges were compliant, then we would expect broader distributions of step size. However, our evidence shows very little cross-bridge compliance. We found (cf. Pollack, Muscles and Molecules, Ebner & Sons, Seattle, 1990) that cross-bridges actually cross-link adjacent thick filaments: the two opposing S-1s link to one another, and thereby form a permanent bridge. Such ring-like links between thick filaments should have very little compliance. Therefore, cross-bridge compliance should not broader the step size histogram peaks.

Maughan: Is there enough flexibility in the thin (actin) filament to accommodate your model?

Pollack: The model does not rely on actin-filament flexibility. It is based on propagated local shortening along the actin filament. There is appreciable evidence for the actin filament shortening (see Pollack, Cells Gels and The Engine of Life, Ebner and Sons, Seattle, 2001).

Maughan: I didn't mean to convey the impression that there is flexibility in actin. But I convey the notion that some aspects of the thin filament structure changes accommodate the jump to next actin.

CROSSBRIDGE FORMATION DETECTED BY STIFFNESS MEASUREMENTS IN SINGLE MUSCLE FIBRES

Barbara Colombini, Maria Angela Bagni, Rolando Berlinguer Palmini and Giovanni Cecchi*

1. INTRODUCTION

The first mechanical change induced by the stimulation of a muscle fibre, is an increase in stiffness that begins few milliseconds after the stimulus, before force development, and continues throughout all the tension rise up to the tetanus plateau (Ford et al., 1986; Claflin et al., 1990; Bagni et al., 1994). Stiffness increase is due to the formation of crossbridges between actin and myosin filaments in the overlap regions. In relaxed fibres the myofilaments are free to slide one past the other and sarcomeres can be elongated by applying only the very small force required to stretch the passive elasticity. When crossbridge form they link mechanically the two sets of filaments and, compared to the relaxed fibre, a much greater force is needed to elongate the sarcomere because the filament sliding now require the distortion of the attached crossbridges. An activated fibre is therefore much stiffer than a relaxed fibre. If filament compliance (the reciprocal of stiffness) is negligibly small, sarcomere elasticity is located mainly in the crossbridges which can be assumed to act in parallel. This means that sarcomere stiffness is the sum of all the individual crossbridge stiffness, hence stiffness is a measure of the number of attached crossbridges. An example of the usefulness of stiffness measure comes from experiments made during isotonic shortening. In principle, the reduction of force during the shortening, could be due to three different mechanisms: 1) a reduction of crossbridge number, or 2) a reduction of individual crossbridge force or 3) the presence of crossbridges generating negative force (opposing shortening) counteracting crossbridge generating positive force. Stiffness measurements allows to select among these possibilities. It was found that stiffness decreased with increasing shortening speed (lowering the isotonic force) but less than in proportion to the steady tension (Julian and Morgan, 1981; Ford et al., 1985; Griffiths et al., 1993). At shortening velocity equal to

* All Authors are from: Dipartimento di Scienze Fisiologiche, Università degli Studi di Firenze, Viale G.B. Morgagni 63, Florence I-50134, Italy.

V_{max}, in spite of the zero tension developed by the fibre, sarcomere stiffness was still about 30% of the isometric value. Assuming that crossbridge compliance is linear, this means that at V_{max} about 30% of the isometric crossbridges are still attached. The more likely explanation of this finding is that during the shortening the positive tension generated by a group of crossbridges is exactly counteracted by the negative tension generated by another group of crossbridges, in agreement with the predictions of the Huxley's model (1957).

2. STIFFNESS MEASUREMENTS

Muscle fibres stiffness has been commonly measured by applying to one fibre end a fast length change (*dl*) and measuring the resulting change in force (*df*) at the other end. The ratio *df/dl* is a measure of the stiffness. To compare stiffness data in different preparation it is some time useful to express stiffness as Young modulus (E) in which case *df* is normalized for the cross section area of the fibre while *dl* is normalized for the fibre length. The most common shape of the length change used is the step although a number of experimenters have also used sinusoidal length changes (Julian and Sollins, 1975; Cecchi et al., 1982). Other methods used to calculate fibre stiffness were based on the measure of the propagation time of a mechanical pulse (Truong, 1974; Schoenberg and Wells, 1984) or of an high frequency ultrasonic acoustic wave (Hatta et al., 1988) along the preparation.

Fig. 1 illustrates the method commonly used to measure stiffness in single muscle fibres. The sudden length change (a stretch in this case) applied at tetanus plateau, produces a series of characteristic force changes known as " tension transient" or isometric transient (Ford et al., 1977), constituted mainly by an fast rise of tension followed by a quick force recovery (time constant around 1 ms; Ford et al., 1977) towards the initial state. The stiffness corresponds to the ratio of the fast tension change over the length change. It is clear that the presence of the quick force recovery, which proceeds during the length change itself, tends to truncate the force change therefore reducing the measured stiffness. The longer the duration of the length change (or the faster the quick recovery) the greater would be the force truncation and the lower will be the stiffness measured. To illustrate the importance of this point it can be recalled that the amplitude of the length change necessary to drop the tension to zero, y_0 (a measure of fibre compliance corresponding to tetanic tension (P_0) over fibre stiffness) in frog muscle fibre, was 8 $nm.hs^{-1}$ when measured with a step length change of 1 ms duration but reduced to less than 5 $nm.hs^{-1}$ when the step duration was reduced to 0.2 ms (Ford et al., 1977). This means that the stiffness measured with the slowest length change was underestimate by about 40% respect to that measured with faster length change. The effect of the quick recovery on stiffness can be corrected with a method that allows to calculate the amount of truncation of the force change (Ford et al., 1977). Alternatively, stiffness should be measured with a very fast length change such that the quick recovery has not time to proceed significantly during the change itself, thus eliminating force truncation. However, increasing the speed of the length change tends to introduce other stiffness artefacts. This is because inertial forces from the fibre mass, as well as viscous forces, become important when the fibre is stretched or released with high speeds and great accelerations. These passive forces add to the elastic force altering the results of stiffness measurements.

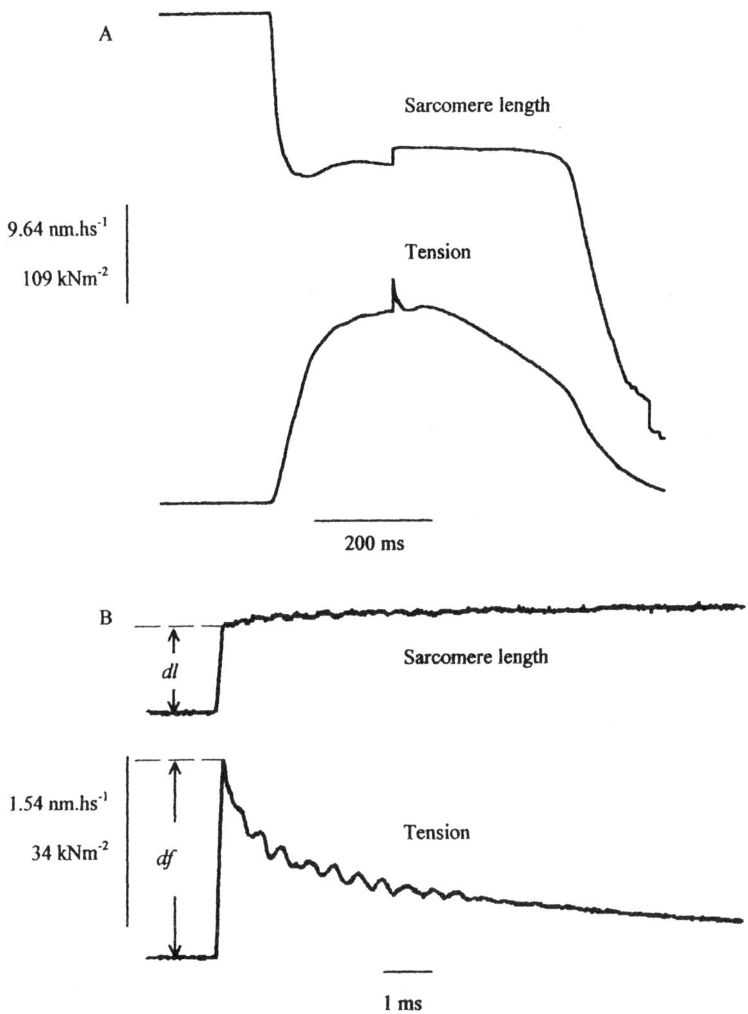

Figure 1. Force response to a small step stretch of a single muscle fibre at tetanus plateau, at slow (A) and fast (B) time base. Upper traces, sarcomere length; lower traces, tension. Sarcomere stiffness can be calculated by the ratio *df/dl*.

Some other effects occurs: the muscle fibre is an elongated structure and the mechanical wave travels along the fibre with a finite velocity (Truong, 1974). If the length change is very fast, it can be completed before the mechanical wave had the time to distribute uniformly along the fibre. This effect too would produce a change on the measured stiffness. Furthermore, since the force transducer and the stretcher are much stiffer than the fibre, the mechanical wave travelling along the fibre would be reflected back and forth from one fibre end to the other. If the dumping factors of the system (internal fibre viscosity, external viscosity and quick recovery) are not great enough, the wave will not be efficiently dumped and the fibre will tend to mechanically resonate at a frequency depending on fibre length (l_0) and fibre Young modulus according to the formula: $Fr = 1/2l_0 \sqrt{(E/density)}$ (viscous force are neglected). Fibre resonance has been directly shown in experiments using high frequency sinusoidal length changes and very long fibres (around 10-12 mm) from ileofibularis frog muscle (Cecchi et al., 1986). Fibre resonance too would produce further artefacts on stiffness measurements. In addition, very fast force transducer need to be used or their response should be corrected for frequency response limitation. It is clear therefore that a precise instantaneous stiffness measurements with very fast length changes require a series of careful considerations regarding all the possible artefacts and limitations.

Stiffness has been measured also by using sinusoidal length changes rather than step length changes. Oscillations method is particularly useful when stiffness has to be followed during a period of time since a stiffness measure can be obtained every cycle of oscillations. One drawback with sinusoidal oscillations is that, due to the non linear dependence of the quick recovery speed from direction and amplitude of the length change, (Ford et al., 1977) it is more complicated to correct for force truncation. The effect of the quick recovery is to produce a phase lead of force respect to length (similarly to the effect of a viscosity) while fibre inertia produces the opposite effect. To reduce as much as possible all the problems related to fibre inertia, resonance and force truncation, very short fibres (about 2 mm) from frog lumbricalis muscle and length oscillations at frequency up to 9 kHz have been used (Cecchi et al., 1982).

Measure of the propagation speed of the mechanical wave along a muscle (Truong, 1974) or fibre bundle (Schoenberg and Wells, 1984) have also been performed in the past to calculate stiffness. These were based on the consideration that the velocity of propagation (V) is equal to $\sqrt{(E/fibre\ density)}$. The main problem with this kind of measure is that the formula above is valid only for an uniform rod with no viscosity. If viscosity is present, as in the fibre, it should be considered since it affects significantly the transmission velocity (Truong, 1974; Ford et al., 1977).

Measurements considered up to now were often made on whole muscle or fibres including tendon attachment. To obtain sarcomere stiffness it is necessary to eliminate or to correct for the effect of tendons compliance which, even in a single carefully dissected and mounted fibre, contributes significantly to the total stiffness (Cecchi et al., 1987). This has been done by measuring sarcomere length changes directly using a laser diffraction technique (Bagni et al., 1994) or devices like the spot follower or striation follower (Ford et al., 1977; Cecchi et al., 1987). Probably the best way to account for all the possible artefacts on stiffness measurements arising from fibre inertia, internal and external viscosity, quick recovery and tendons effects, is the way described by Ford et al. (1981). They developed an electrical model constituted by a delay line simulator which reproduced the mechanical delay line behaviour of the single fibre. The delay line was fed with electrical signals representing the length changes applied to the fibre and the various parameters representing fibre elasticity, viscosity and quick recovery process, were adjusted to fit the experimental responses. The most important result obtained with this analysis was the demonstration that crossbridge compliance accounted for almost all (at least 80-90%) the sarcomere compliance. It was concluded therefore that sarcomere stiffness was a valid measure of crossbridge number (Ford et al., 1977).

3. STIFFNESS DURING TETANUS RISE AND RELAXATION

According to the crossbridge theory (Huxley, 1957) the development of tension in a tetanic contraction is determined by the progressive increase in the number of the attached crossbridges. Accordingly, sarcomere stiffness should rise linearly in proportion to tension. Measurements made using relatively slow length changes showed that stiffness was actually almost proportional to tension, although it was suggested a possible deviation from linearity (Bressler and Clinch, 1974). It should be pointed out, that measurements of stiffness during the tetanus rise are truncated more than at plateau because the speed of the quick recovery is faster at low tension either during the tetanus rise or isotonic shortening (Bagni et al., 1985; Ford et al., 1985; 1986). This effect tends to alter the shape of the stiffness tension relation, stressing even more the necessity of using very fast length changes in making stiffness measurements. In fact, data obtained with high frequency length oscillations (up to 9 kHz) on very short fibres (less than 2 mm) from frog lumbricalis muscle (Cecchi et al., 1982) showed that the relation between tension and stiffness during the tetanus rise deviates substantially from linearity, stiffness being relatively higher than tension (see Fig. 2A).

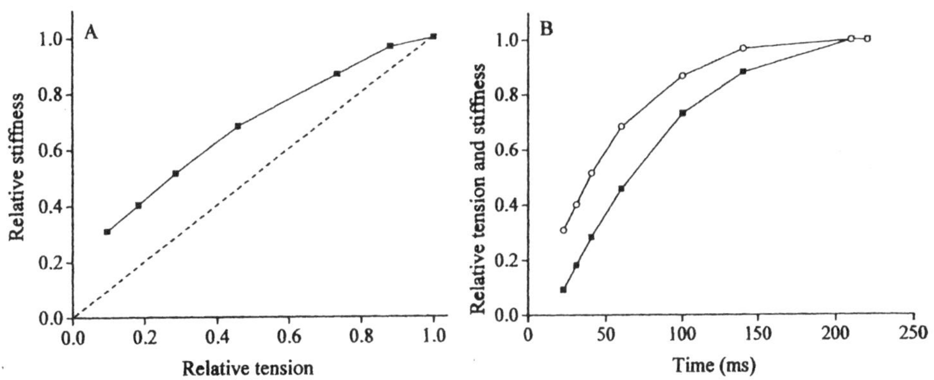

Figure 2. (A) tension-stiffness relation during the tetanus rise. Values are expressed relatively to the plateau values. (B) development of stiffness (filled squares) and tension (empty circles) during the tetanus rise. Time zero indicates the start of stimulation.

At tension of 0.5 P_0, relative stiffness was about 0.65. At lower oscillation frequencies the deviation decreased as expected from the greater force truncation on the tetanus rise. If tension and stiffness are plotted as function of time after the stimulation, it can be seen (Fig. 2B) that stiffness leads tension development by some 15-20 ms (Cecchi et al., 1984). These experiments were made on whole fibres, however it was shown that tendon compliance could not be responsible for the deviation from linearity (Cecchi et al., 1982). Successive experiments using the spot follower device and the delay line simulator (Ford et al., 1981), or striation follower (Cecchi et al., 1987) confirmed these data, demonstrating that the non linearity between stiffness and tension (or the leading of stiffness over force) was a property of the contractile apparatus. Similar experiments

made on the tetanus relaxation showed that the fall of stiffness was lagging behind the fall of force by about 80-100 ms, however, in spite of this difference, the tension-stiffness relations was similar to that found on the tetanus rise (Cecchi et al., 1984).
More recent experiments made with a technique that eliminate the effects of fibre inertia and viscosity, showed that the deviation from linearity is probably smaller than previously found (Bagni et al., 1999). To explain the non linearity of the tension-stiffness relation, Cecchi et al. (1982) and Ford et al. (1986) suggested that during the tetanus rise (or relaxation) a small fraction of attached crossbridges could be in a configuration developing less (or no) force than at plateau contributing more to sarcomere stiffness than to force. This idea was consistent with x-ray diffraction data showing that during the tetanus rise, the changes in the intensity of the 1,1 1,0 equatorial reflections, though to arise from movement of S1 towards actin filament, leaded tension development (Kress et al., 1986; Cecchi et al., 1991). A model able to explain both the non linearity of the tension-stiffness relation and the increase of quick recovery speed at low tensions, was put forward by Bagni et al. (1988).

4. FILAMENT STIFFNESS

The conclusions above were based on the assumption that sarcomere stiffness was proportional to the number of attached crossbridge as demonstrated by the observation that filament compliance was almost negligible (Ford et al., 1981). However, successive high resolution x-ray diffraction data revealed a significant tension-dependent increase of the axial periodicity of both actin and myosin filament showing that both filaments are much more compliant than previously assumed. Calculation showed that filament compliance accounts for at least 40-50% of the sarcomere compliance (Huxley et al., 1994; Wakabayashi et al., 1994). Qualitatively similar observations came from mechanical experiments in which sarcomere compliance was measured at the plateau region of the length-tension relationship in isolated fibres from frog muscles (Bagni et al., 1990). By knowing that stiffness changes in this sarcomere length region can only be attributed to the different amount of actin filament free from crossbridges (the effective overlap and attached crossbridge number remain constant), it was possible to calculate actin filament compliance. These experiments showed that filament compliance was about 20% of the sarcomere compliance, about one half of the value found with x-ray diffraction. Similar measurements in rabbit and frog muscle fibres in rigor made with sarcomere length control, gave higher values, around 30-50% of the total sarcomere stiffness (Higuchi et al., 1995; Linari et al., 1998). Further data from experiments on isolated actin or actin/tropomyosin filament confirmed that thin filament compliance contribute for more than 50% to the sarcomere compliance (Kojima et al., 1994).
It is clear that these findings have a major impact on the assumption that crossbridge number can be calculated from sarcomere stiffness. To determine the crossbridge stiffness in fact, sarcomere stiffness should be corrected for the contribution of filament stiffness. This can be done by means of the calculation proposed by Thorson and White (1969) and further elaborated by other investigators (Ford et al., 1981; Mijailovich al., 1996). A slightly different model, more accurate when the number of attached crossbridges is relatively small as during the initial part of the tetanus rise, was proposed by Forcinito et al. (1997). These methods calculated the crossbridge stiffness considering correctly that filament and crossbridge compliances are arranged in a series-parallel

network. It should be said however, that the error made by assuming that filament compliance is simply in series with crossbridges is almost negligible (less than few % under most conditions; Mijailovich et al., 1996).

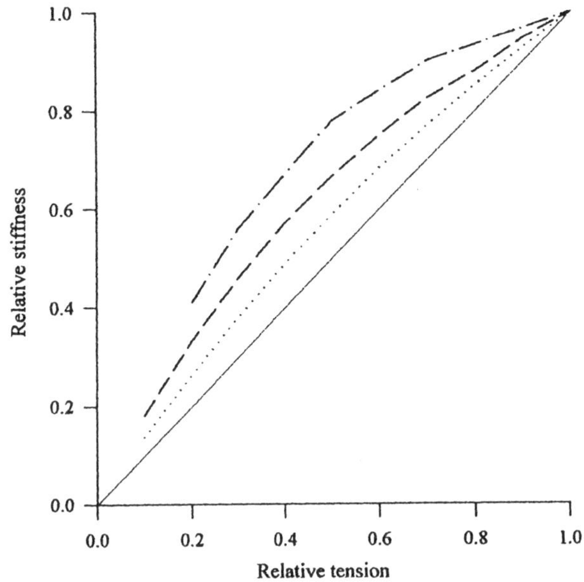

Figure 3. Effects of various amounts of filament compliance on the calculated tension-stiffness relation during the tetanus rise. Filament and crossbridge compliances are assumed to be in series and linear. Filament compliance was 30% (dotted line) or 50% (dashed line) of sarcomere compliance. Point–dashed line is calculated from the model proposed by Mijailovich et al. (1996, see text), assuming 50% filament compliance.

The presence of the filament compliance offers an alternative explanation to the observation that stiffness was relatively higher than tension during the tetanus rise, not implying additional crossbridge states (Goldman and Huxley, 1994). A simple example may be useful to illustrate this point. Suppose that at the tetanus plateau, filament compliance, Cf, is equal to crossbridge compliance, Cb (an assumption made on the bases of the x-ray results reported above) and assume that filament compliance is simply in series with crossbridges. The sarcomere compliance at plateau would be equal to 2Cb. Suppose now that only one half of crossbridges are attached as for example on the tetanus rise when the tension is $0.5P_0$. Since crossbridge number is halved their compliance increases to 2Cb, doubling respect to the plateau. Total compliance would then be Cf + 2Cb = 3Cb and stiffness would be 1/3Cb. At plateau the stiffness was 1/2Cb. The ratio of the stiffness on the rise/stiffness at plateau is therefore (1/3Cb) x 2Cb = 0.666 and relative stiffness is 33% higher than relative tension. The relationship between stiffness and tension would be concave upward, similarly to that found experimentally, with a deviation from the linearity progressively increasing at lower tensions. An example of the calculated effects of various amount of filament compliance on tension-stiffness relation is shown in Fig. 3.

It should be pointed out that the simple calculation above is valid only if filament compliance, as well as crossbridge compliance, are both Hookean, i.e. independent of tension. If actin (or myosin) filament have a non-linear compliance (compliance decreasing with tension) the calculation above can give a quite different result. It can be shown that for a given amount of compliance at plateau, the greater is the non-linearity of filament compliance the smaller is the deviation introduced on the tension-stiffness relation. A particular case occurs when the stiffness of the actin filament is assumed to rise proportionally to force. In this case no deviation from the linearity can be accounted for by the filament compliance. In the example above, Cf would increase at low tensions exactly as Cb and the relation between tension and stiffness would be linear (Bagni et al., 1999). Actually, x-ray diffraction and mechanical experiments of fibres in rigor seems to show that actin filament is somewhat non linear (Huxley et al., 1994; Wakabayashi et al., 1994; Higuchi et al., 1995) but the amount of non linearity is still unclear. In absence of precise information about the stress-strain relations of crossbridge and myofilaments, stiffness measurements would still be affected by this uncertainty.

5. AN ALTERNATIVE METHOD TO MEASURE CROSSBRIDGE NUMBER

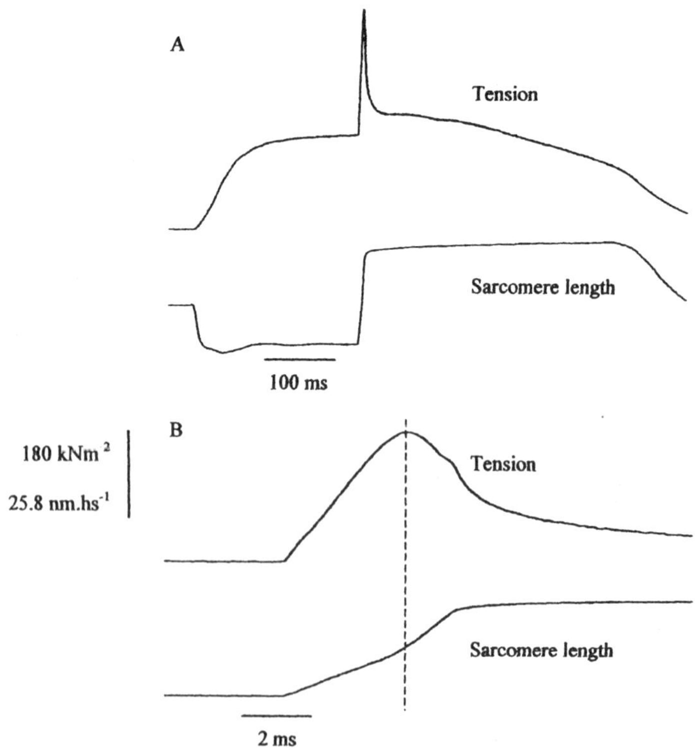

Figure 4. Force response to a fast ramp stretch applied on a single muscle fibre at tetanus plateau at slow (A) and fast time base (B). The vertical dashed line on trace B indicated the time of the "sarcomere give" (lower trace) at which tension suddenly starts to fall. This is caused by the forced crossbridges detachment occurring when sarcomere elongation reaches about 11 $nm.hs^{-1}$.

The results of our recent experiments made to investigate the properties of the actomyosin interaction show that a measure of attached crossbridge number can be obtained with a method that is not affected by either the quick recovery mechanism or filament compliance. This is a variation of the method used in the past by Flitney and Hirst (1978) to investigate the "give" phenomena (Katz, 1939).
It is well known that upon stretching of an activated fibre, tension rises steeply to a point (tension break, P_b) at which the resistance to elongation is greatly diminished and the tension either remains constant or falls to a lower steady level depending upon the stretching speed. At the same time of the tension break, as consequence of the quick reduction of fibre stiffness, sarcomere elongation increases steeply (sarcomere "give"). The break of force is caused by the forced crossbridge detachment imposed by the stretch and it occurs when the sarcomere elongation reaches a critical value (*lc*) of about 12-15 $nm.hs^{-1}$. If the stretching speed is very high as in our experiments the break of tension is quite sharp (see Fig. 4).

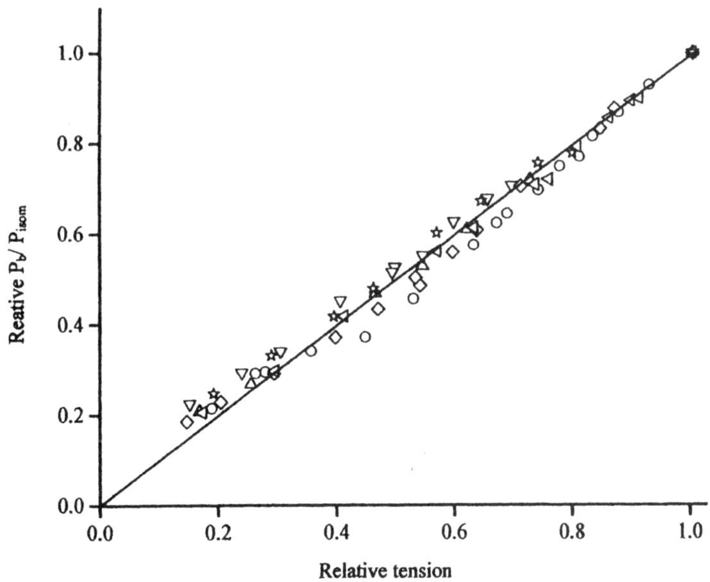

Figure 5. Break tension/isometric tension ratio during the rise of a tetanic contraction. Data from 7 fibres. Note the linearity of the relation suggesting that crossbridge number is proportional to tension.

Since P_b was directly proportional to the overlap between actin and myosin filaments, Flitney and Hirst (1978) suggested that break tension is directly proportional to the number of attached crossbridges. Therefore measuring the breaking force can be a way to assess crossbridge number. The advantage of this methods over stiffness measurements is that P_b is independent of the compliance in series with the crossbridges and of the quick recovery speed.
We have examined the force response to fast stretches during various phases of fibre activity.

Single fibres dissected from tibialis anterior muscle of the frog were tetanically stimulated at 14°C at sarcomere length of about 2.2 μm. Ramp stretches (20-40 $nm.hs^{-1}$ amplitude and 0.6-2 ms time duration) were applied at various tension levels during the tetanus rise, relaxation and isotonic shortening. Force was measured with a fast force transducer (resonance frequency 30-50 kHz) and length changes were measured at sarcomere level using a striation follower device. Fig. 5 shows the results from 5 fibres on the tetanus rise. It can be seen that the break force is directly proportional to the tension developed, suggesting that the number of attached crossbridges is proportional to tension. The mean value of ratio P_b/isometric tension (P_{isom}) at tetanus plateau, on 7 fibres, was 2.37 ± 0.133 (standard error) and it remained constant throughout the tetanus rise being independent of tension.

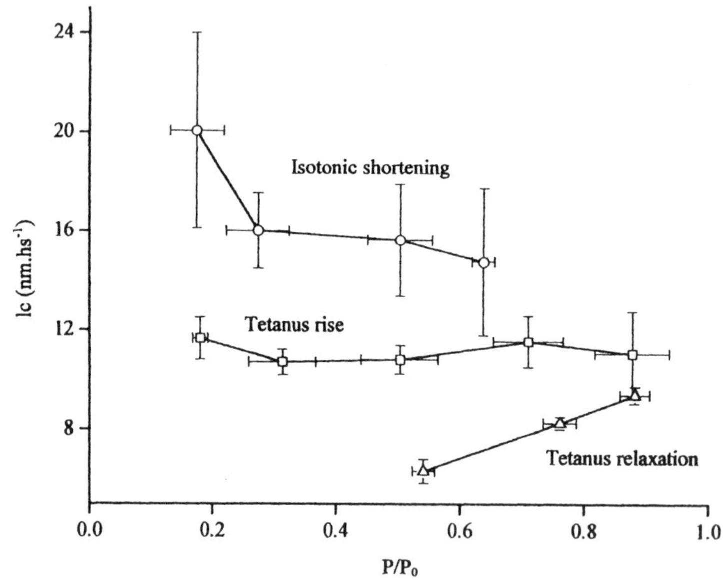

Figure 6. Critical elongation (*lc*) at various tension during isotonic shortening (circles), during tetanus rise (squares) and relaxation (triangles). Note that *lc* is independent of tension only during the tetanus rise.

Taking into account the higher stretching speeds used here, this values is in good agreement with previous values reported in literature. Interestingly, we found that the critical elongation was independent of the tension developed by the fibre (Fig. 6). The mean value found in 5 fibres at a mean tension of 0.50 ± 0.05 P_0 during the tetanus rise, was 10.83 ± 0.57 $nm.hs^{-1}$. This effect is somewhat unexpected since at low break tensions, filament compliances absorb a smaller portion of the length change applied than at plateau, therefore *lc* should be reduced. This result may be explained if filament compliance were non linear, however this explanation would require an opposite non linearity from crossbridges compliance to explain the relative linearity of the tension increase during the ramp stretch.

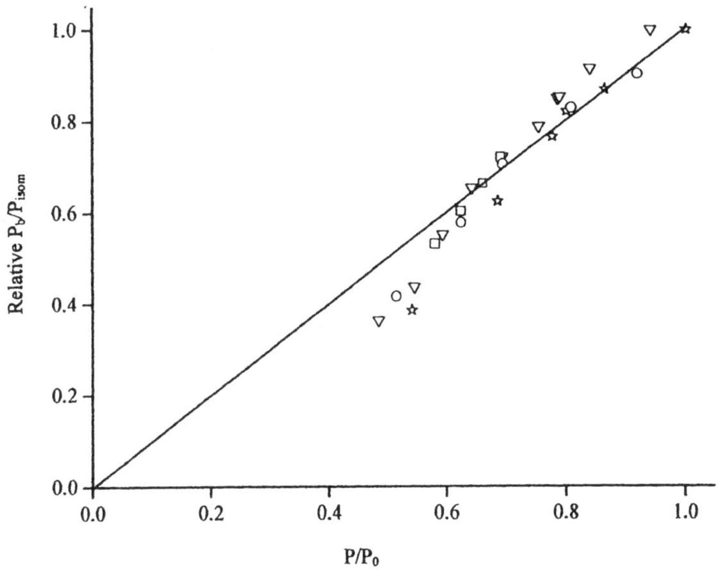

Figure 7. Break tension versus isometric tension during relaxation. Data from 4 fibres.

The same experiments made on the tetanus relaxation showed slightly different results. Break force measurements were made exclusively on the isometric phase of relaxation during which sarcomere length along the fibre remains uniform. Contrary to tetanus rise, the ratio P_b/P_{isom} decreases with tension during relaxation becoming smaller than during the rise (Fig. 7). This suggests that at a given tension fewer crossbridges developing a greater individual force, are present on the relaxation compared to the tetanus rise. In agreement with this idea, the critical elongation is smaller during relaxation (Fig. 6) suggesting that crossbridges have a greater mean extension. Thus, the higher individual mean crossbridge force developed during relaxation could be accounted for by the greater crossbridge mean extension, possibly caused by the stretch imposed on the sarcomeres by the progressive elastic recoil of tendon elasticity as tension falls.

The same stretching technique was also applied during isotonic shortening at various velocities. The averaged results from 5 fibres, are reported in Fig. 8.

At mean isotonic tension (P_{isot}) of 0.45 ± 0.03 P_0 the relative ratio P_b/P_{isot} was 0.644 ± 0.043, 43% higher than during the tetanus rise and plateau.

The higher the shortening velocity and the lower the isotonic tension, the higher is the P_b/P_{isot} ratio. This finding suggests that at given force, during isotonic shortening crossbridge number is greater than during isometric contraction, in agreement with Huxley's model (1957) and with the stiffness measurements reported previously. Therefore the mean force developed by crossbridges during isotonic shortening is smaller that under isometric conditions. In analogy with relaxation data, a lower individual crossbridge force should be associated with a smaller crossbridge extension. Consequently, a greater sarcomere elongation should be needed to reach the tension break.

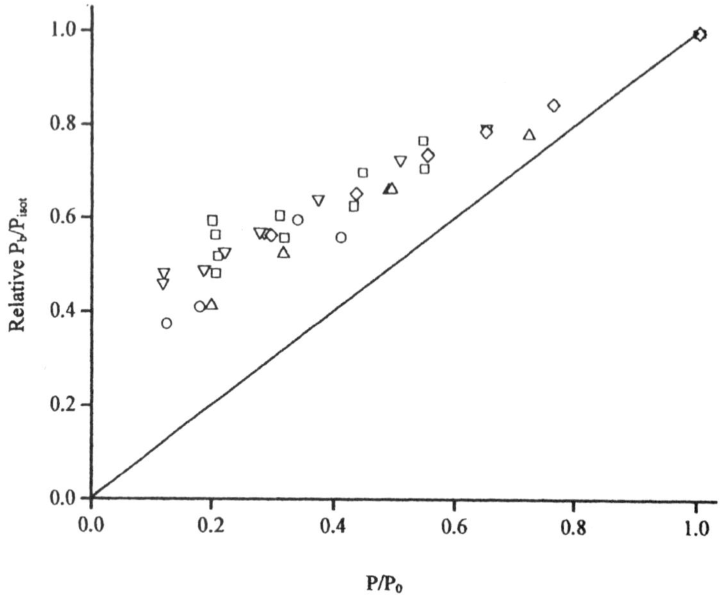

Figure 8. Break tension versus isotonic tension during a constant velocity shortening. Data from 5 fibres.

The data reported in Fig. 6 shows that this is indeed the case: critical elongation measured at isotonic tension of 0.50 ± 0.05 $P_{0,}$ was 15.65 ± 2.26 $nm.hs^{-1}$ about 44% greater that during isometric tetani.

These preliminary results show that the relative crossbridge number can be estimated by measuring the tension at break during fast stretching.

In conclusion the data reviewed in this paper show that the evaluation of attached crossbridge number through stiffness measurements is affected by various possible artefacts and it is more complicated than initially thought mainly because of the filament compliance. In addition it is also necessary to make assumption about the shape of the stress-strain relation of both crossbridges and myofilaments. In spite of these limitations, with appropriate corrections, stiffness measurements can be still considered an useful way to estimate the number of attached crossbridge. We have presented here an alternative way to measure attached crossbridge number which may have less problems than stiffness measurements.

6. ACKNOWLEDGMENTS

The authors wish to thank the University of Florence and the Ente Cassa di Risparmio di Firenze for financial support.

7. REFERENCES

Bagni, M.A., Cecchi, G., Colombini, B., and Colomo, F., 1999, Sarcomere tension-stiffness relation during the tetanus rise in single frog muscle fibres. *J. Muscle Res. Cell Motil.* **20**:469-476.

Bagni, M.A., Cecchi, G., and Colomo, F., 1985, The velocity of the "quick tension recovery" during the rise of tension in an isometric tetanus in frog single muscle fibres, *J. Physiol.* **365**:68P.

Bagni, M.A., Cecchi, G., Colomo, F., and Garzella, P., 1994, Development of stiffness precedes cross-bridge attachment during the early tension rise in single frog muscle fibres, *J. Physiol.* **481**(2):273-278.

Bagni, M.A., Cecchi, G., Colomo, F., and Poggesi, C., 1990, Tension and stiffness of frog muscle fibres at full filament overlap, *J. Muscle Res. Cell Motil.* **11**:371-377.

Bagni, M.A., Cecchi, G., and Schoenberg, M., 1998, A model of force production that explains the lag between crossbridge attachment and force after electrical stimulation of striated muscle fibers, *Biophys. J.* **54**:1105-1114.

Bressler, B.H., and Clinch, N.F., 1974, The compliance of contracting skeletal muscle, *J. Physiol.* **237**:477-493.

Cecchi, G., Bagni, M.A., Griffiths, P.J., Ashley, C.C., and Maeda Y., 1990, Detection of radial crossbridge force by lattice spacing changes in intact single muscle fibers, *Science* **250**:1409-1411.

Cecchi, G., Colomo, F., Lombardi, V., and Piazzesi, G., 1987, Stiffness of frog muscle fibers during rise of tension and relaxation in fixed-end or length-clamped tetani, *Pflügers Arch.* **409**:39-46.

Cecchi, G., Griffths, P.J., Bagni, M.A., Ashley, C.C., and Maeda Y., 1991, Time-resolved changes in equatorial X-ray diffraction and stiffness during rise of tetanic tension in intact length-clamped single muscle fibres. *Biophys. J.* **59**:1273-1283.

Cecchi, G., Griffiths, P.J., and Taylor, S., 1982, Muscular contraction: kinetics of crossbridge attachment studied by high-frequency stiffness measurements, *Science* **217**:70-72.

Cecchi, G., Griffiths, P.J., and Taylor, S., 1984, The kinetics of cross-bridge attachment and detachment studied by high frequency stiffness measurements, *Adv. Exp. Med. Biol.* **170**:641-655.

Cecchi, G., Griffiths, P.J., and Taylor, S.,1986, Stiffness and force in activated frog skeletal muscle fibers. *Biophys. J.* **49**(2):437-451.

Claflin, D.R., Morgan, D.L., and Julian, F.J., 1990, Earliest mechanical evidence of cross-bridge activity after stimulation of single skeletal muscle fibers, *Biophys. J.* **57**(3):425-432.

Flitney, F.W., and Hirst, D.G., 1978, Cross-bridge detachment and sarcomere 'give' during stretch of active frog's muscle, *J. Physiol.* **276**:449-465.

Forcinito, M., Epstein, M., and Herzog, W., 1997, Theoretical considerations on myofibril stiffness, *Biophys. J.* **72**:1278-1286.

Ford, L.E., Huxley, A.F., and Simmons, R.M., 1977, Tension responses to sudden length changes in stimulated frog muscle fibres near slack length, *J. Physiol.* **269**:441-515.

Ford, L.E., Huxley, A.F., and Simmons, R.M., 1981, The relation between stiffness and filament overlap in stimulated frog muscle fibres, *J. Physiol.* **311**:219-249.

Ford, L.E., Huxley, A.F., and Simmons, R.M., 1985, Tension transients during steady shortening of muscle fibres, *J. Physiol.* **361**,:131-150.

Ford, L.E., Huxley, A.F., and Simmons, R.M., 1986, Tension transients during the rise of tetanic tension frog muscle fibres, *J. Physiol.* **372**:595-609.

Goldman, Y.E., and Huxley, A. F., 1994, Actin compliance: are you pulling my chain? Editorial *Biophys. J.* **67**:2131-2133.

Griffiths, P.J., Ashley, C.C., Bagni, M.A., Maeda Y., and Cecchi, G., 1993, Cross-bridge attachment and stiffness during isotonic shortening of intact single muscle fibers, *Biophys. J.* **64**:1150-1160.

Hatta, I., Sugi, H., and Tamura, Y., 1988, Stiffness changes in frog skeletal muscle during contraction recorded using ultrasonic waves, *J. Physiol.* **403**:193-209.

Higuchi, H., Yanagida, T., and Goldman, Y.E., 1995, Compliance of thin filaments in skinned fibers of rabbit skeletal muscle, *Biophys. J.* **69**:1000-1010.

Huxley, A.F., 1957, Muscle stiffness and theories of contraction., *Prog. Biophys. Biophys. Chem.* **7**:255-318.

Huxley, H.E., Stewart, A., Sosa, H., and Irving, C.T.,1994, X-ray diffraction measurements of the extensibility of actin and myosin filaments in contracting muscle, *Biophys. J.* **67**:2411-2421.

Julian, F.J., and Morgan, D.L., 1981, Tension, stiffness, unloaded shortening speed and potentiation of frog muscle fibres at sarcomere length below optimum, *J. Physiol.* **319**:205-217.

Julian, F.J., and Sollins M.R., 1975, Variation of muscle stiffness with force at increasing speeds of shortening, *J. Gen. Physiol.* **66**:287-302.

Katz, B., 1939, The relation between force and speed in muscular contraction, *J. Physiol.* **96**:45-64.

Kojima, H., Ishijima, A., and Yanagida, T., 1994, Direct measurement of stiffness of single actin filaments with and without tropomyosin using in vitro nano-manipulation. *Proc. Natl. Acad. Sci. USA* **91**:12962-12966.

Kress, M., Huxley, H.E., Faruqi, A.R., and Hendrix, J., 1986, Structural changes during activation of frog muscle studied by time-resolved X-ray diffraction, *J. Mol. Biol.* **188**:325-342.

Linari, M., Dobbie, I., Reconditi, M., Koubassova, N., Irving, M., Piazzesi, G., and Lombardi, V., 1998, The stiffness of skeletal muscle in isometric contraction and rigor: the fraction of myosin heads bound to actin, *Biophys. J.* **74**:2459-2473.

Mijailovich, S.M., Fredberg, J.J., and Butler, J.P., 1996, On the theory of muscle contraction: filament extensibility and the development of isometric force and stiffness, *Biophys. J.* **71**:1475-1484.

Schoenberg, M., and Wells, J.B., 1984, Stiffness, force and sarcomere shortening during a twitch in frog semitendinosus muscle bundles, *Biophys. J.* **45**:389-397.

Thorson, J.W., and White, D.C.S., 1969, Distributed representations for actomyosin interaction in the oscillatory contraction of muscle, *Biophys. J.* **9**:360-390.

Truong, X.T., 1974, Viscoelastic wave propagation and rheologic properties of skeletal muscle, *Am. J. Physiol.* **226**:256-264.

Wakabayashi, K., Sugimoto, Y., Tanaka, H., Ueno, Y., Takezawa, Y., and Amemiya, Y., 1994, X-ray diffraction evidence for the extensibility of actin and myosin filaments during muscle contraction, *Biophys. J.* **67**:2422-2435.

DISCUSSION

Huxley: Why the A.F. Huxley 1981 model (Ford et al., J. Physiol. 269: 441-515, 1981) was wrong?

Cecchi: The only thing I can say is that mechanical measurements are not very sensitive.

Brenner: A comment on Huxley's question: In the delay-line simulation by Ford et al., the delay-line properties were assumed for simulating muscle properties. But their non-linear properties were not considered that may have shown discrepancies.

Cecchi: Yes, you may be right. Actually electric elements of the delay line are linear, while we know that fiber has some nonlinearity. This may introduce some significant differences.

NON CROSS-BRIDGE STIFFNESS IN SKELETAL MUSCLE FIBRES AT REST AND DURING ACTIVITY

Maria Angela Bagni, Barbara Colombini, Francesco Colomo, Rolando Berlinguer Palmini, and Giovanni Cecchi*

1. INTRODUCTION

Biochemical studies on the mechanism of muscle contraction have shown that myosin subfragment 1 (S1) containing ATP or ADP + inorganic phosphate (Pi) at its active sites can attach to actin to form cross-bridges even under relaxing conditions (Chalovich et al., 1981). These bridges are characterized by a fast equilibrium between attached and detached states and are commonly referred to as weakly binding bridges. A rapid stretch applied to a skinned fiber bathed in a relaxing medium at low ionic strength produced a significant force response, indicating that passive fiber possesses a significant stiffness (Brenner et al., 1982; 1984). The stiffness increased with the speed of the applied stretch and decreased with the sarcomere length approximately in the same way as the overlap between actin and myosin filaments. This suggested that a population of non generating cross-bridges may be present in the relaxed fibers in the absence of calcium. The dependence of the stiffness on the stretching speed can be explained by assuming that the cross-bridges are in rapid equilibrium between attached and detached states, similarly to the weakly binding bridges found in solution. If the stretch applied is slow, most of the bridges will have time to detach during the stretch itself and the force decreases. Therefore the response of the fiber will increase with speed until the stretch is so fast that no bridges have time to detach during its application. At this point, since all the bridges are picked up by the stretch, the force response and the stiffness will be maximum and constant. The speed at which the force becomes maximum will depend upon the rate constant for the detachment of the weakly binding bridges. As shown by Schoenberg (1985), the kinetics of weakly binding bridges are equivalent to those of a viscoelastic system (a pure viscous element in series with an elastic element) having a relaxation time nearly equal to the reciprocal detachment rate constant.
Several reports in literature have shown that the first mechanical effect of stimulation of a muscle fiber is a stiffness increase which precedes the development of tension (Cecchi et al., 1982; Ford et al., 1986). The lead of stiffness over tension is established early during

* All authors are from: Dipartimento di Scienze Fisiologiche, Università degli Studi di Firenze, Viale G.B. Morgagni, 63, I-50134, Firenze, Italy.

the latent period and it is maintained throughout the whole tension rise in a twitch or in a tetanic contraction. Since most of the fiber stiffness is attributable to cross-bridge attachment (Ford et al., 1981) this finding has led to the suggestion that the initial cross-bridge attachment to actin could occur in a non-force or low force generating state that is converted into a high force state only after a significant delay. According to Brenner (1990) the weakly binding bridges are present also in intact fibers at normal ionic strength and their number may increase following the increase in the myoplasmic calcium concentration that follows the stimulation. This effect may result in an increase of the fiber stiffness without a corresponding increase in force, and could explain the leading of stiffness over force found during the tension development.

The first experiments reported in this paper were made to investigate if weakly binding bridges are present under physiological conditions both in resting intact or skinned frog fibers and if the stiffness increase that lead the tension development after stimulation may be due to an increased number of weakly binding bridges (Bagni et al., 1992; 1994; 1995).

It was found that: passive force responses to ramp stretches at various velocities and amplitudes at different sarcomere lengths in intact and in skinned frog muscle fibers are composed by at least three components but none of them was found to have the properties expected from the weakly binding bridges. The fiber responses to stretches applied at progressively increasing times after a single stimulus showed that part of the stiffness increase caused by activation is not attributable to cross-bridges formation. This component has been termed "static stiffness" because its characteristics are similar to those of an elasticity. Static stiffness increases after the stimulus following a time course similar to Ca^{2+} concentration, starting to rise few milliseconds after the stimulus and falling to zero before the peak of twitch tension.

Steady state static stiffness properties were studied on tetanic contractions at different sarcomere lengths (Bagni et al., 2002). Fibers were normally bathed in Ringer solution containing 2,3-Butanedione-2-monoxime (BDM), an agent that greatly reduces tension without altering static stiffness (Horiuti et al., 1988). The results showed that the characteristics of static stiffness are equivalent to those expected from a linear elasticity located in parallel with cross-bridges. As in the twitch, static stiffness changed characteristically with activation following a time course distinct from that of tension and roughly similar to that of internal Ca^{2+}, suggesting that the static stiffness could arise from an unknown sarcomere structure whose stiffness changes in a Ca^{2+} dependent way. This idea was confirmed by the results of a series of experiments in which static stiffness was measured in presence of a series of agents that inhibited tension either by reducing the Ca^{2+} release by the sarcoplasmatic reticulum or directly by inhibiting actomyosin interaction. It was found that static stiffness was reduced only by the agents that reduced at the same time calcium release (Bagni et al., 2004).

2. METHODS

All the experiments reported in this review were made on single intact or skinned frog muscle fibers isolated from the tibialis anterior or on the shorter lumbricalis digiti IV muscle of the frog (*Rana esculenta*). Frogs were killed by decapitation followed by destruction of the spinal cord. Single intact fibers were mounted by means of aluminum foil clips between the lever arms of a force transducer (40-60 kHz natural frequency) and an electromagnetic motor (minimum ramp time 40 μs) to apply ramp length changes to the fiber. The experimental chamber was provided with a glass floor for ordinary and laser light illumination. The temperature was thermostatically maintained constant ± 0.2°C in a range of 5-15°C in the different series of experiments. Resting sarcomere length (l_0) was set at about 2.1 μm but in some experiments data were also collected at longer lengths.

Force responses to stretches in activated fibers were studied on both twitch and tetanic contractions. A single stimulus or a tetanic volley was applied to a pair of platinum-plate electrodes running parallel to the fiber. Tetanic stimulation consisted of brief (250-600 ms duration) volleys at 3 minute intervals using the minimum frequency necessary to obtain fused tetani (50-60 Hz). After a test of fiber viability and a measure of the isometric tetanic tension (P_0) in normal Ringer solution, all the experiments on tetanic contractions were made in Ringer added with BDM at concentration between 6 and 10mM. This was done to reduce active tension and to reduce the probability that stretches could damage the fiber. Sarcomere length changes were measured using a laser diffractometer (Bagni et al., 1992; 1995) or a striation follower device (Bagni et al., 2002; 2004) in a fiber segment selected for striation uniformity in a region as close as possible to the force transducer. This eliminated the effects of tendon compliance on stiffness measurements allowing to attribute the results directly to the sarcomere structure. In some experiments the output of the diffractometer was electronically differentiated to obtain the instantaneous velocity of sarcomere length changes.

Stiffness was measured by applying ramp stretches (amplitude, 2-70 $nm.hs^{-1}$ and velocity, $2x10^3$-$280x10^3$ $nm.hs^{-1}s^{-1}$) to one fiber end and measuring the force response at the other end. To measure the static stiffness the activated fiber was rapidly stretched and hold for a period longer than the stimulation time. The tension transient produced by the stretch was followed by a period during which the tension settled to an almost constant level which exceeded the isometric force. This level, subtracted by the tension developed at the time of the stretch and by the passive response of the fiber to the same stretch, is the "static tension". The ratio between the static tension and the sarcomere length elongation produced by the stretch represents the "static stiffness" of the sarcomere. In order to follow the time course of the static stiffness development following the activation, stretches were applied at rest and at different times after the start of stimulation. Usually three records were taken for each measure: isometric, isometric with stretch and passive response to the stretch. The isometric and the passive response were subtracted from the isometric record with stretch to obtain the subtracted trace on which measurements were made. By subtracting the isometric record we could measure the static tension always on a flat baseline even when the stretch was applied on tension rise or relaxation. By subtracting the passive response we corrected for the resting tension and stiffness of the relaxed fibers.

Resting fiber length (l_f), fiber cross-sectional area and resting sarcomere length were measured under ordinary light illumination using a 10x or 40x dry objective and 25x eyepieces. The normal Ringer solution had the following composition (mM): 115 NaCl, 2.5 KCl, 1.8 $CaCl_2$, 3 phosphate buffer at pH 7.1. Different Ringer solutions were obtained by adding different agents at the appropriate concentration to the normal Ringer solution.

Skinned fibers were obtained by immersion in skinning solution (3 min at 15°C) of the intact fiber in the experimental chamber, without altering its tendon attachments. The measurements were repeated at high and low ionic strength. The composition of the relaxing solution was as follows. High ionic strength (mM): 95 potassium propionate, 5.3 magnesium propionate, 20 N-tris(hydroxymethil)methyl-2-aminoethane sulphonic acid (Tes), 5 EGTA, 5 ATP at pH 7, ionic strength 170mM. Low ionic strength solution (mM): 1 potassium propionate, 3 magnesium propionate, 10 Tes, 1 EGTA, 1 ATP at pH 7, ionic strength 20mM. Skinning solution as relaxing at high ionic strength solution plus 0.5% Triton X-100 (v/v).

Force, fiber length and sarcomere length signals were measured with a digital oscilloscope (4094 Nicolet, USA), stored on floppy disks and transferred to a personal computer for further analysis.

3. RESULTS AND DISCUSSION

3.1. Response to Stretch of Passive Intact Fibers

Fig. 1 shows a typical force response of a passive intact fiber to a constant velocity stretch (Bagni et al., 1992). The force transient can be divided at least into 3 components which have been defined as P_1, P_2, P_3. P_1 is a fast response that occurs synchronously with the acceleration phase of the stretch during which the stretching velocity at sarcomere level reaches the steady state speed. The second components, P_2, gives the greater contribution to the response and coincides with the period of stretching at constant velocity. P_3 is the component responsible for the resting tension developed by the fiber during a steady elongation.

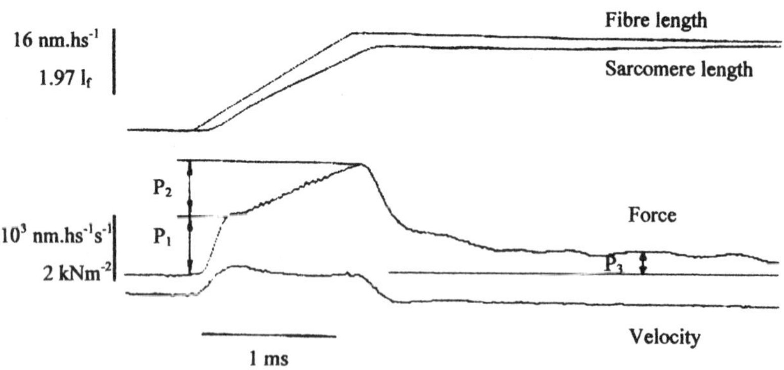

Figure 1. Typical experimental records obtained applying a rapid stretch at constant speed to an intact passive fiber from lumbricalis frog muscle. Stretching velocity, 24x10^3 nm.hs^{-1}s^{-1}; sarcomere length, 2.18 μm

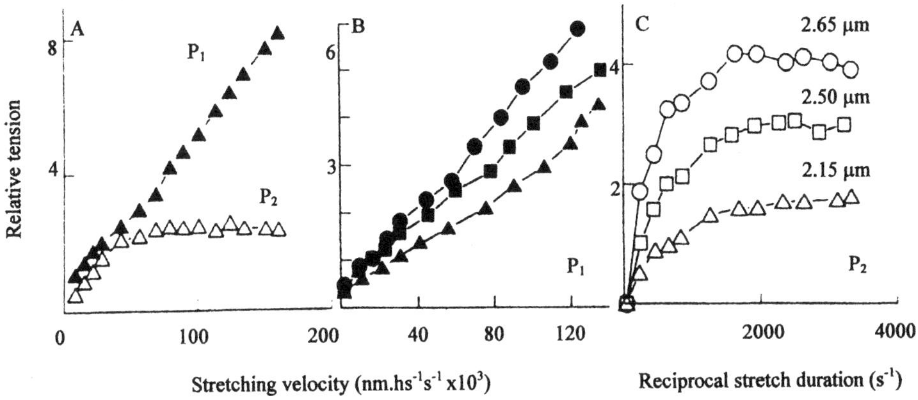

Figure 2. A: P_1 and P_2 responses versus stretching velocity. The extrapolated value of P_1 at zero speed is not zero; this is very likely due to the superposition on the P_1 response of the tension developed by the short range elastic component that at low speeds is higher than the P_1 response. P_2 response is corrected for the resting tension measured 50 ms after the end of the stretch. Stretching amplitude: 39 $nm.hs^{-1}$; sarcomere length 2.13 μm. B,C: P_1 versus stretching velocity and P_2 versus reciprocal stretch duration at different sarcomere lengths. Stretching amplitude, 45 $nm.hs^{-1}$.

Fig. 2A shows the relation between P_1 and P_2 amplitudes and stretching velocity. It can be seen that P_1 response is linearly correlated with the velocity in the whole range of velocities tested. P_2 amplitude instead increases initially with the velocity, but reaches a plateau at velocities around $55x10^3$ $nm.hs^{-1}$. These results indicate that P_1 and P_2 could be the responses of a viscous and a viscoelastic element respectively. The linear relationship between stretching velocity and P_1 amplitude, however, does not exclude that P_1 is the viscoelastic response of a system (a viscosity in series with an elasticity) having a relaxation time much shorter than the minimum stretch duration (230 μs) used in these experiments. Since P_1 probably corresponds to the force response studied in skinned fibers and attributed to the weakly binding bridges, it is interesting to further investigate its nature. In a simple viscoelastic system stretched at constant acceleration (as during the initial part of the stretch) the force response lags the speed by a time corresponding to the relaxation time. In our experiments the delay between force and speed during the acceleration period was never greater than 5 μs, suggesting that P_1 is more likely a viscous response. It should be pointed out that a viscous response is inconsistent with weakly binding bridges kinetics described previously (Brenner et al., 1982; Schoenberg, 1988) since it would correspond to a detachment rate constant of infinite value. Even a viscoelastic system with 5 μs relaxation time would mean a detachment rate constant of $2x10^5s^{-1}$ that is one order of magnitude higher than the maximum value proposed for weakly binding bridge kinetics in skinned fibers.

The dependence of P_2 amplitude from the stretching velocity suggests that P_2 corresponds to the response of a viscoelastic system and it could be due, in principle, to the stretching of weakly binding bridges. However, the relaxation time associated with this phase was about 1 ms which, in terms of cross-bridges kinetics, corresponds to a detachment rate constant of 10^3s^{-1}, one order of magnitude smaller than that reported for skinned fibers (Brenner et al., 1982; Schoenberg, 1988).

The possibility that the force response to stretch could be attributed to weakly binding bridges was further investigated by measuring the stretch responses at various sarcomere length. If P_1 and P_2 were due to cross-bridges, their amplitude should scale in parallel with the degree of the overlap between the myofilaments. However, as shown in Fig. 2B and C the result is exactly the opposite: increasing the sarcomere length from 2.15 to 2.65 μm greatly increased both P_1 and P_2 suggesting strongly that these responses are not arising from cross-bridges.

P_1 response could be due to the viscous resistance to the sliding of myofilaments during the rapid stretching, while P_2 could be related to the presence of the titin filament that joins myosin filaments to Z discs.

3.2. Force Response to Ramp stretch of Relaxed Skinned Fibers

Fig. 3 shows typical force responses to a ramp stretch applied to the same fiber before and after the skinning. The rising phase of the force response of the skinned fibers shows the same two phases, characteristics of intact fibers, but their amplitude is modulated by the ionic strength. Compared to intact fibers, P_1 and P_2 responses are strongly reduced at high ionic strength (170mM), but they increase again at low ionic strength (20mM). The analysis made at both high or low ionic strength (Fig. 4) shows that: P_1 increases linearly with the stretching velocity, as in intact fibers, but the slope of the relationship (the viscosity coefficient) decreases substantially after the skinning at normal ionic strength to rise again at low ionic strength. This suggests that, similarly to intact fibers, P_1 has a viscous nature.

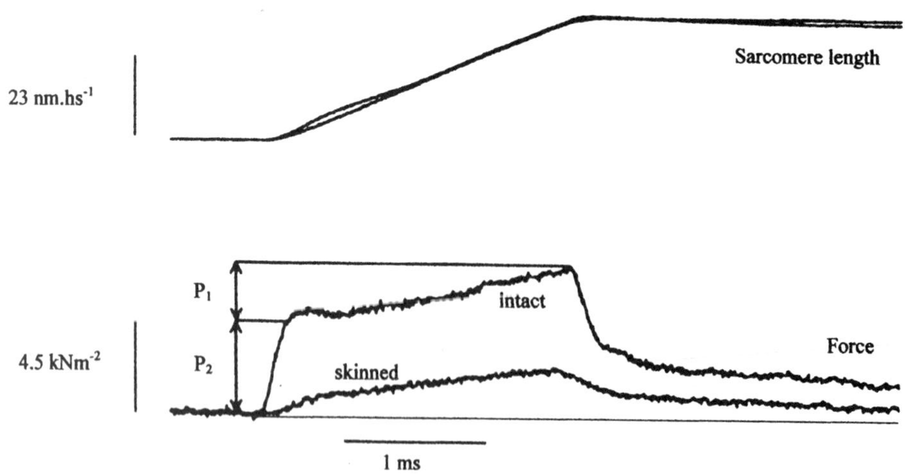

Figure 3. Force response to a ramp stretch of the same relaxed muscle fiber before and after skinning in high ionic strength solution. Initial sarcomere length, 2.15 μm; stretch amplitude, 32 nm.hs^{-1}; stretching velocity 23x10^3 nm.hs^{-1}s^{-1}.

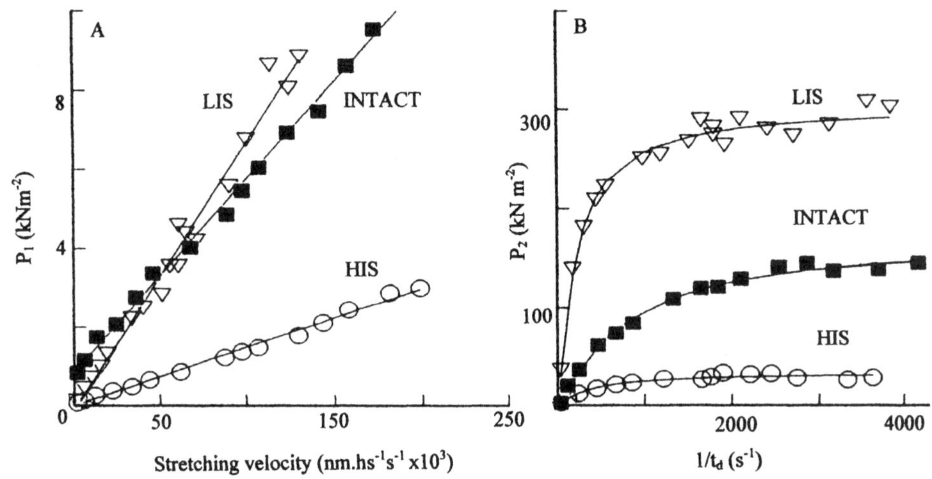

Figure 4. A: P_1 amplitude versus stretching velocity and B: P_2 amplitude versus reciprocal stretch duration in a fiber before and after the skinning at high and low ionic strength.

At both ionic strengths skinning increased the relaxation time of P_2, which however, remained an order of magnitude smaller than that expected from weakly binding bridges kinetics (Bagni et al., 1995). The response at various sarcomere lengths, confirmed the results obtained in intact fibers: both P_1 and P_2 increased substantially with sarcomere length in the range between 2.12-3.27 μm.

Another important result common to all skinned fiber conditions was that, as in intact fibers, the force response to the stretch did not show any sign of "give" (cross-bridge detachment) even when the stretch amplitude was far greater than the elastic range of the cross-bridges (about 12 $nm.hs^{-1}$). This finding too suggests a non cross-bridge nature of the force response to stretch of passive fibers.

All the results reported above strongly support the idea that the force response to stretch of passive fibers either intact or skinned is not arising from weakly binding bridges, but it represents the response of the passive mechanical structures of the fibers.

3.3. Response to Stretch of Activated Intact Fibers

The possible presence of weakly binding bridges in activated fibers was investigated by studying the force responses to ramp stretches applied during the latent period and the early phases of tension development after a single stimulus (Bagni et al., 1994). The comparison between the force response at rest and during the latent period (Fig. 5A) shows that activation increases mainly the P_2 component while P_1 is almost unchanged.

An interesting effect of the activation was that the force transient, at the end of the stretch, did not drop quickly as in passive fibers, but it was followed by a period of some milliseconds during which the tension remained at an approximately constant value. This effect is clearly seen in the subtracted traces (Fig. 5A) reporting the differences between active and passive force responses.

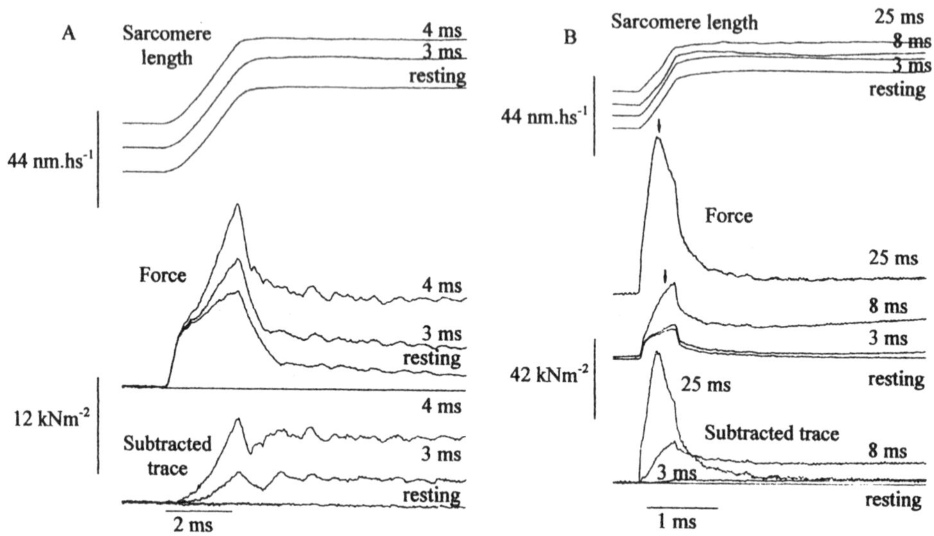

Figure 5. Superposition of force responses to ramp stretches in resting and activated fibers. A: subtracted force traces are the difference between twitch tension and resting response. Sarcomere length, 2.13 μm; stretch amplitude, 38 $nm.hs^{-1}$; stretching velocity, $64x10^3$ $nm.hs^{-1}s^{-1}$. B: Arrows on the force transient mark the point of the tension "give". Sarcomere length, 2.15 μm; stretch amplitude, 28 $nm.hs^{-1}$; stretching velocity, $55x10^3$ $nm.hs^{-1}s^{-1}$. Numbers on the traces represent the delay between the stimulus and the stretch.

The excess of tension following the stretch, is the "static tension". The structure responsible for this tension was assumed to be elastic with a "static stiffness" equal to the static tension divided by the sarcomere length changes. Stretches applied when the fiber was developing active tension (Fig. 5B), showed the presence of tension "give", due to the forced cross-bridge detachment. Nevertheless, static tension was still present at the end of the stretch.

The time course of the static stiffness measured from 0 to 40 ms delay after the stimulus is reported in Fig. 6A. It can be seen that static tension peaks at about 8 ms delay and drops to zero at about 20 ms when the twitch tension is much smaller than the peak value. It is interesting that there is no correlation between the force developed by the fiber and the static tension. The effect of sarcomere length on the time course of static stiffness is shown in the same figure. It is clear that static stiffness increases when the overlap between actin and myosin decreases. This result and the absence of correlation with tension, suggest that static stiffness is not arising from cross-bridges. This hypothesis has been confirmed by experiments made on fibers bathed with Ringer-BDM (5mM), as shown in Fig 6B. Ringer-BDM leaves the response of the resting fiber unaltered, but strongly reduces the force transient of the stimulated fiber. The subtracted traces show that, with respect to normal Ringer, BDM reduces the peak amplitude of the force transient and abolishes the "give", while the static tension level is the same in both conditions.

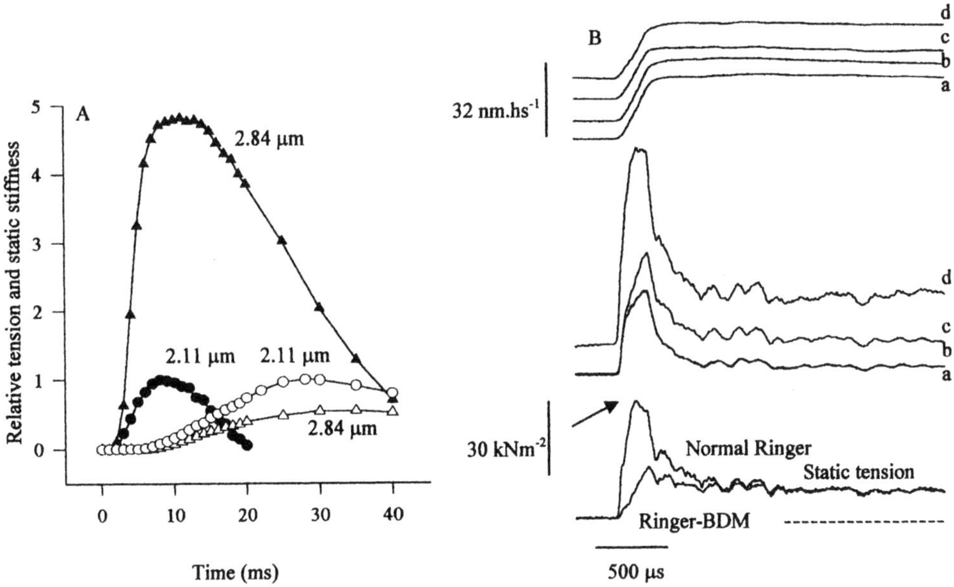

Figure 6. A: Effect of sarcomere length on twitch tension (empty symbols) and static stiffness (filled symbols). Tension and stiffness are expressed relatively to their maximum value at 2.11 µm. Stretch amplitude, 28nm.hs^{-1}; stretching velocity, 45 x10^3 nm.hs^{-1}s^{-1}. B: Stretches applied at rest and 12 ms after one stimulus in normal and Ringer-BDM (5mM). Resting response: (a) BDM, (b) normal Ringer; activated response: (c) BDM, (d) normal Ringer. Sarcomere length, 2.13 µm; stretch amplitude, 26 nm.hs^{-1}; stretching velocity, 85 x10^3 nm.hs^{-1}s^{-1}.

The time course of static stiffness was similar to that of intracellular calcium concentration (Claflin et al., 1994) suggesting that Ca^{2+} could be responsible of the static stiffness increases following activation.

3.4. Static Stiffness During Tetanic Contractions

The study reported above were all made on twitch contraction. To study static stiffness behavior in condition of full activation, ramp stretches (amplitude, 2-40 nm.hs^{-1}; velocity, 2-70x10^3 nm.hs^{-1}s^{-1}) were applied during the rise, the plateau, and the relaxation phase of tetanic contractions. BDM was used throughout all these experiments in order to reduce the tension developed therefore reducing both the possibility of fiber damage and the cross-bridge contribution to the force transient evoked by the stretch, thus isolating the components arising from other structures of the fiber.

Fig. 7A shows the effect of a stretch applied at the tetanus plateau in a fiber bathed in Ringer-BDM (6mM) and illustrates the procedure followed to measure the static tension and the static stiffness. In this particular experiment the static tension is about two times greater than the isometric tension and static stiffness corresponds to about 2% of the total fiber stiffness (S_0) at tetanus plateau in normal Ringer (Bagni et al., 2002).

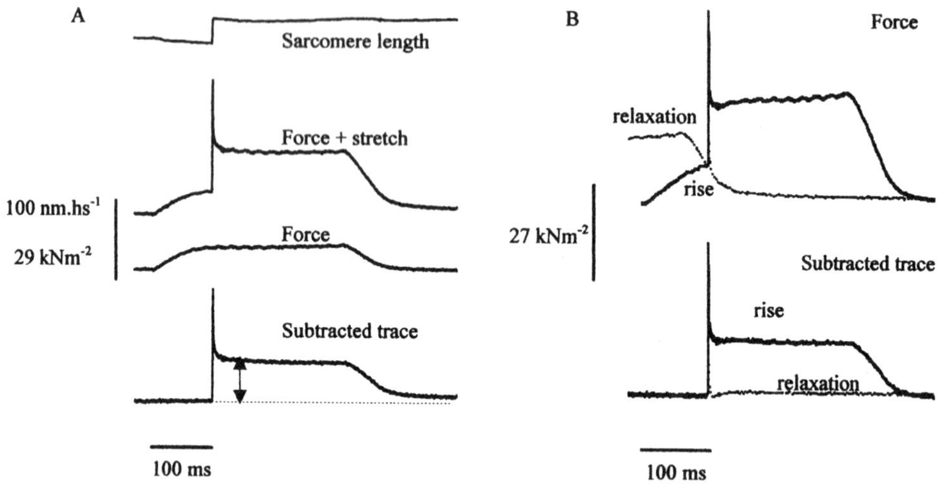

Figure 7. A: Force response to a stretch applied at the tetanus plateau (amplitude 29.7 nm.hs^{-1}, duration 1.2 ms) in a fiber bathed in Ringer-BDM at 6mM concentration. The subtracted trace is obtained by subtracting the passive (not shown), active and plus stretch force responses. The static tension is measured on this trace after the end of the fast transient as indicated by the arrow. The sarcomere length change during the isometric tetanus is not presented. B: Force responses to a stretch (amplitude, 33.7 nm.hs^{-1}; duration 0.6 ms) applied at the same tension level (about 50% of the maximum) during the tetanus rise (thick lines) and the relaxation (dotted thin lines) in two tetanic contractions on the same fiber. Note that the static tension is clearly present on the tetanus rise but it is almost absent on the relaxation.

The experiments on tetanic contractions showed that: 1) confirming previous results on twitches, BDM had a great effect on active tension but did not alter the static tension; 2) the relationship between static tension and stretch amplitude (Fig. 8A) was highly linear, showing that static stiffness is independent of the stretching amplitude; 3) static stiffness was independent of stretching velocity at least in the range tested (Fig. 8B); 4) the comparison of static stiffness and tension time courses shows that stiffness development precedes active tension during tetanus rise (Fig. 8C) and leads active tension during relaxation; 5) static stiffness increased substantially at longer sarcomere length from 2.1 μm up to 2.8 μm, to decrease again at longer length (Fig. 8C).

These properties suggest that static stiffness arises from a passive structure of the fiber in parallel with the cross-bridge behaving like a linear spring. However this is not a simple passive structure since its stiffness increases with activation independently of force and follows a time course similar to intracellular calcium concentration.

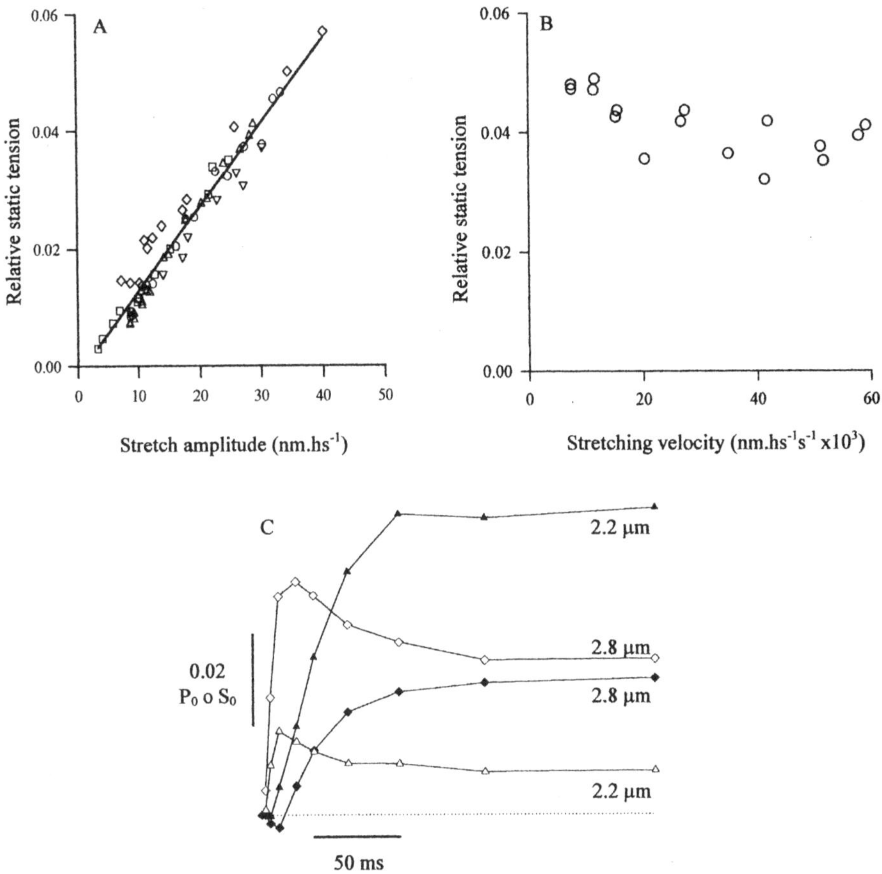

Figure 8. A: Relationships between stretch amplitude (different amplitudes but same time duration, range 2-40 nm.hs^{-1}) and static tension; different symbols refer to 5 different fibers. Static tension is expressed relative to P_0. Stretch duration 0.5 ms. The regression line fitted on the pooled data (r = 0.97) is represented by the following equation: Static tension = $-1.44 \times 10^{-3} + 1.42 \times 10^{-3}$ x Stretch amplitude. The angular coefficient represents the static stiffness. B: Relationships between stretching velocity and static tension in a single fiber. Range tested 2×10^3-70×10^3nm.hs^{-1}s^{-1}; amplitude 31 nm.hs^{-1}C: Time course of isometric tetanic tension (filled symbols) and static stiffness (empty symbols) at different sarcomere lengths.

3.5. Effects of Intracellular Ca^{2+} on Static Stiffness

The possible dependence of static stiffness from Ca^{2+} concentration, suggested by the experiments reported above, was investigated by measuring the force responses of single intact frog muscle fibers under various conditions in which isometric tension was inhibited either by reducing the Ca^{2+} release by the sarcoplasmatic reticulum or directly by inhibiting actomyosin interaction. We analyzed the static stiffness in presence of BDM, Dantrolene (DAN), Deuterium Oxide (D_2O), Methoxyverapamil (D600) and hypertonic solutions. All these agents inhibit twitch tension but they have a different action mechanism: Dantrolene (Helland et al., 1988; Takauji et al., 1975), D_2O (Allen et al., 1984; Sato and Fujino, 1987) and D600 (Eisenberg et al., 1983; Morgan et al., 1997) all depress force development mainly by reducing the Ca^{2+} release, while BDM (Horiuti et al., 1988) and hypertonic solution (Parker and Zhu, 1987) inhibit tension generation mainly by affecting actomyosin interaction so as to reduce cross-bridge formation with little or no effect on calcium release. The comparison of static stiffness in fibers in which similar degrees of force inhibition were obtained with and without inhibition of Ca^{2+} release, allowed us to isolate the effects of intracellular Ca^{2+} on static stiffness.

In Fig. 9 are reported the effects of the various tension inhibitors on twitch tension and static stiffness. In all solutions static stiffness development leaded the tension rise reaching the peak value 8-12 ms after the stimulus when twitch tension was still very low and falling to zero before the twitch peak. Static stiffness was almost unaffected by BDM (Fig. 9A) and hypertonic solution (Fig. 9B) but greatly reduced by D600 (Fig. 9A), Dantrolene (Fig. 9C) and D_2O (Fig. 9D). Twitch tension is slightly reduced by BDM and hypertonic solution but it is strongly decreased by D600, Dantrolene and D2O. Fig. 9C shows that the effects of Dantrolene are reversed by the calcium release potentiatior Nitrate (Sun et al., 1996).

In summary, all the agents tested decreased substantially twitch tension, however only those reducing Ca^{2+} release reduced at the same time the static stiffness. These effects are consistent with previous data showing the similarity between intracellular Ca^{2+} time course and static stiffness time course and consistent with the hypothesis that static stiffness is Ca^{2+} dependent.

4. CONCLUSIONS

In conclusion the results reported in this paper show that the force response of resting intact and skinned frog fibers to fast stretches arises from a viscous and a viscoelastic passive component and is not attributable to attached cross-bridges.

Force responses to fast ramp stretches of various amplitude and velocity, applied during twitch or tetanic contractions on fibers bathed in normal or Ringer-BDM solutions show the presence of a component of the fiber stiffness, the static stiffness, not attributable to cross-bridges. This stiffness was independent of the active tension developed by the fiber, and of stretch amplitude and stretching velocity; it increased with sarcomere length in the range 2.1-2.8 μm to decrease again at longer lengths.

Our results do not provide information about the structure responsible for the static stiffness, however they show that Ca^{2+}, in addition to promote cross-bridge formation, increases the stiffness of some unknown sarcomere structure. This effect could be due to a Ca^{2+} dependent titin-actin interaction (Kellermayer and Granzier, 1996), or to a Ca^{2+} dependent change in titin elasticity (Tatsumi et al., 2001).

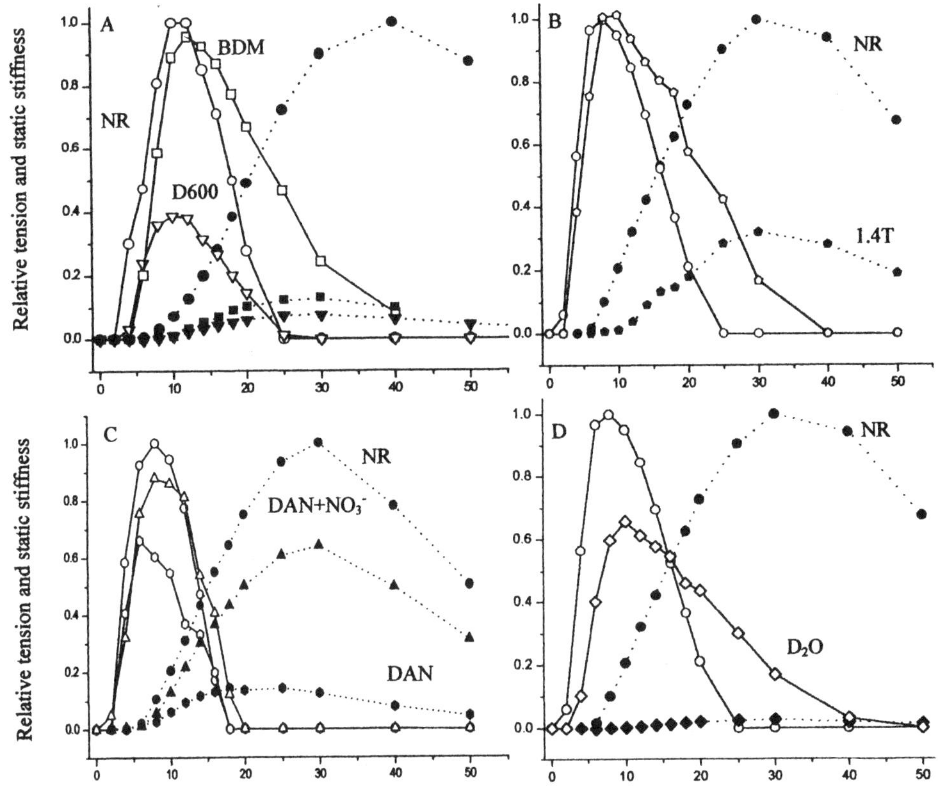

Figure 9. Effects of BDM (2.5mM), D600 (20μM), Dantrolene (6.25μM), Nitrate-Ringer (92mM of NaCl substituted by $NaNO_3$) plus Dantrolene; D_2O Ringer (98% of water substituted with D_2O) and hypertonic solution (1.4 T, obtaining by adding 50mM NaCl to the normal Ringer), on twitch tension and static stiffness time course. Dotted lines and filled symbols represent tension, continuous lines and open symbols represent static stiffness; circles, normal Ringer; squares, BDM; down triangles, D600; pentagons, hypertonic solution, diamonds, D_2O; hexagons, Dantrolene; up triangles, Nitrate-Ringer plus Dantrolene. Tension is greatly reduced by all agents, but static stiffness is inhibited only by D600, Dantrolene and D_2O.

A recent work (Labeit et al., 2003), in which calcium effects on titin properties were studied both on skinned fibers and at molecular level, strongly supports the second possibility. In agreement with our data on intact fibers, it was shown that Ca^{2+} increased the titin based force response to stretch in skinned fibers and decreased the persistence length of the elastic PEVK titin segment.

The calcium dependent increases of static stiffness, especially if the titin hypothesis should be confirmed, could be an important factor in maintaining the sarcomere functionality under condition of sarcomere length instability as in the first phases of activation or at long sarcomere lengths.

6. REFERENCES

Allen, D.G., Blinks, J.R., and Godt, R.E., 1984, Influence of deuterium oxide on calcium transients and myofibrillar responses of frog skeletal muscle, *J. Physiol.* **354**:225-51.

Bagni, M.A., Cecchi, G., Colombini, B., and Colomo, F., 2002, A non-cross-bridge stiffness in activated frog muscle fibres, *Biophys. J.* **82**(6):3118-27.

Bagni, M.A., Cecchi, G., Colomo, F., and Garzella, P., 1992, Are weakly binding bridges present in resting intact muscle fibres? *Biophys. J.* **63**:1412-1415.

Bagni, M.A., Cecchi, G., Colomo, F., and Garzella, P., 1994, Development of stiffness precedes cross-bridge attachment during the early tension rise in single frog muscle fibres, *J. Physiol.* **481**(2):273-278.

Bagni, M.A., Cecchi, G., Colomo, F., and Garzella, P., 1995, Absence of mechanical evidence for attached weakly binding cross-bridges in frog relaxed muscle fibres, *J. Physiol.* **482**(2):391-400.

Bagni, M.A., Colombini, B., Geiger, P., Berlinguer Palmini, R., and Cecchi, G., 2004, A non cross-bridge calcium-dependent stiffness in frog muscle fibers, *Am. J. Physiol. Cell Physiol.* in press.

Brenner, B., Schoenberg, M., Chalovich, J.M., Greene, L.E., and Eisenberg, E., 1982, Evidence for crossbridges attachment in relaxed muscle at low ionic strength, *PNAS* **79**:7288-7391.

Brenner, B., Yu., L.C., and Podolsky, R.J., 1984, X-ray diffraction evidence for cross-bridge formation in relaxed muscle fibres at various ionic strengths. *Biophys. J.* **46**:299-306.

Brenner, B., 1990, Muscle mechanism and biochemical kinetics, in: *Molecular Mechanism in Muscular Contraction*, J.M: Squire, ed., Macmillan Press Ltd, Southampton, UK, pp. 77-149.

Cecchi, G., Griffiths, P.J., and Taylor, S., 1982, Muscular contraction: kinetics of crossbridge attachment studied by high-frequency stiffness measurements, *Science* **217**:70-72.

Chalovich, J.M., Chock, P.B., and Eisenberg, E., 1981, Mechanism of action of troponin-tropomyosin inhibition of actomyosin ATPase activity without inhibition of myosin binding to actin. *J. Biol. Chem.* **256**:575-578.

Claflin, D.R., Morgan, D.L., Stephenson, D.G., and Julian, F.J., 1994, The intracellular Ca^{2+} transient and tension in frog skeletal muscle fibres measured with high temporal resolution,. *J. Physiol.* **475**(2):319-325.

Eisenberg, R.S., McCarthy, R.T., and Milton, R.L., 1983, Paralysis of frog skeletal muscle fibres by the calcium antagonist D-600, *J. Physiol.* **341**:495-505.

Ford, L.E., Huxley, A.F., and Simmons, R.M., 1981, The relation between stiffness and filament overlap in stimulated frog muscle fibres, *J. Physiol.* **311**:219-249.

Ford, L.E., Huxley, A.F., and Simmons, R.M., 1986, Tension transients during the rise of tetanic tension frog muscle fibres, *J. Physiol.* **372**:595-609.

Helland, L.A., Lopez, J.R., Taylor, S.R., Trube, G., and Wanek, L.A., 1988, Effects of calcium "antagonists" on vertebrate skeletal muscle cells, *Ann. N. Y. Acad. Sci.* **522**:259-68.

Horiuti, K., Higuchi, H., Umazume, Y., Konishi, M., Okazaki, O., and Kitohara, S., 1988, Mechanism of action of 2,3-butanedione 2-monoxime on contraction of frog skeletal muscle, *J. Muscle Res. Cell Motil.* **9**:156-164.

Kellermayer, M.S.Z., and Granzier, H.L., 1996, Calcium-dependent inhibition of in vitro thin-filament motility by native titin, *FEBS Lett.* **380**:281-286.

Labeit, D., Watanabe, K., Witt, C., Fujita, H., Wu, Y., Lahmers, S., Funck, T., Labeit, S., and Granzier, H., 2003, Calcium-dependent molecular spring elements in the giant protein titin, *PNAS* **100**(23):13716-13721.

Morgan, D.L., Claflin, D.R., and Julian, F.J., 1997, The relationship between tension and slowly varying intracellular calcium concentration in intact frog skeletal muscle, *J. Physiol.* **500**.1:177-92.

Parker, I., and Zhu, P.H., 1987, Effects of hypertonic solutions on calcium transients in frog twitch muscle fibres, *J. Physiol.* **383**:615-27.

Sato, Y., and Fujino, M., 1987, Inhibition of arsenazo III Ca transient with deuterium oxide in frog twitch fibres at a resting sarcomere length, *Jpn. J. Physiol.* **37**(1):149-53.

Schoenberg, M., 1985, Equilibrium muscle crossbridge behavior: theoretical considerations, *Biophys. J.* **48**:467-475.

Schoenberg, M., 1988, Characterization of the myosin adenosine triphospate (M.ATP) crossbridge in rabbit and frog skeletal muscle fibres. *Biophys. J.* **54**, 135-148.

Sun, Y.B., Lou, F., and Edman, K.A., 1996, The relationship between the intracellular Ca^{2+} transient and the isometric twitch force in frog muscle fibres, *Exp. Physiol.* **81**(5):711-724.

Takauji, M., Takahashi, N., Nagai, T., 1975, Effect of dantrolene sodium on excitation-contraction coupling in frog skeletal muscle, *Jpn. J. Physiol.* **25**(6):747-58.

Tatsumi, R., Maeda, K., Hattori, A., and Takahashi, K., 2001, Calcium binding to an elastic portion of connectin/titin filaments, *J. Muscle Res. Cell Motil.* **22**(2):149-62.

DISCUSSION

Pollack: Your hypotheses are related to titin. One suggestion is to do experiments on a preparation that contains no titin. I would like to invite you to our laboratory, where we are doing experiments on dynamics of one thick and one thin filament, sliding past one another.

Bagni: Thanks for the suggestion and many thanks for your invitation.

Brenner: Have you examined the skinned frog fibers by X-ray diffraction to see whether the skinned fiber equatorials change as ionic strength is lowered?

Bagni: No, but it is a good idea. We will try.

Gonzalez-Serratos: Have you measured the two types of stiffness that you described in your skinned preparation as a function of calcium concentration?

Bagni: We are now doing experiments with skinned frog fibers at different calcium concentrations, but the results are not enough to be presented here.

NATURE'S STRATEGY FOR OPTIMIZING POWER GENERATION IN INSECT FLIGHT MUSCLE

David Maughan[1] and Jim Vigoreaux[2]

1. INTRODUCTION

The most important mechanical function of the actomyosin motor in striated muscles is to generate power, particularly in systems that operate in an oscillatory manner. Since all striated muscles have the same basic motor protein (myosin II), the question arises as to what structural modifications of the sarcomere have evolved to allow oscillatory work and power generated by the motor proteins to not only be amplified, but also to be so well tuned to the operating frequency of the muscle system.

In this chapter we discuss key structural features of the myofilament lattice that have evolved to enhance oscillatory work and power output, using *Drosophila* indirect flight muscle (IFM) as an example. Insect flight muscle is particularly endowed with power-enhancing features since insect flight requires higher power output per gram body weight than other forms of animal locomotion (Marden, 2000; Tregear, 1983). These features include 1) strengthening the weak links of the sarcomere, 2) orienting the head of the myosin molecule for optimum power generation, and 3) modifying the kinetics of the myosin head. We focus on alternative forms of myosin II and proteins associated with the thick filament that serve these functions (**Fig. 1**), particularly those in which mutations have been generated and whose muscle properties have been examined *in vivo* and *in vitro*.

2. PROTEINS AND UNIQUE SEQUENCES THAT MODULATE POWER OUTPUT

Table 1 summarizes thick filament proteins of *Drosophila* IFM in which mutations have been generated and studied by sinusoidal length perturbation analysis (Dickinson et al., 1997; Maughan et al., 1998). The effect of each mutation on the resonance frequency of the flight system (i.e., the wing beat frequency) and muscle power are indicated, as are the primary mechanisms most likely responsible for modifying wing beat frequency (WBF) and muscle power. While WBF and frequency of maximum muscle power output are correlated in native systems (Molloy et al., 1987), this is not always the case in situations where one part of the system has been re-engineered genetically. For example, one mutant may show significant reductions in both WBF and muscle power (RLC $Mlc2^{S66A,S67A}$), while another may show

[1]Molecular Physiology and Biophysics, University of Vermont, Burlington VT 05405 USA
[2]Department of Biology, University of Vermont, Burlington VT 05405 USA

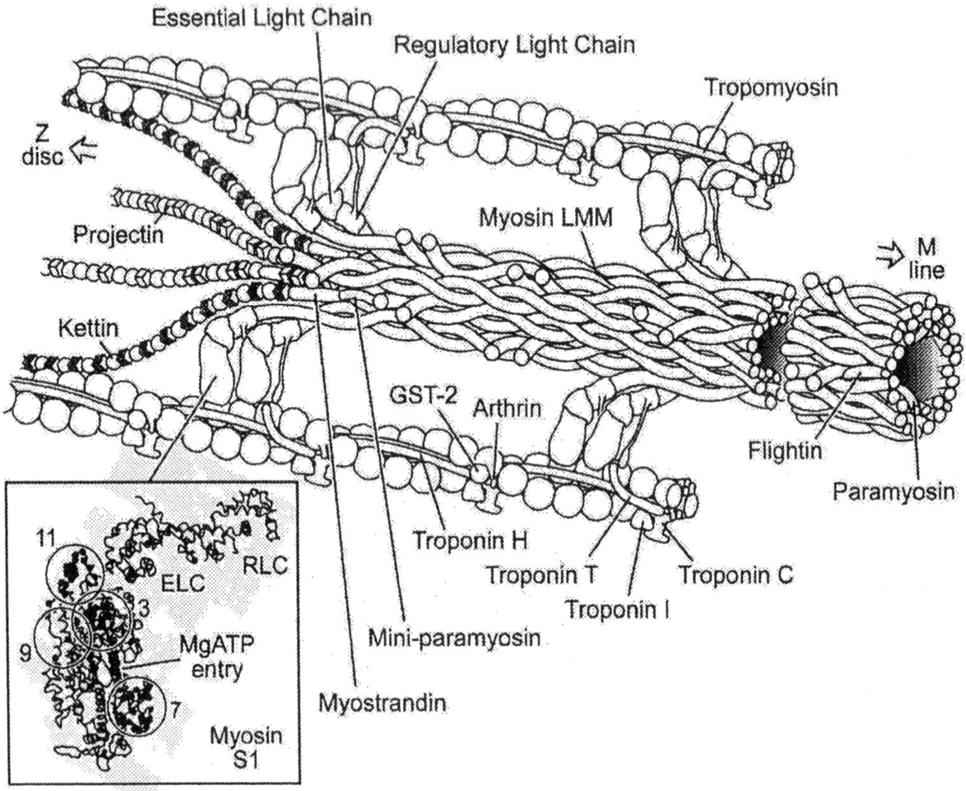

Figure 1. Schematic illustration of a portion of the *Drosophila* myofilament lattice at the A-I junction. Aspects of the lattice (including positions of myostrandin, arthrin, GST-2, and flightin) are highly speculative. Flightin is shown located in a segment of the thick filament further from the end, as determined by immuno-electronmicroscopy (Reedy et al., 2000). Call out: the closely related 3D structure of chicken myosin (Bernstein and Milligan, 1997), where the circles denote the alternatively spliced regions encoded by exons 3, 7, 9 and 11 of the *Drosophila Mhc* gene.

elevated WBF with reduced muscle power (paramyosin pm^{S18A}). The different outcomes reflect different mechanisms that come into play, depending on the protein and site of the mutation and the fly's strategy, voluntary or otherwise, in responding to the challenges of the mutation.

Myosin. Consider first the *myosin heavy chain* mutants, specifically those located in subunit 1 (S1), the motor head. What structural regions of the head determine power output? This question can be addressed in *Drosophila* by taking advantage of the fact that there is only one *Mhc* gene (compared to up to 9 separate MHC genes in mammals). The one gene consists of 19 exons, 5 of which are alternatively spliced (exon 3, 7, 9, 11 and 15: George et al., 1989) and one, exon 18, that is either included or excluded. Four of these exons encode regions of the S1 head which, is the the region of the molecule most likely to determine kinetics (Rayment et al., 1993).Thus the question of optimizing power output can be initially reduced to addressing which regions, or combination of regions, contribute to the differences in myosin isoform functional properties.

Drosophila melanogaster is an exceptionally useful experimental preparation for myosin structure/function studies. The availability of MHC null mutants and powerful molecular genetic tools allow for the transgenic replacement of specific amino acids or regions of the molecule in the flight and jump muscles, often with dramatic results (Bernstein et al., 1993). Whole isoform substitutions produce particularly striking outcomes (Swank et al., 2002a,b; Littlefield et al., 2003. Substituting the embryonic isoform (EMB) for the native indirect flight muscle isoform (IFI) transforms the IFM from a high power-generating muscle that works optimally at high oscillation frequencies to a lower power-generating muscle that works optimally at low oscillation frequencies.

Single region substitutions also produce large changes in fiber kinetics, although the differences are less striking than observed for whole isoform substitutions. The region encoded by exon 11 (Fig. 1, inset) includes the "converter", an important structural link between the catalytic part of the molecule and the light chain-draped lever arm that conveys the torque to the myosin rod. Skinned fiber experiments show that exchanging alternative versions of this region affects power production (*P*) and the frequency of maximum power production (*f*). Expressing the EMB-IC myosin chimera (made by exchanging the IFI converter into EMB myosin) in IFM fibers increased *P* and *f* 2-fold compared to IFM fibers expressing EMB (i.e., *EMB-IC* vs. EMB: Table 1). Conversely, fibers expressing the chimera made by the opposite exchange (replacing the native converter with the EMB converter) reduced *P* by a third and *f* by one half compared to native IFM (i.e., *IFI-EC* vs *IFI*: Table 1). Preliminary experiments suggest other variable regions do not play as great a role in determining myofibril kinetics (Miller et al., 2003; Swank et al., 2002a; Swank et al., 2004), supporting the hypothesis that the converter region is a strategic location.

Light chains. The myosin ***essential*** and ***regulatory light chains*** hug the _-helical backbone of S1, reinforcing and stiffening the lever that extends from the catalytic head piece (Fig. 1). A stiffer lever enhances transmission of torque and power to the myofilaments. While little is known of the essential light chain's role in *Drosophila* outside this function, much is known about the modulatory features of the regulatory light chain (RLC). *Drosophila* RLC has two conserved myosin light chain kinase-dependent phosphorylation sites. Disruption of these sites ($Dmlc2^{S66A,S67A}$: Table 1) markedly alters flight ability by reducing oscillatory power output of the IFM (Tohtong et al., 1995). Sinusoidal length perturbation analysis (Dickinson et al., 1997) combined with *in vivo* X-ray diffraction studies (Irving and Maughan, 2000) established that the reduced power output is due to reduced recruitment of cross-bridges into the work-producing pool rather than to changes in the kinetics of actively cycling cross-bridges. The key observations are the relatively high I_{20}/I_{10} intensity ratio observed in X-ray diffraction patterns from live flies at rest (indicating the close proximity of the myosin heads to the thin filament), the relatively small increase in I_{20}/I_{10} intensity ratio observed during the wing beat (indicating little additional lateral movement of the heads toward the thin filament), and the significant reduction in I_{20}/I_{10} intensity ratio in the phosphorylation site mutant (indicating a shift in myosin heads toward the thick filament backbone). It is tempting to speculate that the phosphorylated RLC maintains the cross-bridges in an extended position (roughly perpendicular to the thick filament axis (AL-Khayat et al., 2003), thereby helping to maintain a critical inter-filament spacing or head position for optimum force and oscillatory work production, as proposed for phosphorylated RLC in vertebrates (Sweeney et al., 1993).

IFM oscillatory power output in *Drosophila* is also enhanced by an N-terminal extension of the RLC that has a counterpart in vertebrate ELC (Bhandari et al., 1986; Fewell et al., 1998; Sweeney, 1995). In flies lacking nearly all 48 amino acids of the N-terminal extension (*Dmlc2*$^{\Delta 2\text{-}46}$: Table 1), skinned IFM near *in vivo* lattice spacing exhibit significantly reduced net oscillatory work output at submaximal calcium activation (Irving et al., 2001), a deficit that could account for an observed flight impairment (Moore et al., 2000). Preliminary nuclear resonance experiments using bacterially-expressed *Drosophila* RLC (referred to in Irving et al., 2001) provide evidence for a N-terminus RLC-actin interaction similar to that reported earlier for ELC (Prince et al., 1981; Trayer et al., 1987), shown by others to be between specific basic residues near the ELC N-terminus and C-terminal residues of the actin monomer (Andreev et al., 1999; Sutoh, 1982). Again, it is likely that, like the phosphorylation of RLC residues 66 and 67, the N-terminal extension helps maintain a critical inter-filament spacing and/or pre-positions the head for optimum force production and oscillatory work output, as proposed for the ELC extension in vertebrate cardiac muscles (Irving et al., 2001; Sweeney, 1995).

Light meromyosin and other myosin-associated proteins. Apart from myosin S1 and the light chains, the rod portion of myosin and other proteins associated with the myosin rod play major roles in determining power output, especially the efficiency with which power is transmitted to the ends of the sarcomere. Because the myofilaments are reinforced by special proteins (or special adaptations of conserved proteins), the filaments are very stiff and optimized for efficient force transmission, yet compliant enough to offer effective elastic energy storageThe insect IFM thick filament is thicker than its vertebrate counterpart, with 4-fold rotational symmetry and 4 double-headed myosins per crown (of which 2 are shown in the plane of Fig. 1) compared with 3 in vertebrates. The primary structural element of the thick filament is ***light meromyosin*** (LMM), i.e., the long coiled-coil segment of the myosin molecule, with periodic charges that electrostatically bind one LMM alongside another. The greater stiffness of insect thick filaments follows simply from the fact, stated in engineering terms, two spring-like viscoelastic elements side-by-side are stiffer than one, especially if they are cross-linked.

Additional stiffening of the thick filament results from annealing adjacent LMMs along their length, just as a set of parallel short springs in series is stiffer than a set of parallel long springs. ***Flightin*** (Fig. 1) may be responsible for annealing, or spot-welding LMMs to one another, by bridging myosin pairs by ionic bonds to form subfilaments, and/or by electrostatic interactions with the S2 hinge to reinforce this particularly flexible part of the myosin rod (Reedy et al., 2000; Vigoreaux et al., 1998). Binding of flightin to LMM and the susceptibility of the S2 hinge to proteolysis in IFM with mutations that prevent flightin accumulation support this view (Kronert et al., 1995; Reedy et al., 2000; Ayer and Vigoreaux, 2003). A flightin null (*fln*^{0}: Table 1) and myosin rod mutant (*Mhc*^{13}: Table 1) wit

Table 1. Thick filament proteins of *Drosophila* IFM in which mutations have been studied by sinusoidal analysis.

Protein	Mutant	Mutation	ΔWBF @22°C (mutant/control)	Δ muscle power (*P*)*	Δ muscle kinetics (*f*)	Δ muscle stiffness (*Y*)*	Reference(s)
Projectin	$bt^D/+$	C-terminal truncation	1.10	n.s.d.	1.44	0.63	Vigoreaux et al., 2000
MHC S1	*EMB*	EMB myosin substitution	none	0.23	0.13	1.53**	Swank et al., 2002b;
	IFI-EC	EMB converter in IFI	n.s.d.	0.68	0.49	1.67**	Littlefield et al., 2003
	EMB-IC	IFI converter in EMB	none	0.52	0.29	1.76**	
MHC rod	Mhc^6	R1559H	N.A.	0.12	0.61	0.76	Henkin et al., 2004
	Mhc^{13}	E1554K	N.A.	0.03	0.38	0.18	Henkin et al., 2004
Paramyosin	pm^{S18A}	S18A	1.07	0.58	n.s.d.	0.69	Hao et al., 2004
	$pm^{S\text{-}A4}$	S9A,S10A,S13A,S18A	n.s.d.	0.72	n.s.d.	0.75	Liu et al. 2002
RLC	$Mlc2^{S66A,S67A}$	S66A,S67A	0.83	0.09	n.s.d.	0.12**	Tohtong et al., 1995
	$Mlc2^{\Delta 2\text{-}46}$	Δ2-46	n.s.d.	0.55	n.s.d.	0.60	Moore et al., 2000
Flightin	$Df(3L)fln^1/+$	deficiency	n.s.d.	n.s.d.	1.78		Vigoreaux et al., 1998
	fln^0/fln^0	null	none	virt. none	0.64	0.39	Henkin et al., 2004

n.s.d. = no significant difference
Power (*P*) and dynamic stiffness (*Y*) measured at the frequency of maximum work output (*f*).
Fractions refer to value of index of mutant divided by that of wild type control at pCa 5.
N.A. = flies ~ _ hr old to avoid time-dependent degradation of the IFM; not capable of flight so soon after eclosion.
* adjusted for any change in number of myofilaments per cross-sectional area of skinned fiber.
** ratio of viscous moduli measured at frequencies of maximum work output (~ *Y* ratio)

low levels of flightin accumulation show marked reductions in passive and dynamic stiffness, commensurate with an almost complete loss of stretch activation and power output. Notably, another myosin rod point mutant with reduced levels of phosphorylated flightin (*Mhc*^{6}: Table 1) exhibits no significant reduction in dynamic elastic modulus, but does show a significant reduction in dynamic viscous modulus that accompanies a sharp drop in oscillatory power output (Henkin et al., 2004). This result raises the possibility that oscillatory power production is modulated by changes in flightin phosphorylation. However, one cannot exclude the possibility that the reduced power output in both *Mhc*^{1} and *Mhc*^{6}, listed as myosin rod mutations in Table 1, is partly due to a structural defect in the rod itself.

Actin and actin-associated proteins. Like the thick filament, various proteins that bind to filamentous actin reinforce the thin filament. It stands to reason that the longitudinal strength and stiffness of the thin filament lattice must be more or less equivalent to that of the thick filament lattice, else the thin filament would represent a weak link in the system. Although the thick filament is the focus of this review, it is important to mention those attributes of the thin filament lattice that allows it to match to the strength and stiffness of the thick filament. First, the ratio of thin to thick filaments is 3:1 in insect flight muscle, so each thin filament does not have to support as much tension as each thick filament. Second, the I-bands of the IFM sarcomere are very narrow (~0.5 _m), representing about 15% of the sarcomere length (~3.2 _m). A short, stiff connection from the ensemble of working cross-bridges to the Z-band (Fig. 1) provides a crucial link for effective power transmission.

Proteins associated with filamentous actin play an ancillary but key role in further stiffening the thin filament. In addition to its calcium-dependent regulatory role, ***tropomyosin*** (TM, Fig. 1) stiffens the thin filament stiffness by continuously running alongside the actin monomers, 'spot welded' to them via electrostatic interactions, within each groove of the double-stranded actin filament (Cammarato et al., 2004). *Drosophila* ***Troponin H*** (TnH, Fig. 1), unlike its *Lethocerus* counterpart that forms part of the troponin complex, is an elongated TM fusion protein that is roughly two-thirds longer than the standard TM, either lies alongside or periodically replaces TM at a molar ratio of ~ 1:1 (Clayton et al., 1998). IFM from mutants that lack the standard tropomyosin are structurally weaker, are less stiff, and produce significantly less power that wild type flies or flies rescued with the wild type standard TM gene (Kreuz et al., 1996; Miller et al., 1993; Molloy et al., 1993). IFM from mutants with reduced expression of TnH are also compromised structurally, producing less power than control flies, but the reduction is not nearly as severe as that of the TM mutant at comparable levels of expression (Kreuz et al., 1996). Ultrastructural analysis of TM and TnH mutants indicate both proteins are required for normal myofibril assembly.

Filamentous proteins of the supertitin family also contribute to the stiffness of the sarcomere (Fig. 1) via their interactions with the myofilaments. ***Projectin*** (Saide, 1981) and ***kettin*** (Kulke et al., 2001) link thick filaments to Z-disc proteins. Both contribute to passive stiffness (Kulke et al., 2001; Vigoreaux et al., 2000). Projectin, like titin, appears to be involved in enhancing stretch-activation (Granzier and Wang, 1993; Vigoreaux et al., 2000) and it is likely kettin is similarly involved (Kulke et al., 2001). Both bind to Z-disc proteins, and both probably share a close association with the thin filament in the I-band region before attaching to the thick filament as has been demonstrated for titin in vertebrate muscle

(Granzier et al., 1997). IFM from flies with a portion (~15%) of the wild type projectin replaced by a mutant C-terminally truncated form (*bent^D/+*: Table 1) show reduced oscillatory work output, consistent with a loss of contact with actin or myosin, resulting in a more compliant link to the power-producing actomyosin cross-bridges. But *bent^D/+* IFM also exhibit enhanced kinetics (an increase in frequency at which maximum work is generated), the net result of which is unchanged power output. Enhanced kinetics, which cannot be simply explained on the basis of increased thin filament compliance (Wang et al., 1999), is probably related to the removal of a putative kinase domain (Weitkamp et al., 1998) within the deleted sequence. The substrate may be RLC (Dickinson et al., 1997), flightin (Vigoreaux and Perry, 1994), troponin T (Domingo et al., 1998), troponin I (Weitkamp et al., 1998), or projectin itself (Maroto et al., 1992). A change in the phosphorylation state of any one of these possible substrates of projectin kinase may affect the contractile properties of the IFM.

4. SUMMARY AND CONCLUSION

Table 1 summarizes the primary mechanisms most likely responsible for modifying wing beat frequency (WBF) and muscle power in the *Drosophila* mutants discussed above. The different outcomes reflect different mechanisms that come into play, depending on the protein and site of the mutation. For example, the reduced muscle power and WBF of the RLC phosphorylation site mutant *Mlc2^{S66A,S67A}* reflect the reduced number of myosin heads available to form working cross-bridges and the concomitant reduction in muscle stiffness. The mixed results of the other mutants are more difficult to explain. For example, while the reduced muscle stiffness of the paramyosin rod mutant *pm$^{S1\&4}$* and the projectin mutant *bent^D/+* may in part reflect mutation-related increases in compliance of the thick filaments (*pm$^{S1\&4}$*) or connecting filaments (*bent^D/+*), the elevated WBF is unexpected because one would expect reduced muscle stiffness to lower WBF rather than raise it[1]. Other aspects of the results are equally baffling. In the case of *pm$^{S1\&4}$*, e.g., myofilament kinetics are enhanced, opposite to what one would predict from reduced myofilament stiffness (Wang et al. 1999), but consistent with a direct effect of the mutation on cross-bridge kinetics. It is tempting to speculate that the fly increases the resonance frequency of its flight system, perhaps even over-compensating, as a mechanism for bringing the optimum frequency of power output of the flight system in line with the optimum frequency of power output of the myofilaments in order to achieve flight. The fly might accomplish this by voluntarily activating flight control muscles that change the stiffness and shape of the thoracic box (Tu and Dickinson, 1996), thereby significantly changing the basal stiffness of the resonance system. This effective

---

[1] The flight system of *D. melanogaster* is essentially a resonance system, whose fundamental frequency, in its simplest formulation, is directly proportional to the square root of the dynamic stiffness of the indirect flight muscles (Hyatt and Maughan, 1994; Molloy et al. 1993). Of course, other factors modify wing beat frequency, such as the length of the wings and their mass, as well as central nervous system recruitment of voluntary (synchronous) direct flight muscles (Lehmann and Gotz, 1996).

strategy would serve to tune flight system kinetics to that of the actomyosin motor for optimum power transmission.

Notably, of the four thick filament mutations listed in Table 1 produce no significant changes in wing beat frequency, three exhibit reduced muscle power, so these flies must make other adjustments to maintain flight competency. These may be additional cases in which the effects of marked changes in cross-bridge kinetics (MHC *IFI-EC*), cross-bridge deployment ($Mlc2^{\Delta 2\text{-}46}$), or sarcomere (thick filament) stiffness ($pm^{S\text{-}A4}$ and $Df(3L)fln^1/+$) are ameliorated by the intervention of direct flight muscles. In summary, it may well be that the fly's general response to mutations that alter one component of the flight system is to alter another in order to maintain optimum transmission of power and flight competency. That is, nature's strategy for optimizing power generation throughout the flight system is probably the same as that at the level of the myofibril: that is, strengthen weak links, orient parts for optimum power production, and modify power train proteins through isoform switches or post-translational modifications to assure all components are in tune with one another.

5. ACKNOWLEDGEMENTS

The authors thank Doug Swank, Mark Miller, Brad Palmer, Josh Henkin, Bill Barnes, Yuan Wang, and Joan Braddock for their help and advice, and Gary Nelson for the artwork. This work was supported by NIH R01 HL68034, R01 AR49425, NSF MCB-0090768 and MCB-0315865.

6. REFERENCES

AL-Khayat H. A., L. Hudson, M. K. Reedy, T. C. Irving, and J. M. Squire. 2003. Myosin head configuration in relaxed insect flight muscle: X-ray modeled resting cross-bridges in a pre-powerstroke state are poised for actin binding. *Biophysical Journal* 85: 1063-79.

Andreev O. A., L. D. Saraswat, S. Lowey, C. Slaughter, and J. Borejdo. 1999. Interaction of the N-terminus of chicken skeletal essential light chain 1 with F-actin. *Biochemistry* 38: 2480-5.

Ayer G. and J. O. Vigoreaux. 2003. Flightin is a myosin rod binding protein. *Cell Biochem.Biophys* 38: 41-54.

Bernstein S. I. and R. A. Milligan. 1997. Fine tuning a molecular motor: the location of alternative domains in the Drosophila myosin head. *J.Mol.Biol.* 271: 1-6.

Bernstein S. I., P. T. O'Donnell, and R. M. Cripps. 1993. Molecular genetic analysis of muscle development, structure, and function in *Drosophila*. *Internat Rev Cytol* 143: 63-152.

Bhandari D. G, B. A. Levine, I. P Trayer, and M. C Yeadon. 1986. 1H-NMR study of mobility and conformational constraints within the proline-rich N-terminal of the LC1 alkali ligth chain of skeletal myosin. Correlation with similar segments in other protein systems. *Eur.J.Biochem.* 160: 349-56.

Cammarato A., V. Hatch, J. Saide, R. Craig, J. C. Sparrow, L. S. Tobacman, and W. Lehman. 2004. Drosophila muscle regulation characterized by electron microscopy and three-dimensional reconstruction of thin filament mutants. *Biophys J.* 86: 1618-24.

Clayton J. D., R. M. Cripps, J. C. Sparrow, and B. Bullard. 1998. Interaction of troponin-H and glutathione S-transferase-2 in the indirect flight muscles of Drosophila melanogaster. *J.Muscle Res.Cell Motil.* 19: 117-27.

Dickinson M. H., C. J. Hyatt, F-O Lehmann, J. R. Moore, M. C. Reedy, A. Simcox, R. Tohtong, J. O. Vigoreaux, H. Yamashita, and D. W. Maughan. 1997. Phosphorylation-dependent power output of transgenic flies: an integrated study. *Biophys.J.* 73: 3122-34.

Domingo A., J. Gonzalez-Jurado, M. Maroto, C. Diaz, J. Vinos, C. Carrasco, M. Cervera, and R. Marco. 1998. Troponin-T is a calcium-binding protein in insect muscle: in vivo phosphorylation, muscle-specific isoforms and developmental profile in Drosophila melanogaster. *J.Muscle Res.Cell Motil.* 19: 393-403.

Fewell J. G., T. E. Hewett, A. Sanbe, R. Klevitsky, E Hayes, D Warshaw, D. Maughan, and J. Robbins. 1998. Functional significance of cardiac myosin essential light chain isoform switching in transgenic mice. *J.Clin.Invest.* 101: 2630-9.

George E. L., M. B. Ober, and C. P. Emerson. 1989. Functional domains of the *Drosophila melanogaster* muscle myosin heavy-chain gene are encoded by alternatively spliced exons. *Mol.Cell.Biol.* 9: 2957-74.

Granzier H., M. Kellermayer, M. Helmes, and K. Trombitas. 1997. Titin elasticity and mechanism of passive force development in rat cardiac myocytes probed by thin-filament extraction. *Biophys J.* 73: 2043-53.

Granzier H. L. M. and K. Wang. 1993. Interplay between passive tension and strong and weak binding cross-bridges in insect indirect flight muscle: A functional dissection by gelsolin-mediated thin filament removal. *J.Gen.Physiol.* 101: 235-70.

Hao Y., M. S. Miller, D. M. Swank, H. Liu, S. I. Bernstein, D. W. Maughan, and G. H. Pollack. 2004. Mutation of paramyosin phosphorylation sites affects the passive stiffness of Drosophila indirect flight muscle. *Biophys J.* 86: 184a

Henkin J. A., D. W. Maughan, and J. O. Vigoreaux. 2004. Mutations that affect flightin expression in Drosophila alter the viscoelastic properties of flight muscle fibers. *Am.J.Physiol Cell Physiol* 286: C65-C72

Hyatt C. J. and D. W. Maughan. 1994. Fourier analysis of wing beat signals: Assessing the effects of genetic alterations of flight muscle structure in *Diptera*. *Biophys.J.* 67: 1149-54.

Irving T., S. Bhattacharya, I. Tesic, J. Moore, G. Farman, A. Simcox, J. Vigoreaux, and D. Maughan. 2001. Changes in myofibrillar structure and function produced by N-terminal deletion of the regulatory light chain in Drosophila. *J.Muscle Res.Cell Motil.* 22: 675-83.

Irving T. C. and D. W. Maughan. 2000. In vivo X-ray diffraction of indirect flight muscle from Drosophila melanogaster. *Biophys J.* 78: 2511-5.

Kreuz A. J., A. Simcox, and D. Maughan. 1996. Alterations in flight muscle ultrastructure and function in *Drosophila* tropomyosin mutants. *J.Cell Biol.* 135: 673-87.

Kronert W. A., P. T. O'Donnell, A. Fieck, A. Lawn, J. O. Vigoreaux, J. C. Sparrow, and S. I. Bernstein. 1995. Defects in the Drosophila myosin rod permit sarcomere assembly but cause flight muscle degeneration. *J.Mol.Biol.* 249: 111-25.

Kulke M., C. Neagoe, B. Kolmerer, A. Minajeva, H. Hinssen, B. Bullard, and W. A. Linke. 2001. Kettin, a major source of myofibrillar stiffness in Drosophila indirect flight muscle. *J.Cell Biol.* 154: 1045-57.

Lehmann F. O. and K. G. Gotz. 1996. Activation phase ensures kinematic efficacy in flight-steering muscles of Drosophila melanogaster. *J.Comp Physiol [A]* 179: 311-22.

Littlefield K. P., D. M. Swank, B. M. Sanchez, A. F. Knowles, D. M. Warshaw, and S. I. Bernstein. 2003. The converter domain modulates kinetic properties of Drosophila myosin. *Am.J.Physiol Cell Physiol* 284: C1031-C1038

Liu H, Swank, D. M., Miller, A., Mardahl-Dumesnil M, Sweeney ST, O'Kane CJ, Maughan DW, and Bernstein SI. Mol.Biol.Cel 13S:320a. 2002.

Marden J. H. 2000. Variability in the size, composition, and function of insect flight muscles. *Annu.Rev.Physiol* 62: 157-78.

Maroto M., J. Vinos, R. Marco, and M. Cervera. 1992. Autophosphorylating protein kinase activity in titin-like arthropod projectin. *J.Mol.Biol.* 224: 287-91.

Maughan D., J. Moore, J. Vigoreaux, E. Blanchard, and L. Mulieri. 1998. Work production and work absorption in muscle strips from vertebrate cardiac and insect flight muscle fibers. In *"Mechanism of work production and work absorption in muscle"*; edited by Sugi and Pollack, 471-89 (Plenum Press, New York).

Maughan D. W. and R. E. Godt. 2001. Protein osmotic pressure and the state of water in frog myoplasm. *Biophys J.* 80: 435-42.

Miller B. M., M. Nyitrai, S. I. Bernstein, and M. A. Geeves. 2003. Kinetic analysis of Drosophila muscle myosin isoforms suggests a novel mode of mechanochemical coupling. *J.Biol.Chem.* 278: 50293-300.

Miller R. C., R. Schaaf, D. W. Maughan, and T. R. Tansey. 1993. A non-flight muscle isoform of *Drosophila* tropomyosin rescues an indirect flight muscle tropomyosin mutant. *J.Mus.Res.Cell Motil.* 11: 203-15.

Molloy J., A. Kreuz, R. Miller, T. Tansey, and D. Maughan. 1993. Effects of tropomyosin deficiency in flight muscle of Drosophila melanogaster. In *"Mechanism of Myofilament Sliding in Muscle Contraction"* Edited by Sugi and Pollack, 165-72 (Plenum Press, New York): Plenum Press. 165-72.

Molloy J. E., V. Kyrtatas, J. C. Sparrow, and D. C. S. White. 1987. Kinetics of flight muscles from insects with different wing beat frequencies. *Nature* 328: 449-51.

Moore J. R., M. H. Dickinson, J. O. Vigoreaux, and D. Maughan. 2000. The effect of removing the N-temrinal extension of the Drosophila myosin regulatory light chain upon flight ability and the contractile dynamics of indirect flight muscle. *Biophys J.* 78: 1431-40.

Prince H. P, H. R. Trayer, G. D. Henry, I. P. Trayer, D. C. Dalgarno, B. A. Levine, P. D. Cary, and C. Turner. 1981. Proton nuclear-magnetic-resonance spectroscopy of myosin subfragment 1 isoenzymes. *Eur.J.Biochem.* 121: 213-9.

Rayment I., H. M. Holden, M. Whittaker, C. B. Yohn, M. Lorenz, K. C. Holmes, and R. A. Milligan. 1993a. Structure of the actin-myosin complex and its implications for muscle contraction. *Science* 261: 58-65.

Rayment I., W. R. Rypniewski, K. Schmidt-Base, R. Smith, D. R. Tomchick, M. M. Benning, D. A. Winkelmann, G. Wesenberg, and H. M. Holden. 1993b. Three-dimensional structure of myosin subfragment-1:A molecular motor. *Science* 261: 50-8.

Reedy M. C., B. Bullard, and J. O. Vigoreaux. 2000. Flightin is essential for thick filament assembly and sarcomere stability in Drosophila flight muscles. *J.Cell Biol.* 151: 1483-500.

Saide J. D. 1981. Identification of a connecting filament protein in insect fibrillar flight muscle. *J.Mol.Biol.* 153: 661-79.

Sutoh K. 1982. Identification of myosin binding sites on the actin sequence. *Am.Chem.Soc.* 21: 3654-61.

Swank D. M., S. I. Bernstein, and D. W. Maughan. 2002a. Alternative N-terminal regions of Drosophila myosin heavy chain tune crossbridge kinetics for optimal muscle power output. *Biophys J.* 82: 191a

Swank D. M., A. F. Knowles, J. A. Suggs, F. Sarsoza, A. Lee, D. W. Maughan, and S. I. Bernstein. 2002b. The myosin converter domain modulates muscle performance. *Nat.Cell Biol.* 4: 312-6.

Swank D. M., W. A. Kronert, D. W. Maughan, and S. I. Bernstein. 2004. Alternative versions of the myosin relay loop influence Drosophila muscle kinetics. *Biophys J.* 86: 656a

Sweeney H. L. 1995. Function of the N terminus of the myosin essential light chain of vertebrate striated muscle. *Biophys.J.* 68: 112s-19s.

Sweeney H. L., B. F. Bowman, and J. T. Stull. 1993. Myosin light chain phosphorylation in vertebrate striated muscle: regulation and function. *Am.J.Physiol.* 264: C1085-C1095

Takahashi S., H. Takano-Ohmuro, and K. Maruyama. 1990. Regulation of *Drosophila* myosin ATPase activity by phosphorylation of myosin light chains. I. Wild-type fly. *Comp.Biochem.Physiol.* 95B: 179-81.

Tohtong R., M Yamashita, M. Graham, J. Haeberle, A. Simcox, and D. Maughan. 1995b. Impairment of flight ability and flight muscle function caused by mutations of phosphorylation sites of myosin regulatory light chain in Drosophila. *Nature* 374: 650-3.

Trayer I. P., H. R. Trayer, and B. A. Levine. 1987. Evidence that the N-terminal region of A1-light chain of myosin interacts directly with the C-terminal region of actin: A proton magnetic resonance study. *Eur.J.Biochem.* 164: 259-66.

Tregear R. T. 1983. Physiology of insect flight muscle. *Handbook Physiol.* 10: 487-506.

Tu M. S. and M. H. Dickinson. 1996. The control of wing kinematics by two steering muscles of the blowfly (Calliphora vicina). *J.Comp Physiol [A]* 178: 813-30.

Vigoreaux J. O., C. Hernandez, J. Moore, G. Ayer, and D. W. Maughan. 1998. A genetic deficiency that spans the flightin gene of Drosophila melanogaster affects the ultrastructure and function of the flight muscles. *J.Exp.Biol.* 201: 2033-44.

Vigoreaux J. O., J. R. Moore, and D. W. Maughan. 2000. Role of the elastic protein projectin in stretch activation and work output of Drosophila flight muscles. *Adv.Exp.Med.Biol.* 481: 237-47.

Vigoreaux J. O. and L. M. Perry. 1994. Multiple isoelectric variants of flightin in Drosophila stretch-activated muscles are generated by temporally regulated phosphorylations. *J.Mus.Res.and Cell Motility* 15: 607-16.

Wang G., W. Ding, and M. Kawai. 1999. Does thin filament compliance diminish the cross-bridge kinetics? A study in rabbit psoas fibers. *Biophys.J.* 76: 978-84.

Weitkamp B., K. Jurk, and G. Beinbrech. 1998. Projectin-thin filament interactions and modulation of the sensitivity of the actomyosin ATPase to calcium by projectin. *The J. Biol. Chem.* 273: 19802-8.

DISCUSSION

Cecchi: What is the Nature's strategy for optimizing myosin power output in insect flight muscle?

Maughan: It is an interesting common feature that thick filament proteins, paramyosin and flightin have four phosphorylation sites, and C-protein has three or four phosphorylation sites, myosin light chain has at least two phosphorylation sites, and light chain extension has four or more phosphorylation sites. You can change the stiffness of sarcomeres quite dramatically by changing the phosphorylation by glueing one protein to another.

III. MUSCLE ENERGETICS

FROM CROSSBRIDGES TO METABOLISM: SYSTEM BIOLOGY FOR ENERGETICS

Martin J. Kushmerick*

INTRODUCTION

AV Hill's studies of muscle mechanics and heat production defined muscle energetics. Myothermal analysis relied on the fact that the total energy change between two states of a muscle is equal to the sum of the heat output and the work done. To interpret this application of the first law of thermodynamics as chemical changes in muscle one must identify the metabolic reactions, know their molar enthalpies, quantify their extents of reaction, and of course measure the heat produced and the work done (32). ATP splitting was suspected as the chemical driving force for these interactions that produced mechanical and thermal outputs, but changes in ATP content could not be measured directly in living muscle. Earlier Lundsgaard showed decreases in phosphorylcreatine (PCr) did occur in muscle but that PCr was not a substrate for myosin. Biochemical and physiological information appeared to be in conflict. By inhibiting creatine kinase, Davies's lab did show directly a splitting of ATP into ADP and inorganic phosphate in a single contraction of frog muscle (3). Understanding of muscle rapidly evolved with analyses of heat, work, chemical changes, and economy and efficiency brought to fruition by a number of laboratories within the context originally defined by Hill (reviewed in (33)).

The first goal of this paper is to define a set of biochemical rules that are equivalent to those of thermodynamics in the sense of accounting for muscle energetics. These rules are not a priori predictable from first principles of thermodynamics because the mechanisms involved follow chemical principles; the pathways and reactions used were selected by evolution. Our analysis reveals biochemical-based rules and integrates them quantitatively into a comprehensive model useful for basic and translational research. This work forms an outline for a "system biology for energetics" meaning that a synthesis of mechanisms from the molecular to tissue and organ is feasible. This synthesis is quantitatively given by a set of rules and equations. The burden of this

* Martin J. Kushmerick, University of Washington, Seattle, Washington, 98195.

approach is to quantify the rate and extent of the critical reactions, measure their stoichiometry, quantify their fluxes and describe the mechanisms quantitatively by kinetic equations. We will see that the central reaction of the crossbridge, chemomechanical splitting of ATP, is necessarily and mechanistically connected to the regulation of metabolic pathways generating ATP; this regulation in turn influences crossbridge function through the operation of metabolic pathways, which alter the concentrations of ATP, ADP, Pi and pH and thereby influence crossbridge force and kinetics. These interacting mechanisms are central features of the system biology approach outlined here.

The evolution of these ideas began with the work of Lipmann (21). He used the phrase "high energy phosphate bond" to point out that, with all the intricacy of metabolic pathways, there are very few molecules involved in energy transducing mechanisms, the most significant being ATP. From our vantage point of energy balance, metabolic pathways are biochemical structures whose regulation controls the ATP synthesis. Energy transductions in muscle are carried out by molecular electro-chemical and chemo-osmotic machines and chemo-mechanical motors schematically shown as modules connected through a common intermediate ATP (Fig. 1). These modules synthesize or dissipate ATP, the cell's source of chemical potential energy, in a way that couples chemical energy to the metabolic, electrical, osmotic and mechanical work. Although a small amount of ATP is stored in the cell (on the order of 6-8 mM), ATP must be produced by metabolism in a steady state of work performance. The reason is that typical rates of ATP utilization in active muscle (on the order of 0.5 mM per sec) would rapidly exhaust the muscle's source of chemical potential energy in ~ 10 sec. We know that steady states of metabolic activity are possible during exercise over many minutes to hours. The question is what are the rules and mechanisms that match supply to demand?

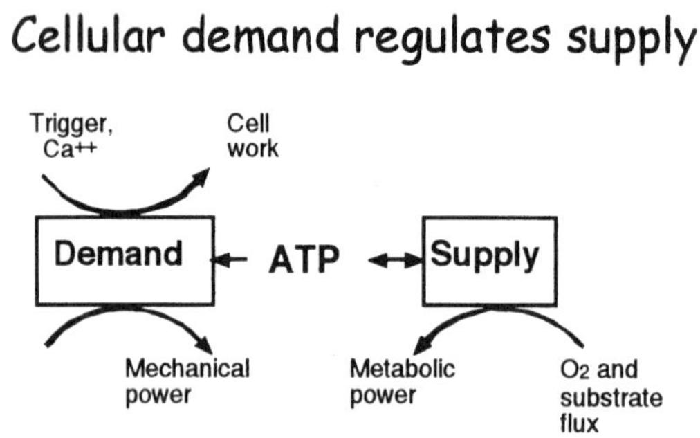

Figure 1: Diagram of the two major components in energy balance in muscle: demand of ATP utilization and supply by its synthesis. Both modules are coupled by functions of ATP.

My second goal here is to demonstrate methods we developed to reveal the mutual interconnections between energy metabolism and crossbridge function. Instead of studying the details of myothermal balances (which remain a valuable experimental approach with high time resolution and accuracy), we bring analyses of chemical fluxes coupled to mechanical output and the regulation of metabolic ATP synthesis to the forefront. I envision an integration of mechanisms from the kinetics of ATP hydrolysis in a cycle of attachment and detachment of crossbridges to the regulation of mitochondrial oxidative phosphorylation and cytoplasmic glycogenolysis in full biochemical detail.

To build this integrated view we need to begin with the essential reactions that are involved and focus on their stoichoimetric relations. Understanding of the implications of these reactions lead us to novel methods for measuring critical chemical fluxes. These reactions produce or take up Pi, ADP and H^+ by stoichiometric rules of chemistry. It is important to recognize that these molecules serve to modulate crossbridge activity and kinetics as well as control metabolic fluxes. Pi affects kinetics of strong attachment transitions. H^+ effects are distributed, and may be modeled best as influencing the strength of myosin to actin interaction because the major part of the binding site energy is ionic. Hence, in the muscle, cell crossbridge function cannot be divorced from intermediary metabolic pathways involved in the regeneration of ATP as depicted in Figure 1.

CONTRIBUTIONS TO ENERGETICS BY 31P NMR SPECTROSCOPY

Much progress toward the goal of a non-invasive measure of muscle cell energetics has been achieved by applications of ^{31}P magnetic resonance spectroscopy (for reviews see (23, 28)). Typical results are displayed in Fig 2 which are taken from our work on human forearm muscle (1). With the onset of mechanical activity in human and other animal muscles, PCr content decreases (down arrow on PCr peaks), and the content of inorganic phosphate (Pi) increases (up arrow on Pi peaks). During aerobic recovery from moderate activity, there can be an apparently simple exponential time course of re-synthesis of PCr and decrease in Pi with a return of normal intracellular pH (25). These observations are consistent with feedback control of cellular respiration by ADP concentration. Thus measurements of PCr decline and re-synthesis can provide a means for characterizing the components of the muscle energy balance to define directly the functional state of the muscle. These studies forced a revision of metabolic control by these metabolites because:

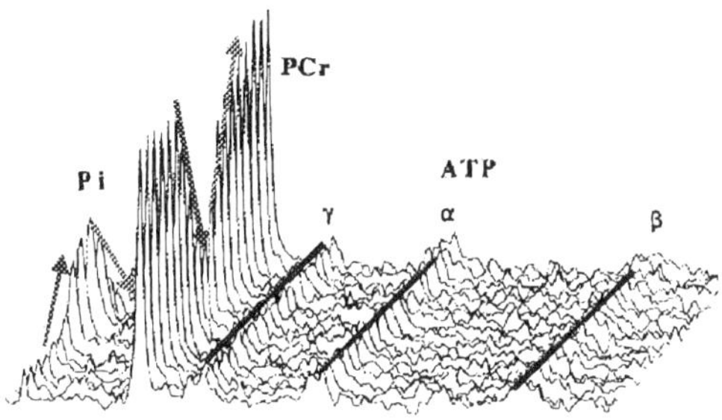

Figure 2: Plot of every 10th spectrum obtained serially during a ^{31}P NMR experiment. The peak identities from the left are: Pi, PCr, and the three peaks of ATP. Down and up arrows indicate changes measured during muscle activity.

- the concentration of Pi is on the order of 1-3 mM, not 10 mM or higher, as defined by prior analyses of muscle extracts,
- the concentration ADP is on the order of 0.02 mM, not the 0.4 to 0.6 mM as measured in muscle extracts that include a large amount of ADP bound to actin and other proteins, and
- the range of pH can be as large as a full pH unit in the acid direction from a normal pH of ~ 7.

We defined an experimental approach for separating ATP-utilization from ATP-synthesis by a quantitative metabolic stress test in normal volunteers (1, 2). This test used a comparison between normal and ischemic contraction and common aerobic recovery protocol. This approach allowed separation and quantification of the ATP-utilization phase from ATP re-synthesis in the human muscles. In the ischemic protocol the initial rate of PCr decline equals the ATPase flux for that experiment. Release of the pressure cuff allowed aerobic recovery by restoring blood flow and oxygen delivery with initiation of mitochondrial oxidative phosphorylation. The time course of PCr resynthesis is accurately fitted by a monoexponential curve which describes the capacity for mitochondrial ATP synthesis by oxidative phosphorylation (24). PCr content returns to the resting, pre-stimulation level. Recovery kinetics of PCr following exercise are in accord with a simple model of respiratory control (4).

RULES FOR STOICHIOMETRY OF METABOLIC REACTIONS

Here we consider some details of the chemistry of relevant reactions in energetics; full details are available (17). The first kind of stoichiometric relationships are those defined by the laws of chemical combination and mass conservation. The fact that they are found in muscle is the product of evolution rather than deductive reasoning from first principles. ATP splitting is tightly coupled by crossbridge motor to perform work; other ATPases are similarly coupled to electrical and osmotic work of ion transport:

ATP hydrolysis: $ATP = ADP + Pi + \alpha H^+ + work$ Eqn 1

The large Gibbs energy for this reaction is due largely to proton binding under intracellular physiological conditions (10). The value of the stoichiometric coefficient α is 0.74 at pH 7.0.

The second reaction of importance is that catalyzed by creatine kinase:

$PCr + ADP + \beta H^+ = ATP + Cr$ Eqn 2

Note the term βH^+ is on the left which means that protons are taken up as the reaction advances in the direction written. The value of β at pH 7.0 is 0.93 in the cytoplasm. The absolute magnitude of β decreases significantly at lower pH values such as occurs in heavy muscular exercise or tissue hypoxia. Of course both reactions occur in the cell, so that their combined action is given by their summation, which is known as the Lohmann reaction:

$PCr = Cr + Pi + \gamma H^+$ Eqn 3

The algebraic sum $\alpha + \beta$ defines the proton stoichiometric coefficient γ. The value of γ at pH 7.0 is -0.19; this means that net PCr breakdown throughout the course of contraction causes the cytoplasm to become alkaline. The reverse occurs in ATP resynthesis. Notice the stoichiometric coefficients (Greek symbols) for H^+ are not unity nor are they constant; they depend on the concentration of cations and pH of the cytoplasm (17).

APPLICATIONS OF STOICHIOMETRY FOR MEASUREMENT OF PATHWAY FLUX

These stoichiometric rules allow the time course of chemical changes to be made in terms of specific components of energy balance and can extend the interpretation. The first example is the creatine kinase reaction. Because it is coupling to ATPase by the Lohmann reaction, these stoichiometric relationships establish the validity of the interpretation of PCr changes as ATPase and ATP synthesis fluxes. A second example is the ability to measure the intracellular buffer capacity (5, 6). During the first several seconds of the abrupt onset of exercise the intracellar pH rises from 7.0 to ~ 7.1. During the initial ~ 20 sec at the onset of contractile activity, these stoichiometric relationships predict an alkalinization, provided that the only reactions occurring are those given in Eqn 1 – 3 as depicted schematically in Fig. 3. The upper curve of the left panel shows the predictions for the Lohmann reaction with alkalinization occurring until the onset of recovery. The lower curve of the left panel depicts the observed pH change. The difference indicated by the vertical lines represents the contribution to glycolysis. We can use these stoichiometric relationships to assess pH buffer capacity by the measurements depicted in the right panel. The decrease in PCr while pH goes alkaline are defined by the dashed lines in the figure. Because the only net reaction occurring at this time is PCr breakdown to Pi and Cr at a significantly higher rate than at rest, the stoichiometry of this reaction results in uptake of H^+. The buffer capacity, defined as the moles of H^+ produced per unit pH change, is measured from these relationships; standard units are

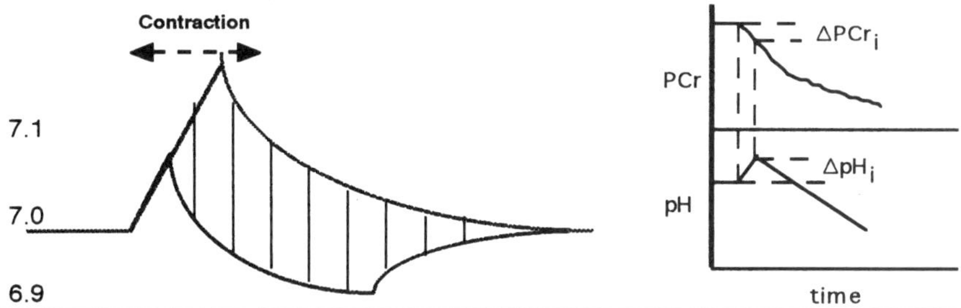

Figure 3: Diagram for method of measuring intracellular buffer capacity. The left panel shows simulations of expected pH calculated from the stoichiometric relationships in text; in the lower curve the measured pH is depicted and the vertical bars represent the difference. The right panel depicts the measurement of initial change in PCr and pH used to define intracellular buffer capacity; see text.

Slykes, mM H+ added per unit pH change; it is measured by the following equation:

Non-Pi buffer capacity = $\gamma \bullet \Delta PCr_{initial}/\Delta pH_{initial}$ Eqn 4

The total buffer capacity is the sum of the non-Pi buffer capacity and that due to the changing Pi concentration. This strategy was used to measure glycogenolytic flux and to analyze the turning on and off of glycolysis and glycogenolysis (7, 8). This method uses the stoichiometric coefficients as described above and full details are published (5, 6). Validation of this method was shown by direct measurements of lactate produced in muscle (14).

RULES FROM MASS AND ENERGY BALANCE

Energy balance can be described by a few simple rules on conservation of energy expressed in terms of biochemical changes observed by ^{31}P NMR spectroscopy and application of these stoichiometric rules:

1. ATP provides the energy for all forms of muscle work.

This rule is evident from the prior discussion and Lipmann's insight into the central role of ATP.

2. Creatine kinase functions as a biochemical capacitor.

A biochemical capacitor is a generalization of a buffer reaction analogous to electronic circuits where capacitors store energy. The creatine kinase reaction is a biochemical capacitor for ATP and ADP and, as such, this reaction is centrally important for muscle energetics as already described (19, 26, 31, 34). The evidence shows the CK reaction operates near equilibrium in muscle cells (15, 22, 29). The function of a biochemical capacitor is to buffer ATP and ADP concentrations against larger changes in concentration that would otherwise occur due to bursts of ATPase or ATP synthesis fluxes. In so far as ADP is a metabolic control signal, buffering also attenuates the magnitude of the signal and slows the response of the system. Creatine kinase does depart transiently from equilibrium at the onset of contraction but not more than ~ 25% from its equilibrium value; the free energy cost of this is small, ~RTln(1.25).

3. The sum of the coupled ATPases sets the demand side of the balance and defines energetic states.

The major extent of cellular ATPases occur during the mechanical output. The sum of all of these defines the energy demand side of energy balance.

4. The products of ATPases provide control signals for energy balance.

ADP and Pi are products of ATPases with ADP being strongly buffered as described above. Both molecules are substrates for mitochondrial oxidative phosphorylation with ADP thought to be the primary controller as discussed above. The results obtained from animal and human skeletal muscle are quantitatively consistent with an ADP feedback model. Thus ATPase signals its own resynthesis and this regulation is central to the integral coupling of crossbridge function to metabolic generation of ATP.

5. The products of ATPases and ATP synthesis modulate crossbridge function.

Concentrations of ATP, Pi, ADP and pH all influence crossbridge function. As indicated above major changes in concentration occur in Pi, typically from ~ 2 to 25 mM in strenuous activity. pH changes of 0.5 unit or more during exercise (3-fold in concentration of H+) are common. Changes in ADP occur from ~ 20 to 200 μM. These variations in metabolite concentrations are sufficient to decrease force by altering the population of the AMADPPi and AMADP states of the crossbridge (11, 12, 27). Moreover differences in sensitivity of muscle phenotype to pH are known (9).

QUANTITATIVE MODEL FOR ENERGY BALANCE

It is straightforward to translate the stoichiometric principles and the rules for energy balance into a simple model of muscle bioenergetics (16, 18). Such a scheme is given in Fig 4 that gives details on the constituent mechanisms involved grouped as interacting modules; fig. 4 is a more detailed view of fig 1. The left lobe of this diagram represents the ATPase, depicting the predominant crossbridge ATPase. Activity begins by neural signals and transient Ca^{2+} increase. The right lobe represents the biochemical capacitors, vertical lobes represent the ATP synthesizing modules: oxidative phosphorylation at the top and glycolysis and glycogenolysis at the bottom. All of the molecules in the central ring are critical for these interactions as reactants, products and signals.

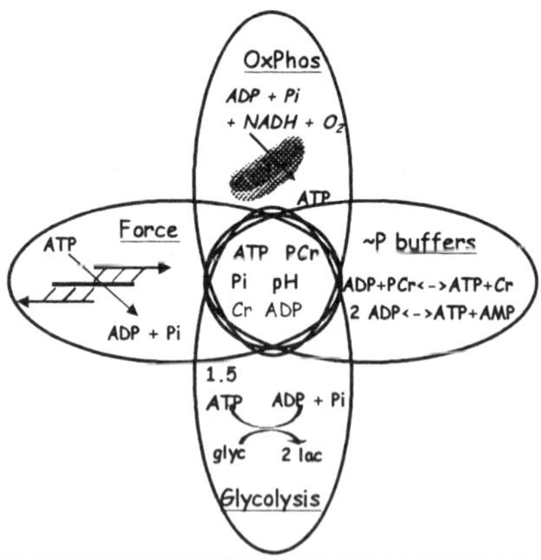

Figure 4: Diagram of intersections for components of energy balance; see text.

Energy balance can be stated mathematically by the following differential equation, which has only three terms defining the rate of change of ATP during contractile activity: a lumped ATPase, an empirical term for control of mitochondrial oxidative phosphorylation, and a term for the creatine kinase buffer. In the discussion that follows, for simplicity we will ignore the details of glycolysis which are available (20). The kinetic function for oxidative phosphorylation is a Hill function of ADP determined empirically (13). Creatine kinase activity is determined solely by the concentration of its substrates and products; a full set of kinetic relationships, based on *in vitro* kinetic constants, were applicable to muscle (Schimerlik & Cleland, 1973) as we reported (McFarland *et al.*, 1994).

$$\frac{dATP(t)}{dt} = -kATP(t) + OxPhos(t) - \frac{dPCr(t)}{dt} \qquad \text{Eqn 5}$$

Simulations like those in Fig. 5 result by integrating Eqn 5 with parameters obtained from energy balance experiments on human limb muscle (18). The simulation in Fig 5 begins with k at a value for resting muscle. At 210 sec, the rate increases as measured for twitch stimulation and is steady until 300 sec when it returns to resting. The simulation was made by numerical approximation of these relationships: for each time step, ATPase decreased [ATP] according to the first order term, -k•ATP(t). CK flux was adjusted and new concentrations of ATP, PCr, ADP and creatine were calculated. ATP synthesis for that time step was calculated and the process was iterated for small time steps. The simulation reproduced the time courses of chemical changes observed.

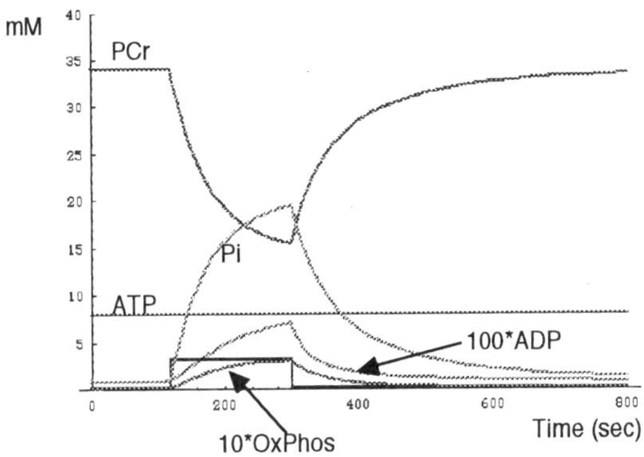

Figure 5: Simulations of chemical changes predicted by a model for energy balance. Ordinate scale is concentration of metabolites in mM or flux in mM/sec; abscissa is time in sec. 100 sec of rest is followed by 200 sec of elevated ATPase flux, then by recovery. The ATPase curve is the rectangle. The curve for ADP is multiplied by 100; the curve for oxidative phosphorylation is multiplied by 10.

FITTING EXPERIMENTAL DATA TO THE ENERGY BALANCE MODEL

This model for energy balance can be used for data analysis (30). Two further examples from consecutive measurements on one subject is displayed in Fig 6. The panel on the left represents a series of ischemic isometric twitches and the panel on the right shows the normoxic series of twitches. For each experiment, integration of spectral peaks like that shown in Fig 2 gives the data shown in the closed symbols. The lines drawn are the model fits using the equations and parameters used in the simulation in Fig 5 with the following adjustable parameters: resting and active ATPase and ADP concentration for half maximum oxidative phosphorylation. Thus model is adequate to explain the data for rest-contraction-recovery transitions in ischemic and normoxic contractions in human muscle. For both experiments the mitochondrial capacity for oxidative phosphorylation was 0.45 mM ATP/sec, resting ATPase was 0.03 mM ATP/sec, and active ATPase was 0.41 mM ATP/sec. The apparent "Km" of ADP was 15 μM. The time course of the component metabolic fluxes is also calculable from the fitted results. The conclusion is that the

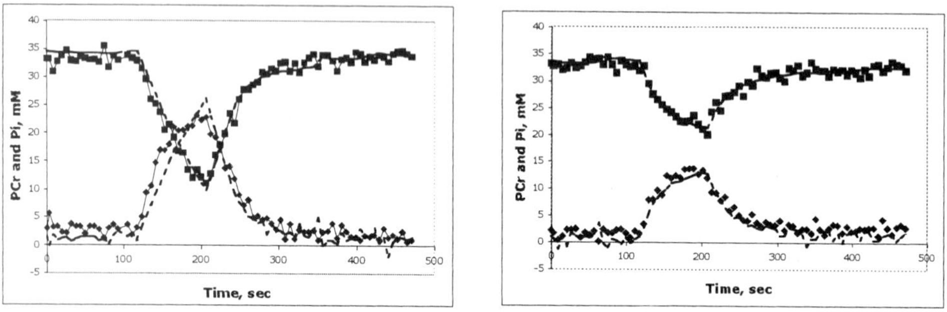

Figure 6. Examples of data and model fits using an energy balance model. Two serial experiments on one subject are displayed. The symbols are data obtained from ^{31}P spectra from the anterior compartment muscle of the human leg; PCr boxes and Pi diamonds. The protocol was 120 sec of rest; 90 sec of twitch contraction followed by recovery. The muscle was ischemic during the twitches in the left panel and normoxic in the right panel. For both, recovery occurred with normal blood flow. The lines drawn through or near the data points are the curves fitted by an energetic model (30).

model, obviously incomplete in many details including the absence of glycolytic ATP synthesis, nonetheless does provide an adequate description of the chemical events in brief bouts of exercise. These simple analyses show that this basic explanation of the reactions and components of biochemical energy balance depicted in Figs. 1 and 4 is largely correct, because the equations used, which embody current understanding of these processes, correctly account for experimental data.

CONCLUSIONS

The pathways for metabolic generation of ATP and crossbridge and other molecular machines that use ATP are necessarily connected because both sets of molecular modules share common metabolites, specifically ATP, ADP, Pi and pH. These metabolites act to modulate the flux in the mechanisms for energy demand and supply. The products of the ATPases are signals regulating the flux of ATP synthesis. The results of ATPase and ATP synthesis modulate metabolites that influence crossbridge kinetics and force generation. There are only a small number of critical chemical reactions that are necessary to advance quantitative understanding of these processes. The stoichiometric inter-relationships, defined by chemical rules, thus enforce a degree of simplification onto the system. The reason is the constraints imposed by the constancy of the internal composition, mass balances, stoichiometric rules and shared pH environment must follow ordinary conservation rules for mass and charge balances. We argue that these principles permit the obtaining of basic information concerning the organization of muscle energetics and the discerning of simple yet surprisingly powerful rules governing function. Moreover, they offer bases from which strong inferences can be made to integrate mechanisms as diverse as molecular crossbridge motors, metabolic fluxes generating ATP, cell respiration, and muscle oxygen consumption.

Acknowledgements:
This work was supported by grants from the US National Institutes of Health (AR 41928 and AR 36281) and from the National Space Biomedical Research Institute (MA00212).

REFERENCES

1. **Blei ML, Conley KE, and Kushmerick MJ.** Separate measures of ATP utilization and recovery in human skeletal muscle. *J Physiol Lond* 465: 203-222, 1993.
2. **Blei ML, Conley KE, Odderson IB, Esselman PC, and Kushmerick MJ.** Individual variation in contractile cost and recovery in a human skeletal muscle. *Proc Natl Acad Sci U S A* 90: 7396-7400, 1993.
3. **Cain DF and Davies RE.** Breakdown of adenosine triphosphate during a single contraction of working muscle. *Biochem Biophys Res Commun* 8: 361-366, 1962.
4. **Chance B and Williams GR.** The respiratory chain and oxidative phosphorylation. *Adv Enzymol* 17: 65-134, 1956.
5. **Conley KE, Blei ML, Richards TL, Kushmerick MJ, and Jubrias SA.** Activation of glycolysis in human muscle in vivo. *Am J Physiol* 273: C306-C315, 1997.
6. **Conley KE, Kushmerick MJ, and Jubrias SA.** Glycolysis is independent of oxygenation state in stimulated human skeletal muscle in vivo. *J Physiol Lond* 511: 935-945, 1998.
7. **Crowther GJ, Carey MF, Kemper WF, and Conley KE.** The control of glycolysis in contracting skeletal muscle. I. Turning it on. *Am J Physiology Endocrnol Metab* 282: E74 - E79, 2002.
8. **Crowther GJ, Kemper WF, Carey MF, and Conley KE.** The control of glycolysis in contracting skeletal muscle. II. Turing it off. *Am J Physiol Endocrinol Metab* 282: E67 - E73, 2002.
9. **Donaldson SKB, Hermansen L, and Bolles L.** Differential, direct effects of H+ on Ca2+-activated force of skinned fibers from the soleus, cardiac and adductor magnus muscles of rabbits. *Pfluegers Archiv* 376: 55-65, 1978.
10. **George P and Rutman RJ.** The "High Energy Phosphate Bond" Concept. *Prog Biophys Biophys Chem* 10: 2 - 53, 1960.
11. **He Z, Stienen GJ, Barends JP, and Ferenczi MA.** Rate of phosphate release after photoliberation of adenosine 5'-triphosphate in slow and fast skeletal muscle fibers. *Biophys J* 75: 2389-2401, 1998.
12. **Homsher E, Lacktis J, and Regnier M.** Strain-dependent modulation of phosphate transients in rabbit skeletal muscle fibers. *Biophysical Journal* 72: 1780-1791, 1997.
13. **Jeneson JAL, Wiseman RW, Westerhoff HV, and Kushmerick MJ.** The signal transduction function for oxidative phosphorylation is at least second order in ADP. *J Biol Chem* 271: 27995-27998, 1996.
14. **Kemper WF, Lindstedt SL, Hartzler LK, Hicks JW, and Conley KE.** Shaking up glycolysis: Sustained, high lactate flux during aerobic rattling. *Proc Natl Acad Sci U S A* 98: 723-728., 2001.
15. **Kingsley-Hickman PB, Sako EY, Mohanakrishnan P, Robitaille PML, From AHL, Foker JE, and Ugurbil K.** 31P NMR studies of ATP synthesis and hydrolysis kinetics in the intact myocardium. *Biochemistry* 26: 7501-7510, 1987.
16. **Kushmerick MJ.** Energy balance in muscle contraction: A biochemical approach. In: *Current Topics in bioenergetics*, edited by Sanadi R. New York: Academic Press, 1977, p. 1-37.
17. **Kushmerick MJ.** Multiple equilibria of cations with metabolites in muscle bioenergetics. *Am J Physiol* 272: C1739-1747, 1997.
18. **Kushmerick MJ.** Skeletal muscle: A paradigm for testing principles of bioenergetics. *J Bioenergetics and Biomembranes* 27: 555-569, 1995.
19. **Kushmerick MJ, Meyer RA, and Brown TR.** Regulation of oxygen consumption in fast- and slow-twitch muscle. *Am J Physiol* 263: C598-606, 1992.
20. **Lambeth MJ and Kushmerick MJ.** A computational model for glycogenolysis in skeletal muscle. *Ann Biomed Eng* 30: 808-827., 2002.
21. **Lipmann F.** Metabolic generation and utilization of phosphate bond energy. In: *Advances in Enzymology*, edited by Nord FF and Werkman CH. New York: Intersciences Publishers Inc., 1941, p. 99-162.
22. **McFarland EW, Kushmerick MJ, and Moerland TS.** Activity of creatine kinase in a contracting mammalian muscle of uniform fiber type. *Biophys J* 67: 1912-1924, 1994.

23. **Meyer R, Kushmerick M, and Brown T.** Application of ^{31}P-NMR spectroscopy to the study of striated muscle metabolism. *Am J Physiol* 242: C1-C11, 1982.
24. **Meyer RA.** A linear model of muscle respiration explains monoexponential phosphocreatine changes. *Am J Physiol* 254: C548-C553, 1988.
25. **Meyer RA and Foley JM.** Cellular Processes Integrating the Metabolic Response to Exercise. In: *Handbook of Physiology.*, edited by Rowell LB and Shepherd JT. New York: Oxford University Press, 1996, p. 841-869.
26. **Meyer RA, Sweeney HL, and Kushmerick MJ.** A simple analysis of the "phosphocreatine shuttle." *Am J Physiol* 246: C365-C377, 1984.
27. **Pate E and Cooke R.** A model of crossbridge action: the effects of ATP, ADP and Pi. *Journal of Muscle Research and Cell Motility* 10: 181-196, 1989.
28. **Radda GK.** The Use of NMR Spectroscopy for the Understanding of Disease. *Science* 233: 640-645, 1986.
29. **Ugurbil K, Kingsley-Hickman PB, Sako EY, Zimmer S, Mohanakrishnan P, Robitaille PML, Thoma WJ, Johnson A, Foker JE, and From AHL.** 31P NMR studies of the kinetics and regulation of oxidative phosphoylation in the intact myocardium. *Ann N Y Acad Sci* 508: 265-286, 1987.
30. **Vicini P and Kushmerick MJ.** Cellular energetics analysis by a mathematical model of energy balance: estimation of parameters in human skeletal muscle. *Am J Physiol Cell Physiol* 279: C213-224., 2000.
31. **Wallimann T, Wyss M, Brdiczka D, Nicolay K, and Eppenberger HM.** Intracellular compartmentation, structure and function of creatine kinase isoenzymes in tissues with high and fluctuating energy demands: the 'phosphocreatine circuit' for cellular energy homeostasis. *Biochem J* 281: 21-40, 1992.
32. **Wilkie DR.** Thermodynamics and the interpretation of biological heat measurements. *Prog Biophys & Biophys Chem* 10: 260-298, 1960.
33. **Woledge RC, Curtin NA, and Homsher E.** *Energetic Aspects of Muscle Contraction.* New York: Academic Press, 1986.
34. **Wyss M, Smeitink J, Wevers RA, and Wallimann T.** Mitochondrial creatine kinase: A key enzyme of aerobic energy metabolism. *Biochim Biophys Acta Bio Energetics* 1102: 119-166, 1992.

DISCUSSION

Rall: You stated that contracting human skeletal muscle in vivo has an ATPase during contraction 15 fold higher than at rest. But in frog skeletal muscle at low temperature, the rate of energy liberation during contraction increases more than 100 fold. Why is there the difference between human and frog muscle?

Kushmerick: The range of contraction to resting flux referred to the specific examples in my presentation, where there was a 15-fold ATPase during contraction as compared to the resting state. Bear in mind that my experiments used isometric twitches at 3 per sec, not tetanic stimulation as the frog muscle to which you refer were stimulated. Twitch tension is about 10% of a maximal voluntary contraction, which is comparable to a tetanus. Moreover the initial heat plus work rate in frog muscle tetanic stimulation is higher than the steady state rate whereas my results referred to an average rate over 90 sec of the twitch stimulation. The energy fluxes are rather comparable, and I did show a table showing that my results with intact human leg muscle (crossbridge and ion pumping ATPases) were about twice those from skinned human fibers (He et al., Biophys. J. 79: 945-961, 2000; Stienen et al., Physiol. 493: 299-307, 1996) when compared on the basis of integral of tension times time per unit strain.

Gonzalez-Serratos: Have you considered in your model that a substantial amount of the ATPase activity is due to the sarcoplasmic reticulum ATPase, Na^+ / K^+ ATPase pumps etc?

Kushmerick: The model and simulations described lump all ATPases together, including the Ca^{2+} transport in SERCA. Best estimates suggest that ion transport ATP costs are on the order of 40% of the total. I consider the total ATPase occurs during the mechanical output of the twitch because that is the time the crossbridges are splitting ATP and because of the evidence from mouse muscle that cytoplasmic free Ca^{2+} is returned to baseline before the peak of twitch force (Hollingworth et al., J. Gen. Physiol. 108: 455 - 469, 1996). Precise alignment of the time courses of the separate ATPase mechanisms matters little because they occur on a sub-second time scale whereas the relevant time scale for metabolism is on the order of tens of seconds.

T. Yamada: How about pH_i changes based on the chemical shift of P_i? Under the resolution of spectra you obtained, do you see any correlation between the metabolic and mechanical changes you mentioned and the pH_i changes?

Kushmerick: Due to time limitations, I did not discuss any results measuring pH, but as described in the text these occur and have consequences. pH can be measured with a precision of better than 0.1 pH unit and sometimes with a precision of ~ 0.02 pH unit. The text of my paper discusses some of the more important metabolic consequences. The major one is acidification influences the position of the creatine kinase equilibrium, and for the same change in PCr reduces the concentration of ADP as compared to what it would have been without the pH decrease. Another consequence derives from the initial pH rise, due to net PCr splitting without other confounding reaction. This transient pH rise allows us to calculate the pH buffer capacity for each experiment, and to use that information to compute the extent of glycogenolytic and glycolytic ATP production. These and other effects are complex, and this is one of the reasons why it is necessary to make quantitative models of the processes occurring, including Pi, ADP and pH effects on crossbridges, on Ca^{2+} transport and in signaling metabolic pathways to advance our understanding of the effects of pH change.

Brenner: What about effects of compartments in cells/tissues as effects of Pi /ADP etc on crossbridges are derived?

Kushmerick: The notion of compartmentalization of low molecular weight metabolites in the cytoplasm has become popular again, but I maintain the importance of metabolite compartmentalization is over-emphasized and the effects are likely smaller than some imagine for the following reasons. This statement of course does not apply to organelles surrounded by semi-permeable membranes, the definition use for compartmentalization. Many enzymes are bound to cellular structures or are large enough to suffer restrictions to diffusion, but the small metabolites including solvent water have been found to have diffusivities on the order of one quarter to one half that of simple solutions at comparable ionic strength. We argued the case for PCr, ATP, ADP and creatine in great detail (Meyer et al., Am. J. Physiol. 246: C365-C377, 1984). Creatine kinase is definitely localized in most interesting manners (Wallimann et al., Biochem. J. 281: 21-40, 1992), but that does not mean that its reactants and products are not freely diffusible. This statement is based on NMR methods, diffusive mobility and electrophoretic mobility of solutes injected into cells and the fact that biochemically measured enzyme activities and concentrations account for measured fluxes in cells. Most metabolites and solutes of interest have concentrations much larger than the concentration of enzyme sites to which they bind. Definitely Ca^{2+} is a major exception to this statement because of its low concentration (even at the height of its release), high concentration of binding sites and active transport into SERCA. There must be concentration gradients, for example between the myofibril and mitochondria, because there is a finite flux, on the order of 0.1 to 1 mM/sec in active muscle, but remember that the distances involved are less than a few microns. My rule of thumb for ordinary small molecular weight solutes that diffusion equilibration occurs over 1 μm in about a millisecond, and this rule suggests only a few or a few tens of μM gradients are possible, but this does not justify the notion of compartments of metabolites.

ENERGETICS, MECHANICS AND MOLECULAR ENGINEERING OF CALCIUM CYCLING IN SKELETAL MUSCLE

Jack A. Rall*

1. INTRODUCTION

During skeletal muscle contraction, Ca^{2+} moves through an intracellular cycle. Ca^{2+} is released from the sarcoplasmic reticulum (SR) in response to stimulation and then diffuses to bind to troponin C (TnC) on the thin filaments to activate the contractile machinery. Relaxation ensues when stimulation ceases and Ca^{2+} dissociates from TnC in response to Ca^{2+} re-accumulation into the SR driven by the SR Ca^{2+} ATPase. This review will trace the development of thought with regard to three aspects of this Ca^{2+} cycle: a) the energetic sequelae of Ca^{2+} cycling, b) the role of the intracellular Ca^{2+} binding protein parvalbumin (PA) in the Ca^{2+} cycle and c) the consequences of altering the Ca^{2+} exchange kinetics with TnC on muscle relaxation rate. This review is not comprehensive but rather selective with emphasis on topics of interest to our laboratory over the years.

2. ENERGETICS OF CALCIUM CYCLING

In 1949 A.V. Hill[1] defined the "activation heat" as the "...a 'triggered' reaction setting the muscle in a state in which it can shorten and do work." After the discovery of the role of Ca^{2+} in contraction and relaxation[2], the activation heat came to be regarded as the "sum of the thermal accompaniments of the liberation of calcium into the sarcoplasm, its movement to and from the myofibrillar binding sites, and its return to its storage site by an ATP-dependent transport process in the sarcoplasmic reticulum.[3] In a completely reversed cycle, all other processes should thermally cancel, so that the net activation heat would result from the hydrolysis of ATP associated with the cycling of calcium.[3] The problem then became how to separate the energy utilization by the cross-bridges from that associated with the cycling of Ca^{2+}. Two developments, one of great conceptual importance and the other entirely practical, were pivotal in isolating and investigating the energetics of the Ca^{2+} cycling pathway. In 1954, A.F. Huxley and R. Niedergerke[4] and

* Jack A. Rall, Department of Physiology and Cell Biology, Ohio State University, Columbus, OH 43210.

H.E. Huxley and J. Hanson[5] proposed the sliding filament model of muscle contraction. This theory suggested that a muscle stretched beyond myofilament overlap would generate no isometric force. Thus the energetics of Ca^{2+} cycling could be separated from the energetics of force and work production assuming that the Ca^{2+} cycling process was not inhibited by muscle stretch. Unfortunately the classic skeletal muscle preparation, the frog sartorius, could not be stretched to this extent without damage. But in 1972 E. Homsher[3] working in W.F.H.M. Mommaert's laboratory and I.C.H. Smith[6] working in R.C. Woledge's laboratory discovered that the frog semitendinosus muscle could be reversibly stretched to long strengths. They performed comprehensive studies that form the foundation of our present day understanding of the energetics of Ca^{2+} cycling. The basic experiment is illustrated in Fig. 1. Isometric twitch force and accompanying energy liberation at optimum myofilament overlap is superimposed upon results at an extreme muscle length where twitch force is nearly zero (Fig. 1A). When energy liberation is plotted versus isometric twitch force at stretched muscle lengths, a straight line results with an intercept that is about 30% of the energy liberated in a maximum twitch (Fig. 1B). Similar results were observed in isometric tetani[3]. Furthermore, the net activation heat over repeated contraction-relaxation cycles was produced in proportion to the hydrolysis of ATP. This result suggested that the activation heat was ultimately derived from ATP hydrolysis by the SR Ca^{2+} pump. Thus Homsher *et al* and Smith concluded that about 30% of the energy liberated, or ATP utilized, during a maximum isometric twitch in amphibian muscle at low temperature was associated with the cyclic movement of Ca^{2+}.

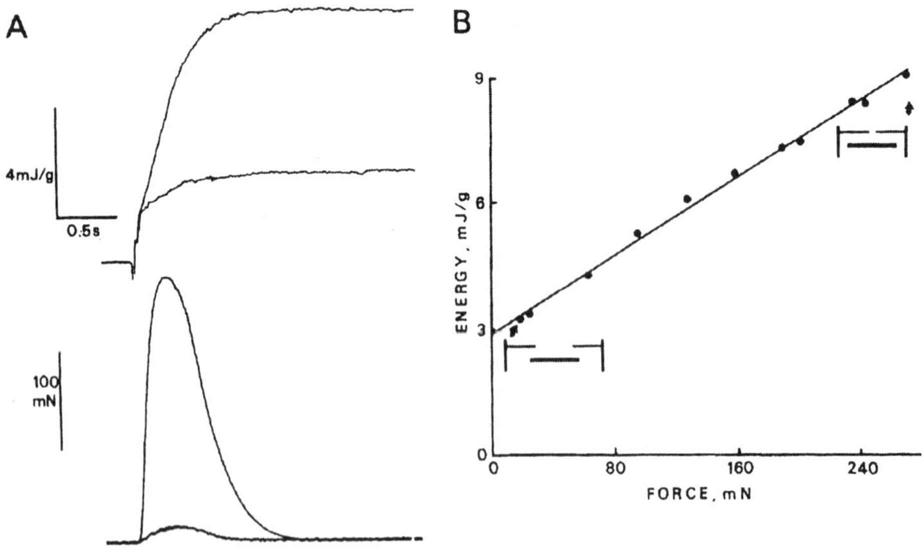

Figure 1. Measurement of activation heat in an isometric twitch. From reference 12 with permission.

It was of interest to test the generality of these results by measuring activation heat or Ca^{2+} cycling energetics in a variety of muscles under different conditions. In 1973, B. Schottelius and I[7] characterized the activation heat in isometric twitches of phasic and tonic skeletal muscles from the chicken at room temperature. The phasic posterior latissimus dorsi muscle exhibited kinetics of isometric contraction and relaxation and isotonic shortening that were 4 to 6-fold faster than that observed for the tonic anterior latissimus dorsi muscle. Yet in both muscles the activation heat represented about 20 – 25% of the energy liberation in a maximum isometric twitch. In the early 1970s, C. Gibbs and his laboratory were the first to extensively study the energetics of contraction and relaxation in mammalian fast- and slow-twitch muscle. They found that the energy associated with Ca^{2+} cycling was about 35 – 40% of the steady rate of energy liberation in a tetanus in rat fast- and slow-twitch muscle.[8,9] In 1983, M.T. Crow and M.J. Kushmerick[10] found that the activation energetics represented about 30 to 40% of the total energy utilization in tetanic contractions of mammalian fast- and slow-twitch muscle of the mouse at 20 °C. L. Rome has shown the even in the fastest contracting and relaxing muscle known, the toadfish swimbladder muscle, the fraction of the energy utilization attributed to Ca^{2+} cycling is about 25 to 40% of the total energy utilized.[11] Thus across the animal kingdom there is a significant fraction of the total energy utilized in a twitch or tetanus that is devoted to Ca^{2+} cycling.

These experiments led to four fundamental conclusions. First, during a complete contraction-relaxation cycle, the amount of activation heat reflects the amount of ATP hydrolyzed to move Ca^{2+} through the cycle and thus the total amount of Ca^{2+} moved. Second, during an isometric twitch or tetanus of skeletal muscle, a surprisingly large fraction of the ATP (20 to 40%) utilized was devoted to pumping Ca^{2+} back to the SR and not to producing force and/or work. Third, since the fraction of ATP utilized to pump Ca^{2+} was approximately constant in a twitch and a tetanus where the total amounts of ATP utilized were vastly different, Ca^{2+} must continually move through its cycle during the maintenance of tetanus force. Fourth, since the fraction of ATP utilized to cycle Ca^{2+} was approximately the same in fast and slow contracting muscles in twitches and tetani, the rate of Ca^{2+} accumulation by the SR must be "tuned" to match the rate of cross-bridge cycling. In other words, it would be pointless to have a fast contracting muscle that relaxed slowly, since it could not contract rapidly again until it relaxed.

Using the amplitude of the activation heat produced in a twitch as an indirect measure of the amount of Ca^{2+} cycled has led to insights into the mechanism of action of various perturbations of muscle contraction (see review by Rall[12] and the references therein). For example, the activation heat amplitude or amount of Ca^{2+} cycled in a twitch is: a) increased by agents that potentiate excitation-contraction (E-C) coupling such as Zn^{2+}, NO_3^-, caffeine and UO_2^{2+}, b) decreased by agents that inhibit E-C coupling such as elevated CO_2, strongly hypertonic solutions, D_2O and dantrolene, c) unaltered by temperature in the range of 0 to 20 °C, and d) independent of muscle length. One interesting feature of the activation heat is its characteristic repriming time.[13] When a stimulus is given at varying time intervals after a twitch or tetanus, it is possible to map the repriming of the activation heat amplitude, and presumably the amount of Ca^{2+} cycled, as a function of stimulus interval. This measure provides insight into the time it takes Ca^{2+} to return to a releasable state in the SR. The activation heat in frog muscle at low temperature reprimes as a single exponential after a twitch but as a double exponential after a tetanus. In general the activation heat produced at any time after a tetanus is depressed to a greater extent than after a twitch. Activation heat also has been measured in cardiac muscle but other approaches have been taken since it is not possible to stretch cardiac muscle to long sarcomere lengths.[14] More recently techniques have been developed to directly measure the steady rate of ATP hydrolysis by the SR Ca^{2+}

pump in single skinned muscle fibers.[15] In general the activation heat has proven to be a fundamental aspect of the muscle contraction-relaxation cycle and an indirect measure of the amount of Ca^{2+} cycled during contraction and relaxation.

Even though this history has focused on a narrow aspect of muscle energetics, no review of the history of muscle energetics would be complete without mention of two classic monographs in the field. In 1965 A.V. Hill published a book entitled "Trails and Trials in Physiology".[16] This book contains an annotated bibliography of his contributions to the field of muscle mechanics and energetics, a field which he dominated for over fifty years. It was considered to be a "bible" to those of us learning muscle energetics. The second monograph, published in 1985, entitled "Enegetic Aspects of Muscle Contraction" is a remarkable *tour de force* by R.C. Woledge, N.A. Curtin and E. Homsher.[17] This monograph is: a) a depository of virtually all studies on muscle energetics up to that time, b) a primer on enzyme and actomyosin kinetics, and c) an advanced treatise on muscle mechanics and energetics with integration, interpretation and theoretical speculation. These books are still valuable for novel ideas, history and inspiration.

3. ROLE OF PARVALBUMIN IN RELAXATION

The description of the activation heat above concentrated on its net amount which is proportional to the ATP hydrolyzed to pump Ca^{2+} back to the SR and thus also proportional to the amount of Ca^{2+} cycled during contraction and relaxation. But it must be remembered that the myothermal technique is nonspecific. Thus any reaction that liberates or absorbs heat will affect the time course of the energy liberation. It is only over a complete cycle that these events would be reversed and thus would be thermally neutral. Indeed during the time course of an isometric tetanus more energy is liberated than can be explained by the accompanying ATP hydrolysis.[18,19] N.A. Curtin and R.C. Woledge showed that some of this "unexplained" energy or enthalpy is associated with Ca^{2+} cycling (the rest is associated with cross-bridge cycling).[20]

C. Gerday and J.M. Gillis in 1976[21] and J.-F. Pechere, J. Derancourt and J. Haiech in 1977[22] proposed that the intracellular Ca^{2+} binding protein parvalbumin (PA) might promote relaxation in skeletal muscle by binding Ca^{2+} in parallel with the SR. At about this time, Curtin and Woledge[23] speculated that Ca^{2+} binding to PA might be the heat producing reaction that could account for the component of the unexplained energy associated with the activation heat time course. Indeed Ca^{2+} binding to PA was shown to be a heat producing reaction.[24,25] PA is a soluble, intracellular Ca^{2+} binding protein found in high concentration in fast contracting and relaxing skeletal muscles across the animal kingdom.[26] The affinity of Ca^{2+} for PA is higher than the affinity of Ca^{2+} for the regulatory sites of TnC.[21] Since PA concentration in some fish muscles can be ten-fold greater than the TnC concentration, speculation arose as to how these muscles could contract at all. The answer, as reviewed in detail by Rall,[27] is that PA also competitively binds Mg^{2+}, albeit with a 10^4 fold lower affinity than exists for Ca^{2+}. But since the free Mg^{2+} concentration in a muscle fiber at rest is about 10^4 higher than the free Ca^{2+} concentration, PA would be largely in the Mg^{2+} bound form, i.e., Mg•PA. Ca^{2+} cannot bind to PA until Mg^{2+} dissociates and this is a relatively slow process. On the other hand, the regulatory sites of TnC do not contain Mg^{2+} and Ca^{2+} binds to these sites in a manner than is limited only by the rate of diffusion of Ca^{2+} in the sarcoplasm. Thus muscle activation is not impaired by the presence of PA. Gerday and Gillis[21] further showed that the SR has a higher affinity for Ca^{2+} than does PA. Computer simulations showed that PA could, in principle, act to shuttle Ca^{2+} from TnC to the SR and thus promote relaxation in the process.[28,29] Figure 2 shows a scheme indicating how PA might promote skeletal muscle relaxation.

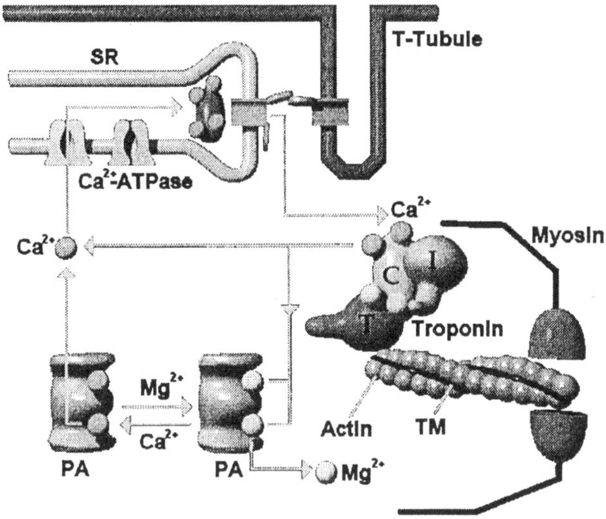

Figure 2. Proposed role of parvalbumin in promoting skeletal muscle relaxation. From reference 27 with permission.

Our laboratory tested and verified this scheme for the role of PA in promoting muscle relaxation in frog skeletal muscle at 0 and 10 °C.[30-32] Frog skeletal muscle was chosen to study because it has a high concentration of PA. Furthermore experiments were conducted at low temperatures because under these conditions PA would be expected to be more effective in promoting relaxation relative to the SR Ca^{2+} pump. This speculation follows from the fact that the Ca^{2+} pump ATPase exhibits a higher temperature sensitivity than does the exchange of Ca^{2+} for Mg^{2+} on PA. According to the scheme shown in Fig. 2, the effect of PA in promoting relaxation rate should diminish with increasing tetanus duration as PA becomes saturated with Ca^{2+} during the tetanus and is thus unable to further bind Ca^{2+} during relaxation. Furthermore the time course of diminution of the PA effect on relaxation should be determined by the rate of Mg^{2+} dissociation from PA. This prediction was tested by comparing the time course of the slowing of relaxation rate in muscle with the time course of Mg^{2+} dissociation from PA at the same temperature. Relaxation rate slowed progressively in frog skeletal muscle at 0 °C with increasing tetanus duration with a rate constant of 1.2 s^{-1}. After a four second tetanus, the relaxation rate reached a final value of about 40% of the maximum rate of relaxation. During this time there is little fatigue of maximum force development. In parallel experiments in solution, the Mg^{2+} dissociation rate from PA was 0.9 s^{-1}, a value similar to the rate of slowing of relaxation. Also, consistent with these results, the rate of fall of the Ca^{2+} transient declines with increasing tetanus duration in parallel with the decline of relaxation rate.[33]

A second prediction of the hypothesis is that the time course of recovery of the ability of PA to promote relaxation after a tetanus should reflect the time course of Mg^{2+} re-association to PA which is determined by the time course of Ca^{2+} dissociation from PA. In frog skeletal muscle at 0 °C, the depressed rate of relaxation after an isometric tetanus was accelerated with a rest period with a rate constant of 0.12 s^{-1}. This value is similar to the rate constant of Ca^{2+} dissociation from isolated PA of 0.19 s^{-1}. These same predictions also were verified at 10 °C.[31] A final prediction is that PA should be able to cause relaxation even when the SR Ca^{2+} pump is inhibited but only from a brief tetanus. This prediction was tested in frog skeletal muscle at 10 °C when the SR Ca^{2+} pump was poisoned selectively with 2,5-di(tert-butyl)-1,4-benzohydroquinine (TBQ).[32] If the muscle was stimulated to produce a tetanus that would be predicted to saturate PA with Ca^{2+} at 10 °C (1.1 second) in the presence of TBQ, there was little change in the rate of contraction or the amplitude of contraction but the fall of the Ca^{2+} transient and relaxation required nearly five minutes instead of the usual 1 second. If instead the tetanus was brief (0.3 second), full relaxation occurred, albeit at a rate that was eight-fold slower than in the un-poisoned control muscle. Taken together, these results are consistent with the hypothesis that PA can promote relaxation in a brief contraction by binding Ca^{2+} in parallel with the SR.

Recent studies have employed genetic techniques to directly test the role of PA in promoting relaxation in mammalian muscle. PA gene inactivation[34] or gene knockout[35] led to a decrease in relaxation rate in fast-twitch muscles of the mouse. Direct gene transfer of PA cDNA into rat slow-twitch muslce[36] or rat cardiac myocytes[37], which normally do not express PA, resulted in an acceleration of relaxation rate. Over-expression of PA in slow-twitch muscle of the mouse not only resulted in an acceleration of relaxation but also in other alterations in the muscle phenotype, including changes in the profile of metabolic enzymes.[38] It now is generally accepted that PA can promote relaxation in fast contracting and relaxing skeletal muscle.

4. CALCIUM EXCHANGE WITH TROPONIN C AND RELAXATION

The observation that PA modulates relaxation by transiently binding Ca^{2+} implies that the rate of Ca^{2+} sequestration must limit the rate of relaxation, especially at low temperatures in frog skeletal muscle. But what is the rate limiting process at higher temperatures? The effect of PA in frog skeletal muscle diminishes at higher temperatures, e.g., it is half as large at 10 °C as at 0 °C.[31] In fact, J.D. Johnson and Y. Jiang made the provocative observation that even though the rate of the fall of the Ca^{2+} transient and rate of relaxation in frog skeletal muscle increased in the presence of a cell permeable form of the Ca^{2+} chelator EGTA at 10 °C, there was no change at 20 °C.[39] It appeared that at 20 °C the rate of relaxation was not limited by Ca^{2+} sequestration but rather by some myofibrillar process(es). They further noted that the rate of relaxation was similar to the rate of Ca^{2+} dissociation from isolated whole troponin (Tn). Thus they proposed that at higher temperatures, relaxation rate might be limited by the rate of Ca^{2+} dissociation from TnC.

This conclusion was a surprise since it had been assumed for many years that Ca^{2+} dissociation from TnC in the muscle was much more rapid than the rate of mechanical relaxation and thus was not involved in controlling the rate of relaxation. This assumption was based on the fact that the Ca^{2+} dissociation rate from purified TnC is more than twenty-fold faster than the rate of muscle relaxation at the same temperature.[40]

But Ca^{2+} dissociation from isolated whole Tn (TnC-TnI-TnI) is much slower and similar to the rate of muscle relaxation.[41] Thus it appears that under some conditions the rate of Ca^{2+} dissociation from whole Tn may be slow enough to limit the rate of muscle relaxation.

In order to test if the rate of Ca^{2+} dissociation from TnC is a factor in determining the rate of muscle relaxation, we have: a) generated mutants of TnC with varying Ca^{2+} affinities and Ca^{2+} dissociation rates, b) developed a simplified *in vitro* system that mimics isolated whole Tn to screen mutant properties and c) begun to measure the effects of some of these TnC mutants on the rate of muscle relaxation. J.P. Davis and S.B. Tikunova in our laboratory have generated mutants of TnC by substituting hydrophobic residues in the regulatory domain of TnC with glutamine.[42] These TnC mutants exhibit a wide range of Ca^{2+} affinities and Ca^{2+} dissociation rates. Also they have shown that the rate of Ca^{2+} dissociation from a complex of TnC and a peptide of TnI (TnI_{96-148}) is similar to the rate of Ca^{2+} dissociation from isolated whole Tn.[43] Furthermore Figure 3 shows that the rate of Ca^{2+} dissociation from the TnC-TnI_{96-148} complex is much slower than the rate of Ca^{2+} dissociation from isolated TnC but similar to the rate of muscle relaxation.[43] Thus the TnC-TnI_{96-148} complex was utilized to screen TnC mutants for variations in Ca^{2+} dissociation rates (>30-fold) and Ca^{2+} affinities (>240-fold). Interestingly, mutations that affected Ca^{2+} sensitivity of isolated TnC but not the TnC-TnI_{96-148} complex had little affect on the Ca^{2+} sensitivity of muscle force development. Thus, the effect of mutations on the Ca^{2+} sensitivity of force development could be better predicted from the change in Ca^{2+} affinity of the TnC-TnI_{96-148} complex than from the change in affinity of the isolated TnC alone.

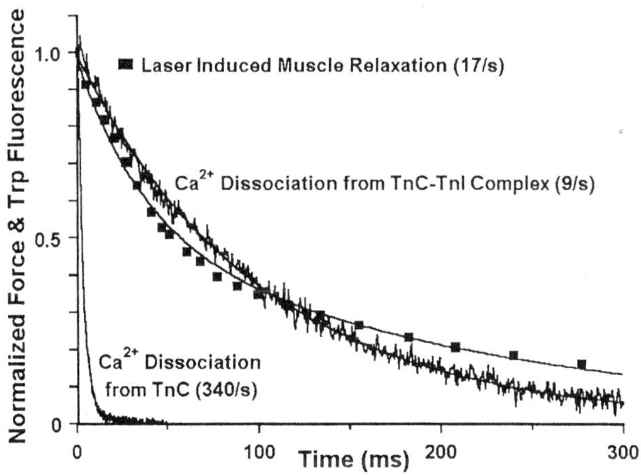

Figure 3. Ca^{2+} dissociation from the TnC-TnI complex, but not from isolated TnC alone, is similar to the rate of relaxation in skinned psoas muscle fibers at 15 °C. Ca^{2+} dissociation was measured from changes in fluorescence of trp engineered into TnC^{F29W}. Relaxation was induced by flash photolysis of the Ca^{2+} chelator diazo-2. From reference 43 with permission.

In collaboration with L. Smillie, our laboratory has determined the effects of TnC mutants which exhibited a higher or lower Ca^{2+} dissociation rate from the TnC-TnI_{96-148} complex on the rate of muscle relaxation.[44] TnC mutants were incorporated into fast-twitch mammalian skinned psoas fibers after extraction of endogenous TnC. The skinning process destroyed the SR in these fibers. Contraction was induced by elevating the free Ca^{2+} concentration and relaxation was induced by flash photolysis of the photolabile Ca^{2+} chelator, diazo-2. A TnC mutant with a higher Ca^{2+} affinity and a two-fold slower Ca^{2+} dissociation rate as measured in solution, TnC^{M82Q}, increased the Ca^{2+} sensitivity of isometric force production and decreased the rate of relaxation in muscle fibers by two-fold. A TnC mutant with a lower Ca^{2+} affinity and 1.5-fold faster Ca^{2+} dissociation rate, TnC^{NHDel} (deletion of residues 1-11), decreased the Ca^{2+} sensitivity of force production but did not significantly accelerate muscle relaxation. The rate of cross-bridge detachment was slowed by reducing the inorganic phosphate concentration. This perturbation also slowed relaxation rate. Thus slowing of either the Ca^{2+} dissociation rate from TnC or cross-bridge detachment rate slowed muscle relaxation but acceleration of one rate, Ca^{2+} dissociation from TnC, had little effect on relaxation rate. These results are consistent with the interpretation that the rate of Ca^{2+} dissociation from TnC and the rate of cross-bridge detachment are comparable and contribute equally to determining the rate of relaxation.

In order to test this idea more fully, mutants of TnC have been constructed with a wider range of Ca^{2+} dissociation rates and these mutants will be tested in muscle fibers. Since at present, it is not possible to measure the rate of Ca^{2+} dissociation from TnC in a muscle fiber, these mutants are being incorporated into isolated myofibrils to test for their influence on Ca^{2+} dissociation rate from TnC in a system intermediate in structure between TnC-$TnI_{96\text{-}148}$ and the skinned fiber.[45] To date, we have found that the Ca^{2+} dissociation rate from the isolated myofibrils is similar to the rate of Ca^{2+} dissociation from the TnC-$TnI_{96\text{-}148}$ complex and the rate of muscle relaxation. Also, we recently have constructed mutants of cardiac TnC with varying Ca^{2+} dissociation rates in order to determine their effects in cardiac muscle relaxation and to compare the results to those observed in skeletal muscle.[46] In cardiac muscle, the phosphorylation of TnI is thought to accelerate relaxation by increasing the rate of Ca^{2+} dissociation from TnC.[47,48] This idea will be tested directly with the cardiac TnC mutants. In conclusion, the notion that Ca^{2+} dissociation from TnC plays a role in determining the rate of skeletal muscle relaxation under certain conditions seems probable but is still speculative and more work will need to be done in this area.

5. SUMMARY

During muscle contraction and relaxation, Ca^{2+} moves through a cycle. About 20 to 40% of the ATP utilized in a twitch or a tetanus is utilized by the SR Ca^{2+} pump to sequester Ca^{2+}. Parvalbumin is a soluble Ca^{2+} binding protein that functions in parallel with the SR Ca^{2+} pump to promote relaxation in rapidly contracting and relaxing skeletal muscles, especially at low temperatures. The rate of Ca^{2+} dissociation from troponin C, once thought to be much more rapid than the rate of relaxation, is likely to be similar to the rate of cross-bridge detachment and to the rate of muscle relaxation under some conditions. During the past fifty years, great progress has been made in understanding the Ca^{2+} cycle during skeletal muscle contraction and relaxation. Nonetheless, there are still mysteries waiting to be unraveled.

6. REFERENCES

1. A.V. Hill, The heat of activation and the heat of shortening in a muscle twitch, *Proc. Roy. Soc. B* **136**, 195-211 (1949).
2. S. Ebashi, M. Endo, and I. Ohtsuki, Control of muscle contraction, *Quat. Rev. Biophys.* **2**, 351-384 (1969).
3. E. Homsher, W.F.H.M. Mommaerts, N.V. Ricchiuti, and A. Wallner, Activation heat, activation metabolism and tension-related heat in frog semintendinosus muscles, *J. Physiol.* **220**, 601-625 (1972).
4. A.F. Huxley and R. Niedergerke, Structural changes in muscle during contraction, *Nature* **173**, 971-973 (1954).
5. H.E. Huxley and J. Hanson, Changes in the cross-striations of muscle during contraction and stretch and their structural interpretations, *Nature* **173**, 973-976 (1954).
6. I.C.H. Smith, Energetics of activation in frog and toad muscle, *J. Physiol.* **220**, 583-599 (1972).
7. J.A. Rall and B.A. Schottelius, Energetics of contraction in phasic and tonic skeletal muscles of the chicken, *J. Gen. Physiol.* **62**, 303-323 (1973).
8. I.R. Wendt and C.L. Gibbs, Energy production of rat extensor digitorum longus muscle, *Am. J. Physiol.* **224**, 1081-1086 (1973).
9. C.L. Gibbs and W.R. Gibson, Energy production of rat soleus muscle, *Am. J. Physiol.* **223** 864-871 (1972).
10. M.T. Crow and M.J. Kushmerick, Correlated reduction of velocity of shortening and the rate of energy utilization in mouse fast-twitch muscle during a continuous tetanus, *J. Gen. Physiol.* **82**, 703-720 (1983).
11. L.C. Rome and A.A. Klimov, Superfast contractions without superfast energetics: ATP usage by SR-Ca^{2+} pumps and crossbridges in toadfish swimbladder muscle, *J. Physiol.* **526**, 279-286 (2000).
12. J.A. Rall, Energetics of Ca^{2+} cycling during skeletal muscle contraction, *Fed. Proc.* **41**, 155-160 (1982).
13. J.A. Rall, Effects of previous activity on the energetics of activation in frog skeletal muscle, *J. Gen. Physiol.* **75**, 617-631 (1980).
14. N.R. Alpert, E.M. Blanchard, and L.A. Mulieri, Tension-independent heat in rabbit papillary muscle, *J. Physiol.* **414**, 433-453 (1989).
15. G.J.M. Stienen, R. Zaremba, and G. Elzinga, ATP utilization for calcium uptake and force production in skinned muscle fibres of *Xenopus laevis*, *J. Physiol.* **482**, 109-122 (1995).
16. A.V. Hill, *Trails and Trials in Physiology* (The Williams & Wilkins Company, Baltimore, 1965).
17. R.C. Woledge, N.A. Curtin, and E. Homsher, *Energetics Aspects of Muscle Contraction* (Academic Press, London, 1985).
18. N.A. Curtin and R.C. Woledge, Chemical change and energy production during contraction of frog muscle: how are their time courses related? *J. Physiol.* **288**, 353-366 (1979).
19. E. Homsher, C.J. Kean, A. Wallner, and V. Garibian-Sarian, The time-course of energy balance in an isometric tetanus, *J. Gen. Physiol.* **73**, 553-567 (1979).
20. N.A. Curtin and R.C. Woldege, The effect of muscle length on energy balance in frog skeletal muscle, *J. Physiol.* **316**, 453-468 (1981).
21. C. Gerday and J.M. Gillis, Possible role of parvalbumin in the control of contraction, *J. Physiol.* **258**, 96-97P (1976).
22. J.-F. Pechere, J. Derancourt, and J. Haiech, The participation of parvalbumins in the activation-relaxation cycle of vertebrate fast skeletal-muscle, *FEBS Letters* **75**, 111-114 (1977).
23. N.A. Curtin and R.C. Woledge, Energy changes and muscular contraction, *Physiol. Rev.* **58**, 690-761 (1978).
24. S.J. Smith and R.C. Woledge, Thermodyamic analysis of calcium binding to frog parvalbumin, *J. Muscle Res. Cell Mot.* **6**, 757-768 (1985).
25. M. Tanokura and K. Yamada, A calorimetric study of Ca^{2+} binding to two major isotypes of bullfrog parvalbumin, *FEBS* **185**, 165-169 (1985).
26. C.W. Heinzman, Parvalbumin, an intracellular calcium-binding protein; distribution, properties and possible roles in mammalian cells, *Experientia Basel* **40**, 910-921 (1984).
27. J.A. Rall, Role of parvalbumin in skeletal muscle relaxation, *News Physiol. Sci.* **11**, 249-255 (1996).
28. J.M. Gillis, D. Thomason, J. Lefevre, and R.H. Kretsinger, Parvalbumins and muscle relaxation: a computer simulation study, *J. Muscle Res. Cell Mot.* **3**, 377-398 (1982).
29. M. Cannell and D.G. Allen, Model of calcium movements during activation in the sarcomere of frog skeletal muscle, *Biophys. J.* **45**, 913-925 (1984).
30. T. Hou, J.D. Johnson, and J.A. Rall, Parvalbumin content and Ca^{2+} and Mg^{2+} dissociation rates correlated with changes in relaxation rate of frog muscle fibres, *J. Physiol.* **441**, 285-304 (1991).
31. T. Hou, J.D. Johnson, and J.A. Rall, Effect of temperature on relaxation rate and Ca^{2+}, Mg^{2+} dissociation rates from parvalbumin of frog muscle fibres, *J. Physiol.* **449**, 399-410 (1992).

32. Y. Jiang, J.D. Johnson, and J.A. Rall, Parvalbumin relaxes frog skeletal muscle when the sarcoplasmic reticulum Ca-ATPase is inhibited, *Am. J. Physiol.* **270**, C411-C417 (1996).
33. M.B. Cannell, Effect of tetanus duration on the free calcium during the relaxation of frog skeletal muscle fibres, *J. Physiol.* **376**, 203-218 (1986).
34. J.M. Raymackers, P. Gailly, M.C. Schoor, D. Pette, B. Schwaller, W. Hunziker, M.R. Celio, and J.M. Gillis, Tetanus relaxation of fast skeletal muscles of the mouse made parvalbumin deficient by gene inactivation, *J. Physiol.* **527**, 355-364 (2000).
35. B. Schwaller, J. Dick, G. Dhoot, S. Carroll, G. Vrbova, P. Nicotera, D. Pette, A. Wyss, H. Bluethmann, W. Hunziker, and M.R. Celio, Prolonged contraction-relaxation cycle of fast-twitch muscles in parvalbumin knockout mice, *Am. J. Physiol.* **276**, C395-C403 (1999).
36. M. Muntener, L. Kaser, J. Weber, and M.W. Berchtold, Increase of skeletal muscle relaxation speed by direct injection of parvalbumin cDNA, *Proc. Nat. Acad. Sci.* **92**, 6504-6508 (1995).
37. P.A. Wahr, D.E. Michele, and J.M. Metzger, Parvalbumin gene transfer corrects diastolic dysfunction in diseased cardiac myocytes, *Proc. Nat. Acad. Sci.* **96**, 11982-11985 (1999).
38. E.R. Chin, R.W. Grange, F. Viau, A.R. Simard, C. Humphries, J. Shelton, R. Bassel-Duby, R.S. Williams, and R.N. Michel, Alterations in slow-twitch muscle phenotype in transgenic mice overexpressing the Ca^{2+} buffering protein parvalbumin, *J. Physiol.* **547**, 649-663 (2003).
39. J.D. Johnson, Y. Jiang, and M. Flynn, Modulation of Ca^{2+} transients and tension by intracellular EGTA in intact frog muscle fibers, *Am. J. Physiol.* **272**, C1437-C1444 (1997).
40. J.D. Johnson, S.C. Charlton, and J.D. Potter, A fluorescence stopped flow analysis of Ca^{2+} exchange with troponin C, *J. Biol. Chem.* **254**, 3497-3502 (1979).
41. J.D. Potter and J.D. Johnson, Troponin, in: Calcium and Cell Function, Vol II, edited by W.Y. Cheung (Academic Press, New York, 1982), pp. 145-172.
42. S.B. Tikunova, J.A. Rall, and J.P. Davis, Effect of hydrophobic residue substitutions with glutamine on Ca^{2+} binding and exchange with the N-domain of troponin C, *Biochem.* **41**: 6697-6705 (2002).
43. J.P.Davis, J.A. Rall, C.Alionte, and S.B. Tikunova, Mutations of hydrophobic residues in the N-domain of troponin C affect calcium binding and exchange with the troponin C-troponin $I_{(96\text{-}148)}$ complex and muscle force production, *J. Biol. Chem*, in press (2004).
44. Y. Luo, J.P. Davis, L.B. Smillie, and J.A. Rall, Determinants of relaxation rate in rabbit skinned skeletal muscle fibres, *J. Physiol.* **545**, 887-901 (2002).
45. J.P. Davis, S.B. Tikunova, D.R. Swartz, and J.A. Rall, Measurement of Ca^{2+} dissociation rates from troponin C (TnC) in skeletal myofibrils, *Biophys.* J. **86**, 218a (2004).
46. S.B. Tikunova, J.P. Davis, and J.A. Rall, Engineering cardiac troponin C (cTnC) mutants with dramatically altered Ca^{2+} dissociation rates as molecular tools to study cardiac muscle relaxation, *Biophys. J.* **86**, 394a (2004).
47. S.P. Robertson, J.D. Johnson, M.J. Holroyde, E.G. Kranias, J.D. Potter, and R.J. Solaro, The effect of troponin I phosphorylation on the Ca^{2+}-binding properties of the Ca^{2+}-regulatory site of bovine cardiac troponin, *J. Biol. Chem.* **257**, 260-263 (1982).
48. R. Zhang and J.D. Potter, Cardiac troponin I phosphorylation increases the rate of cardiac muscle relaxation, *Circ. Res.* 76, 1028-1035 (1995).

HEAT, PHOSPHORUS NMR AND MICROCALORIMETRY IN RELATION TO THE MECHANISM OF FILAMENT SLIDING

Kazuhiro Yamada*

SUMMARY

For muscle heat measurements the methods available are sensitive and rapid, and the heat is related to the chemical changes in a manner that provides a firm outline for understanding the mechanism of contraction. For example linear dependence of the shortening heat on the sarcomere length has shown that the rate of turnover of cross-bridges increases during shortening. However, heat is bound to lack *specificity*.

In order to cope with this problem, various methods such as rigorous chemical analyses, phosphorus NMR and microcalorimetry have been introduced. As a result of ultra-rapid freezing and chemical analysis by D. R. Wilkie (Gilbert, Kretzchmar, Wilkie and Woledge, 1971), the energy balance discrepancy between (heat + work) and the amount of phosphocreatine (PCr) split emerged, i.e. the unexplained enthalpy.

Calcium ions move from the sarcoplasmic reticulum to the calcium-receptive proteins in the sarcoplasm during contraction. In an attempt to find the cause of the unexplained enthalpy, microcalorimetry of calcium binding to calcium-receptive proteins has been performed. The results have shown that calcium ions dislocated between sites within the sarcoplasm on activation may produce about 1/3 of the unexplained heat. In addition calcium pump should operate by consuming PCr to relocate the calcium after the contraction.

Time-resolved phosphorus NMR has also shown that a certain amount of PCr splitting continues during early minute of recovery period after the contraction without Pi released. This delayed splitting of PCr is most likely caused by the kinetic properties of the contractile proteins and can explain another 1/3 ofthe unexplained enthalpy.

The mechanism of how muscle is regulated is another important question. Studies of calcium binding to calcium-receptive proteins in the sarcoplasm by using titration

* Department of Physiology, University of Oita Faculty of Medicine, Oita 879 -5593 Japan

microcalorimetry has shown that troponin C has a characteristic single calcium-binding site that is most likely to be involved in the regulation of contraction.

INTRODUCTION

In 1959 A. V. Hill (see Hill, 1965) wrote "chemical methods are usually extremely slow and insensitive... particularly since most chemical estimations in living cells involve the destruction of the material itself. For measuring heat, however, the methods available are very sensitive and can be made very rapid. The heat is related to the chemical exchanges, not always indeed in a specific way, but at least in a manner which provides a firm outline that must not be overstepped and can be filled in as knowledge accrues."

Following A. V. Hill much effort has been made to improve sensitivity in heat measurements. However heat measurements are bound to lack *specificity*. In order to cope with this problem, various methods have been introduced. Wilkie introduced, in addition to a rigorous chemical approach, phosphorus NMR (Dawson, Gadian and Wilkie, 1979), and Woledge introduced microcalorimetry in the field (Woledge, 1973). Since then several novel findings and associated advances in understanding have been attained. I would like to discuss what problems have been solved and what problems remain.

ENERGY BALANCE STUDIES AND THE EMERGENCE OF THE UNEXPLAINED ENTHALPY

Wilkie introduced a novel hammer apparatus for freezing muscles rapidly. This was to cope with the above non-specificity problem by using rigorous chemical analysis. As a result a gap emerged in energy balance (the unexplained enthalpy, Gilbert et al., 1971).

The result was that the break-down of PCr was not large enough to account for the heat produced during the first few s of isometric contraction. By the end of a 15 s tetanus as much as 42 mJ/g remained unaccounted for. Gilbert et al. (1971) have suggested some process which could produce about 42 mJ/g during the early part of a tetanus and which is not reversed during a whole tetanus. One is binding of calcium ions to proteins, and the conformational changes of the proteins. 42 mJ/g muscle is equivalent to 105 kJ per mole of moved calcium, by assuming 0.4 μmole calcium moved between sites within the sarcoplasm per g muscle (Winegrad, 1970).

Break-down of PCr continues during the minute following relaxation, but without heat production. Similarities have been noticed between the unexplained enthalpy and the labile part of the maintenance heat production, and are discussed, together with differences, by Woledge, Curtin and Homsher (1965).

It has been hypothesized that the unexplained energy is produced by the binding of calcium ions to parvalbumins (see for example Woledge, Curtin and Homsher, 1985). Concentration of parvalbumins in the sarcoplasm (0.4 μmole/g, Gosselin-Rey and Gerday, 1977) is considerably higher than that of troponin (0.07 μmole/g, Ebashi, Endo and Ohtsuki, 1969). Calcium binding to parvalbumins of frog muscle produces heat of 28 kJ/mole (average of two major isotypes at 5°C, Tanokura and Yamada, 1987), therefore calcium binding to parvalbumins would produce heat of 22 mJ per g muscle. This amount of heat corresponds to about a half of the unexplained enthalpy.

A comparison of total myoplasmic calcium and total calcium-binding capacity indicates that about a half of the calcium-binding sites remain unsaturated during a prolonged tetanus (Hou, Johnson and Rall, 1991). Moreover, kinetic analyses of calcium movement in the sarcoplasm during contractile activation have shown that the calcium pump of the sarcoplasmic reticulum may operate in parallel with the calcium binding to parvalbumins (Hou, Johnson and Rall, 1993). Therefore the above figure is likely to be an overestimate.

Another possible explanation of the unexplained enthalpy is delayed PCr splitting. PCr splitting that continues after contraction has ended has long been known to exist during the minute after relaxation (see Gilbert et al., 1971). It is very likely that the post-contractile splitting of PCr is coupled to the reversal of an unidentified exothermic process occurring early in the contraction. A quantitative estimate of the amount of the delayed splitting of PCr by time-resolved phosphorus NMR in living muscles is described in the later section of this article.

ACTIVATION HEAT

Heat production during isometric tetanic contractions depends both on time after the contraction is initiated, and on muscle lengths (see Woledge et al., 1985). The rate of heat production that is proportional to filament overlap is most certainly derived from actin-myosin interaction.

The activation heat, measured in twitch contractions with muscles stretched to lengths where active force development is nearly abolished, is very likely to be related to activation processes by calcium (Smith, 1972; Homsher, Mommaerts, Ricchiuti and Wallner, 1972). It can be deduced from studies on repriming of the activation heat that this component of heat most likely represents calcium release from the sarcoplasmic reticulum and subsequent interaction with troponin on the thin filament to activate contraction (Rall, 1980).

Calcium release process from the sarcoplasmic reticulum is most likely thermo-neutral because heat production during the ATP-driven calcium uptake by sarcoplasmic reticulum is small except for that of ATP hydrolysis (Kodama, Kurebayashi and Ogawa, 1980). Therefore, the fastest part of the activation heat at least should be caused by calcium binding to troponin.

Enthalpy change of calcium binding to calcium-specific sites of troponin is 74 kJ/mole calcium bound (Yamada, Mashima and Ebashi, 1976, see Fig. 1). Yamada et al. (1976) have discussed that 0.1 μmole of calcium ions per g of muscle, which is closely involved on contractile activation (Weber and Herz, 1963), would produce 7.4 mJ per g of muscle. This is more than necessary to explain the activation he at, which is about 4 mJ/g muscle maximally. Therefore the activation heat most likely represents the amount of calcium released by the sarcoplasmic reticulum on activation.

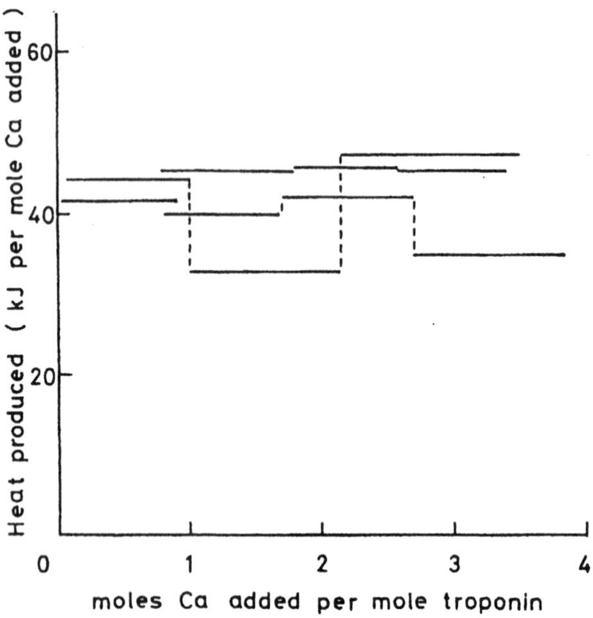

Fig. 1. Heat produced per mole of Ca added to troponin (rabbit skeletal) in 1 mM Mg (10°C). Reproduced with permission from Yamada, Mashima and Ebashi, 1976.

DEPENDANCE OF SHORTENING HEAT ON SARCOMERE LENGTH

During shortening the heat rate is higher than in the periods when the muscle is isometric. The shortening heat is thought to represent a greater rate of cross-bridge turnover during the shortening. The concept of independent force-generator indicates that the shortening heat is expected to be directly proportional to the extent of filament overlap. At longer muscle lengths allowance has to be made for the thermoelastic heat absorption on releasing pre-stretched muscles, which is likely to be caused by parallel elastic structures. It has been shown that the shortening heat at a rapid isovelocity releases linearly depended on the sarcomere lengths as the isometric tension did in the sarcomere length range between 2.2 and 3.7 μm (Irving and Woledge, 1981 ; Kometani and Yamada, 1983).

It has also been shown that in the shortening at near V_{max} there is a significant discrepancy between the observed and the explained enthalpy (Homsher, Yamada, Wallner and Tsai, 1984). This discrepancy disappears after the shortening and is not seen in the shortening at a slower speed (1/2 V_{max}).

Kodama and Yamada (1979) proposed a model based on results from kinetic and calorimetric studies of myosin and actomyosin ATPase reaction, and showed that could account for such separation of the energy output, i.e. (heat + work), from the high-energy phosphate splitting by shortening muscles. The model predicts that a significant amount of heat is produced, in addition to that from ATP splitting, when isometrically contracting muscle is allowed to shorten and the transition from isometric to isotonic contraction occurs (see Figs. 2 and 3).

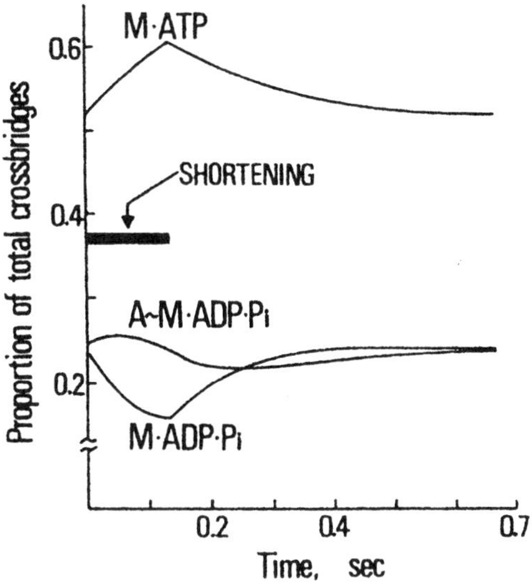

Fig. 2. Change in proportion of cross-bridge states during shortening followed by isometric contraction based on a kinetic model of actomyosin ATPase reaction. Reproduced with permission from Kodama and Yamada, 1979.

TIME-RESOLVED PHOSPHORUS NMR

Phosphorus NMR has become available for the non-destructive measurements of phosphorus compounds in intact muscle early in 1970s. Wilkie and colleagues have developed the method of using NMR further, for the study of phosphorus compounds in living muscle during rest, contraction and recovery (Dawson et al., 1977). This was inspired by serious doubt on the then current biochemical description of the situation based on studies with disrupted tissue, even if ultra-rapid freezing was employed as described above.

We have developed a method to improve the time-resolution from 7 min (Dawson et al., 1977) to 16 s to study living muscle contraction (Yamada and Tanokura, 1983; Kawano, Tanokura and Yamada, 1988). Contraction and recovery of bull-frog skeletal muscles were studied by using the methods described above in contractions of various durations at 5°C. The amount of PCr split coincided well with Pi appeared except for the initial few minutes following relaxation. During the early recovery period ΔPCr was significantly smaller than ΔPi (temporal separation between ΔPi and ΔPCr). The difference was 0.35 mmole/kg immediately after 2 s and longer tetanic contractions (Fig. 4). It is notable that this figure is comparable to the number of myosin heads in the muscle (0.28 mmole/kg, Ebashi, Endo and Ohtsuki, 1969).

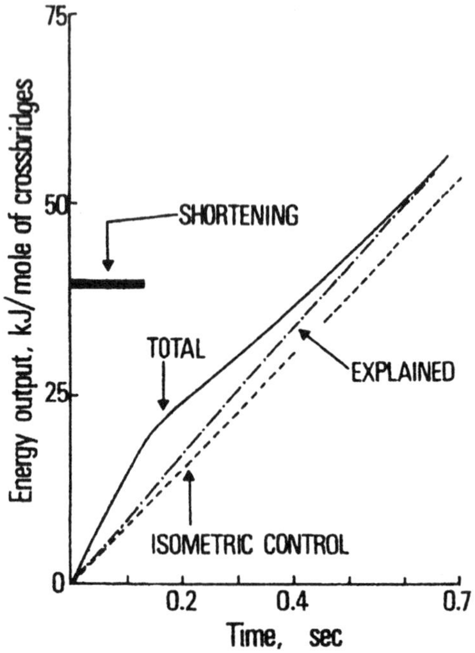

Fig. 3. Time course of energy output by muscle that is allowed to shorten and then returned to isometric contraction based on the enthalpy profile of actomyosin ATPase reaction. Reproduced with permission from Kodama and Yamada, 1979.

This phenomenon may be related to the repeated observations that certain amount of high-energy phosphate is utilized after contraction (see Gilbert et al., 1971). It may thus be related to the results of energy balance studies in which a significant fraction of the (heat + work) cannot be explained by PCr splitting (unexplained enthalpy, Gilbert et al., 1971; see Woledge et al., 1985). Operation of calcium pump may not produce such a separation of Pi and PCr.

This phenomenon suggests a delayed release of ADP from myosin heads in the early recovery period. This possibility seems compatible with the fact that the rate of release of ADP from rabbit myosin S1 is known to be markedly temperature dependent and can be very slow at low temperatures (Bagshaw, 1975).

Enthalpy change of PCr splitting has been estimated to be –34 kJ per mole by *in vitro* calorimetry under the conditions inside muscle cells (Woledge, 1973). Enthalpy change of PCr in muscle has been reported to be –46 kJ per mole (Gilbert et al, 1971). Therefore, the delayed splitting of PCr (0.35 mmole/kg) would produce heat of 12–16 mJ per g muscle. Because PCr break-down continues without heat production after contraction (Gilbert et al., 1971), the same amount of heat should have been produced without PCr splitting early during contraction (the unexplained enthalpy), and is reversed during early recovery period associated with PCr splitting without heat exchanges. The

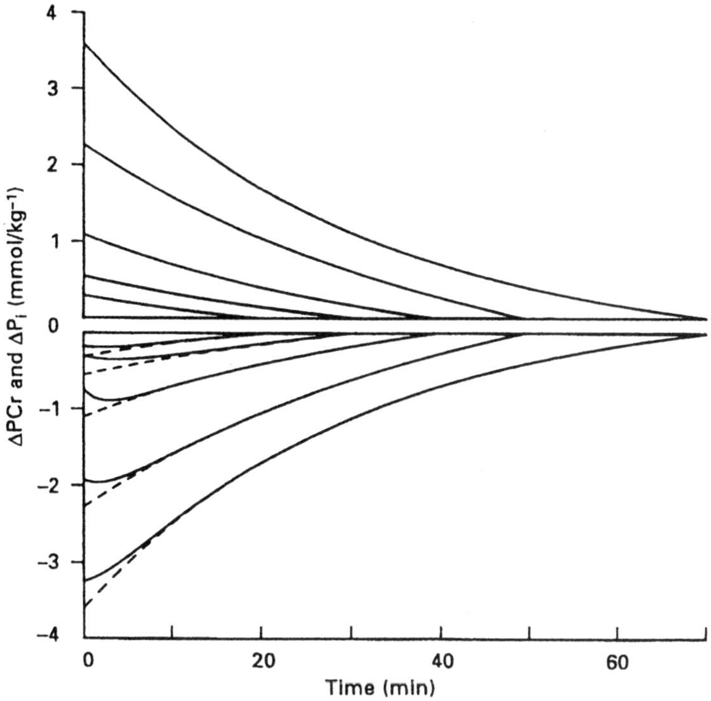

Fig. 4. Results summarizing phosphorus NMR studies on the effect of duration of contraction in bull-frog semitendinosus muscles. Changes in Pi and PCr immediately after the contraction, and time course of their recovery are shown in tetanic contractions of 0.2, 0.5, 2, 5 and 10 s duration at 5oC. Reproduced with permission from Kawano, Tanokura and Yamada, 1988.

unknown exothermic process producing 12-16 mJ of heat per g muscle early during isometric contraction corresponds to 3/10 – 4/10 of the unexplained enthalpy.

According to the kinetic model proposed by Kodama and Yamada (1979) as described above, there could also be a shift in the cross -bridge states during the transition from rest to contraction, producing a significant amount of heat without ATP splitting. According to Kodama and Yamada (1979), a shift of cross-bridge states from M. ADP. Pi to AM. ADP. Pi and M. ATP during an isometric contraction of 1 s would produce heat of about 10 mJ/g muscle (see Fig. 2).

MICROCALORIMETRIC APPROACH TO THE CALCIUM BINDING OF TROPONIN C AS A PRIMARY STEP OF THE REGULATION OF CONTRACTION

One of the problems in the mechanism of filament sliding is its regulation. Kodama and Woledge (1976) introduced titration microcalorimetry for studying the mechanism of ATP hydrolysis by myosin, and Yamada and Kometani (1983) for studying calcium binding of troponin C associated with the regulatory mechanism.

Contraction in all muscles is triggered by an increase in the calcium concentration to 100 ~ 1000 times the resting level, and this calcium signal is conveyed to the actomyosin system. The calcium sensitivity of the actomyosin system results from the presence of the troponin complex and tropomyosin (Ebashi, 1963). The binding of calcium to troponin C, the target protein of the complex, initiates a series of events involving the regulatory proteins of the I filament.

Troponin C is the highly specialized calcium-binding protein of the troponin complex. Troponin C has two high-affinity sites that bind calcium and magnesium competitively, and two low-affinity sites which selectively bind calcium over magnesium (Ebashi and Endo, 1968; Potter and Gergely, 1975; Potter, Seidel, Leavis, Lehrer and Gergely, 1976). Since the change in free magnesium concentration did not affect calcium sensitivity of the myofibrillar ATPase activity (Potter and Gergely, 1975), it is reasonable to assume that the low-affinity sites are the ones related to the regulation of contraction.

When troponin C of the rabbit skeletal muscle was titrated with calcium in a microcalorimeter a substantial amount of heat is produced in several phases according to the site involved (Yamada and Kometani, 1983). In addition we noticed an anomalous phase where heat produced becomes very small or even heat is absorbed. This heat is also characterized by an increase in heat capacity, which is obtained by the dependence of ΔH on temperature. This characteristic phase of positive enthalpy and heat capacity changes should be involved in the mechanism of regulation of contraction because these anomalous phases are associated exclusively with the proteins having regulatory functions (Yamada, 2004). We may call these regulatory calcium sites as "active" calcium sites of troponin C. Thus microcalorimetric titrations of EF-hand calcium-binding proteins resulted in the notion that calcium binding to the "active" calcium sites induce characteristically anomalous enthalpy and heat-capacity changes (Yamada, 2004).

In rabbit skeletal muscle the characteristic phase of positive enthalpy and heat capacity changes is associated with calcium binding to one of the two high-affinity calcium sites of the C-terminal domain (Yamada and Kometani, 1982). Calorimetric studies have also shown that, in frog skeletal troponin C, one of the N-terminal low-affinity calcium-binding sites is the "active" calcium site (Imaizumi and Tanokura, 1990). This feature is in good agreement with the results of structural studies of the N-terminal domain of skeletal troponin C originated from the reconstituted chicken troponin C (Slupsky, Kay, Reinach, Smillie and Sykes, 1995). The structure of the calcium-saturated N-terminal domain is "open" and the hydrophobic pocket is exposed to the solvent compared with the apo N-terminal domain. The frog troponin C has the same number of amino-acid residues as chicken troponin C, which has three amino-acid residues more than rabbit skeletal troponin (van Eerd, Canopy, Ferraz and Pechere, 1978).

The hydrophobic effect is an important factor in the organization of the molecules in biology (Tanford, 1973). A group of hydrophobic residues exposed to the solvent on calcium binding to the "active" site of troponin C may interact with the hydrophobic NH_2-terminal regions of troponin I (see Yamada, 1999).

Anomalously positive enthalpy change (heat absorption) has also been detected in ATP cleavage reaction on myosin (Kodama and Woledge, 1979; Kodama, 1981; also see

Kodama, 1985). In this case, however, entropy change is positive and heat-capacity change is negative, indicating that hydrophobic residues are retracted from the surface to the interior of the molecule.

Microcalorimetric studies of various kinds of calcium-binding proteins that have specific functions in muscle cells have indicated that there could be a fundamental difference in the mechanism of regulation of contraction by the calcium-troponin-tropmyosin system among skeletal muscles of different species and between cardiac and skeletal muscle (Yamada, 2004). Specific features attained by structural studies should fit in with the outline provided by calorimetry.

ACKNOWLEDGEMENTS

I express my sincere gratitude to my former colleagues who participated in the studies cited in this article.

REFERENCES

Bagshaw, C. R., 1975, The kinetic mechanism of the manganous ion-dependent adenosine triphosphatase of myosin subfragment 1. FEBS Letters **58**; 197.

Dawson, M. J., Gadian, D. G., and Wilkie, D. R., 1977, Contraction and recovery of living muscles studied by ^{31}P nuclear magnetic resonance. J. Physiol. (London) **267**; 703.

Ebashi, S., 1963, Third component participating in the superprecipitation of actomyosi n. Nature **200**; 1010.

Ebashi, S., and Endo, M., 1968, Calcium ion and muscle contraction. Progr. Biophys. Molec. Biol. **18**; 125.

Ebashi, S., Endo, M., and Ohtsuki, I., 1969, Calcium ion and muscle contraction. Q. Rev. Biophys. **2**; 351.

Van Eerd, J.-P., Cannopy, J.-P., Ferraz, C., and Pechere, J.-P., 1978, The amino-acid sequence of troponin C from frog skeletal muscle. Eur. J. Biochem. **91**; 231.

Gilbert, C., Kretzchmar, K. M., Wilkie, D. R., and Woledge, R. C., 1971, Chemical change and energy output during muscular contraction. J. Physiol. (London) **218**; 163.

Gosselin-Rey, C., and Gerdsay, C., 1977, Parvalbumins from frog skeletal muscle. Isolation and characterization. Structural modifications associated with calcium binding. Biochim. Biophys. Acta **492**; 53.

Hill, A.V., 1965, Trails and Trials in Physiology, Edward Arnold, London. p. 15.

Homsher, E., Mommaerts, W. F. H. M., Ricchiuti, N. V., and Wallner, A., 1972. Activation heat, activation metabolism and tension-related heat in frog semitendinosus muscles. J . Phyisiol. (London), **220**; 601.

Homsher, E., Yamada, T., Wallner, A., and Tsai, J., 1984, Energy balance studies in frog skeletal muscles shortening at one-half maximal velocity. J. Gen. Physiol. **84**; 347.

Hou, T.-T., Johnson, J. D., and Rall, J. A., 1991, Parvalbumin content and Ca^{2+} and Mg^{2+} dissociation rates correlated with changes in relaxation rate of frog muscle fibres, J. Physiol. **441**; 285.

Hou, T., Johnson, J. D. and Rall, J. A., 1993, Role of parvalbumins in relaxation of frog skeletal muscle. Adv. Exp. Med. Biol. **332**; 141.

Imaizumi, M., and Tanokura, M., 1990, Heat capacity and entropy changes of troponin C from bullfrog skeletal muscle. Eur. J. Biochem. **192**; 275.

Irving, M., and Woledge, R. C., 1981, The dependence on extent of shortening of the extra energy liberated by rapidly shortening frog skeletal muscle. J. Physiol. (London) **321**; 411.

Kawano, Y., Tanokura, M., and Yamada, K., 1988, Phosphorus nuclear magnetic resonance studies on the effect of duration of contraction in bull-frog skeletal muscles. J. Physiol. (London) **407**; 243.

Kodama, T., and Woledge, R. C., 1976, Calorimetric studies of the interaction of myosin with ADP. J. Biol. Chem. **251**; 7499.

Kodama, T., 1985, Thermodynamic analysis of muscle ATPase mechanism. Physiol. Rev. **65**; 467.

Kodama, T., and Yamada, K., 1979, An explanation of the shortening heat based on the enthalpy profile of the myosin ATPase mechanism. In: Cross-bridge Mechanism in Muscle Contraction. H. Sugi and G. H. Polack, ed., Univ. Tokyo Press, Tokyo, pp. 481-488.

Kodama, T., Kurebayashi, N., and Ogawa, Y., 1980, Heat production and proton release during the ATP -driven Ca uptake by fragmented sarcoplasmic reticulum from bullfrog and rabbit skeletal muscle. J. Biol. (Tokyo) **88**; 1259.

Kometani, K., and Yamada, K., 1983, Dependence of shortening heat on sarcomere length in frog muscle and fiber bundles. Jpn. J. Physiol. **33**; 895.

Potter J. D., and Gergely J., 1975, The calcium and magnesium binding sites on troponin and their role in the regulation of myofibrillar adenosine triphosphatase. J. Biol. Chem. **250**; 4628.

Potter J. D., Seidel, J. C. Leavis, P., Lehrer, S. S., and Gergely, J., 1976, Effect of Ca^{2+} binding on troponin C. Changes in spin label motility, extrinsic fluorescence, and sulfhydryl reactivity. J. Biol. Chem. **251**; 7551.

Rall, J. A., 1980, Effects of previous activity on the energetics of activation in frog skeletal muscle. J. Gen. Physiol. **75**; 617.

Slupsky, C. M., Kay, C. H., Reinach, F. C., Smillie, L. B., and Sykes, B. D., 1995, Calcium-induced dimerization of troponin C: Mode of interaction and use of trifluoroethanol as a denaturant of quarternary structure. Biochemistry **34**; 7365.

Smith, I. C. H., 1972, Energetics of activation in frog and toad muscle. J. Physiol. (London), **220**; 583.

Tanford, C., 1973, The Hydrophobic Effect. Formations of Micelles and Biological Membranes. John Wiley & Sons, New York.

Tanokura, M., and Yamada, K., 1987, Heat capacity and entropy changes of the two major isotypes of bullfrog (Rana catesbeiana) parvalbumins induced by calcium binding. Biochemistry. **26**; 7668.

Weber, A., and Herz, R., 1963, The binding of calcium ions to actomyosin systems in relation to their biological activity. J. Biol. Chem. **238**; 599.

Winegrad, S., 1970, The intracellular site of calcium activation of contraction in frog skeletal muscle. J.Gen. Physiol. **55**; 77.

Woledge, R. C., 1973, In vitro calorimetric studies relating to the interpretation of muscle heat experiments. Cold Spring Harbor Symp. Quant. Biol. **37**, 629.

Woledge, R. C., Curtin, N. A. and Homsher, E., 1985, Energetic Aspects of Muscle Contraction. Academic Press, London.

Yamada, K., and Kometani, K., 1982; The changes in heat capacity and entropy of troponin C induced by calcium binding. J. Biochem. (Tokyo) **92**; 1505.

Yamada, K., Mashima, H., and Ebashi, S., 1976, The enthalpy change accompanying the binding of calcium to troponin relating to the activation heat production of muscle. Proc. Japan Acad. **52**; 252.

Yamada, K., and Tanokura, M., 1983, Post-contractile phosphocreatine splitting in muscle as revealed by time-resolved ^{31}P nuclear magnetic resonance. Jpn. J. Physiol. **33**; 909.

Yamada, K., 1999, Thermodynamic analyses of calcium binding to troponin C, calmodulin and parvalbumins by using microcalorimetry. Molec. Cell Biochem. **190**; 39.

Yamada, K., 2004, Calcium binding to troponin C as a primary step of the regulation of contraction. A microcalorimetric approach. In: Molecular and Cellular Aspects of Muscle Contraction. H. Sugi, ed., Kluwer/Plenum, New York, pp. 203-213.

DISCUSSION

Rall: You call the calcium binding site in calcium binding proteins that exhibits anomalous heat production you have as the "active" site. What do you mean by active site?

K. Yamada: By "active site" I mean, in a very tentative sense, the single Ca-binding site of troponin C which has an thermodynamically anomalous nature, and is most likely to be involved in the mechanism of regulation. It should be the same as the "regulatory" site.

Kushmerick: Does the delayed splitting of PCr caused the ADP coming from slow creatine kinase reaction.

K. Yamada: The delay in creatine kinase reaction is unlikely to occur during contraction at low temperature. This is because 1) muscles develop force which is very well maintained even with long tetanus, and 2) the amount of delayed PCr splitting does not increase with the duration of contraction longer than about 1 s.

HOW TWO-FOOT MOLECULAR MOTORS MAY WALK

Kazuhiko Kinosita, Jr.[1], M. Yusuf Ali[2], Kengo Adachi[1], Katsuyuki Shiroguchi[1], and Hiroyasu Itoh[3,4]

1. INTRODUCTION

Myosins and kinesins each constitute a large family of linear molecular motors that track along a filamentous rail, myosins along an actin filament and kinesins along a microtubule. These motors are powered by free energy derived from ATP hydrolysis, and the mechanisms of chemo-mechanical conversion in these motors have been under intensive study (Kinosita et al., 1998; Vale and Milligan, 2000; Mehta, 2001; Vale, 2003; Schliwa and Woehlke, 2003; Endow and Barker, 2003). Most of myosins and kinesins have two globular domains that bind to the filamentous rail and that hydrolyze ATP in a rail-dependent manner. The two domains are usually called "heads" or "motor domains" and are connected via a neck-like structure to a common stalk. Some of the two-headed motors are processive, in that a single molecule moves along a rail for many ATPase cycles without detaching from the rail. These processive motors appear to "walk," using the two heads alternately in a hand-over-hand fashion, as has recently been demonstrated for myosin V (Yildiz et al., 2003) and conventional kinesin (Asbury et al., 2003; Yildiz et al., 2004). Unlike walking of a human being, molecular motors cannot rely on inertia, which is negligible for biological molecules that work in water or membranes. A human has a right and a left foot, but the two heads of a molecular motor are identical (Fig. 1*a*). How, then, do they walk? In this article, we focus on the mechanism(s) that warrant forward, not backward, stepping. Because we discuss walking mechanisms in this article, we call the heads "feet" and necks "legs."

[1]Okazaki Institute for Integrative Bioscience, National Institutes of Natural Sciences, Higashiyama 5-1, Myodaiji, Okazaki 444-8787, Japan. [2]Department of Physics, Faculty of Physical Sciences, Shahjalal University of Science and Technology, Sylhet-3114, Bangladesh. [3]Tsukuba Research Laboratory, Hamamatsu Photonics KK, and [4]CREST "Creation and Application of Soft Nano-Machine, the Hyperfunctional Molecular Machine" Team 13*, Tokodai, Tsukuba 300-2635, Japan.

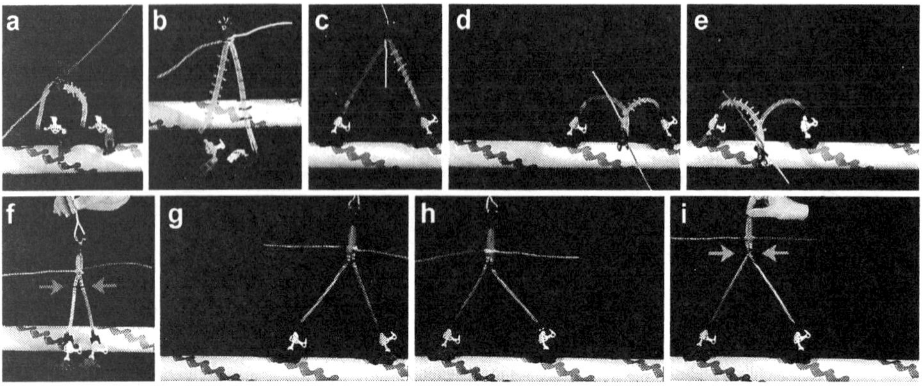

Figure 1. Walking with two identical feet. The legs are made of elastic vinyl pipes, which can accommodate both torsional and flexural strains. (**a**) Two feet of a molecular motor are identical, because they are coded by the same gene. (**b**) The two legs are joined in twofold symmetry. Thus the two feet would be oriented in opposite directions, one toe facing east if the other faces west, unless the legs are flexible. (**c**) When two feet land simultaneously during walking, both toes must be oriented forward, because the landing sites on the rail are all identical. Elastic legs would thus severely distorted. In the configuration shown in this image (*b*), most of the strain is in torsion of the legs (note the directions of the small bars on the legs) whereas bending is relatively small. (**d**) By turning one of the feet by 360°, much of the torsional strain is relieved, resulting in severe bending. The total strain is the same as in *c*. (**e**) When the motor moves forward by one step and attains the same posture as in *d*, a cargo (pink and blue long bars) attached to the body rotates by 180°. A 180° rotation also accompanies every step if the motor steps with the posture in *c*. (**f**) Flexible joints in the legs (arrows) allow the two feet to adopt the same orientation. (**g**) With flexible joints, landing with two feet does not introduce strain. (**h**) But, because of the basic twofold symmetry, the body still tends to rotate 180° every time the motor steps. (**i**) With flexible joints, though, the 180° rotations can be prohibited by an external force without introducing too much strain in the legs. Note that the two legs are now crossed near the junction (arrows); a small torque suffices to bring about this configuration.

2. HINTS ON WALKING MECHANISMS

2.1. Walking with Two Identical Feet

The two legs of a myosin or kinesin molecule emerge from a coiled coil of α-helices (the stalk, or "body"), each helix extending into a leg (see Figs. 2*c* and *d*). Because the coiled coil is twofold symmetric, as confirmed for kinesin (Kozielski et al., 1997) and myosin (Li et al., 2003), the two identical legs are also arranged, basically, in the same twofold symmetry: if there were no flexibility, the two feet would be oriented in opposite directions, making it extremely difficult to walk on landing spots that are unidirectionally oriented on a rail (Figs. 1*b-e*). In fact, each leg of kinesin is a single polypeptide chain and is considered flexible over the entire length, unless it is "docked" onto the foot (Fig. 2*d*); in a crystal structure of kinesin dimer where both legs were docked (Kozielski et al., 1997), the two feet were not arranged in twofold symmetry, indicating that the short undocked portion is already flexible. Myosin's legs are reinforced with light chains (Figs. 2*a* and *b*), and thus are probably semi-rigid. Flexibility likely resides at the leg-body junction. Indeed, analyses of rotational Brownian motion of myosin II (conventional myosin) indicated the presence of a flexible joint near the leg-body junction (Kinosita et al., 1984; Ishiwata et al., 1987), and electron micrographs of myosin

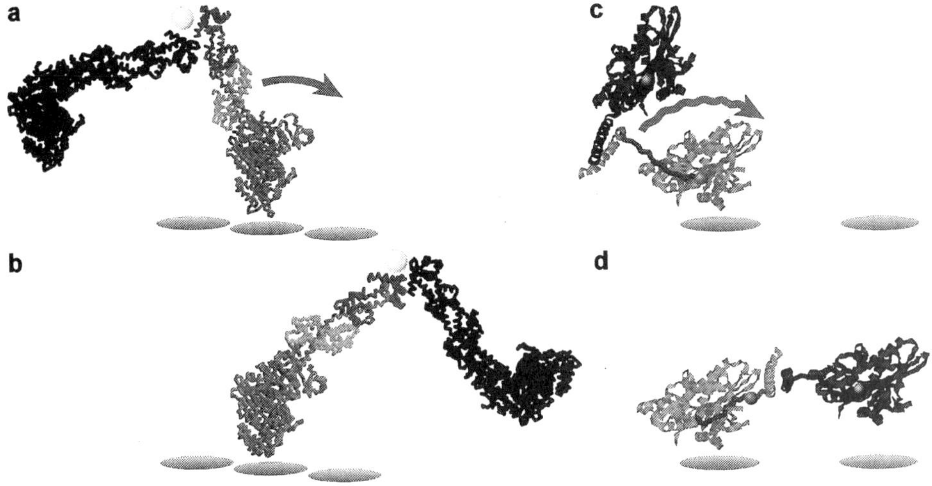

Figure 2. Postulated lever action in two-foot motors. The motors in this figure move toward right. (**a, b**) Scallop myosin. This myosin is non-processive and thus does not "walk." Nevertheless, we show possible walking postures in order to indicate lever action (arrow) in myosin: the orange leg (the long α-helix) rotates between *a* and *b*. Flexible joints likely exist around the location indicated by the yellow ball, beyond which the two α-helices form a coiled coil (the "stalk"; not shown in the figure). Thus, the red leg presumably undergoes rotational diffusion around the yellow ball. The lever action of the orange leg biases the Brownian motion of the red foot forward. Each figure is composed of two structures of scallop myosin subfragment 1 (S1), arranged arbitrarily. The orange foot/leg in *a* represents the structure of S1 binding MgADP·VO_4, and that in *b* is S1 without a nucleotide (Houdusse et al., 2000). The red foot/leg is S1 binding MgADP, which mimics a structure after detachment from actin (Houdusse et al., 1999). Cyan/blue, essential light chains; green/dark green, regulatory light chains. The gray disks with a diameter of 5.5 nm represent binding sites on an actin filament. (**c, d**) Conventional kinesin. The blue and green legs ("neck linkers") between the balls are presumably flexible, and thus the red foot undergoes Brownian motion. The flexible green leg in *c* "docks" onto the orange foot, in an ATP-dependent process, through Brownian motion (wavy arrow). After docking (*d*), the Brownian motion of the red foot occurs around the green ball, and thus is biased toward the forward binding site. The docking is equivalent with the lever action in myosin. The gray disks with a diameter of 4 nm represent binding sites on a microtubule. The figures are constructed from a structure of a dimeric kinesin (Kozielski et al., 1997) by assigning arbitrary structures to the neck linkers (except the green one in *d*) and orienting the central coiled coil (stalk) arbitrarily.

V revealed the angle between the two legs to be quite variable (Walker et al., 2000). In a crystal of the coiled coil portion of myosin II, the end next to the legs was disordered (Li et al., 2003).

The flexibility in the undocked legs of kinesin and that at the body-leg junction of myosin allows the two feet to adopt the same orientation, which is a required posture when the two feet simultaneously land during walking. Because the two feet and legs are identical, the landed posture with least strain must be the same whether one or the other foot leads, although the flexibility may allow many other postures without imposing much strain. If the motor reaches the unique, most stable posture after every step, then symmetry dictates that the body must turn around its axis by 180° as the leading and trailing feet swap (Howard, 1996; also see Figs. 1*g* and *h*). Hua et al. (2002) sought for, but did not observe, the 180° reorientations that would accompany every step in kinesin. They thus suggested inchworm walking in which one foot always leads. We also failed to observe the 180° reorientations in myosin V (Ali et al., 2002) and myosin VI (Ali et al., 2004). Recent studies, however, have established that myosin V and kinesin walks in a hand-over-hand fashion in which the two feet alternate in the lead. A single fluorophore attached to one of the two feet of myosin V (Yildiz et al., 2003) or kinesin (Yildiz et al., 2004) advanced, for every two ATP hydrolysis cycles, over a distance twice the motor's step size which is the distance traveled by the body of the motor in one ATP cycle; inchworm walking would advance the fluorophore by a distance equaling the step size. Also, in some mutant kinesins, dwells between steps were alternately long and short, indicating an asymmetric hand-over-hand walking with two different conformations for two-foot landing (Asbury et al., 2003). Why, then, have the 180° reorientations not been detected so far? Previous observations were made by attaching a large probe, a microtubule or micron-sized bead, to the motor. The attached probe may well have broken the symmetry of the motor, as in the mutations, and impeded the reorientation. With the assumed flexibility, the torque needed to break the symmetry is small (Fig. 1*i*). Genuine motion of these motors, without a cargo, might still involve 180° reorientations, although the flexible joints likely obscure the rotation.

2.2. Lever Action and Biased Diffusion

Until recently, researchers working on myosin and kinesin had somewhat different views. Because myosin's legs are likely stiff, prevailing theory states that a landed leg acts as a lever (Huxley, 1969; Holmes and Geeves, 2000): when a landed ankle is bent forward, the leg leans forward, carrying the body forward (Figs. 2*a* and *b*). The tilting is best evidenced for myosin V (Walker et al., 2000; Moore et al., 2001; Veigel et al., 2001; Burgess et al., 2002; Forkey et al., 2003). The lifted leg thus easily finds a forward landing site. Kinesin's legs, in contrast, are flexible and unlikely to serve as a stiff lever. Instead, a lower part of a landed leg docks onto the landed foot such that the upper leg emerges from a forward part of the foot (Figs. 2*c* and *d*). This biases the Brownian motion of the lifted foot forward, and the foot lands on a forward site. The relatively small bias could be efficient, because kinesin's legs are short and must be fully extended to reach a forward or backward site: of the two sites that are available, only the forward site can be reached after the docking of the landed foot. The docking in kinesin may be regarded as the equivalent of the lever action in myosin, and then the two schemes are apparently quite similar. Myosin researchers, however, tend to stress the lever action whereas kinesin researchers diffusional search. The distinction seems to be coupled to the question of which process produces force: lever action itself in myosin, and landing of the diffusing foot in kinesin. Is this really so?

3. SPIRAL MOTION OF MYOSIN AROUND AN ACTIN FILAMENT

3.1. Observation of Spiral Motion

Myosin V is now one of the best characterized molecular motors. It moves toward the barbed, or plus (fast-growing) end of an actin filament using two long (~23 nm) legs each reinforced with six light chains (Cheney et al., 1993; see Fig. 3). The movement is processive (Mehta et al., 1999; Rief et al., 2000; Sakamoto et al., 2000). The step size

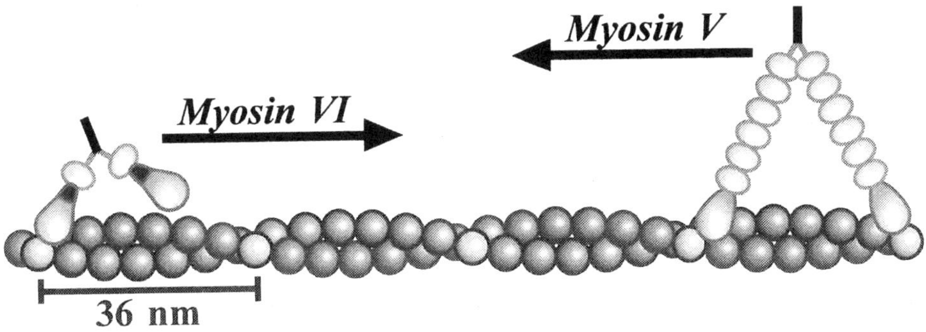

Figure 3. Myosin V and myosin VI. Myosin V has two legs each reinforced with six light chains (orange) and walks on an actin filament (green) toward the barbed, or plus, end (left in the figure). Myosin VI, on the other hand, moves toward the pointed, or minus, end. Previously, myosin VI was thought to bind one light chain per leg, and thus myosin VI was considered short-legged as shown in this figure. The red portion in the feet of myosin VI indicates an extra insert not seen in other myosins. Every 13th monomer of the actin filament is shown in blue. Walking along one side of an actin filament implies an average step size of ~36 nm. Adapted from Ali et al. (2004).

of myosin V has been estimated to be about 36 nm (Mehta et al., 1999; Rief et al., 2000; Rock et al., 2001; Tanaka et al., 2002), which coincides with the helical repeat of an actin filament of 36 nm (Holmes et al., 1990; see Fig. 3). Electron micrographs of myosin V (Walker et al., 2000) have shown that bound feet are separated by ~36 nm on an actin filament, corroborating the step size measurements. In most of these previous studies, however, myosin V could approach an actin filament from only one side, and thus myosin V may have been forced to step on the actin sites that are 36-nm apart (Fig. 3).

To address this possibility, we designed an experiment in Fig. 4*a*, where myosin V could freely rotate around an actin filament that was suspended between two large (4.5 μm) beads. We attached a duplex of 1-μm beads to myosin V to facilitate visualization of rotation. The microscope focus was set such that the bead closer to the observer appeared white and the bead farther from the observer black. As seen in Fig. 4*b*, the bead duplex moved in a left-handed spiral around the right-handed double helix of actin. The rotational pitch was 2.2±0.3 μm per turn, compared to the pitch of the actin double helix of 72 nm. The slight left-handed spiral can be explained if the step size of myosin V is slightly less than the actin helical repeat (half pitch) of 36 nm: in Fig. 4*a*, stepping on blue subunits that are 36-nm apart produces straight motion without spiraling, whereas slightly smaller strides, e.g. on the purple subunits, would lead to a left-handed spiral.

The average step size of myosin V is calculated from the spiral pitch in the following way. If the step size is $(36 - x)$ nm, then myosin V will rotate toward left by $180° \times (x/36)$ per step. To accumulate one turn, myosin V has to make $360° / [180° \times (x/36)] = (72/x)$ steps, or to proceed over a distance of $(72/x) \times (36 - x)$ nm. This latter distance is equal to the spiral pitch of 2200 nm. Thus, $(72/x) \times (36 - x) = 2200$. Solving for x yields $x = 1.1$ nm, and thus the step size is 34.9 nm. (This calculation is slightly different from the one in Ali et al. (2002), but the answer is essentially the same.) With this average step size, the predominant landing site from the position in Fig. 4*a* is on the blue subunit, or the 13th subunit counting both strands. Occasionally myosin V lands on the 11th subunit (purple), and less frequently on the 15th or 9th.

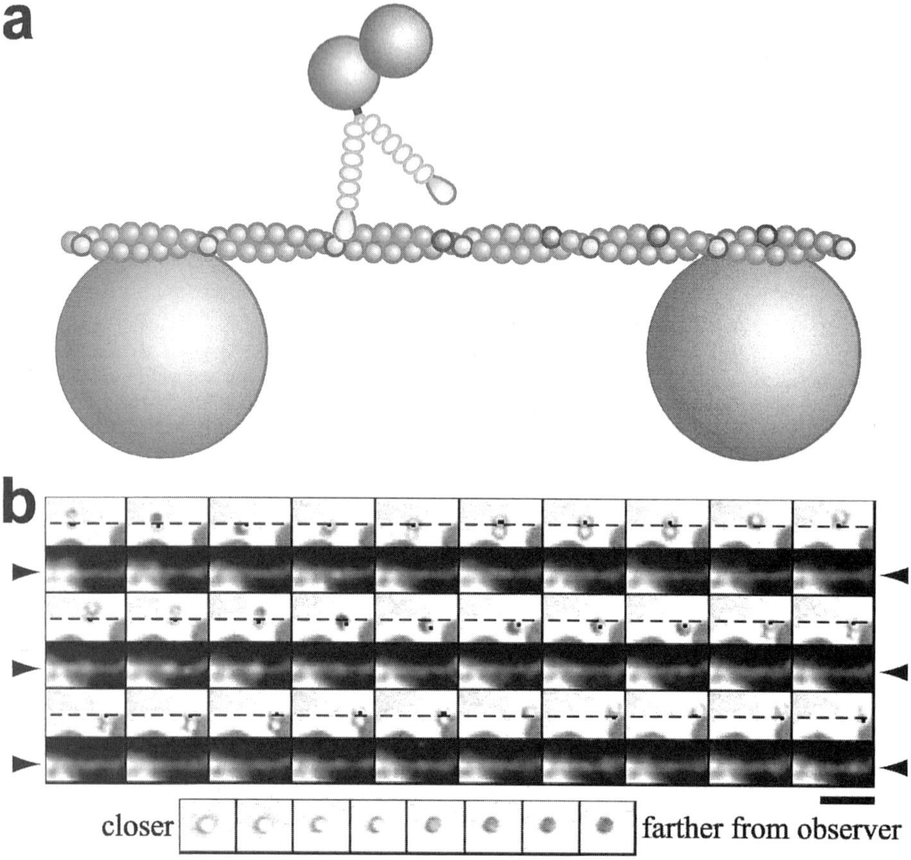

Figure 4. Spiral motion of myosin V around an actin filament. (**a**) Experimental setup. An actin filament, fluorescently labeled, was suspended between beads of diameter 4.5 μm placed on a glass surface. To myosin V, a duplex of 1-μm beads was attached. (**b**) Sequential images, at 1-s intervals, of spiral motion of myosin V at 400 μM ATP. Upper frames in a row show transmitted images of the bead duplex, where a bead closer to the observer appears white and a bead farther away black (see control images at bottom). Lower frames show fluorescence images of the actin filament (between arrow heads). The dashed line in the upper frames show the position of the actin filament deduced from the lower frames. The dot in the upper frames indicates the center of the bead which we judged to be attached to actin. Bar, 5 μm. Adapted from Ali et al. (2002).

The step size we obtained under no load (viscous load is negligibly small for the slow translational motion) is not so different from the previous estimates. Previous steps of ~36 nm were not imposed by experimental geometry. Myosin V naturally walks almost straight, on one side of an actin filament. The implication is that the lever action in myosin V occurs, on the average, almost precisely forward, or in a plane including the actin filament, to let the lifted foot aim from above, not obliquely from side, at the landing sites on the filament. The slight tendency to spiral to the left is explained if forward bias by the lever action is not sufficient to comfortably place the lifted foot on the 13th subunit. Or, the ankles of myosin V may have an intrinsic tendency to bend slightly to the left. When many molecules of myosin II carry an actin filament forward, the filament rotates as a right-handed screw (Nishizaka et al., 1993) with a pitch of ~1 μm (Sase et al., 1997). Although myosin II is not processive and thus does not walk, the right-handed spiral can be explained if myosin II "kicks" actin by bending a landed ankle slightly obliquely to the right. The structure of the foot (motor domain) of myosin V is similar to that of myosin II (Coureux et al., 2003; Holmes et al., 2003), but a small difference in the direction of ankle action is not inconceivable.

3.2. Myosin VI May Move Like Kinesin

Myosin VI moves, unlike most other myosins, toward the pointed, or minus, end of an actin filament (Wells et al., 1999; Homma et al., 2001). It is a processive motor (Rock et al., 2001; Nishikawa et al., 2002), and thus is considered to "walk" using its two feet. Myosin VI has only one typical light-chain binding site per leg (Wells et al., 1999) and was initially considered short-legged (Fig. 3). Nevertheless, its step size was long, ~30 nm (Rock et al., 2001) or ~36 nm (Nishikawa et al., 2002), apparently incompatible with the expected physical size of the legs. Again, these measurements were made in a configuration that allowed myosin VI to approach from one side of an actin filament, raising the possibility that this constraint somehow forced the short-legged myosin VI to stride on the 36-nm spaced landing sites on one side of the filament. We thought that, if we allowed myosin VI to rotate freely around an actin filament, it would spiral more extensively toward left than myosin V. It might even spiral to the right with a pitch of 72 nm, if its natural step size is smaller than 18 nm such that the motor tracks along one strand of the actin double helix. So we tested this idea, again using the actin-bridge assay (Fig. 4*a*).

A result is shown in Fig. 5*a*, where the bead duplex spiraled as a right-handed screw with a pitch of ~1.4 μm. Many more right-handed spirals were observed, the pitch averaging 2.4 μm. The shallow right-handed spiral indicates an average step size slightly greater than 36 nm (e.g., walking on purple subunits in Fig. 5*b*); the average step size calculated as above is 37.1 nm. There were, however, many more bead duplexes (>80%) that failed to show clear rotation (>0.5 revolution) during a processive run of a few μm. If these represent straight motion, the step size would be 36 nm (on blue actin subunits in Fig. 5*b*). Or, these may also be a part of long-pitch right-handed spiral. In any event, the average step size of myosin VI somewhat exceeds 36 nm, being the longest among known molecular motors. Other long steppers include myosin V, and myosin XI with the average step size of 35 nm (Tominaga et al., 2003).

The long strides of myosin VI are obviously incompatible with walking with short legs. In an electron micrograph (Nishikawa et al., 2002), legs of myosin VI were somehow extended, possibly by unzipping of the stalk coiled coil. Another possibility that the insert unique to myosin VI (red part in Fig. 3) may adopt an extended conformation (Rock et al., 2001) has been denied by a recent study (Bahloul et al., 2004) which shows that the insert is actually a second calmodulin (light chain) binding domain. More recent work (B. R. Rami and J. A. Spudich, Stanford University, and H. L. Sweeney and C. Franzini-Armstrong, Pennsylvania University, personal communication) shows that the coiled coil is indeed unzipped over a considerable length and the unzipped portion is flexible (Fig. 5*b*). The long and flexible legs can account for the unconstrained step size of 36-37 nm. Thus, myosin VI seems to walk almost like kinesin, relying on biased diffusion. A lower part of the legs of myosin VI is

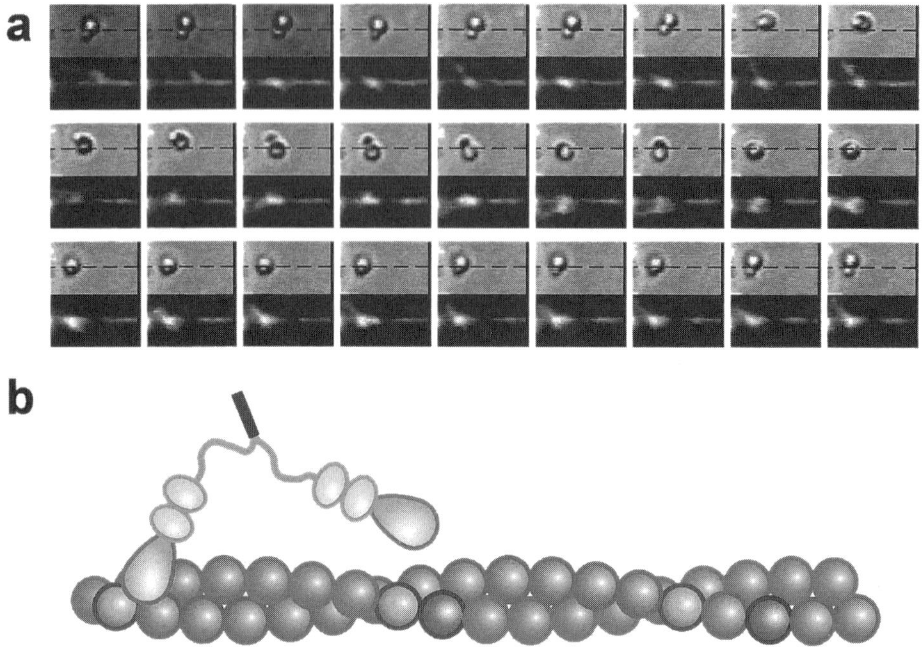

Figure 5. Spiral motion of myosin VI around an actin filament. (**a**) Sequential images, at 0.4-s intervals, of spiral motion of myosin VI at 400 μM ATP. See Fig. 4 for details. This bead duplex (1 and 0.45-μm) carried a short actin filament, which also rotated as seen in the fluorescence images in the lower frames. Of the bead duplexes examined, only ~15% showed right-handed spiral, as seen in these images. Others moved straight without spiraling. (**b**) A recent study (Bahloul et al., 2004) shows that the extra insert in the leg sequence (red part in Fig. 3) actually binds a calmodulin light chain. Also, the coiled coil beyond the light-chain binding region was found to be unzipped. Myosin VI can thus span >36 nm, as required by the right-handed spiraling. Adapted from Ali et al. (2004).

presumably semi-rigid, and its lever action would serve the purpose of biasing the diffusion of a lifted foot forward.

4. TOE UP-DOWN MECHANISM

4.1. Moving Forward against a Backward Force

Myosin and kinesin both appear to walk forward by biasing the diffusion of the lifted foot by an action of the landed ankle, either lever action or docking. Does this mechanism really warrant that the lifted foot lands on a forward, not backward, site? Such a mechanism, alone, would not work properly when the body is pulled backward, particularly when the legs are flexible as in kinesin and myosin VI. If the legs are flexible over a large part, diffusion will even be biased backward (Fig. 6*a*). In kinesin, the free energy difference between the docked and undocked states is small (Rice et al., 2003), implying that docking would fail when the body is pulled back by a load. Then, there will be little bias for diffusion, or backward diffusion may be preferred (Fig. 2*c*).

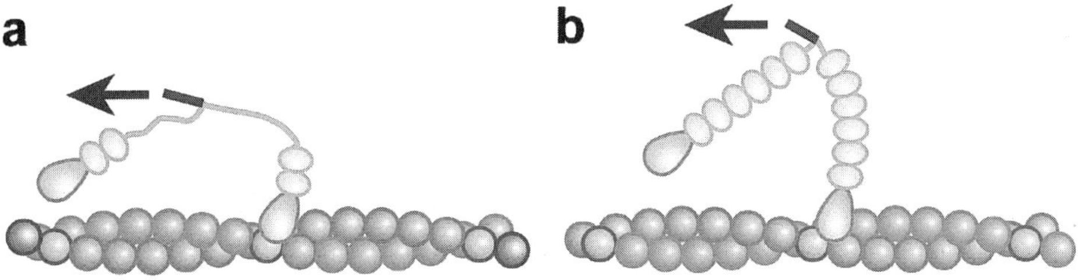

Figure 6. How to move forward against a backward force (arrows). (**a**) If legs are flexible over a considerable length, diffusion would move the lifted foot toward a backward site. (**b**) Under a high backward force, the presumably semi-rigid leg of myosin V may also be bent backward, such that the diffusion is biased backward.

Myosin V might appear to be free from these problems, but its legs cannot be perfectly rigid. Veigel et al. (2002) have estimated its leg stiffness to be 0.2 pN/nm, implying that the body would be pulled back by 10 nm under 2 pN of backward force (Fig. 6*b*). Also, the ankle of the landed foot may yield under the external force, because sustaining 2 pN at the end of the 23-nm leg requires a torque of 46 pN·nm, comparable to the torque of the powerful rotary motor F_1-ATPase (Yasuda et al., 1998; Kinosita et al., 2004). In spite of these problems, all these motors are known to move forward under a few pN of backward load. An additional mechanism, other than biasing the translational diffusion of the lifted foot by moving the effective pivot forward, must be involved.

4.2. Toe Up-Down Mechanism

The mechanism we propose (Ali et al., 2004) is the ankle action in the lifted foot: proper up-down motion of the lifted toe will orient the sole correctly such that landing on a forward site is favored over a backward site even if the body is pulled backward. The toe up-down mechanism warrants forward motion even if legs are completely flexible.

To illustrate the mechanism, let us again use the toy model introduced in Fig. 1. First we deal with the case of semi-rigid legs as in myosin V (Fig. 7). We assume the presence of a flexible joint near the leg-body junction (Fig. 1*f*), and thus there will be little torsional strain in the two legs. Bending strain, however, cannot be relieved by the flexible joint. In Fig. 7, a posture with a bent leg is attained only at the expense of a high energetic cost, or is rarely attained (unless coupled to a reaction accompanying a large drop in free energy).

Figure 7*a*1 shows a posture of the motor after the trailing foot (green) is lifted from the rail. The ankle of the leading foot (red) is already bent forward (lever action), such that the pivot for rotary diffusion of the lifted leg is forward of the landed foot (red) and is ready to bring the green foot forward. However, if the lifted toe is up, as in Fig. 7*a*1, the sole will be misoriented after a forward swing (Fig. 7*a*2), making forward landing difficult. Forced landing would bend the leg severely (Fig. 7*a*3). Natural landing sites for the toe-up foot will be next to the landed foot (Fig. 7*a*4), front or back, and thus the motor will hardly walk forward.

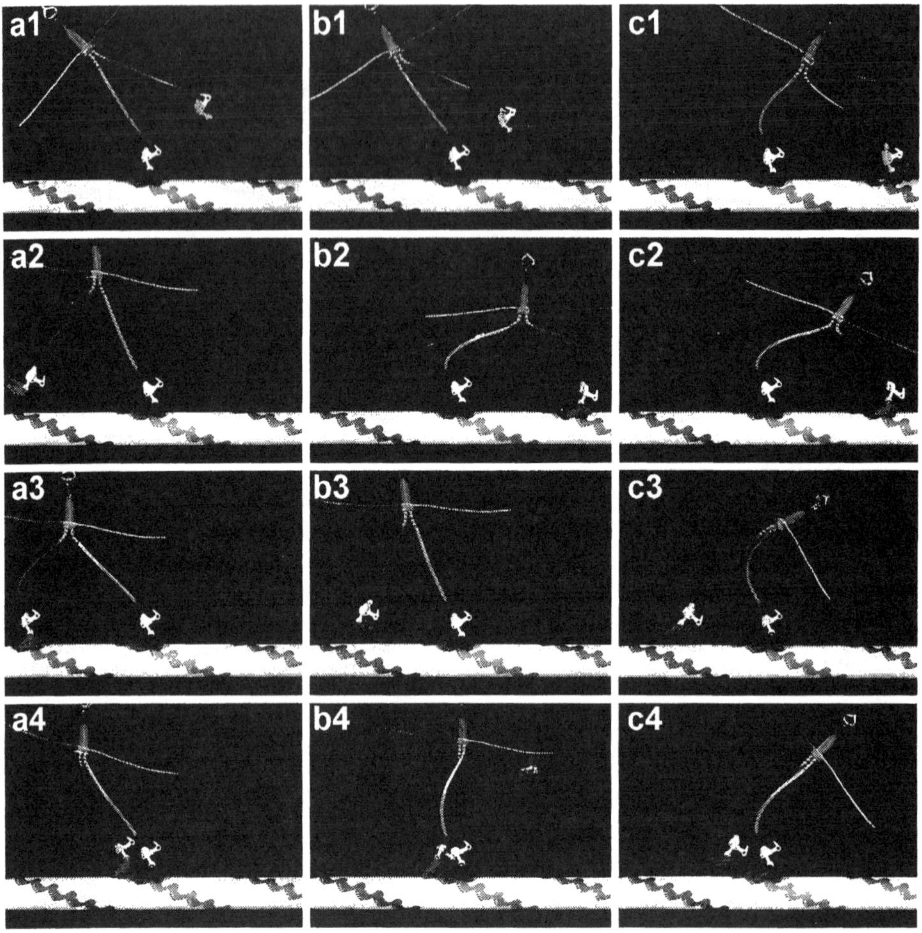

Figure 7. Toe up-down mechanism for the case of semi-rigid legs. (**a**) If the lifted toe is up, landing on a forward site is difficult, and natural landing would be on a nearby site (*a*4), if one is available. (**b**) If the toe goes down upon lifting of the foot, landing with the least strain will be on a distant forward site. (**c**) A backward force will pull the pivot backward (*c*1), but, as long as the lifted toe is down, landing on a distant forward site is favored. Under a high force, landing on a nearby site is equally probable (*c*4). Note that, in *c*4, bending of the red leg is caused by the external force. Adapted from Ali et al. (2004).

If, on the other hand, the lifted toe goes down, as in Fig. 7*b*1, re-landing on the backward site will be prohibited (Fig. 7*b*2), whereas landing on a distant forward site will be smooth and natural (Fig. 7*b*3). Landing on a nearby site would result in a strain (Fig. 7*b*4) and thus is much less likely.

A nice feature of the toe up-down mechanism is that it operates properly even in the presence of a backward force that tends to pull back the pivot beyond the position of the landed foot (Fig. 7*c*1). Thanks to the sole orientation, backward landing is still prohibited (Fig. 7*c*2), whereas a small thermal agitation will allow landing on a distant forward site (Fig. 7*c*3). Landing on a nearby site is also allowed under a high backward force (Fig. 7*c*4), but this is the condition where the motor begins to stall.

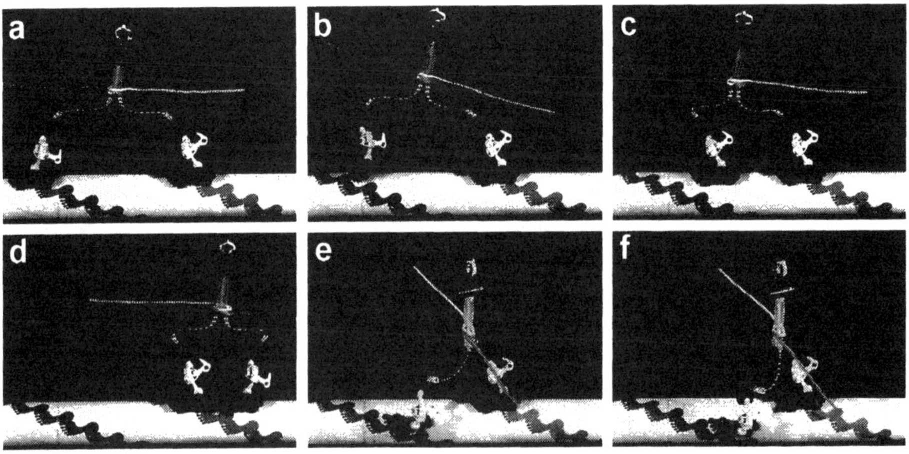

Figure 8. Toe up-down mechanism for the case of completely flexible legs. The legs in the model are made of metal chains. (**a**) The size of the legs are such that landing on a distant forward site is allowed only when the legs are fully extended and the leading toe is down and trailing toe up. (**b, c**) With the green toe up, forward landing is impossible. (**d**) As long as the green toe is down, there is no hope in backward landing. (**e, f**) If a lower part of legs is stiff, oblique landing on a side of the cylindrical rail is also prohibited. The postures here represent the lowest possible position for the green foot, which fails to reach a blue foot mark below.

The toe up-down mechanism also works with flexible legs, as long as landing sites are widely separated such that landing requires almost full extension of the legs. This is because, on a fully extended leg, the orientation of the sole is restricted, in the same way as in a rigid leg. In Fig. 8, the legs of the toy model are made of flexible chains. Suppose, for the moment, that only those landing sites (foot marks) that are at the top of the rail cylinder are available. This situation applies to kinesin on a microtubule. The green foot can land on the forward site if its toe is down (Fig. 8*a*), but it cannot land if the toe is up (Figs. 8*b* and *c*). When the green toe is down, forward movement is warranted because landing on the backward site is impossible (Fig. 8*d*). The preferential forward landing also operates in the presence of backward force. In Fig. 8*d*, even if the red ankle yields to a backward force and bends backward (kinesin "undocks" by backward pull), the green foot cannot reach the backward site as long as its toe remains down. Forward landing (Fig. 8*a*) would still be possible, when the body happens to move forward by Brownian motion (a rare event when the backward force is high).

On an actin filament along which myosin VI moves, all foot marks in Fig. 8 are basically available for landing. Unless the legs are too long, however, landing on a distant backward site is prohibited when the green toe is down (Fig. 8*d*). In addition, the presumably stiff lower portion of the legs of myosin VI opposes oblique landing on a side of the filament (Figs. 8*e* and *f*). Thus, toe-down landing will be made primarily on a distant forward site (Fig. 8*a*), and occasionally on sites next to the landed foot. The frequency of landing on a nearby site depends critically on the make of the legs and ankles (less easy if shanks are long and ankles are bent acutely), and also on the backward load. The large variation in the size of individual steps of myosin VI (Rock et al., 2001; Nishikawa et al., 2002) may result from mixing of ~36-nm strides with short ones (±5.5 nm for the adjacent landing sites).

4.3. How to Walk Forward

In our view, at least four mechanisms are needed to let a two-foot motor walk forward: (1) the lever action or docking in the landed foot, (2) diffusion of the lifted foot, (3) toe up-down in the lifted foot, and (4) preferential detachment of the rear foot. (1) The lever action has been considered, by many myosin researchers, to play the central role in throwing the lifted leg forward. Its true significance, however, might be in preventing the pivot from pulled back too much by a backward force (Fig. 6). (2) Diffusion around the pivot is the only way by which the lifted foot move, whether forward or backward. It is quite unlikely that ATP-dependent conformational changes alone, in the landed and/or lifted feet, carry the lifted foot all the way onto a landing site. Passive diffusion must be involved, which assists forward and backward movements equally. (3) Toe up-down in the lifted foot, we propose, plays the major role in selecting a forward landing site over backward ones. (4) When both feet land, it must be the rear foot that is lifted first. This is accomplished by a strain dependence of the nucleotide kinetics in the feet. For kinesin, for example, pulling a landed foot forward has been shown to increase the affinity of that foot for ADP (Uemura and Ishiwata, 2003). This would take place in the rear foot, and the affinity for a microtubule binding site is weakened when kinesin's foot binds ADP.

In somewhat different contexts, a human being also uses the four mechanisms above when he/she walks. The major differences, apart from the distinction between the right and left feet, are that a human relies on inertia, particularly in (2) and (4), and that gravity helps in (2) and (3).

5. ACKNOWLEDGMENTS

We thank M. Fukatsu for the toy model, and S. Ishiwata, S. Uemura, and the members of Kinosita lab for discussion. This work was supported in part by Grants-in-Aid from the Ministry of Education, Culture, Sports, Science and Technology of Japan. M. Y. Ali was, and K. Shiroguchi is, a Research Fellow of the Japan Society for the Promotion of Scicnce.

6. REFERENCES

Ali, M. Y., Uemura, S., Adachi, K., Itoh, H., Kinosita, K. Jr., and Ishiwata, S., 2002, Myosin V is a left-handed spiral motor on the right-handed actin helix, *Nat. Struct. Biol.* **9**:464-467.

Ali, M. Y., Homma, K., Iwane, A. H., Adachi, K., Itoh, H., Kinosita, K. Jr., Yanagida, T, and Ikebe, M., 2004, Unconstrained steps of myosin VI appear longest among known molecular motors, *Biophys. J.* **86**, in press.

Asbury, C. L., Fehr, A. N., and Block, S. M., 2003, Kinesin moves by an asymmetric hand-over-hand mechanism, *Science* **302**:2130-2134.

Bahloul, A., Chevreux, G., Wells, A. L., Martin, D., Nolt, J., Yang, Z., Chen, L.-Q., Potier, N., Dorsselaer, A. V., Rosenfeld, S., Houdusse, A., and Sweeney, H. L., 2004, The unique insert in myosin VI is a structural calcium-calmodulin binding site. *Proc. Natl. Acad. Sci. USA.* **101**:4787-4792.

Burgess, S., Walker, M., Wang, F. J., Sellers, R., White, H. D., Knight, P. J., and Trinick, J., 2002, The prepower stroke conformation of myosin V, *J. Cell Biol.* **159**:983-991.

Cheney, R. E., O'shea, M. K., Heuser, J. E., Coelho, M. V., Wolenski, J.S., Espreafico, E. M., Forscher, P., Larson, R. E., and Mooseker, M. S., 1993, Brain myosin-V is a two-headed unconventional myosin with motor activity, *Cell* **75**:13-23.

Coureux, P.-D., Wells, A. L., Ménétrey, J., Yengo, C. M., Morris, C. A., Sweeney, H. L., and Houdusse, A., 2003, A structural state of the myosin V motor without bound nucleotide, *Nature* **425**:419-423.

Endow, S. A., and Barker, D. S., 2003, Processive and nonprocessive models of kinesin movement, *Annu. Rev. Physiol.* **65**:161-175.

Forkey, J. N., Quinlan, M. E., Shaw, M. A., Corrie, J. E. T., and Goldman, Y. E., 2003, Three-dimensional structural dynamics of myosin V by single-molecule fluorescence polarization, *Nature* **422**:399-404.

Holmes, K. C., Popp, D., Gebhard, W., and Kabsch, W., 1990, Atomic model of the actin filament, *Nature* **347**:44–49.

Holmes, K. C., and Geeves, M. A., 2000, The structural basis of muscle contraction, *Phil. Trans. R. Soc. B.* **355**: 419-431.

Holmes, K. C., Angert, I., Kull, F. J., Jahn, W., and Schröder, R. R., 2003, Electron cryo-microscopy shows how strong binding of myosin to actin releases nucleotide, *Nature* **425**:423-427.

Homma, K., Yoshimura, M., Saito, J., Ikebe, R., and Ikebe, M., 2001, The core of the motor domain determines the direction of myosin movement, *Nature* **412**: 831-834.

Houdusse, A., Kalabokis, V. N., Himmel, D., Szent-Györgyi, A. G., and Cohen, C., 1999, Atomic structure of scallop myosin subfragment S1 complexed with MgADP: A novel conformation of the myosin head, *Cell* **97**, 459-470.

Houdusse, A., Szent-Györgyi, A. G., and Cohen, C., 2000, Three conformational states of scallop myosin S1, 2000, *Proc. Natl. Acad. Sci. USA* **97**, 11238-11243.

Howard, J., 1996, The movement of kinesin along microtubules, *Annu. Rev. Physiol.* **58**:703-729.

Hua, W., Chung, J., and Gelles, J., 2002, Distinguishing inchworm and hand-over-hand processive kinesin movement by neck rotation measurements, *Science* **295**:844-848.

Huxley, H. E., 1969, The mechanism of muscular contraction, *Science* **164**:1356-1366.

Ishiwata, S., Kinosita, K. Jr., Yoshimura, H., and Ikegami, A., 1987, Rotational motions of myosin heads in myofibril studied by phosphorescence anisotropy decay measurements, *J. Biol. Chem.* **262**:8314-8317.

Kinosita, K. Jr., Ishiwata, S., Yoshimura, H., Asai, H., and Ikegami, A., 1984, Submicrosecond and microsecond rotational motions of myosin head in solution and in myosin synthetic filaments as revealed by time-resolved optical anisotropy decay measurements, *Biochem.* **23**:5963-5975.

Kinosita, K. Jr., Yasuda, R., Noji, H., Ishiwata, S., and Yoshida, M., 1998, F_1-ATPase: a rotary motor made of a single molecule, *Cell* **93**:21-24.

Kinosita, K., Jr., Adachi, K., and Itoh, H., 2004, Rotation of F_1-ATPase: how an ATP-driven molecular machine may work, *Annu. Rev. Biophys. Biomol. Struct.* **33**:245-268.

Kozielski, F., Sack, S., Marx, A., Thormählen, M., Schönbrunn, E., Biou, V., Thompson, A., Mandelkow, E.-M., and Mandelkow, E., 1997, The crystal structure of dimeric kinesin and implications for microtubule-dependent motility, *Cell* **91**:985-994.

Li, Y, Brown, J. H., Reshetnikova, L., Blazsek, A., Farkas, L., Nyitray, L., and Cohen, C., 2003, Visualization of an unstable coiled coil from the scallop myosin rod, *Nature* **424**:341-345.

Mehta, A. D., Rock, R. S., Rief, M., Spudich, J. A., Mooseker, M. S., and Cheney, R. E., 1999, Myosin-V is a processive actin-based motor, *Nature* **400**:590-593.

Mehta, A., 2001, Myosin learns to walk, *J. Cell Sci.* **114**:1981-1998.

Moore, J. R., Krementsova, E. B., Trybus, K. M., and Warshaw, D. M., 2001, Myosin V exhibits a high duty cycle and large unitary displacement, *J. Cell Biol.* **155**:625-635.

Nishikawa, S., Homma, K., Komori, Y., Iwaki, M., Wazawa, T., Iwane, A. H., Saito, J., Ikebe, R., Katayama, E., Yanagida, T., and Ikebe, M., 2002, Class VI myosin moves processively along actin filaments backward with large steps, *Biochem. Biophys. Res. Commun.* **290**:311-317.

Nishizaka, T., Yagi, T., Tanaka, Y., and Ishiwata, S., 1993, Right-handed rotation of an actin filament in an in vitro motile system, *Nature* **361**:269-271.

Rice, S., Cui, Y., Sindelar, C., Naber, N., Matuska, M., Vale, R., and Cooke, R., 2003, Thermodynamic properties of the kinesin neck-region docking to the catalytic core, *Biophys. J.* **84**: 1844-1854.

Rief, M., Rock, R. S., Mehta, A. D., Mooseker, M. S., Cheney, R. E., and Spudich, J. A., 2000, Myosin-V stepping kinetics: A molecular model for processivity, *Proc. Natl. Acad. Sci. USA.* **97**: 9482-9486.

Rock, R. S., Rice, S E., Wells, A. L., Purcell, T. J., Spudich, J. A., and Sweeney, H. L., 2001, Myosin VI is a processive motor with a large step size, *Proc. Natl. Acad. Sci. USA.* **98**:13655-13659.

Sakamoto, T., Amitani, I., Yokota, E., and Ando, T., 2000, Direct observation of processive movement by individual myosin V molecules, *Biochem. Biophys. Res. Commun.* **272**:586–590.

Sase, I., Miyata, H., Ishiwata, S., and Kinosita, K. Jr., 1997, Axial rotation of sliding actin filaments revealed

by single-fluorophore imaging, *Proc. Natl. Acad. Sci. USA.* **94**:5646-5650.
Schliwa, M., and Woehlke, G., 2003, Molecular motors, *Nature* **422**:759-765.
Tanaka, H., Homma, K., Iwane, A. H., Katayama, E., Ikebe, R., Saito, J., Yanagida, T., and Ikebe, M., 2002, The motor domain determines the large step of myosin-V, *Nature* **415**:192-195.
Tominaga, M., Kojima, H., Yokota, E., Orii, H., Nakamori, R., Katayama, E., Anson, M., Shimmen, T., and Oiwa, K., 2003, Higher plant myosin XI moves processively on actin with 35 nm steps at high velocity, *EMBO J.* **22**:1263-1272.
Uemura, S., and Ishiwata, S., 2003, Loading direction regulates the affinity of ADP for kinesin, *Nat. Struct. Biol.* **10**:308-311.
Vale, R. D., and Milligan, R. A., 2000, The way things move: looking under the hood of molecular motor proteins, *Science* **288**:88-95.
Vale, R. D., 2003, Myosin V motor proteins: marching stepwise towards a mechanism, *J. Cell Biol.*, **163**:445-450.
Veigel, C., Wang, F., Bartoo, M. L., Sellers, J. R., and Molloy, J. E., 2002, The gated gait of the processive molecular motor, myosin V, *Nat. Cell Biol.* **4**:59-65.
Walker, M. L., Burgess, S. A., Sellers, J. R., Wang, F., Hammer, J. A., Trinick, J., and Knight, P. J., 2000, Two-headed binding of a processive myosin to F-actin, *Nature* **405**:804-807.
Wells, A. L., Lin, A. W., Chen, L.-Q., Safer, D., Cain, S. M., Hasson, T., Carragher, B. O., Milligan, R. A., and Sweeney, H. L., 1999, Myosin VI is an actin-based motor that moves backwards, *Nature* **401**:505-508.
Yasuda, R., Noji, H., Kinosita, K. Jr., and Yoshida, M., 1998, F_1-ATPase is a highly efficient molecular motor that rotates with discrete 120° steps, *Cell* **93**:1117-1124.
Yildiz, A., Forkey, J. N., McKinney, S. A., Ha, T., Goldman, Y. E., and Selvin, P. R., 2003, Myosin V walks hand-over-hand: Single fluorophore imaging with 1.5-nm localization, *Science* **300**:2061-2065.
Yildiz, A., Tomishige, M., Vale, R. D., and Selvin, P. R., 2004, Kinesin walks hand-over-hand, *Science* **303**:676-678.

DISCUSSION

Pollack: Could you comment on how single-headed kinesins could advance along a microtubule?

Kinoshita: Some part of kinesin would stick non-specifically to the microtubule, as has been shown for KIF1A, and another part would pull the molecule forward, by a lever action or by extending itself forward by diffusion followed by binding to a microbutule site.

Gonzalez-Serratos: Without strain, on which direction would myosin V move?

Kinoshita: If there were no strain-dependent regulation of nucleotide kinetics, either foot, front or rear, could be lifted from actin. The motor would still move forward, but the efficiency would be low, because lifting the front foot would not contribute to forward motion while ATP is expended.

Pollack: With single filaments sliding past one another, you do see backward steps of n x 2.7 nm.

Sugi: One single question is the biased diffusion you mentioned in your talk is the same as so-callled Brownian ratchet mechanism?

Kinosita: The Brownian ratchet, in my opinion, is a broad term that applies to almost any mechanisms of molecular machines. Moving the pivot forward, by a lever action, to help forward landing is a kind of ratchet. Assuring forward landing by bringing the toe up, without necessarily the pivot motion, is another way of realizing a ratchet.

Kushmerick: You stated the ATP synthesis by F1-ATPase may be nearly 100 % efficient. What is the efficiency of the H+ potonmotive-driven component of FoF1 ATP synthesis? Is the overall FoF1 rotatory motor reversible?

Kinosita: For ATP synthesis, I expect the number efficiency (the number of ATP molecules synthesized per 120-degree rotation) to be close to one, under favorable nucleotide concentrations and at low rotary speeds. The energetic efficiency (free energy for ATP synthesis divided by mechanical work done on the gamma subunit) is critically dependent on the nucleotide concentrations and applied torque, and would be high only near reversal (low torque). Peter Gra[a with umlaut]ber's group, among others, has shown that FoF1 ATP synthase is reversible, and that synthesis/hydrolysis in F1 is balanced by the proton flow through Fo. When the proton motive force across Fo is greater than the free energy needed for ATP synthesis, ATP is synthesized. ATP is hyrdolyzed when the proton motive force is lower. They have shown that synthesis of one ATP molecule requires translocation of about four protons, but, in addition, some protons may leak through the membrane or the Fo motor. Thus, experimental determination of the efficiency of synthesis is very difficult.

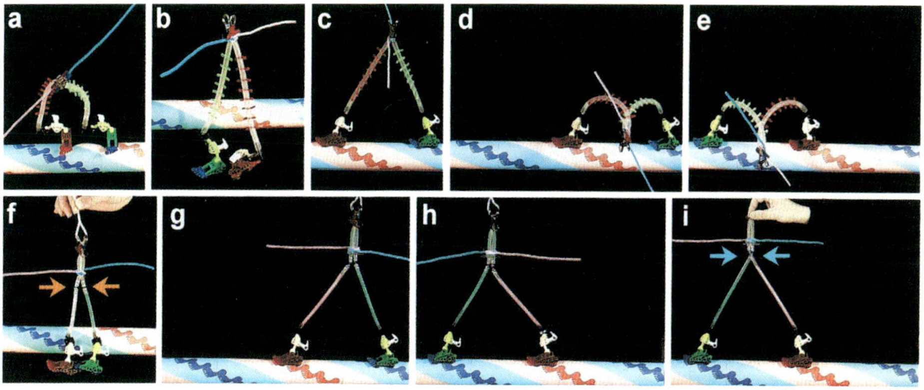

Figure 1. Walking with two identical feet. The legs are made of elastic vinyl pipes, which can accommodate both torsional and flexural strains. (**a**) Two feet of a molecular motor are identical, because they are coded by the same gene. (**b**) The two legs are joined in twofold symmetry. Thus the two feet would be oriented in opposite directions, one toe facing east if the other faces west, unless the legs are flexible. (**c**) When two feet land simultaneously during walking, both toes must be oriented forward, because the landing sites on the rail are all identical. Elastic legs would thus severely distorted. In the configuration shown in this image (*b*), most of the strain is in torsion of the legs (note the directions of the small bars on the legs) whereas bending is relatively small. (**d**) By turning one of the feet by 360°, much of the torsional strain is relieved, resulting in severe bending. The total strain is the same as in *c*. (**e**) When the motor moves forward by one step and attains the same posture as in *d*, a cargo (pink and blue long bars) attached to the body rotates by 180°. A 180° rotation also accompanies every step if the motor steps with the posture in *c*. (**f**) Flexible joints in the legs (arrows) allow the two feet to adopt the same orientation. (**g**) With flexible joints, landing with two feet does not introduce strain. (**h**) But, because of the basic twofold symmetry, the body still tends to rotate 180° every time the motor steps. (**i**) With flexible joints, though, the 180° rotations can be prohibited by an external force without introducing too much strain in the legs. Note that the two legs are now crossed near the junction (arrows); a small torque suffices to bring about this configuration.

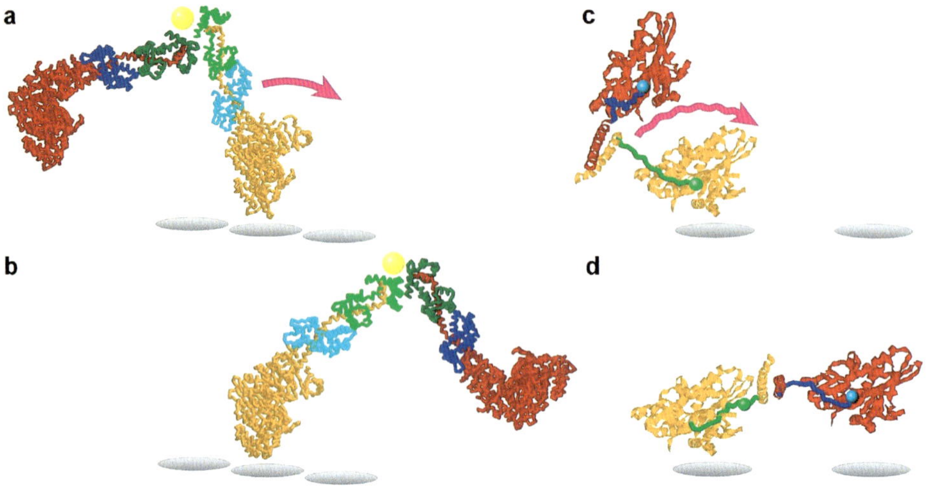

Figure 2. Postulated lever action in two-foot motors. The motors in this figure move toward right. **(a, b)** Scallop myosin. This myosin is non-processive and thus does not "walk." Nevertheless, we show possible walking postures in order to indicate lever action (arrow) in myosin: the orange leg (the long α-helix) rotates between *a* and *b*. Flexible joints likely exist around the location indicated by the yellow ball, beyond which the two α-helices form a coiled coil (the "stalk"; not shown in the figure). Thus, the red leg presumably undergoes rotational diffusion around the yellow ball. The lever action of the orange leg biases the Brownian motion of the red foot forward. Each figure is composed of two structures of scallop myosin subfragment 1 (S1), arranged arbitrarily. The orange foot/leg in *a* represents the structure of S1 binding MgADP·VO_4, and that in *b* is S1 without a nucleotide (Houdusse et al., 2000). The red foot/leg is S1 binding MgADP, which mimics a structure after detachment from actin (Houdusse et al., 1999). Cyan/blue, essential light chains; green/dark green, regulatory light chains. The gray disks with a diameter of 5.5 nm represent binding sites on an actin filament. **(c, d)** Conventional kinesin. The blue and green legs ("neck linkers") between the balls are presumably flexible, and thus the red foot undergoes Brownian motion. The flexible green leg in *c* "docks" onto the orange foot, in an ATP-dependent process, through Brownian motion (wavy arrow). After docking (*d*), the Brownian motion of the red foot occurs around the green ball, and thus is biased toward the forward binding site. The docking is equivalent with the lever action in myosin. The gray disks with a diameter of 4 nm represent binding sites on a microtubule. The figures are constructed from a structure of a dimeric kinesin (Kozielski et al., 1997) by assigning arbitrary structures to the neck linkers (except the green one in *d*) and orienting the central coiled coil (stalk) arbitrarily.

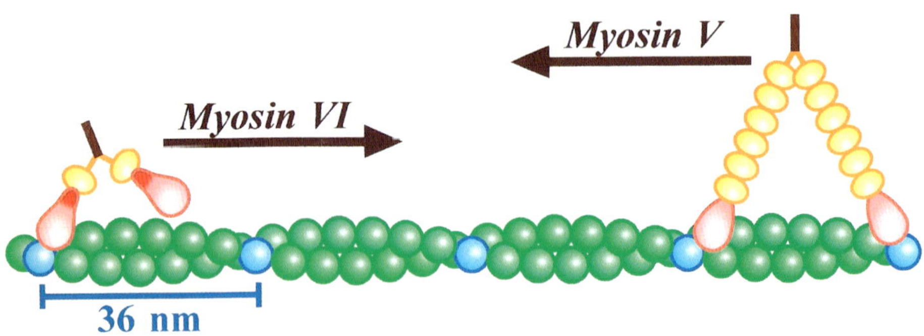

Figure 3. Myosin V and myosin VI. Myosin V has two legs each reinforced with six light chains (orange) and walks on an actin filament (green) toward the barbed, or plus, end (left in the figure). Myosin VI, on the other hand, moves toward the pointed, or minus, end. Previously, myosin VI was thought to bind one light chain per leg, and thus myosin VI was considered short-legged as shown in this figure. The red portion in the feet of myosin VI indicates an extra insert not seen in other myosins. Every 13th monomer of the actin filament is shown in blue. Walking along one side of an actin filament implies an average step size of ~36 nm. Adapted from Ali et al. (2004).

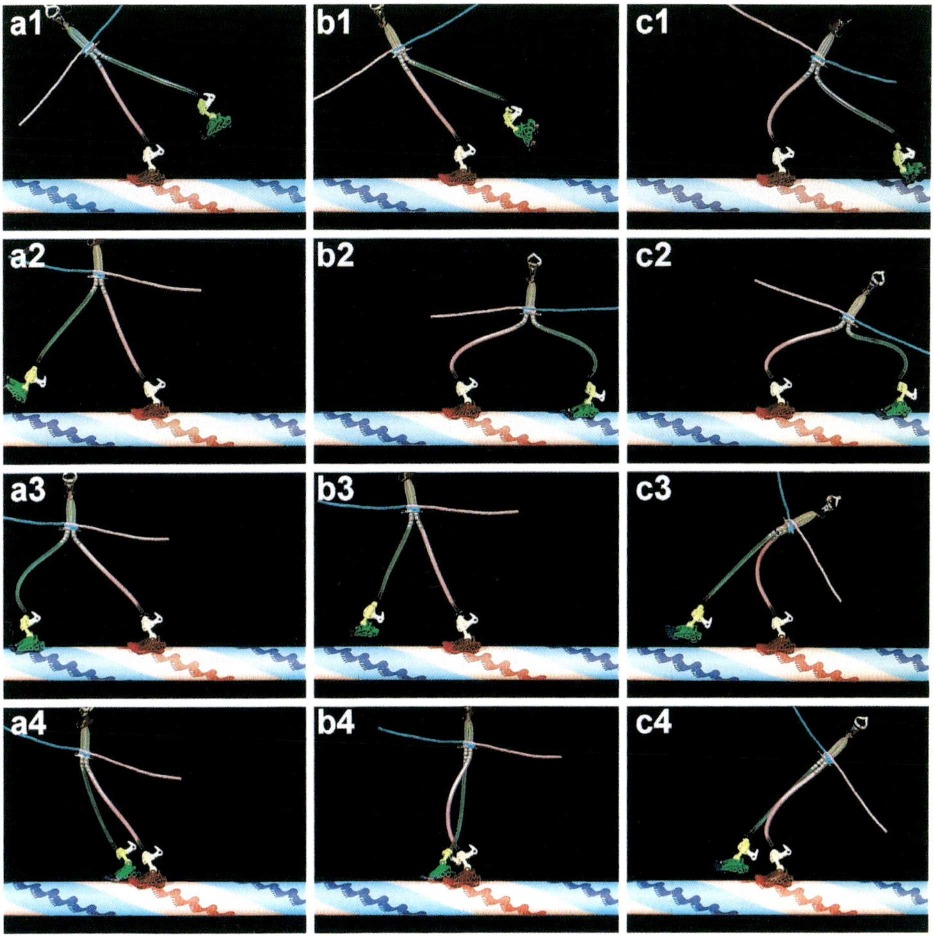

Figure 7. Toe up-down mechanism for the case of semi-rigid legs. **(a)** If the lifted toe is up, landing on a forward site is difficult, and natural landing would be on a nearby site (*a*4), if one is available. **(b)** If the toe goes down upon lifting of the foot, landing with the least strain will be on a distant forward site. **(c)** A backward force will pull the pivot backward (*c*1), but, as long as the lifted toe is down, landing on a distant forward site is favored. Under a high force, landing on a nearby site is equally probable (*c*4). Note that, in *c*4, bending of the red leg is caused by the external force. Adapted from Ali et al. (2004).

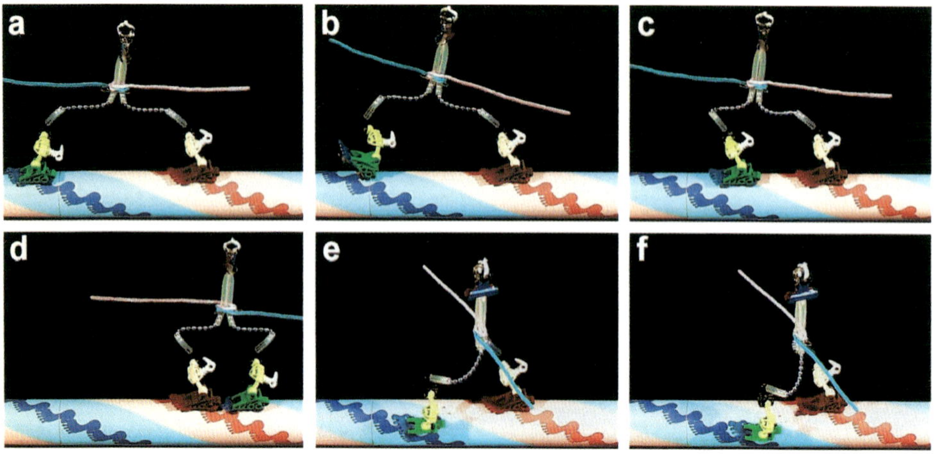

Figure 8. Toe up-down mechanism for the case of completely flexible legs. The legs in the model are made of metal chains. (**a**) The size of the legs are such that landing on a distant forward site is allowed only when the legs are fully extended and the leading toe is down and trailing toe up. (**b, c**) With the green toe up, forward landing is impossible. (**d**) As long as the green toe is down, there is no hope in backward landing. (**e, f**) If a lower part of legs is stiff, oblique landing on a side of the cylindrical rail is also prohibited. The postures here represent the lowest possible position for the green foot, which fails to reach a blue foot mark below.

IV. REGULATORY MECHANISM AND E-C COUPLING

MOLECULAR BASIS OF CALCIUM REGULATION OF STRIATED MUSCLE CONTRACTION

I. Ohtsuki

INTRODUCTION

It is a great pleasure to participate in a symposium commemorating the fiftieth anniversary of the sliding filament theory of muscle contraction. It was almost a decade after the proposal of this mechanism that specific regulatory proteins were found to be required for physiological muscle contraction. A Ca^{2+}-sensitizing protein factor in actomyosin first isolated from muscle was called native tropomyosin because of its similarity in amino acid composition to the previously isolated tropomyosin (Ebashi and Ebashi, 1964; Bailey, 1948). This protein factor was then separated into tropomyosin and a new protein named "troponin" (Ebashi and Kodama, 1965; Ebashi et al., 1968). In the absence of Ca^{2+}, the contractile interaction between myosin and actin is depressed by troponin and tropomyosin. When Ca^{2+} ion acts on troponin, this depression is removed and the contractile interaction is activated. Immunoelectron microscopic investigation revealed that troponin distributes along the entire length of a thin filament at regular intervals of 38 nm (Ohtsuki et al., 1967). This finding led to the construction of a molecular model of thin filament as an ordered assembly of troponin-tropomyosin-actin, in which two end-to-end filaments of fibrous tropomyosin molecules, each attached by troponin at its specific region, lie in the grooves of actin double strands (Ebashi et al., 1969; Ohtsuki, 1974; Ebashi, 1974). The action of Ca^{2+} on troponin is transmitted to actin molecules via tropomyosin and/or neighboring actin molecules. These studies have established the molecular basis of the Ca^{2+}-regulatory mechanisms of striated muscle contraction in vertebrates.

TROPONIN COMPONENTS

Troponin is a complex of three different components: troponins C, I, and T (cf. Ohtsuki et al., 1986). Troponin C binds Ca^{2+} ions. Troponin I inhibits the contractile interaction between myosin and actin in the presence of tropomyosin, and troponin T

Department of Physiology, The Jikei University School of Medicine, Minato-ku, Tokyo 105-8461, Japan binds to tropomyosin. Without the troponin ternary complex, the contractile interaction between myosin and actin-tropomyosin stays in the activated state regardless of Ca^{2+} concentrations. Troponin I inhibits the contractile interaction, and the addition of troponin C releases the inhibition by troponin I regardless of Ca^{2+} concentrations. In the presence of troponin T, in addition to troponins I and C, the contractile interaction of myosin-actin-tropomyosin becomes sensitive to Ca^{2+} and the maximum contraction level becomes slightly higher than that in the absence of the troponin complex. Functionally troponin T represents a regulatory component of the troponin. Several aspects of the structure and function of troponin components will be discussed in the following section.

The molecular arrangement of troponin components in thin filament was shown by immunoelectron microscopic investigations. It was first found that anti-troponins C and I formed the narrow striations in each 38-nm period along the thin filament bundle, whereas the anti-troponin T formed a wide band that is comprised of a pair of narrow striations (Ohtsuki, 1975). It was then demonstrated that troponin T is split into two subfragments, troponin T_1 and troponin T_2, by mild chymotryptic digestion (Ohtsuki, 1979). Immunoelectron microscopic investigation of these troponin T subfragments revealed that each of the striation pair of anti-troponin T is derived from a specific portion of the molecule; the striation on the filament-top side is formed by troponin T_2 and the striation on the Z-line side by troponin T_1. The anti-troponin T_2 striation is located at the same position as anti-troponins C and I, while the anti-troponin T_1 striation is approximately 13 nm apart toward the Z-line direction. These findings led to the conclusion that troponin T molecules are distributed along thin filament in such a way that the troponin T_1 region and the troponin T_2 region of the molecule are arranged on the Z-line side and the filament-top side, respectively.

Troponin T from rabbit skeletal muscle, a peptide of 259 amino acid residues, is split into two soluble subfragments, troponin T_1 and troponin T_2, by mild treatment with chymotrypsin (Tanokura et al., 1982; Ohtsuki, 1980; Ohtsuki and Nagano, 1982). Troponin T_1 is a subfragment of the N-terminal 158 residues and binds only tropomyosin strongly, whereas troponin T_2, a C-terminal subfragment of 101 residues, interacts with tropomyosin, troponin I and troponin C. The Ca^{2+}-regulating action of troponin T is retained only in the smaller subfragment, troponin T_2, but not in troponin T_1. However, a fragment of troponin T_2 that is devoid of the C-terminal 17 residues of troponin T_2, shows a greatly decreased Ca^{2+}-sensitizing action and decreased affinity to tropomyosin, indicating that the Ca^{2+} regulatory activity of troponin T is closely correlated with the tropomyosin-binding activity of the C-terminal region in troponin T_2 (Ohtsuki et al., 1981; Tanokura et al., 1983). The Ca^{2+}-regulating activity of troponin T_2 is stabilized as a whole through the strong binding of troponin T_1 region to tropomyosin.

The troponin isolated from muscle is almost freely soluble in water. An aqueous solution of troponin immediately after its isolation from minced muscle is usually transparent and liquid or slightly viscous. When frozen and stored, the troponin solution becomes more viscous and occasionally turbid. The high Ca^{2+}-sensitizing activity of fresh troponin gradually declines during the course of storage. An electron microscopic examination, using a rotary shadow technique, has shown that fresh preparations consist of globular particles of 13x9 nm, whereas, in the frozen stored preparations, troponin particles have the shape of a smaller globular head and a thin bent tail of various lengths extending from the lateral side of the head portion (Ohtsuki et al., 1988). Troponin particles eventually take the shape of a small globular head with a thin straight tail extending from its center (head-tail type particles). The axial size of the head portion is 8 nm and the straight tail length is 17 nm. The shape and axial size of the head-tail type particle of 25 nm are in accord with those first reported by Flicker et al. (1984).

In the fresh globular particles, the entire tail portion is attached around the globular head portion. During the course of storage, certain changes occur in the interaction between the tail and the head within the globular particles, with the result that the tail portion gradually unfolds from the head region. Troponin T has the shape of a filamentous particle of 17 nm in length. Since the length of about 17 nm coincides with the tail length of the head-tail type particle, the tail portion would be mostly formed by a troponin T molecule and consequently the globular head portion is mostly composed of troponin C and troponin I. The filamentous nature of troponin T and the globular nature of troponin C·I have also been demonstrated by viscosity measurements (Ohtsuki et al., 1988). The prediction of a quaternary structure of globular troponin was attempted (Nagano and Ohtsuki, 1982). Within the crystal structure of the core domain in human cardiac troponin C·I·T_2, troponin T_2 forms a coiled-coil structure with troponin I of about eight nm length, which constitutes the backbone of the helix-rich troponin core domain (Takeda et al., 2003), suggesting that most of the troponin T_2 region of the troponin T molecule is contained within the head region of the head-tail type troponin particle. A detailed examination of the structure of troponin components in native thin filament will be required to establish the physiological molecular configuration of this Ca^{2+}-receptive protein complex.

Concerning the structural aspect of the inhibitory interaction of troponin I with actin-tropomyosin, the axial location of troponin I in the thin filament has been generally considered to be determined by the binding of troponin T to the specific region of tropomyosin. At the same time, it has been shown that, even in the absence of troponin T, troponin I shows its full inhibitory action on actomyosin in the presence of tropomyosin, whereas troponin I has only weak inhibitory activity in the absence of tropomyosin (Perry, 1999). This finding has raised the possibility that troponin I itself has a certain structural relationship with tropomyosin along the actin double strands, though no significant affinity has yet been detected between troponin I and tropomyosin (Ohtsuki et al., 1986). A recent immunoelectron microscopic investigation has shown that troponin I binds to actin at the specific region of tropomyosin molecules along the tropomyosin-actin filament (Ohtsuki and Shiraishi, 2002). Therefore, each troponin I should form a complex with both actin and tropomyosin axially at the specific region of each tropomyosin. In other words, the seven actin molecules in each 38-nm period are by no means equal in terms of their affinity to troponin I, but a specific actin molecule has a higher affinity to troponin I than other actins in the 38-nm period, through its interaction with a specific region of tropomyosin.

In order to explore the regulatory mechanisms of troponin under physiological conditions, it is critical to remove and reconstitute troponin components in a myofibrillar lattice. The troponin C·I·T complex was found to be replaced by troponin T in glycerinated muscle fibers and myofibrillar preparations (Hatakenaka and Ohtsuki, 1991, 1992; Shiraishi et al., 1992). In the presence of an excess of troponin T in the incubating solution, the intrinsic troponin C·I·T complex, as a whole, is exchanged by the externally added troponin T in skinned muscle fibers, and as a result, both troponin C and I are removed from the fibers. The force generation of the troponin CI-depleted fibers (or troponin T-treated fibers) stays activated at all Ca^{2+} concentrations at the level of about 70% of the maximum tension of the intact fibers; the force generation is inhibited by the reconstitution with troponin I, and the original Ca^{2+}-sensitive force generation is recovered after further reconstitution with troponin C. The changes in the Ca^{2+}-activated contraction profiles of skinned fibers during the troponin T-treatment and subsequent reconstitution with troponin I and C are explained by the basic regulatory actions of troponin components, *i.e.*, the inhibitory action of troponin I, the neutralizing (or de-inhibitory) action of troponin C and the activating action of the troponin C·I·T -complex.

This procedure, including the troponin T treatment, has made it possible to exchange all three components of troponin in skinned fibers by the added troponin isoforms or mutants (Ohtsuki, 1995). The intrinsic ternary troponin complex in myofibrillar lattice can also be exchanged with the added ternary troponin complex (Shiraishi et al., 1993). The troponin-exchange procedure has been successfully applied for detecting delicate alterations in the Ca^{2+}-activated contraction profiles caused by disease causing-mutations in genes for human cardiac troponin components, as described in the following section.

GENETIC DISORDER OF CARDIAC TROPONIN AND ITS FUNCTIONAL CONSEQUENCES

Inherited cardiomyopathies, such as hypertrophic cardiomyopathy (HCM) and dilated cardiomyopathy (DCM), are caused by mutations of cardiac sarcomeric proteins (Fatkin and Graham, 2002). HCM is an autosomal dominant heart disease with characterized by ventricular hypertrophy and diastolic dysfunction. Genetic analyses of this disorder have revealed that the HCM is caused by mutations in genes for various cardiac sarcomeric proteins, including β-myosin heavy chain, myosin light chain, troponin T, troponin I, α-tropomyosin, myosin-binding protein C, and titin/connectin. About one half of cases of HCM are caused by β-myosin heavy chain mutations and about fifteen percent are due to troponin T mutations. Studies on the functional consequences of the genetic disorders of troponin components have revealed that an increase in the Ca^{2+}-sensitivity of contraction is the critical functional consequence of these HCM-causing mutations, while a decrease in the Ca^{2+}-sensitivity is caused by the DCM-causing mutations.

Functional characteristics of two missense mutations of human cardiac troponin T (Ile79Asn, Arg92Gln) were first investigated on skinned cardiac muscle fibers by employing the *in situ* troponin-exchange technique, including treatment with excess troponin T, as described in the preceding section. It was found that both mutations had a Ca^{2+}-sensitizing effect without affecting the maximum force or cooperativity (Morimoto et al., 1998). Succeeding examinations of fourteen mutations have revealed that all the mutations except one showed the Ca^{2+}- sensitizing effect on the skinned fiber tension as well as myofibrillar ATPase activity (cf. Ohtsuki et al., 2003). The exceptional mutation (Phe110Ile), associated with patients with benign prognosis, did not show the

Ca^{2+}-sensitization, but showed a marked potentiation of the maximum contraction (Yanaga et al., 1999; Nakaura et al., 1999). Essentially the same results have been reported in subsequent studies from several laboratories (Szczezesna et al., 2000; Tobacman et al., 1999; cf. Knollman and Potter, 2001, Fatkin and Graham, 2002). The HCM-causing mutations of cardiac troponin I also showed the Ca^{2+}-sensitizing effect without any sign of changes in the maximum force but mostly accompanied with a slight increase in the resting level, indicating that the inhibitory action of troponin I is impaired by these mutations (Takahashi-Yanaga et al., 2000, 2001).

Troponin T from human cardiac muscle is a peptide with 288 amino acid residues, along which several interacting regions are distributed. Although the HCM-causing mutation sites distribute along the entire sequence of this molecule, more than half of the examined mutations are located in the two tropomyosin-binding regions, *i.e.*, the helical region of troponin T_1 and the C-terminal region of troponin T_2, suggesting that the Ca^{2+}-sensitization caused by these mutations is closely related to the impaired interaction of troponin T with tropomyosin. In the case of the mutations of troponin I, most mutations distribute in the inhibitory region and its C-terminal side sequence.

At the same time, two mutations in the cardiac troponin T gene (ΔLys210, Arg141Trp) were recently reported to be associated with a different type of cardiomyopathy, *i.e.*, dilated cardiomyopathy (DCM). Functional analyses indicated that these mutations cause a Ca^{2+}-desensitiziation of the skinned fiber contraction and myofibrillar ATPase activity, while the cooperativity and the maximum or resting contraction levels are not affected (Morimoto et al., 2002; Lu et al., 2003). The Arg141Trp mutation, located within the strong tropomyosin-binding helical region of troponin T_1, was found to increase the affinity of troponin T to tropomyosin, whereas any significant change in the interaction with tropomyosin was not detected by the Lys210-deletion mutation located in the troponin T_2 region .

There is a third type of inherited cardiomyopathy, called restrictive cardiomyopathy, which is the least common and characterized by impaired diastolic filling of the left ventricle. In a recent genetic analysis, six mutations of cardiac troponin I were detected in the genetic disorders of five affected families (Morgensen et al., 2003). This has raised the possibility that this type of cardiomyopathy is a specific disorder caused by cardiac troponin I mutations, though more comprehensive surveys will be needed before reaching a definitive conclusion. In preliminary analyses, increased Ca^{2+}-sensitivity of the force generation was observed in skinned cardiac muscle into which these troponin I mutants were incorporated by the use of the troponin-exchange procedure (unpublished results).

The core domain structure of the Ca^{2+}-bound human cardiac troponin C·I·T $_2$ complex was crystallographically determined recently (Takeda et al., 2003). In the core domain, two regions, an IT-arm and a regulatory head, are connected by a flexible linker. The regulatory head is composed of the N-domain of troponin C and the second troponin C-binding helix (H3(I)) of troponin I. On Ca^{2+}-binding to the N-domain of troponin C, the hydrophobic pocket embedded in the molecule is exposed and interacts with the H3-helix of troponin I, rendering the inhibitory region of troponin I unable to exert its inhibitory interaction with actin-tropomyosin. In the IT-arm region, the C-domain of troponin C forms a complex with both the N-terminal helix (H1(I)) of troponin I and the troponin T_2 in the coiled-coil region with troponin I (H2(T_2)), creating a backbone structure of the core domain that is fixed to tropomyosin through the two regions of troponin T, *i.e.*, the helical region of troponin T_1 and the C-terminal region of troponin T_2. It should be noted that most mutation sites in troponin T and I are not included in the core domain but rather are distributed in the regions interacting with the other thin filament constituents, actin and tropomyosin (Fig. 1). Detailed examinations of the interacting

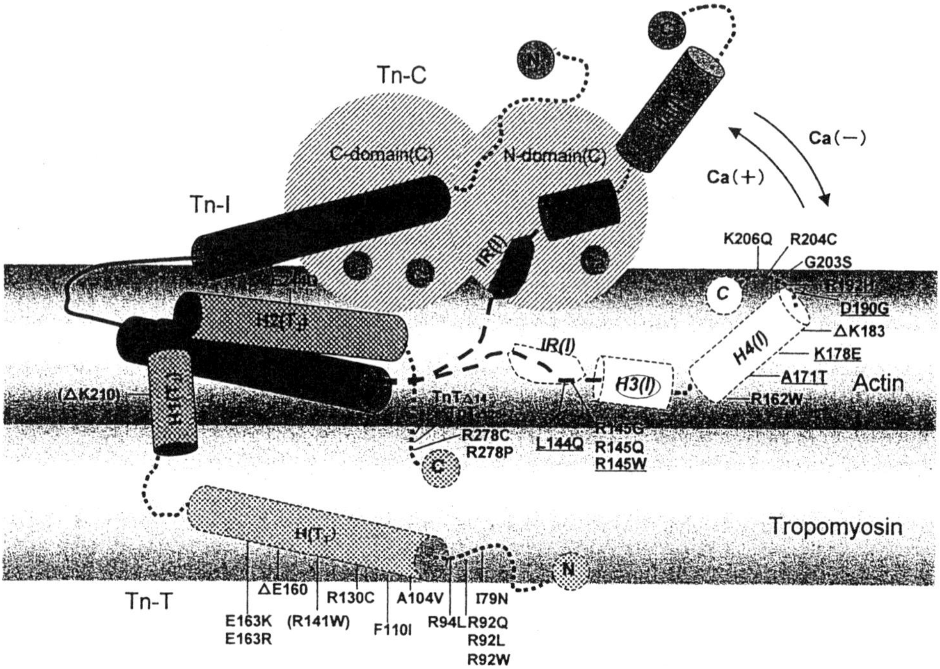

Fig. 1. Schematic illustration showing the distribution of Ca^{2+}-sensitizing and Ca^{2+}-desensitizing mutations associated with inherited cardiomyopathies within cardiac troponin C·I·T -complex.

The position of HCM-causing mutations is indicated along the sequences of TnT and TnI. Two DCM-causing mutations are shown by letters in parentheses. RCM-causing mutations are indicated by underlined letters. The Ca^{2+}-sensitizing effect is produced by the HCM- and RCM-causing mutations, whereas the Ca^{2+}-desensitizing effect is produced by the DCM-causing mutations. The Phe110-Ile mutation of TnT, which is HCM-causing mutation without the Ca^{2+}-sensitizing effect, shows potentiating effect on the maximum contraction.

Molecular arrangement of troponin components is based on the core domain structure of Ca^{2+}-bound human cardiac troponin C·I·T_2, which is composed of IT-arm and regulatory head (Takeda et al., 2003). The IT-arm consists of the C domain of TnC, the helix regions of TnI (H1(I),H2(I)) and TnT_2 (H1(T_2), H2(T_2)). A coiled-coil is formed by H2(I) and H2(T_2). The regulatory head is composed of the Ca^{2+}-bound N-domain of TnC and the helical region of TnI (H3(I)). Putative regions in TnI and TnT out of the core domain are indicated by dotted lines. The putative region for the inhibitory ineraction with actin-tropomyosin(IR(I)) is also indicated in the the inhibitory region (broken line) of TnI. The H4(I) region contains the homologous sequence to that of the second actin-binding region of skeletal TnI. The H3(I) helical region of TnI binds to the hydrophobic pocket in the N-domain of TnC exposed in the presence of Ca^{2+}. No mutation has been reported in the H3(I)-helical region. The position of the inhibitory region and its C-terminal sequence of TnI in the absence of Ca^{2+} is indicated by italic letters (*IR(I), H3(I) and H4(I)*). TnT binds to tropomyosin through the tropomyosin-binding helical region in TnT_1 (H(T_1)) and the C-terminal-end region of TnT_2.

properties of troponin components with actin-tropomyosin are therefore critical for clarifying both the pathogenetic mechanisms of these disease-causing mutations and the Ca^{2+}-regulatory mechanisms under physiological conditions.

MOLLUSCAN TROPONIN

It has been established that the troponin-tropomyosin system is the sole and ubiquitous Ca^{2+}-regulatory mechanism of contraction in vertebrate striated muscle. In invertebrate striated muscle, however, there are two types of Ca^{2+}-regulatory mechanisms: myosin-linked and actin (troponin)-linked Ca^{2+} regulations. The myosin-linked Ca^{2+} regulation was first demonstrated in the striated muscle of scallop and has long been considered to be the sole specific Ca^{2+}-regulatory system in scallop muscle (Kendrick-Jones et al., 1970; Szent-Györgyi et al., 1999).

At the same time, troponin and tropomyosin have also been isolated from scallop muscle (Ojima and Nishita,1986). The physiological significance of scallop troponin-tropomyosin was examined on the Ca^{2+}-activated ATPase of scallop myofibrils that had been desensitized to troponin-linked regulation and to myosin-linked regulation, by removing troponin C and the regulatory light chain of myosin, respectively (Shiraishi et al., 1999). The examination of the effect of reconstitution of troponin C or the myosin regulatory light chain clarified that the ATPase is inhibited by the myosin-linked regulation at low Ca^{2+}, while the ATPase is activated by the troponin-linked regulation at high Ca^{2+}. The activating action of troponin, which is a minor action in vertebrate troponin, is therefore a predominant mechanism of scallop troponin. It is interesting to note that scallop troponin C binds only one Ca^{2+} ion at the site IV in the C-terminal domain, whereas, in vertebrate skeletal and cardiac troponins, the Ca^{2+} binding to the N-terminal domain is critical for Ca^{2+} regulation (Ojima et al., 2000; Yumoto et al., 2001; Yumoto et al., 2003).

REFERENCES

Bailey, K., 1948, Tropomyosin: a new asymmetric protein component of the muscle fibril. *Biochem. J.* **43**, 271-279.

Ebashi, S., 1974, Regulatory mechanism of muscle contraction with special reference to the Ca-troponin-tropomyosin system. *Essays in Biochem.* **10**, 1-36.

Ebashi, S., and Ebashi. F., 1964, A new protein component participating in the superprecipitation of myosin B. *J. Biochem.* **55**, 604-613.

Ebashi, S., and Kodama, A., 1965, A new protein factor promoting aggregation of tropomyosin. *J. Biochem.* **58**, 107-108.

Ebashi, S., Endo, M., and Ohtsuki, I., 1969, Control of muscle contraction. *Q. Rev. Biophys.* **2**, 351-384.

Ebashi, S., Kodama, A., and Ebashi, F., 1968, Troponin 1. Preparation and physiological function. *J. Biochem.* **64**, 465-477.

Fatkin, D. and Graham, R. M., 2002, Molecular mechanism of inherited cardiomyopathies. *Physiol. Rev.* **82**, 945-980.

Flicker, P. F., Phillips, G. N., and Cohen, C., 1984, Structure of troponin and its interaction with tropomyosin ; an electronmicroscope study. *J. Mol. Biol.* **162**, 495-501.

Hatakenaka, M., and Ohtsuki, I., 1991, Replacement of three troponin components with cardiac troponin components within single glycerinated skeletal muscle fibers. *Biochem. Biophys. Res. Commun.* **181**,

1022-1027.

Hatakenaka, M., and Ohtsuki, I., 1992, Effect of removal and reconstitution of troponins C and I on the Ca^{2+}-activated tension development of single glycerinated rabbit skeletal muscle fibers. *Eur. J. Biochem.* **205**, 985-993.

Kendrick-Jones, J., Lehman, W., and Szent-Györgyi, A. G., 1970, Regulation in molluscan muscles. *J. Mol. Biol.* **54**, 313-326.

Knollman, B. C., and Potter, J. D., 2001, Altered regulation of cardiac muscle contraction by troponin T mutations that cause familial hypertrophic cardiomyopathy. *Trends in Cardiovas. Med.* 11, 206-212.

Lu, Q-W., Morimoto, S., Harada, K., Du, C-K., Takahashi-Yanaga, F., Miwa, Y., Sasaguri,T., and Ohtsuki. I.,2003, Cardiac troponin T mutation found in dilated cardiomyopathy stabilizes the troponin T-tropomyosin interaction and causes Ca^{2+} desensitization. *J. Mol. Cell. Cardiol.* **35**, 1421-1427.

Morgensen, J., Kubo, T., Duque, M., Uribe, W., Shaw, A., Murphy, R.,J. R. Gimeno, J. R., Elliott, P., and Mckenna, W. J., 2003, Idiopathic restrictive cardiomyopathy is part of the clinical expression of cardiac troponin I mutations. *J. Clin. Invest.* **111**, 209-216.

Morimoto, S., Yanaga, F., Minakami, R., and Ohtsuki, I., 1998, Ca^{2+}-sensitizing effects of the mutations at Ile-79 and Arg-92 of troponin T in hypertrophic cardiomyopathy. *Am. J. Physiol.***275**, C200-C207.

Morimoto, S., Lu, Q-W., Harada, K., Takahashi-Yanaga, F., Minakami, R., Ohta, M., Sasaguri, T., and Ohtsuki, I., 2002, Ca^{2+}desensitizing effect of a deletion mutation delta-K210 in cardiac troponin T that causes familial dilated cardiomyopathy. *Proc. Natl. Acad. Sci. USA* **99**. 913-918 (2002)

Nagano, K., and Ohtsuki, I., 1982, Prediction of approximate quarternary structure of troponin complex. *Proc. Japan Acad.* **58**, Ser.B 73-77.

Nakaura, H., Yanaga, F., Ohtsuki, I., and Morimoto, S., 1999, Effects of missense mutations Phe110Ile and Glu244Asp in human cardiac troponin T on force generation in skinned cardiac muscle fibers. *J. Biochem.* **126**, 457-460.

Ohtsuki, I., 1974, Localization of troponin in thin filament and tropomyosin paracrystal*J. Biochem.* **75**, 753-765.

Ohtsuki, I., 1975, Distribution of troponin components in the thin filament studied by immunoelectron microscopy. *J. Biochem.* **77**, 633-639.

Ohtsuki, I., 1979, Molecular arrangement of troponin T in thin filament. *J. Biochem.* **86**, 491-497.

Ohtsuki, I., 1980, Functional organization the troponin-tropomyosin system. In "Muscle contraction; its regulatory mechanisms (eds. S. Ebashi et al.)" pp237-250.

Ohtsuki, I., 1995, Troponin components and calcium ion regulation of myofibrillar contraction in skeletal muscle. In"*Calcium as Cell Signal*" edited by K. Maruyama, Y, Nonomura, and K. Kohama (Igaku-shoin Tokyo・New York) pp. 36-42.

Ohtsuki, I., and Nagano, K., 1982, Molecular arrangement of troponin-tropomyosin in the thin filament. *Advances in Biophysics*, **15**, 93-130.

Ohtsuki, I., and Shiraishi, F., 2002, Periodic binding of troponin C・I and troponin I to tropomyosin-actin filaments. *J. Biochem.* **131**, 739-743.

Ohtsuki, I., Maruyama, K., and Ebashi, S., 1986, Regulatory and cytoskeletal proteins of vertebrate skeletal muscle. *Adv. Prot. Chem.* **38**, 1-68.

Ohtsuki, I., Morimoto, S., and Takahashi-Yanaga, F., 2003, Several aspects of calcium regulatory mechanisms linked to troponin*Adv. Exp. Med. Biol.* **538**, 221-229.

Ohtsuki, I., Onoyama, Y., and Shiraishi, F., 1988, Electronmicroscopic study of troponin. *J. Biochem.* **103**, 913-919.

Ohtsuki, I., Yamamoto, K., and Hashimoto, K., 1981, Effect of two C-terminal side chymotryptic subfragments on the Ca2+ sensitivity of superprecipitation and ATPase activities of actomyosin. *J. Biochem.* **90**, 259-261.

Ohtsuki, I., Masaki, T., Nonomura, Y., and Ebashi, S., 1967, Periodic distribution of troponin along thin filament. *J. Biochem.* **81**, 817-819.

Ojima,T., and Nishita, K., 1986, Troponin from *Akazara* scallop striated adductor muscle. *J. Biol. Chem.* **261**, 16749-16754.

Ojima, T., Koizumi, N., Ueyama, K., Inoue, A., and Nishita, K., 2000, Functional role of Ca^{2+}-binding site IV of scallop troponin C, *J. Biochem.* **128**, 803-809.

Perry, S. V., 1999, Troponin I: Inhibitor or facilitator. *Mol. Cell. Biochem.* **190**, 9-32.

Shiraishi, F., Kambara, M., and Ohtsuki, I., 1992, Replacement of troponin components in myofibrils, *J. Biochem.***111**, 61-65.

Shiraishi, F., Nakamura, Y., and Ohtsuki. I., 1993, Replacement of troponin in bullfrog skeletal myofibrils by rabbit skeletal and bovine cardiac troponins. *Biomed. Res.* **14(2)**, 93-97.

Shiraishi, F., Morimoto, S., Nishita, K., Ojima, T., and Ohtsuki, I., 1999, Effects of removal and reconstitution of myosin regulatory light chain and troponin C on the Ca^{2+}-sensitive ATPase activity of myofibrils from scallop striated muscle. *J. Biochem.* **126**, 1020-1024.

Szczesna, D., Zhang, R., Zhao, J., Jones, M., Guzman, G, and Potter. J. D., 2000, Altered regulation of cardiac muscle contraction by troponin T mutations that cause familial hypertrophic cardiomyopathy. *J. Biol. Chem.* **275**, 624-630.

Szent-Györgyi, A. G, Kalabokis, V. N., and Perrealt- Micale, C. L., 1999, Regulation by molluscan myosins. *Mol. Cell. Biochem.* **190**, 55-62.

Takahashi-Yanaga, F., Morimoto, S., and Ohtsuki. I., 2000, Effect of Arg145Gly mutation in human cardiac troponin I found in familial hypertrophic cardiomyopathy. *J. Biochem.* **127**, 355-357.

Takahashi-Yanaga, F., Morimoto, S., Harada, K., Minakami, R., Shiraishi, F., Ohta, M., Lu, Q-W., Sasaguri, T., and Ohtsuki, I., 2001,Functional consequences of the mutations in human cardiac troponin I gene found in familial hypertrophic cardiomyopathy. *J. Mol. Cell. Cardiol.* **33**, 2095-2107.

Takeda, S., Yamashita, A., Maeda, K., and Maeda, Y., 2003, Structure of the core domain of human cardiac troponin in the Ca^{2+}-saturated form. *Nature* **424**, 35-41.

Tanokura, M., Tawada, Y., and Ohtsuki, I., 1982, Chymotryptic subfragments of troponin T from rabbit skeletal muscle. I. Determination of the primary structure. *J.. Biochem.* **91**, 1257-1265.

Tanokura, M., Tawada, Y., Ono, A., and Ohtsuki, I., 1983, Chymotryptic subfragments of troponin T from rabbit skeletal muscle. Interaction with tropomyosin, troponin I and troponin C. *J. Biochem.* **93**, 331-337.

Tobacman, L. S., Lin, D., Butters, C., Landis, C., Back, N., Pavlov, D., and Homsher, E, 1999, Functional consequences of troponin T mutations found in hypertrophic cardiomyopathy. *J. Biol. Chem.* **274**, 28363-28370.

Yanaga, F., Morimoto, S., and Ohtsuki, I., 1999, Ca^{2+} sensitization and potentiation of the maximum level of myofibrillar ATPase activity caused by mutations of troponin T found in familial hypertrophic cardiomyopathy. *J. Biol. Chem.* **274**, 8806-8812.

Yumoto, F., Nagata, K., Adachi, K., Nemoto, N., Ojima, T., Nishita, K., Ohtsuki, I., and Tanokura, M., 2003, NMR structural study of troponin C C-terminal domain complexed with troponin I fragment from Akazara scallop. *Adv. Exp. Med. Biol.* **538**, 195-201.

Yumoto, F., Nara, M., Kagi, H.,. Iwasaki, W., Ojima, T., Nishita, K., Nagata, K., and Tanokura, M., 2001, Coordination structure of Ca^{2+} and Mg^{2+} in Akazara scallop troponin C in solution. FTIR spectroscopy of side-chain carboxy group. *Eur. J. Biochem.* **268**. 6284-6290.

MECHANISMS OF CALCIUM RELEASE FROM THE SARCOPLASMIC RETICULUM IN SKELETAL MUSCLE

Makoto Endo*

1. INTRODUCTION

Apart from the molecular mechanism of contraction, the mechanism of excitation-contraction coupling, i.e., how action potential (an electrical response at the surface membrane) triggers contraction (reaction of the contractile protein system packed in the muscle cell), is one of the most important problems in muscle physiology.

The fact that contraction-relaxation cycle is regulated by calcium ion (Ca^{2+}) was established largely through the achievements of S. Ebashi between 1955 and 1968 (cf. Ebashi and Endo, 1968), summarized as follows.

(a) A minute amount of Ca^{2+}, about 1 μM, is essential for the activation of native actomyosin, the *in vitro* contractile protein system extracted together with regulatory proteins.

(b) "Relaxing factor" derived from muscle homogenate that causes relaxation of previously contracted glycerol-extracted muscle is nothing but fragmented sarcoplasmic reticulum (SR), and the mechanism of relaxation is removal of Ca^{2+} from the contractile system through the ATP-dependent Ca^{2+} uptake by the SR.

(c) Control of muscle contraction through the presence and absence of Ca^{2+} is performed by the collaboration of regulatory proteins in the thin filament, tropomyosin and troponin, the latter being a protein discovered by Ebashi.

The largest problem in excitation-contraction coupling remaining after

* Department of Pharmacology, Saitama Medical School, Saitama 350-0495, Japan

Ebashi is how calcium ion is mobilized from the SR upon production of action potential, and I have been engaged on this problem.

2. CONTINUITY BETWEEN T-TUBULE LUMEN AND THE EXTRACELLULAR SPACE

I joined Prof. A. F. Huxley's laboratory in London in 1962. At that time the understanding about the first step of excitation-contraction coupling was as follows. The signal of action potential must somehow be rapidly transmitted deep into the muscle fiber, because heat production experiments by A. V. Hill showed that the contractile reaction of the whole fiber starts soon after action potential. Hill (1949) pointed out that the rapidity excludes the possibility of diffusion of any possible activating substance from the surface to the interior of the fiber. This suggested the presence of some structure that transmits signal of action potential rapidly into the fiber. Huxley and Taylor (1955; 1958) indeed demonstrated by their famous local stimulation experiment on frog muscle fibers that there exists at the level of Z-line, not of A-band, a certain structure through which the localized part of fiber just underneath the electrode responds to electrical stimulation. Such a structure had been observed by histologists already in 19th century, and electronmicroscopists found it as triad structure (Porter and Palade, 1957). The transverse (T) tubular system, the middle component of the triad, was present at the level of Z-line in frog muscle fibers. If the lumen of the T-tubule is open to the extracellular space, action potentials could either conduct or spread electrotonically through the T-tubule membrane into the fiber rapidly enough to explain the rapidity of occurrence of contraction. However, there was no direct evidence for the fact that the transverse tubular system is open to the extracellular space in frog muscle fibers. This could be examined by using a fluorescent dye that cannot penetrate the cell membrane and by detecting under a fluorescence microscope whether or not the dye enters into and diffuses out of the T-tubules. This was the experiment Prof. Huxley suggested to me.

When a non-penetrating fluorescent dye called Lissamine Rhodamine B 200 was applied to a single fiber for some 2 min under a fluorescence microscope and then the dye around the fiber was rapidly washed away, fluorescent striations could be clearly seen soon after the start of wash out but rapidly faded away, indicating diffusing out of the dye. Such an experiment could be repeated over and over again. The position of fluorescent bands was proved to be the center of the I-bands, where the T-tubules are situated (Endo, 1964; 1966). These results clearly showed that at about the position of the T-tubules in the fiber there is a space that is open to the extracellular medium. At about the same time H. E.

Huxley (1964) demonstrated electron-microscopically that ferritin molecules can diffuse into the lumen of the T-tubules.

Some years later, it was shown that the physiological inward spread of electrical signal through the T-tubule is not the electrotonic spread but the conduction of action potentials (Costantin, 1970).

3. CALCIUM-INDUCED CALCIUM RELEASE

3.1 The Discovery of CICR

After coming back to Tokyo from London, I started working on Natori's skinned fiber (Natori, 1954). At first we were more interested in the contractile responses but then turned to the mechanisms of calcium release from the SR.

We happened to examine the action of caffeine on skinned fibers. Caffeine was known to cause calcium release from the SR (Weber and Herz, 1968), and, therefore, we expected that skinned fibers in a weakly Ca^{2+}-buffered solution would reversibly contract in the presence of caffeine. Contrary to the expectation, we obtained a peculiar result as shown in Fig. 1. When a low concentration of caffeine, such as 0.2 mM, was applied to a skinned fiber, a large, nearly maximal contraction was produced after a pretty long latency but it spontaneously relaxed while caffeine was still present. To our further surprise similar transient contraction recurred after several minutes of silence, and it repeated over and over again with about the same interval as far as caffeine was present.

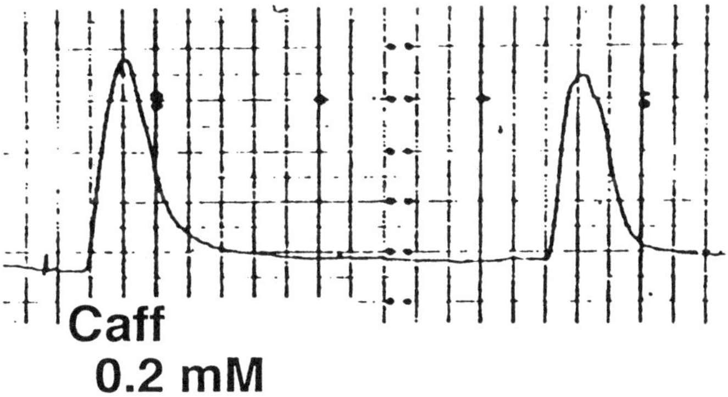

Figure 1. Repeated transient contractions of a skinned fiber under the influence of 0.2 mM caffeine (Endo et al., 1970).

Maximal contractile tension cannot be obtained unless entire part of the fiber contracted synchronously. Therefore, the repeated near maximal contractions with similar time course observed in the presence of caffeine as in Fig. 1 suggested the presence of some positive feedback process through which the contractile response propagates throughout the entire part of the fiber; otherwise, the time course of contraction would have gradually been broadened and peak tension decreased. In order to find out the mechanism of the positive feedback, we examined whether or not products of contractile reaction, ADP or inorganic phosphate, or consequences of contraction such as stretching the fiber cause contraction, but obtained only negative results. We then found that Ca^{2+} itself causes calcium release from the SR (Endo et al., 1970).

Figure 2 shows a piece of evidence for the fact that Ca^{2+} causes calcium release. When free Ca^{2+} concentration of Ca^{2+} buffer surrounding a skinned fiber was raised, in the presence of 2 mM caffeine, from 10^{-8} M to 10^{-6} M, which was below the threshold for contraction by itself (therefore, the fiber developed no tension at the steady state), a transient contraction was evoked if the SR had been preloaded. The fact that the transient contraction was due to calcium release was confirmed because the amount of calcium remaining in the SR after the contraction was less than before. Essentially the same effect of Ca^{2+} could be demonstrated in the absence of caffeine. We called this phenomenon calcium-induced calcium release (CICR) (Endo et al., 1970).

Ford and Podolsky (1970) also discovered CICR at about the same time entirely independently on the somewhat different basis. Soon afterwards, A. Fabiato and F. Fabiato (1972) showed CICR in cardiac muscle.

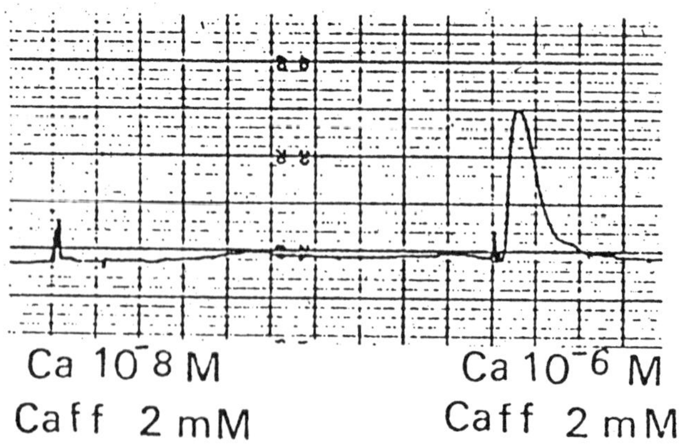

Figure 2. Contraction of a skinned fiber induced by Ca^{2+}(10^{-6}M)-induced calcium release (Endo et al, 1970).

3.2 Properties of CICR

CICR cannot be quantitatively studied by simple monitoring of Ca^{2+} concentration outside the SR, because the increase in Ca^{2+} concentration due to calcium release further stimulates release and at the same time also stimulates calcium pump which works to decrease Ca^{2+} concentration. In order to determine the exact calcium-releasing effect of a certain concentration of Ca^{2+}, we devised the following experimental procedures to cut off the above mentioned positive and negative feedback loops (Endo and Iino, 1988).

First, the SR was loaded with calcium to a fixed level. Then ATP was removed from the medium immersing the fiber so that calcium pump can no longer work. SR was then stimulated with various concentrations of Ca^{2+} that were buffered with a high concentration of Ca^{2+} buffer to minimize the increase in free Ca^{2+} concentration due to calcium release. The amount of calcium released during the period when the calcium stimulus was applied was measured by determining the amount of calcium remaining in the SR after the stimulus (by fully discharging it by applying a high concentration of caffeine) and subtracting it from the amount initially loaded.

In this way it was found that decrease in the amount of calcium in the SR decreased exponentially during continuous stimulation with a fixed concentration of Ca^{2+}, and the rate of decrease, namely the rate of calcium release,

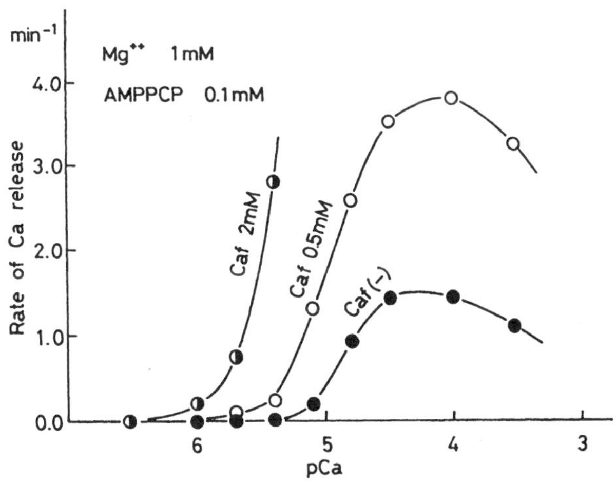

Figure 3. Ca^{2+} concentration dependence of the rate of calcium release and the effect of caffeine. (Endo, 1984)

is dependent on stimulating Ca^{2+} concentration. The dependence of the rate of calcium release on Ca^{2+} concentration is shown in Fig. 3, and as is seen, it is bell-shaped, higher concentrations of Ca^{2+} exerting an inhibitory action as well. Caffeine enhances CICR as is also shown in Fig. 3. The action of caffeine on CICR explains the mechanism of well-known action of this agent on skeletal muscle; at a low concentration it enhances twitch responses and at high concentrations it produces contracture.

Adenine compounds also enhance CICR; while caffeine increases not only the maximal response at the optimal Ca^{2+} concentration but also Ca^{2+} sensitivity of CICR (Fig. 3), adenine compounds do not alter the Ca^{2+} sensitivity of CICR (Endo, 1981).

Mg ion and procaine inhibit CICR; similarly to the relation between caffeine and adenine compounds, Mg ion decreases Ca^{2+} sensitivity of CICR, but procaine does not (Endo, 1981).

3.3 CICR and Physiological Calcium Release in Skeletal Muscle

Does CICR operate as physiological calcium release mechanism? If physiological calcium release is mediated by CICR, an inhibitor of CICR should inhibit physiological calcium release. An inhibitor of CICR, procaine, inhibits caffeine-induced contracture as expected, but it did not inhibit contractions evoked by potassium-induced sarcolemmal depolarization at all (Thorens and Endo, 1975).

However, since procaine also inhibits action potential production, it could not be used to test whether or not it inhibits real physiological contractions triggered by action potentials. To examine real physiological contractions we utilized adenine as an inhibitor of CICR under the physiological condition (Ishizuka and Endo, 1983). Adenine by itself enhances CICR as shown in Fig. 4A. This action is in common with other adenine compounds, adenosine, AMP, ADP, and ATP. Among these compounds the action of ATP is the strongest and the action of adenine is very much weaker. Therefore, in the presence of ATP adenine actually inhibits CICR, because the strong enhancing action of ATP on CICR is replaced by the weak enhancing action of adenine, as shown in Fig. 4B. Thus adenine should inhibit CICR in living muscle where millimolar ATP is present. Adenine indeed inhibited caffeine-induced contracture of an intact muscle fiber but twitches were not inhibited at all (Fig. 5). However, as is also shown in the figure, if twitches are potentiated by caffeine adenine now inhibits twitch tension (although the magnitude of inhibition is rather small because adenine penetrates the cell membrane very slowly), but if the same fiber was potentiated by replacement of chloride in the medium by nitrate, adenine did not inhibit twitches at all (Ishizuka et al., 1983). These results clearly indicate that principal physiological calcium release mechanism in skeletal muscle is not CICR. Whether or not CICR plays a secondary role in physiological EC-coupling is at present somewhat controversial.

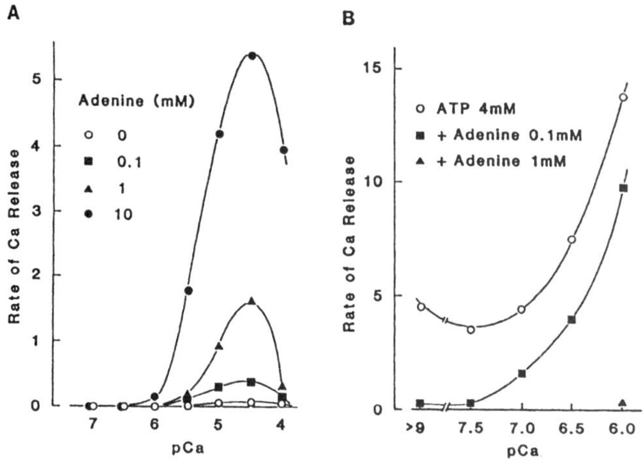

Figure 4. Enhancing effect of adenine alone on CICR (**A**) and inhibitory effect of adenine on CICR in the presence of ATP (**B**) (Ishizuka and Endo, 1983).

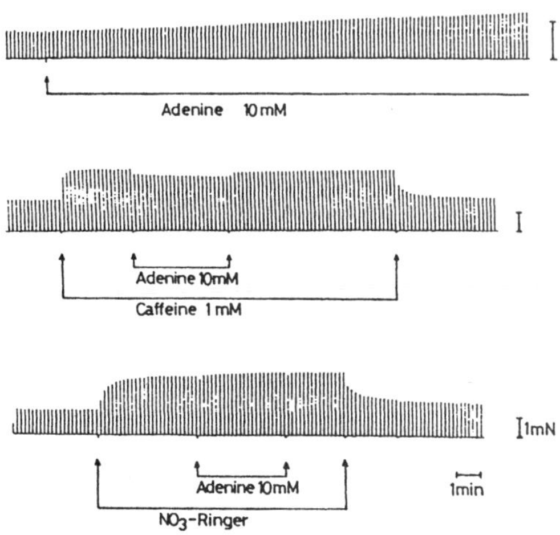

Figure 5. No inhibitory effect of adenine on normal twitch (upper); inhibitory effect of adenine on caffeine-potentiated twitch (middle); no inhibitory effect of adenine on nitrate-potentiated twitch (lower) of a single skeletal muscle fiber (Ishizuka et al., 1983).

3.4 CICR and Malignant Hyperthermia

CICR, however, plays an essential role in a pathophysiological condition, called malignant hyperthermia (MH). Malignant hyperthermia is one of a serious complication of inhalation anesthesia in people with some hereditary disorders (Denborough and Lovell, 1960). The patients develop extremely high fever, over 40 ℃, with skeletal muscle rigidity during inhalation anesthesia without apparent causes. The fever is due to over production of heat through generalized contraction of patients' skeletal muscles. Kalow et al. (1970) showed that in skeletal muscles of MH patients a lower concentration of caffeine was effective in inducing contracture than in normal muscle. This fact suggested us that the hereditary disorder is present in CICR, because the mechanism of action of caffeine is the enhancement of CICR. As shown in Figs. 6 and 7, CICR in skeletal muscles of MH patients indeed shows a higher Ca^{2+} sensitivity and greater maximal response at optimal Ca^{2+} concentration than that of normal muscles (Endo et al., 1983; Kawana et al., 1992). Figure 6 also shows that halothane enhances CICR both in MH and normal muscles. We further found that all the other inhalation anesthetics also enhance CICR (Matsui and Endo, 1986). Based on these findings, we theoretically showed that only if the enhanced CICR of MH patients is further enhanced by inhalation anesthetics, the rate of CICR now exceeds that of calcium uptake by calcium pump at any Ca^{2+} concentrations, and, therefore, calcium release proceed in a more-or-less regenerative manner and MH occurs. Normal muscles under the influence of inhalation anesthetics or MH

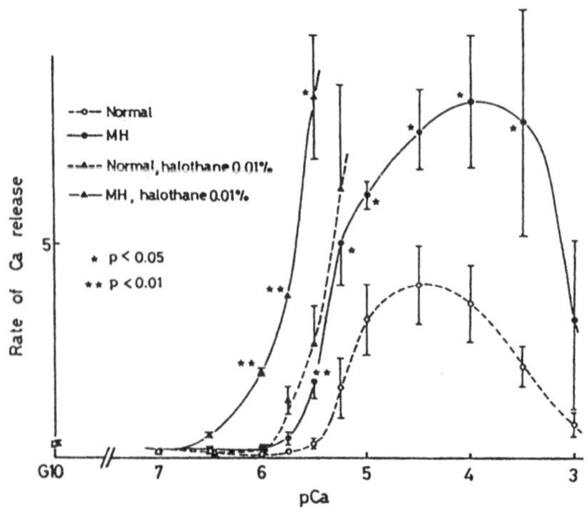

Figure 6. Enhanced rate of CICR in muscle fibers of an MH patient and an enhancing effect of halothane on CICR in normal and patient's muscle fibers (Endo et al., 1983).

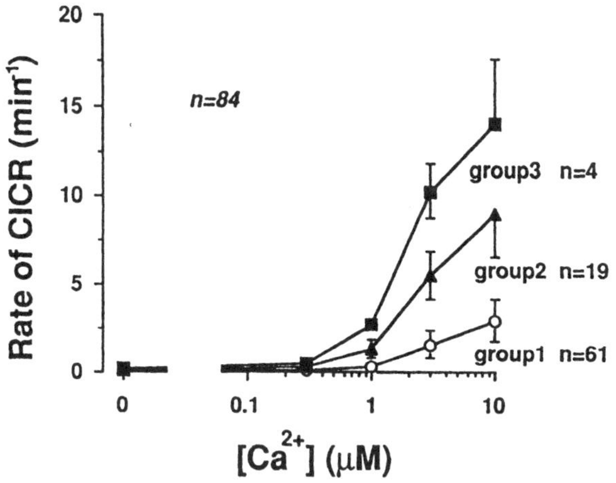

Figure 7. Mean rates of CICR in muscle fibers from groups 1 (normal subjects), 2 and 3 (MH patients). Groups were separated by cluster analysis (Kawana et al., 1992).

muscles without anesthetics were safe because the rate of calcium uptake by SR calcium pump overwhelms the rate of CICR, and Ca^{2+} tends to go back into the lumen of the SR (Endo et al., 1983).

3.5 Molecular Basis of CICR, Ryanodine Receptor

A plant alkaloid, ryanodine, binds to the CICR channels when they are open, and fixes the channels in an open state (Fleischer et al., 1985). This action of ryanodine was very specific, and by utilizing the specific binding of the alkaloid, the calcium release channel protein of the SR was isolated and purified by biochemists as ryanodine receptor (RyR) (Inui et al., 1987; Imagawa et al., 1987; Lai et al., 1988). The RyR, if incorporated into lipid bilayer, shows all the properties of CICR (Lai et al., 1988). It is now known that there are three types of RyRs in mammals (Sorrentino, V. and Volpe, P., 1993).

In mammalian skeletal muscle, the type of RyR expressed is mainly type 1 with a much smaller fraction of type 3. It was demonstrated that if type 1 RyR was knocked out, the animals cannot live after birth because EC-coupling of their respiratory muscles cannot function, although type 3 RyR is still present (Takeshima et al., 1994). EC-coupling of such a muscle can be relieved by reintroduction of type 1 RyR but not of type 2 (Yamazawa et al., 1996). From these results it is concluded that type 1 RyR is *the* physiological calcium release channel. Type 1 RyR behaves as CICR channel as shown in lipid bilayer experiments mentioned above, but physiological calcium release shows properties quite different from CICR as described in 3.3 above and 4 below. Thus, one must conclude that Type 1 RyR operates in two different modes, physiological mode and CICR mode.

4. ACTIVATION OF RYANODINE RECEPTOR BY CLOFIBRIC ACID

Almost all the agents so far known to activate RyR open the channel in the CICR mode. However, we have recently found that clofibric acid, an antihyperlipidemic agent, activates the RyR of skeletal muscle in a mode that is quite different from CICR (Ikemoto and Endo, 2001).

As shown in Fig. 8, 10 mM clofibric acid exerts a strong calcium releasing action on the SR in the practical absence of Ca^{2+} (in the presence of 10 mM EGTA). This calcium release was proven to be the result of RyR activation. The properties of calcium-release induced by clofibric acid are qualitatively different from those of CICR. (1):As shown in Fig. 9, clofibric acid-induced calcium release was *inhibited* by AMP, whereas caffeine-potentiated CICR is of course strongly enhanced by AMP. (2):Clofibric acid-induced calcium release was ***not inhibited*** by procaine, a well-known inhibitor of CICR.

Another difference between clofibric acid-induced calcium release and CICR is the inhibitory action of dantrolene. Dantrolene inhibits CICR at 37 ℃ but not at 20 ℃ (Ohta and Endo, 1986). However, it inhibited clofibric acid-induced calcium release both at 37 ℃ and at 20 ℃. In this sense physiological calcium release was similar to clofibric acid-induced calcium release as it is also inhibited by dantrolene at 20 ℃ as well as at 37 ℃ (Kobayashi and Endo, 1988).

In addition to the inhibitory effect of dantrolene at 20 ℃, there are other qualitative similarities between physiological calcium release and clofibric acid-induced calcium release, which are different from CICR, although the similarities are not complete. Namely, both kinds of release can be evoked in the absence of Ca^{2+}; adenine compounds do not enhance the release; procaine does not inhibit the release, although there are some controversial reports in the case of the effect of procaine on physiological calcium release (Garcia and Schneider, 1995).

From these results clofibric acid appears to be the first example of agents to activate RyR of skeletal muscle not in the CICR mode but in the mode similar to the physiological one, and might, therefore, be useful for elucidating the mechanism of physiological calcium release through the RyR.

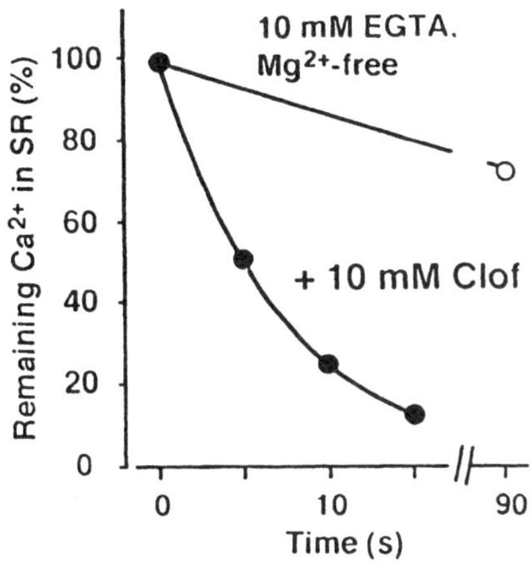

Figure 8. Calcium releasing effect of clofibric acid in the presence of practical absence of Ca^{2+} (Ikemoto and Endo, 2001).

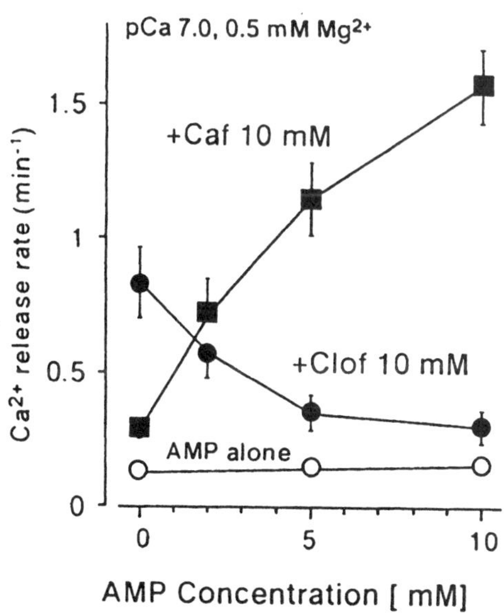

Figure 9. Inhibitory effect of AMP on clofibric acid-induced calcium release vs. potentiating effect of AMP on simple CICR or caffeine-potentiated calcium release (Ikemoto and Endo, 2001).

5. ACTIVATION OF RYANODINE RECEPTOR BY CERIVASTATIN

Clofibric acid and related agents are used for the treatment of hyperlipoproteinemia and one of their rare but serious adverse effect is rhabdomyolysis. Since statins, another class of drugs for hyperlipoproteinemia, also cause rhabdomyolysis, we have examined the effects of statins on skinned fibers and as shown in Fig. 10, we found that cerivastatin also activates RyR to cause calcium release (Inoue et al., 2003). It enhances CICR, but in the practical absence of Ca^{2+} it still substantially increases rate of calcium release.

Cerivastatin-activated calcium release at pCa 6 is inhibited by procaine as expected, but the effect at pCa over 8 was not at all inhibited by procaine. Tetracaine that is known to inhibit both CICR and physiological calcium release inhibits cerivastatin-activated calcium release at pCa over 8 as well as at pCa 6..

While it is not clear at present that rhabdomyolysis caused by these agents is related to their activation of the ryanodine receptor, because it is conceivably possible, we are currently pursuing further along the line.

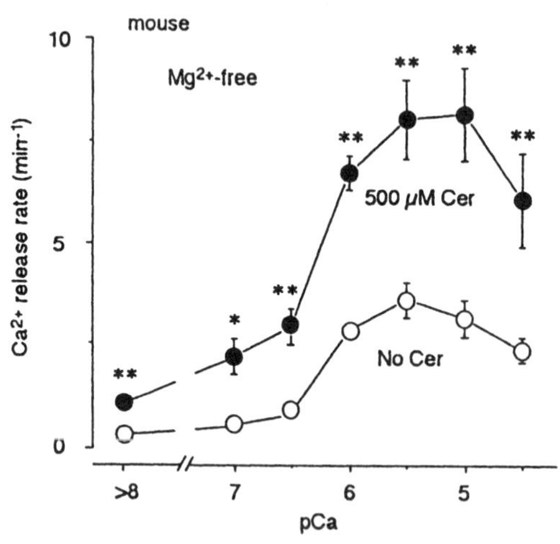

Figure 10. Enhancing effect of cerivastatin on calcium release (Inoue et al., 2003).

6. CONCLUDING REMARKS

Calcium release from the SR in skeletal muscle is brought about by opening of RyR/calcium release channels. RyR can be activated by Ca^{2+} (CICR), but physiological opening of RyR is not mediated by Ca^{2+}. RyR is considered to open physiologically as a result of interaction with voltage sensor protein in the T-tubule membrane, and its opening mode is quite different from that of CICR. The physiological mode of opening is mimicked by clofibric acid and possibly by cerivastatin in the absence of Ca^{2+}. On the other hand, CICR mode of opening of RyR in skeletal muscle plays the major role in MH or during caffeine action. The exact operation of RyRs in physiological calcium release remains to be solved.

REFERENCES

Costantin, L., L., 1970, The role of sodium current in the radial spread of contraction in frog muscle fibers. *J. Gen. Physiol.* **55**: 703-715.

Denborough, M. A. and Lovell, R. R. H., 1960, Anesthetic deaths in a family. *Lancet* II : 45.

Ebashi, S. and Endo, M., 1968, Calcium ion and muscle contraction, in: *Progr. in Biophys. and Mol. Biol.* Vol. **18**: J. A. V. Butler, and D. Noble, eds., Pergamon Press, Oxford, pp. 123-183.

Endo, M., 1964, Entry of a dye into the sarcotubular system of muscle. *Nature* **202**: 1115-1116.

Endo, M., Entry of fluorescent dyes into the sarcotubular system of the frog muscle. *J. Physiol.* **185**: 224-238.

Endo, M., 1981, Mechanism of calcium-induced calcium release in the SR membrane, in: *The Mechanism of Gated Calcium Transport across Biological Membranes* S. T. Ohnishi, and M. Endo, eds., Academic Press, N.Y., pp.257-264.

Endo, M., 1984, Significance of calcium-induced release of calcium from the sarcoplasmic reticulum in skeletal muscle. *Jikeikai Med. J.* **30**: Suppl.1, 123–130.

Endo, M. and Iino, M., 1988, Measurement of Ca^{2+} release in skinned fibers from skeletal muscle. *Methods in Enzymology* **157**: 12-26.

Endo, M., Tanaka, M. and Ogawa, Y., 1970, Calcium-induced release of calcium from the sarcoplasmic reticulum of skinned skeletal muscle fibres. *Nature* **228**: 34-36.

Endo, M., Yagi, S., Ishizuka, T., Horiuti, K., Koga, Y. and Amaha, K., 1983, Changes in the Ca-induced Ca release mechanism in the sarcoplasmic reticulum of the muscle from a patient with malignant hyperthermia. *Biomed. Res.* **4**: 83-92.

Fabiato, A. and Fabiato, F., 1972, Excitation-contraction coupling of isolated cardiac fibers with disrupted or closed sarcolemmas. Calcium-dependent cyclic and tonic contractions. *Circulation Res.* **31**: 293-307.

Fleischer, S. E., Ogunbunmi, E. M., Dixon, M. C. and Fleer, E. A., 1985, Localization of Ca^{2+} channels with ryanodine in junctional terminal cisternae of sarcoplasmic reticulum of fast skeletal muscle. *Proc. Natl. Acad. Sci. USA* **82**: 7256-7259.

Ford, L. E. and Podolsky, R. D., 1970, Regenerative calcium release within muscle cells. *Science* **167**: 58-59.

Garcia, J, and Schneider, M., F., 1995, Suppression of calcium release by calcium or procaine in voltage clamped rat skeletal muscle fibres. *J Physiol.* **485**:437-445.

Hill, A., V., 1949, The abrupt transition from rest to activity in muscle. *Proc. Roy. Soc, (London)*, **Ser. B**, 136: 399-420.

Huxley, A. F. and Taylor, R. E., 1955, Function of Krause's membrane. *Nature*, **176**: 1068.

Huxley, A. F. and Taylor, R. E., 1958, Local activation of striated muscle fibres. *J. Physiol.*, **144**: 426-441.

Huxley, H. E., 1964, Evidence for continuity between the central elements of the triads and extra-cellular space in frog sartorius muscle *Nature* **202**: 1067-1071.

Ikemoto, T. and Endo, M., 2001, Properties of Ca^{2+} release induced by clofibric acid from sarcoplasmic reticulum of mouse skeletal muscle fibres. *Brit. J. Pharmacol.* **134**: 719-728.

Ikemoto, T., Hosoya, T., Aoyama, T., Kihara, Y., Suzuki, M. and Endo, M., 2001, Effects of dantrolene and its derivatives on Ca^{2+} release from the sarcoplasmic reticulum of mouse skeletal muscle fibres. *Brit. J. Pharmacol.* **134**, 729-736.

Imagawa, T., Smith, J. S., Coronado, R. and Campbell, K. P., 1987, Purified ryanodine receptor from skeletal muscle sarcoplasmic reticulum is the Ca^{2+} permeable pore of the calcium release channel. *J. Biol. Chem.* **262**: 16636-16643.

Inoue, R., Tanabe, M., Kono, K., Maruyama, K., Ikemoto, T. and Endo, M., 2003, Ca^{2+}-releasing effect of cerivastatin on the sarcoplasmic reticulum of mouse and rat skeletal muscle fibers. *J. Pharmacol. Sci.* **93**: 279-288.

Inui, M., Saito, A. and Fleischer, S., 1987, Purification of the ryanodine receptor and identity with feet structures of junctional terminal cisternae of sarcoplasmic reticulum from fast skeletal muscle. *J. Biol. Chem.* **262**: 1740-1747.

Ishizuka, T. and Endo, M., 1983, Effects of adenine on skinned fibers of amphibian fast skeletal muscle. *Proc. Japan Acad.* **59**: 93-96.

Ishizuka, T., Iijima, T. and Endo, M., 1983, Effect of adenine on twitch and other contractile responses of single fibers of amphibian fast skeletal muscle. *Proc. Japan Acad.* **59**: 97-100.

Kalow, W., Britt, B. A., Terreau, M. E. and Haist C., 1970, Metabolic error of muscle metabolism after recovery from malignant hyperthermia. *Lancet* Ⅱ: 895-898.

Kawana, Y., Iino, M., Horiuti, K., Matsumura, N., Ohta, T., Matsui, K. and Endo, M., 1992, Acceleration in calcium-induced calcium release in the biopsied muscle fibers from patients with malignant hyperthermia. *Biomed. Res.* **13**: 287- 297.

Kobayashi, T. and Endo, M., 1988, Temperature-dependent inhibition of caffeine contracture of mammalian skeletal muscle by dantrolene. *Proc. Japan Acad.* **64**: 76-79.

Lai, F. A., Erickson, H. P., Rousseau, E., Liu, Q.-Y. and Meissner, G., 1988, *Nature* **331**: 315-319.

Matsui, K. and Endo, M., 1986, Effect of inhalation anesthetics on the rate of Ca release from the sarcoplasmic reticulum of skeletal muscle in the guinea-pig. *Jap. J. Pharmacol.* **40**: Supple. 245P.

Natori, R., 1954, The property and contraction process of isolated myofibrils. *Jikeikai Med.J.* **1**: 119-126.

Ohta, T. and Endo, M., 1986, Inhibition of calcium-induced calcium release by dantrolene at mammalian body temperature. *Proc. Japan Acad.* **62**: 329-332.

Porter, K. R. and Palade, G. E., 1957, Studies on the endoplasmic reticulum. Ⅲ. Its form and distribution in striated muscle cells. *J. Biophys. Biochem. Cytol.*, **3**: 269-300.

Sorrentino, V. and Volpe, P., 1993, Ryanodine receptors: how many, where and why? *Trends Pharmacol. Sci.* 14: 98-103.

Takeshima, H., Iino, M., Takekura, H., Nishi, M., Kuno, J., Minowa, O., Takano, H. and Noda, T., 1994, Excitation-contraction uncoupling and muscular degeneration in mice lacking functional skeletal muscle ryanodine receptor gene. *Nature* 369: 556-559.

Thorens, S. and Endo, M., 1975, Calcium-induced calcium release and "depolarization"-induced calcium release: their physiological significance. *Proc. Japan Acad.* 51: 473-478.

Weber, A. and Herz, R., 1968, The relationship between caffeine contracture of intact muscle and the effect of caffeine on reticulum. *J. Gen. Physiol.* 52: 750-759.

Yamazawa, T., Takeshima, H., Sakurai, T., Endo, M. and Iino, M., 1996, Subtype specificity of the ryanodine receptor for Ca^{2+} signal amplification in excitation-contraction coupling. *EMBO J.* 15, 6172-6177.

DISCUSSION

Ikebe: Does clofibric acid act on all ryanodine receptor (RYR) subtypes? Is the action target of clofibric acid really on RYR, but not other component such as dihydropyridine (DHP)?

Endo: 1) We haven't yet examined the effect of clofibric acid on Type 2 or Type 3 RYR so far, although we are planning to do it. 2) I didn't show that the calcium-releasing effect of clofiblic acid is due to its effect on RYR because of shortage of time. Utilizing the properties of ryanodine that affects calcium-release channels only when they are open and fixes the channels in an open state, ryanodine was added to the skinned fiber in the presence of clofiblic acid in the practical absence of calcium. The result was that the calcium-accumulating capacity of the SR of the skinned fiber was greatly reduced indicating the open fixation of calcium-release channels of the skinned fiber by ryanodine. Without addition of clofiblic acid ryanodine did not reduce the calcium-accumulating capacity at all. This result indicates that clofiblic acid opens RYR.

FROM INWARD SPREAD OF ACTIVATION, ACTIVE ELONGATION TO THE EFFECT OF ORGANIC CALCIUM CHANNEL BLOCKERS IN MUSCLE EXCITATION-CONTRACTION COUPLING.

Gonzalez-Serratos, H*, Ortega A**, Valle-Aguilera R.** and Chang R*

1. OUTLINE

This paper is divided into three sections. The first section is entitled "Perimeter Propagation of an Excitatory Process in Muscle Fibers of Different Species." In this section we describe the remarkable experimental findings of Professor Haruo Sugi[1] published in1974 and those of Sugi & Ochi[2, 3] published in 1967. They showed that in fresh isolated skeletal muscle cells of two different species there is an excitatory process that travels not across the radius of the cells through the transverse tubular system (T-system) but around the periphery of the fibers. The second section is entitled "Is Elongation of Myofibrils During Relaxation Caused by Active Reverse Sliding of Acto-myosin Filaments?" Here we review experiments indicating that when elongation occurs during isotonic relaxation, the actin and myosin filaments always slide back to the same sarcomere length. The third section entitled "Effect of the Organic Ca Channel Blocker Diltiazem on Excitation-Contraction Coupling in Skeletal Muscle." describes recent experiments showing that the organic calcium channel blocker diltiazem has an inhibitory effect on the sarcoplasmic reticulum (SR) calcium pump CERCA-1.

2. PERIMETER PROPAGATION OF AN EXCITATORY PROCESS IN MUSCLE FIBERS OF DIFFERENT SPECIES

By the end of the fifties A. F. Huxley[4], Huxley and Straub[5] and Huxley and Taylor[6, 7] had shown that local depolarizations, with small pipettes of $\leq$ 1.5 μm in

---

*Department of Physiology, School of Medicine, University of Maryland, Baltimore, MD, 1201, USA, **Departamento de Boquimica, Facultad de Medicina, Universidad Nacional de Mexico, Mexico City, 04510, Mexico and ***Departamento de Fisiologia, Facultad de Medicina Universidad de San Luis Postosi, San Luis Postosi, Mexico.

diameter, at the level of the Z line in frogs[4, 6, 7] or the baundery between the I and A bands in lizards[5] produced an inward and localized contraction. The electric currents flowing through the stimulating pipettes were small enough so as not to trigger action potentials.

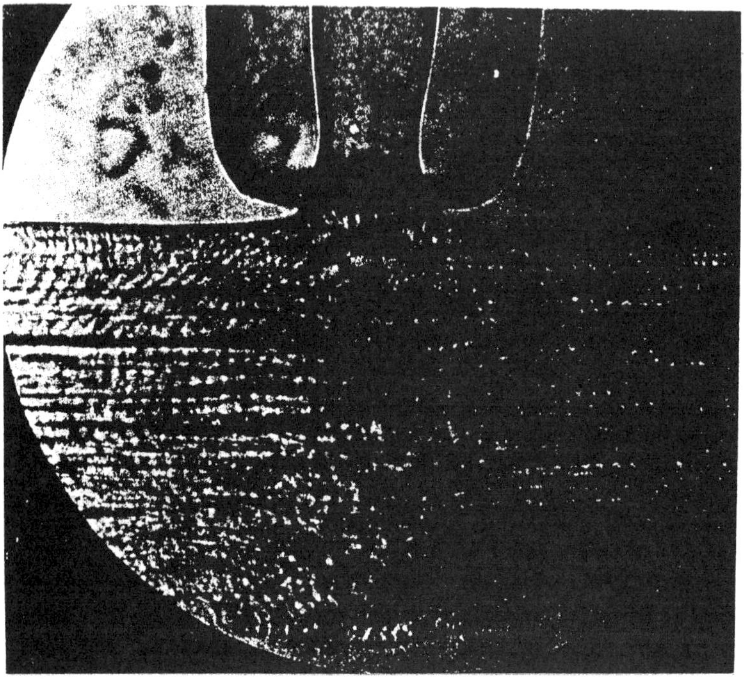

Figure 1. Sample photomicrograph of a muscle fiber dissected from the crayfish during a local contraction that propagates around the fiber's perimeter. The sarcomeres located immediately under the stimulating pipette and on the opposite side of the fiber are contracting strongly. The sarcomeres in the center of the fiber remain unchanged i.e., they are not contracting. From Sugi and Ochi[2].

Increasing the stimulating currents produced further graded penetration of inward spread of contraction of no more than 10 μm, whereas there was no lateral spread of activation. There was no suggestion of an all-or-none behavior, because with a further increase in the current, the fiber went into a full contraction and became damaged. Huxley[4, 8] and Huxley and Taylor[7] proposed that the inward spread of activation occurs through the T-system, the spread is not an all-or-none process like an action potential but is like an electrotonic spread. Gonzalez-Serratos [9, 10] used extracellular stimulating electrodes and high-speed cinemicrophotography to investigate the inward spread of the excitatory process in isolated frog muscle cells. He showed that the inward propagation of the excitatory process traveled all across the muscle fiber diameter at a velocity of 7 cm/s and had a Q_{10} of 2.13. He proposed that this was a regenerative process like an action potential propagating along the transverse tubular system (for a review see Gonzalez-Serratos[11]). Professor Sugi[1] investigated the same problem in single isolated skeletal frog muscle cells. He used a similar approach to the one of Huxley and collaborators but applied depolarizing currents of the surface membrane with stimulating pipettes of 20-40 μm in diameter. He also found an inward spread of the excitatory process that propagated all across the diameter with an average velocity of the same order of magnitude of 1.45 cm/sec and a similar Q_{10} of 2.3. Evidences from several laboratories indicated that the excitatory signal traveling along the T-tubular network was a Na^+ dependent action potential similar to the sarcolemmal action potential [12, 13, 14, 15] (reviewed in [16]).

However, Sugi and Ochi[2] found a remarkable phenomenon in single striated muscle cells isolated from crayfish. They produced local activations of the muscle fibers

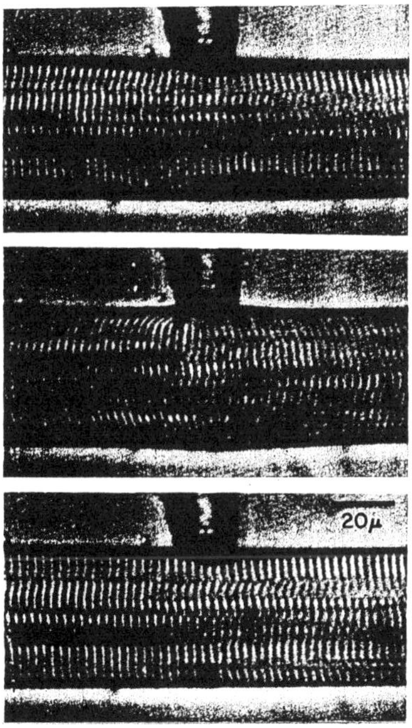

Figure 2. Sample photomicrographs from a muscle fiber dissected from the frog before (A), during (B) and after (C) a local contraction that propagates around the fiber's perimeter. In B, the sarcomeres located immediately under the stimulating pipette and on the opposite side of the fiber are contracting strongly when depolarizing current is applied through the pipette. The sarcomeres in the center of the fiber remain unchanged during the stimulation i.e., they are not contracting. From Sugi and Ochi[3].

by depolarizing the surface membrane with glass pipettes of 20-120 μm tip diameter that was in contact with the surface membrane. The authors applied strong currents that produced depolarizations of >300 mV across a sarcolemmal area, that included several sarcommeres, As shown in Figure 1, the contraction propagated not across the diameter, as described in the previous paragraph, but surprisingly, around the perimeter of the muscle fibers. Moderate depolarizations through the pipettes produced contractions that spread transversely all the way across the crayfish fibers diameter like the ones described in the previous paragraph. The authors further investigated the same phenomenon in single isolated frog skeletal muscle cells[3]. For these experiments they used small diameter pipettes of 3- 40 μm and applied large trans-membrane currents that developed depolarizations of >100 mV. As shown in Figure 2, they discovered that using similar techniques to the ones used for crayfish cells, the contraction in the frog also spread around the whole perimeter of the muscle fibers without propagating through the center of the cells. The authors estimated that this excitatory process spread around the inside perimeter of the muscle fibers at an average velocity of 0.8-6 cm/s which is similar to the

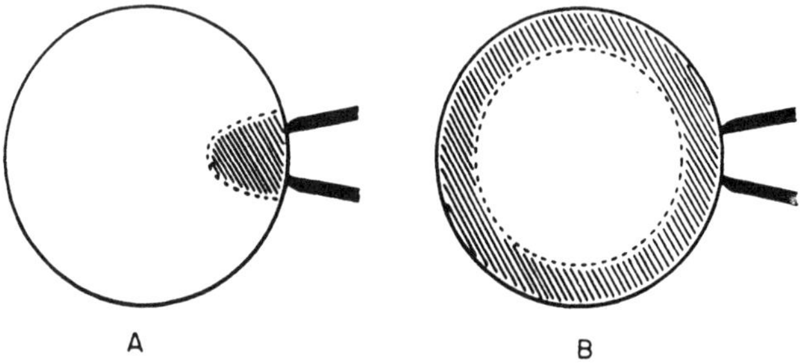

Figure 3. Illustration of the two types of transverse spread of contraction initiated by local activation of skeletal muscle fibers. The outer circles represent the cross section of the fiber. The pipette for local activation is shown to the right-hand side of the circle representing the fiber. Shaded areas represent the contraction regions. A, transverse spread of contraction. B, circumferential spread of contraction. From Sugi and Ochi[2]

one found for the propagation across the diameter of the fibers[1, 16]. A further remarkable finding was that in isolated muscle cells dissected from either crayfish or frog this type of propagation could be elicited in Na^+ free Ringer's solution (substituted with cholin chloride). This led to the conclusion that the contraction spread around the inside perimeter was not a Na^+ dependent process[1, 2, 3] like the tubular action potential[16]. The findings were so unique that they were published soon after their appearance in a leading textbook[17]. One possible explanation put forward[3] was that only the tubules localized in the surface region are capable of conducting this process due to regional differences. This implies that the tubular membranes in the periphery are different in either electrophysiological, morphological or both characteristics compared to the network of transverse tubules inside the cells. The above results showing the two ways of spread of contraction are summarized in Figure 3. Even today, however, the mechanism responsible for this remarkable phenomenon of propagation along the inner perimeter of muscle fibers remains a mystery to be solved. As Professor Sugi pointed out 37 years ago, "Much more experimental work is needed to determine the actual mechanism of...."[3] the above process.

3. IS ELONGATION OF MYOFIBRILS DURING RELAXATION CAUSED BY ACTIVE REVERSE SLIDING OF ACTO-MYOSIN FILAMENTS?

3.1 Introduction

Studies on muscle relaxation have led to contradictory opinions. Ramsey and Street[16] showed that isolated fibers elongate after isotonic contractions. They postulated

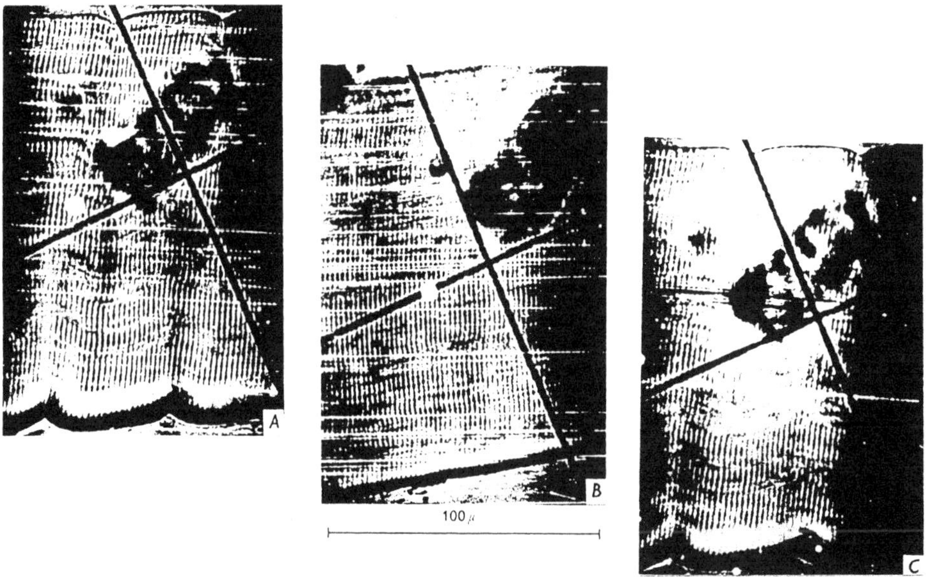

Figure 4. Sample cinehotomicrographs of a contraction-relaxation cycle of a single frog muscle fiber during a propagated twitch. A, wavy myofibrils at rest before stimulation. B, during maximum contraction, the myofibrils have straightened out. C, after the relaxation the myofibrils become wavy again. From Gonzalez-Serratos[9]

that the relaxation was an active process. However, Hill[17] on the basis of the observation that the latent period of a whole sartorius muscle was the same at the slack length and at shorter lengths, argued that elongation was a passive process due to elastic restoring forces set up by the lateral expansion of the fibers. Here we review some experimental findings showing that after shortening upon stimulation, actin and myosin filaments always slide back to the same sarcomere length (sl). The process by which myosin and actin filaments slide back "in reverse" remains unknown and possible mechanisms are discussed.

3.2 Methods and results

The experiments were done in freshly isolated muscle fibers dissected from the semitendinosus muscle of the frog *Rana temporaria*.

If the ends f an isolated muscle fiber were not held as it is resting in the bottom of dish filled with Ringer's solution, its sarcomere length (sl) is about 2.0 μm. When it was held by the end tendons and stretched >2.0 μm/sl and then released, it shortened passively back to 2.0 μm/sl. Passive decrease in the distance between the end tendons below this sl resulted in the fiber sagging while the sarcomeres remained at 2.0 μm/sl and the total fiber length remained constant. When the sl was brought to <2.0 μm, by shortening upon stimulation, the actin and myosin filaments slid pass each other and the

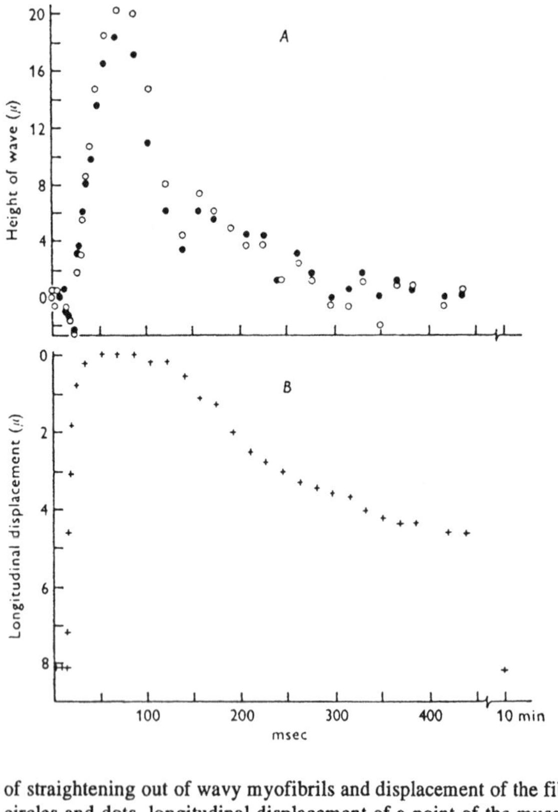

Figure 5. Time course of straightening out of wavy myofibrils and displacement of the fiber and gelatin during a contraction. A, open circles and dots, longitudinal displacement of a point of the muscle fiber surface and a speck of dust in the gelatin respectively. B, height of the wave on a superficial myofibril decreasing from 8 to 0 μm in height. From Gonzalez-Serratos[9].

sagging disappeared. During relaxation, the fiber elongated again and the filaments slid back to 2 μm/sl. The sagging could be prevented from forming by setting the fiber straight in a block of gelatin at an approximate sl of 2.6 μm[9]. Afterwards, if the block of gelatin was compressed along the longitudinal axis of the fiber, the overall length of the muscle fiber (the sarcomere spacing) decreased. When sl was reduced to ≤ 1.98 μm, the fiber as a whole remained straight but the myofibrils within it became wavy[18, 19]. Stimulation of the fiber during contraction sliding of acin and myosin filaments caused the wavy myofibrils to straighten out. However, as shown in Fig. 4, after the contraction the wavy pattern re-appeared during relaxation with the same wave-length, height and sl as before the contraction. Another feature seen during the contraction is that it takes place without appreciable change of fiber diameter. The time course of these processes was followed by high-speed cinemicrophotography. The results from one of these experiments are plotted in Figure 5. Figure 5 also shows a plot of the longitudinal displacement of the fiber and the movement of a speck of dust in the gelatin close to the edge of the fiber. It can be seen that the time course of the movement of the fiber and the gelatin are much faster than the redevelopment of the wavy pattern of the myofibrils.

3.3 Discussion

Ramsey and Street[16] had already found that isolated muscle fibers lying in Ringer's solution re-extended themselves to the original slack length after a twitch or short tetanus. They proposed that relaxation was an active process generated within the contractile machinery itself. Hill[17] reported that the latent period for tension to develop after stimulation of whole sartorii muscles was the same at slack length and shorter lengths. On the basis of this observation he argued that the elongation was a passive process due to elastic restoring forces set up by lateral expansion of the fiber. However, if the latency to the rapid rise in tension observed by Hill is plotted rather than the latent period for the slow development in force, there is indeed a latent period consistent with the idea of a restoration force within the muscle cells[9]. Furthermore, the possibility that the elongation and reformation of wavy myofibrils might be caused by restoring force in the sarcolemma as it expanded, do not operate in the gelatin compressed fibers since the general outline and diameter does not change. The movement of the gelatin has a much faster time course than de re-development of wavy myofibrils. Thus, it is also unlikely the re-development of wavy myofibrils may be due to changes imposed by elastic forces in the gelatin.

From the preceding description, it seems necessary to conclude that an elongation force exists within the myofibrils themselves and causes a reverse of the sliding movement of filaments that had occurred during shortening. There are two possible hypotheses that may explain this reverse sliding of actin and myosin filaments. At sl< 2 μm, when there is a double overlap, the myosin heads in the area of double overlap may engage in reverse the opposite actin sites of attachment pushing back the filaments until the double overlap disappears at 2 μm/sl. This mechanism would imply that the cross-bridges are able to produce negative tension. A. F. Huxley[20] had raised the possibility that cross-bridges are able to exert negative tension. The other possibility is that a cytoskeleton protein could provide the passive elastic force to restore the sarcomeres to 2 μm after shortening. The most plausible cytoskeleon that may provide this passive restoration force is the giant protein filaments, titin, that span between the Z disks and the M lines[21]. Titin is firmly bound to myosin while the part of the protein in the I band is extensible and may provide a passive restoration force when the cells are stretched above 2 μm/sl[22]. However, unless the elastic properties of titin were those of a stiff compression spring that would remain straight and in place during isotonic shortening below 2 μm/sl, this process would not provide the elastic force to explain the sliding back of actin and myosin. Experimental evidence for either mechanism is lacking. Whether the restoring force is an elastic force stored during the contraction in some structure within the fibrils, or an active process developed by the cross-bridges after the contraction, or both, remains unknown. Additional research is needed to test this hypothesis.

4. EFFECT OF THE ORGANIC CA CHANNEL BLOCKER DILTIAZEM ON EXCITATION-CONTRACTION COUPLING IN SKELETAL MUSCLE

4.1 Introduction

One of the steps in the chain of events that brings about the contraction in frog twitch muscle cells is the release of calcium from the terminal cisternae of the sarcoplasmic reticulum (SR). It has been proposed that an influx of Ca ions into twitch muscle cells during an action potential might initiate the contraction[18]. A net calcium influx observed with flux measurements during contraction[24] supported the above proposition. However, when external calcium is lowered with EGTA to 10^{-8} M, isolated frog muscle fibers can produce normal twitches for many minutes[25, 26]. The presence of inward Ca^{2+} currents (I_{Ca}) in phasic skeletal muscle[27] lead to suggest that extracellular Ca^{2+} moving into the cell, due to the I_{Ca}, may be involved in the e-c coupling. This proposition was further supported by the finding that the voltage dependent I_{Ca} channels are localized in the T-tubule membrane[28]. This finding circumvented the problem of the delay due to Ca^{2+} diffusion from the surface membrane into the center of the fibber. The voltage-dependent Ca^{2+} channels can be inhibited by a diverse class of organic compounds, conventionally grouped together as Ca^{2+} antagonists or Ca^{2+}-channel blockers. We have reported[29, 30] that the organic Ca^{2+}-channel blocker D-600 produces twitch potentiation in intact single frog skeletal muscle cells, and an inhibitory effect on Ca^{2+} uptake by isolated SR . We attributed twitch potentiation to a decrease in the rate of Ca^{2+} uptake by the SR. We also described that the interaction with CERCA-1 is specifically in the Ca^{2+}-ATPase Ca^{2+}-binding domain[30]. These observations, along with the fact that Ca^{2+} channel blockers penetrate into cells[31], explain the facilitation of excitation-contraction coupling seen with D-600[29, 30, 32]. We have previously analysed some of the diltiazem effects on the I_{Ca} and the implications that blocking I_{Ca} may have on initiating and sustaining contractions in frog twitches muscle fibers[33]. We decided to use diltiazem as it had been previously shown to be a specific and effective Ca^{2+} channel-blocker at low concentrations[31].

The present study was undertaken to determine whether or not the effect of diltiazem on e-c coupling is caused by a direct interaction with the SR Ca^{2+} ATPase.

4.2. Materials and Methods

The experiments were done in three different types of preparations: intact and split open single muscle fibers and isolated light SR microsoms.

4.2.1. Isolated Single Muscle Fibers.

Single twitch fibers were isolated from the semitendinosus and tibialis anterior muscles of the frog *Rana temporaria* as described previously[34]. Two fiber preparations were used for the first part of this work: 1) intact fresh fibers for contractility studies and 2) single skinned fibers for experiments on SR Ca^{2+} regulation.

4.2.1 a. *Intact Fresh Fibers.* The method used for the contractility experiments have already been published in detail[34]. In summary, after isolation muscle cells remained in the dissecting dish for at least 30 min in Ringer's solution, which consisted of (in mM): NaCl 115, KCl 2.5, $CaCl_2$ 1.8, $MgCl_2$ 0.2 (pH was adjusted with phosphates to 7.2). Next, the fibers were stimulated with single low-voltage electrical shocks. If they responded with brisk twitches and there were no signs of membrane damage, the fibers were used; otherwise, they were discarded. The experiments were carried out at room temperature (2̃022°C).

Each single fiber was transferred to the experimental chamber, which consisted of a 0.3-ml narrow channel where solutions could be changed several times within five seconds. In the chamber one tendon was gripped with a small clamp while the other tendon was attached to the hook of an Ekharts type force transducer (Senso Nor, Horten, Norway). The stimulating electrode was made of platinum plates placed on each side of the fiber. The muscle fiber was then stretched 1.3 times its slack length to reach an average sarcomere length of approximately 2.6 μm. Subsequently, the fiber was stimulated with single electrical pulses of 0.5-msec duration with variable voltage. The voltage was steadily increased until the threshold for contraction was reached. This voltage was then increased by 50%, and the experimental protocol was started. When the muscle cell was depolarised with high external K^+ concentration, the ($[K^+]_o$), the $[K^+]$ $[Cl^-]$ product of the solution was kept constant. Diltiazem was added at the concentrations specified in the results.

4.2.1.b. Split Open Fibers. In order to expose the interior of the muscle fibers to external substances, fibers split along their longitudinal axis were used as described previously[34, 35]. Splitting the fibers longitudinally exposes the interior uniformly to the external solutions including the drug used in this study and to control Ca^{2+} uptake and release by the SR. Single muscle fibers isolated from the tibialis anterior were dissected and bathed with a relaxing solution containing (in mM) 7 potassium Methanosulfonate; 4 MgSO4; 2 ethylen glycol-bis(b-aminoethyl)-N,n,n',n'-tretraacetic acid (EGTA); 20 tris-Malate; 2 CaSO4 and 4 Na-ATP; pH 6.8. Each split fiber was transferred to a glass experimental chamber and was kept at 4-5°C. The experimental chamber had a narrow channel (1.7 cm long, 0.15 cm wide and 0.15 cm deep). In the chamber one tendon was held with micro-dissection forceps and the other tendon was attached to the same type force transducer used for the intact fiber experiments. The average sarcomere length was adjusted to 2.5 μm, the length being directly measured under a light microscope. The SR was loaded with Ca^{2+} by bathing the fibers with a loading solution in the presence of $5X10^{-7}$ M $CaCl_2$. The amount of Ca^{2+} taken up by the SR was estimated from the force developed during exposure of the fibers to a relaxing solutions containing 25 mM caffeine. Experiments were done at 4-5°C.

4.2.2. Isolated Light SR Microsomes

4.2.2.a. Preparation of Light Sarcoplasmic Reticulum (LSR). Microsomes, isolated from rabbit skeletal muscle, were loaded on a discontinuous sucrose gradient composed of three layers of 25%, 27.5%, and 35% (w/v), containing 20 mM Tris-Malate and 1 mM DTT, pH 6.8 that were then centrifuged for 14 hrs at 23,000 rpm. The pellet was washed and loaded on the top of a second discontinuous sucrose gradient of 28%, 32%, 35% and 45% (w/v). The fraction that banded at the 32%/35% interface, which contains the highest specific Ca^{2+} ATPase activity[30], was collected and will now be referred to as light SR (LSR). Heavy SR (HSR) was collected from the pellet (45% sucrose).

4.2.2.b. Calcium Uptake. SR Ca^{2+} uptake was determined using the metallochromic indicator Arsenazo III in a Cl^- free solution containing, in mM: $CaSO_4$, 0.1; $MgSO_4$, 5; KH_2PO_4 , 77; Tris-Malate, 20; Na-ATP, 1, and Arsenazo III 0.7, pH 6.8. Ca^{2+} transport was determined from the change in absorbance at 660 nm. SR protein (0.05 mg/ml) was added to the reaction solution. After a 30 sec incubation the reaction was started by adding 1 mM Mg-ATP.

4.2.2.c. Passive Ca Efflux From Heavy SR. ATP dependent Ca^{2+}-loaded in LSR vesicles was done with a solution containing in (mM) : $CaSO_4$, 0.1; $MgSO_4$, 5; KH_2PO_4 , 77; Tris-Malate, 20; Na-ATP, 1, and 1 μCi $^{45}CaCl_2$, pH 6.8. at a concentration of 100 ug protein/ml. These vesicles were diluted in a solution containing 2 mM EGTA in the absence and presence of diltiazem. The ^{45}Ca remaining in the LSR was determined by filtration. Radioactivity was measured using a scintillation counter.

4.2.2.d. Differential Scanning Calorimetry (DSC). The thermal denaturation profile of LSR membranes was determined by differential scanning calorimetry (DSC) as described previously[30].

4.3 Results

4.3.1 Effect of Diltiazem on E-C Coupling

The fact that diltiazem can inhibit I_{Ca} prompted us to investigate the effect of the drug on e-c coupling. Figure 6 shows the time course of the effect of $5x10^{-3}$ M diltiazem on twitch tension. The increase in twitch tension appeared 10s after the addition of the drug, reached a maximum 180s later, and declined with a half time of 30s after its removal. In addition to the increase in twitch tension, diltiazem also produced a slight increase in the rate of maximal tension development and a substantial decrease in the rate of tension decay during relaxation. The half relaxation time measured from twitches of several experiments under the influence of diltiazem was approximately 1.9 times larger that of the control. Consequently, total duration of the twitches was prolonged. This might be due to a decrease in the rate calcium uptake by the SR pump that would be reflected as an increase in both, the peak of individual twitches and titanic forces produce during semifused tetanii. Experiments done when the fibers were stimulated at rates between 10 and 30 Hz showed that, in the presence of $3x10^{-6}$ M diltiazen, the peak forces as well as the troughs between individual contractions in the partially fused tetanii were substantially increased when compared to those of control experiments. The difference in force development between control experiments and the experiments done in the presence of diltiazem decreased as the stimulation rate was increased. In order to avoid misleading results due to the use dependence effect, the high frequencies of stimulations were preceded by a series of 12 twitches elicited at 8 Hz. The maximum tetanic force elicited when the fibers were stimulated at 80 Hz was similar in the absence and in the presence of diltiazem.

Inward calcium currents have a long latency and a slow time course[27, 28, 33]. It takes approximately 300 ms for a maximal I_{Ca} to reach its peak. The total duration of I_{Ca}

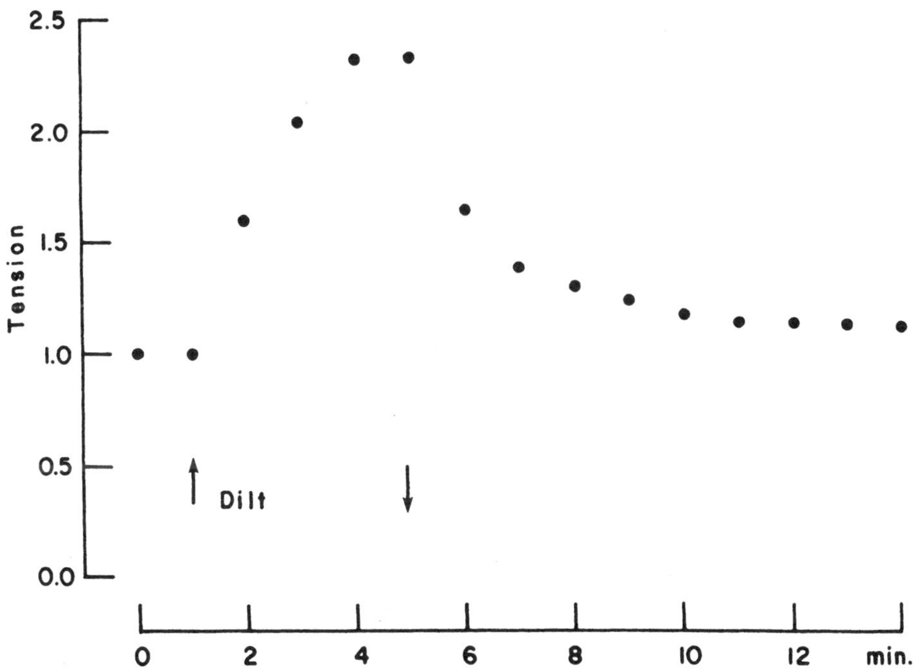

Figure 6. Time course of the effect of $3x10^{-6}$ M diltiazem on the peak twitch tension. Diltiazem was added during the time indicated between the two arrows. Tension is plotted as a fractional change of the control twitch tension before adding diltiazem.

at that temperature is in the order of 1s at 22°C. Therefore, it necessary to impose long depolarization pulses to study them. It may then be possible that I_{Ca} plays an important role in e-c coupling during prolonged depolarizations. This finding and the possibility that the potentiation observed with diltiazem might be due to changes in the threshold for mechanical activation led us to study the influence of the drug on prolonged contractures elicited by K^+ depolarizations. To overcome the use dependence effect and ensure that diltiazem had inhibited I_{Ca}, the muscle cell was stimulated in the presence of diltiazem at 1 Hz for 6 to 8 sec. immediately before each K-contracture. We found that there is practically no difference in the magnitude and time course of the contractures elicited with different $[K]_o$. Also, in the presence of diltiazem the mechanical threshold is similar to the one observed without the drug.

4.3.2. *Effect of Diltiazem on SR Ca^{2+} Uptake and Release in Split Open Fibers*

The experimental results described above led us to suppose that diltiazem may affect SR Ca^{2+} uptake. We explored this possibility by studying the effect of diltiazem on SR Ca^{2+} uptake in split open preparations. One of these experiments is illustrated in Fig.7. It can be seen that Ca^{2+} release with caffeine is depressed when the preceding SR

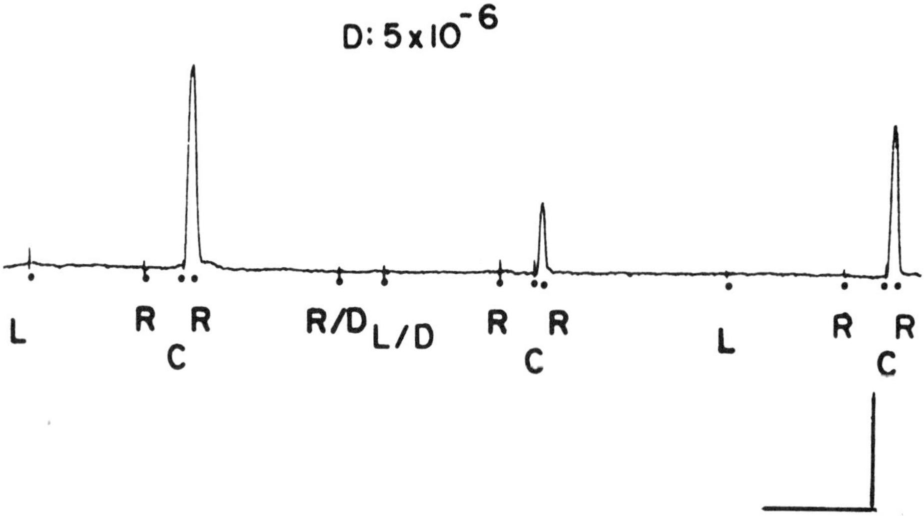

Figure 7. Effect of the presence of $5x10^{-6}$ M diltiazem in the loading solutionon (L/D) on tension development produced by releasing calcium with 25 mM caffeine (C) after the loading period. The loading period in the presence of diltiazem was preceded and followed by loading periods (L) without diltiazem. R indicates the moments when the preceding solution was washed away with relaxing solutions. Time and force calibration bars are 3 min and 60 mg respectively.

Ca^{2+} load was done in the presence of $1x10^{-6}$ M diltiazem as compared with Ca^{2+} in the absence of the drug. The decrease in tension development observed during caffeine exposures following exposures to diltiazem could be the consequence of diltiazem inhibiting either SR Ca^{2+} uptake or SR Ca^{2+} release. To identify if the release was affected by the drug, we did experiments where diltiazem was added to the Ca^{2+} caffeine release solution. We found that the presence of diltiazem not only did not block the tension development but potentiated it as compared to the control.

4.3.3. Effect of Diltiazem on LSR

To determine whether the inhibitory effect of diltiazem on Ca^{2+} uptake by SR is caused by direct interaction with the SR Ca^{2+}-ATPase, we measured Ca^{2+} uptake isolated LSR vesicles. We found that $10x10^{-6}$ M diltizem inhibited calcium uptake by 60 % after 30 min of Ca^{2+} loading and have no inhibitory effect of the ATPase hydrolytic activity at a dose concentration up to 150 μM. To rule out the possibility that diltiazem has an effect on Ca^{2+} uptake caused by Ca^{2+} leakage we measured the effect of 10 μM diltiazem on passive Ca^{2+} efflux from LSR. We found that in the presence of diltiazem, passive Ca^{2+} efflux is reduced as compared to the passive Ca^{2+} efflux of control LSR vesicles. These results indicated that diltiaze interacts with the Ca^{2+}-ATPase. In other experiments we used a rapid filtration technique to assess the effect of diltiazem on the Ca^{2+} release from heavy SR. In these experiments we found that10 μM diltiazem increased the initial rate and total calcium release within one second when compared with control experiments.

4.3.4. Differential scanning calorimetry experiments (DSC)

To test for the effect of diltiazem on the thermal stability of the different domains of the LSR Ca^{2+}-ATPase, LSR membranes were investigated in a temperature dependent manner by DSC scans. The temperature dependence of the excess molar heat capacity of the Ca^{2+}-ATPase between 10 to 100 °C in the absence and presence of diltiazem. The scans were obtained at a heating rate of 1°C/min. Denaturation was irreversible because the transitions are absent during cooling or reheating of the samples. In this study, the heat of transition (ΔH) relates to the heat absorbed during the process of denaturation. Transition *A*, that corresponded to the nucleotide binding domain (NBD) has a Tm at 48.39 ± 0.6°C in the presence of EGTA. In the presence of diltiazem, there is no change either in the T_m or in the heat of transition. However, in the presence of 1mM calcium, a second transition labeled B, related with the transmembrane calcium-binding-domain (CBD)[36]. This transition is destabilized when diltiazem was present. Therefore, this indicates that under this conformation, diltiazem interacts with the transmembrane CBD. The T_m for transition B is 60 ± 0.1°C. In the presence of an increasing concentration of diltiazem (0.10, 0.30 and 0.80 mM), the CBD T_m shifted at 0.30 mM to lower T_m 58 ± 0.1 °C and 57.4 ± 0.4 °C at 0.08 mM diltiazem.

4.4. Discussion

Our results showed: 1) that diltiazem has a twitch-potentiating effect in isolated intact single muscle fibers of the frog at low concentrations. 2) The twitch potentiation is accompanied by a substantial prolongation of the twitch relaxation time rather than by a change in the rate of tension development. 3) In split fibers loaded with Ca^{2+} in the presence of diltiazem, caffeine induced a smaller tan control tension development. 4) After the SR has been loaded with Ca^{2+} in split fibers, adding diltiazem to the caffeine Ca^{2+} release solution potentiated the ensuing contracture. 5) In isolated light SR of the rabbit, diltiazem produced an inhibitory effect of Ca^{2+} uptake and increased Ca^{2+} release.

All of these results lead us to conclude that the organic Ca^{2+} blocker diltiazem has a strong inhibitory effect on the SR Ca^{2+} ATPase, which is the consequence of an intersection with the transmembrane region of the ATPase. As diltiazem penetrates into the myoplasm in intact fibers, it affects the excitation-contraction coupling mechanism by inhibiting the SR Ca^{2+} ATPase. Because of the inhibitory effect of diltiazem on the SR Ca^{2+} ATPase, myoplasmic Ca^{2+} is not completely sequestered into the SR after each twitch. Consequently, cytosolic Ca^{2+} would increase slightly after each activation, and twitches would be potentiated. This also explains the slow time course for the full development of twitch potentiation, since it would take time to reach a new equilibrium between myoplasmic Ca^{2+} and the decreased SR Ca^{2+} uptake. The above proposed mechanism also explains the substantial decrease in the rate of twitch tension relaxation induced by diltiazem.

Experiments with isolated light SR, where the main component is the Ca^{2+} ATPase, have shown that the Ca^{2+} uptake is inhibited by 10 μM diltiazem. Diltiazem in the amount of 10-100 μM has no observed effect on Ca^{2+} uptake when measured in the Cl-reaction solution that is commonly used for kinetic analysis of Ca^{2+} uptake[37]. In the reaction solution for Ca^{2+} uptake we substituted sulfonate for chloride as the primary anion. These conditions seem to emulate the conditions under which diltiazem affects SR Ca^{2+} loading in skinned muscle fibers. The failure of diltiazem to inhibit caffeine-induced calcium release from isolated JSR strongly supports the premise that the Ca^{2+} ATPase and not the Ca^{2+} channels of SR is the target for the calcium antagonist diltiazem.

To determine whether the effect of diltiazem on the Ca^{2+} ATPase is exerted directly on the protein or produced by modifying the membrane environment, thermal analysis of the effect of diltiazem on calcium uptake in light SR was done. The change in thermal stability of the Ca^{2+} ATPase induced by diltiazem is due to an interaction with the protein rather than a general membrane effect.

In an earlier investigation we studied the effect of another organic Ca^{2+} blocker, D-600[24, 25]. We found that D-600 has similar effects to the ones described here. We propose then that the effect of diltiazem on contractility in skeletal muscle fibers can also be explained by a direct effect on SR Ca^{2+} ATPase. It may be that this class of organic Ca^{2+} blockers have a common effect on the transmembrane Ca^{2+} domain.

We thank Mrs. Anne O. Nourse from the University of Maryland for valuable and helpful suggestions in the editing and preparation of the manuscript .

5. REFERENCES

1. H. Sugi, Inward spread of activation in frog muscle fibres investigated by means of high-seed, microcinematography, *J. Physiol.* **242**, 219-235 (1974).
2. H. Sugi and R. Ochi, The mode of transverse spread of contraction initiated by local activation in single crayfish muscle fibres, *J. Gen. Physiol.* **50**, 2167-2176.
3. H. Sugi and R. Ochi, The mode of transverse spread of contraction initiated by local activation in single frog muscle fibres, *J. Gen. Physiol.* **50**, 2145-2166 (1967).
4. A. F. Huxley, Local activation in muscle, *Ann. NY Acad. Sci.* **81**, 446-452 (1959).
5. A. F. Huxley, and R. W. Straub, Local activation and interfibrillar structures in striated muscle, *J. Physiol. London* **143**, 40P-41P (1958).
6. A. F. Huxley and R. E. Taylor, Function of Krause's membrane, *Nature London* **176**, 1068 (1955).
7. A. F. Huxley and R. E. Taylor, Local activation of striated muscle fibres, *J. Physiol. London* **144**, 426-441 (1958).
8. A. F. Huxley, The Croonian Lecture, 1967. The activation of striated muscle and its mechanical response. *Proc. R. Soc. London Ser. B* **178**, 1-27 (1971).
9. H. Gonzalez-Serratos, Inward spread of activation in vertebrate muscle fibres, ***J. Physiol. London* 212** 777-799 (1971).
10. H. Gonzalez-Serratos, Inward spread of activation in twitch skeletal muscle fibers, in: *Handbook of Physiology: Skeletal Muscle* by L.D. Peachey, R. H. Adrian and S. R. Geiger (American Physiological Society, Bethesda, Maryland, 1983).
11. F. C. Bezanilla, C. Caputo, H. Gonzalez-Serratos, and R. A. Venosa, Sodium dependence of the inward spread of activation in isolated twitch muscle fibres of the frog, *J. Physiol. London* **223**, 507-523 (1972).
12. C. Caputo and R. DiPolo, Ionic diffusion delays in the transverse tubules of frog twitch muscle fibres, *J. Physiol. London* **229**, 547-557 (1973).
13. J. Bastian and S. Nakajima, Action potential in the transverse tubules and its role in the activation of skeletal muscle, *J. Gen. Physiol.* **63**, 257-278 (1974).
14. R. H. Adrian and L. D. Peachey, Reconstruction of the action potential of frog sartorius muscle, *J. Physiol. London* **235**, 103-131 (1973).
15. H. Davson, *A Textbook of General Physiology*, (The Williams and Wilkins Co., Baltimore, Marylnd, 1970) pp. 1387-1455.
16. R. W. Ramsey and S. F. , The isometric length-tension diagram of isolated skeletal muscle fibres of the frog, *J. Cell. Comp. Physiol.* **9**, 11-34 (1940).
17. A. V. Hill, Is relaxation an active process? *Proc. R. Soc. B* **136**, 420-425 (1949).
18. L. M. Brown, H. Gonzalez-Serratos and A. F. Huxley, Structural studies of the waves in striated muscle fibres shortened passively below their slack length, *J. Musc. Res. Cell Motility* **5**, 273-292 (1984).
19. L. M. Brown, H. Gonzalez-Serratos and A. F. Huxley, Sarcomere and filament lengths in passive muscle fibres with wavy myofibrils, *J. Musc. Res. Cell Motility* **5**, 293-314 (1984).
20. A. F. Huxley, Review Lecture Muscular Contraction given at the meeting of the Physiological Society at Leeds University on 14-15 December 1973, *J. Physiol.* **243**, 1-43 (1974).
21. K. Wang, R. McCarter, J. Wright, J. Beverly and R. Ramirez-Mitchell, Viscoelasticity of the sarcomere matrix of skeletal muscles: The titin-myosin composite filament is a dual-stagae molecular spring, *Biophys. J.* **64**, 1161-1177 (1993).
22. C. A. Opitz, M. Kulke, M. C. Leake, C. Neagoe, H. Hinssen, R. J. Hajjar and W. A. Linke, Damped elastic recoil of the titin spring in myofibrils of human myocardium, *PNAS* **100**(22), 12688-12693 (2003).
23. A. Sandow, Excitation-contraction coupling in muscular response, *J. Biol. Med.* **25**, 176- 201 (1952).
24. B. A. Curtis, Ca fluxes in single twitch muscle fibers, *J. Gen. Physiol.* **50**, 255-267 (1966).
25. C. M. Amstrong, F. M. Bezanilla and P. Horowicz, Twitches in the presence of ethylen glycol bis (aminoethyl ether)-N,N, tetraacetic acid, *Biocem. Biophys. Acta* **267**, 605-608 (1972).
26. H. Luttgau and W. Spiecker, The effect of calcium deprivation upon mechanical and electrophysiological parameters in skeletal muscle fibers of the frog, *J. Physiol.* **296**, 411-429.
27. J. Sanchez and E. Stefani, Inward calcium current in twitch muscle fibres of the frog, *J. Physiol.* **283**, 197-209 (1978).
28. W. Almers, R. Fuik and P. T. Palade, Calcium depletion in frog muscle tubules: the decline of muscle current under maintained depolarisation, *J. Physiol.* **312**, 177-207 (1981).
29. A. Ortega, H. Gonzalez-Serratos and J. Lepock, Effect of organic calcium channel blocker D-600 on sarcoplasmic reticulum calcium uptake in skeletal muscle, *Am. J. Physiol. (Cell Physiol. 41)* **272,** C310-C317 (1997).

30. A. Ortega, V. M. Becker, R. Alvarez, H. Gonzalez-Serratos and J. R. Lepock, Interaction of D-600 with the tresmembrane domain of the Sarcoplasmic reticulum Ca^{2+} ATPase from skeletal muscle. *Am. J. Physiol. (Cell Physiol.)*(2 000)
31. J. Hescheler, D. Perzer, D. Trube and W. Trautwein, Does the organic calcium channel blocker D-600 act from inside or outside on the cardiac cell membrane? *Pfluegers Arch.* **393,** 287-291 (1982).
32. M. Dorrscheidt-Kafer, M. The action of D-600 on frog skeletal muscle: facilitation of excitation-contraction coupling, ***Pfluegers Arch.*** **369,** 259-267 (1977).
33. H. Gonzalez-Serratos, R. Valle-Aguilera, D. A. Lathrop and M. D. Garcia, M.D. Slow inward calcium currents have no obvious role in frog twitch muscle excitation-contraction coupling, *Nature* **298,** 292-294 (1982).
34. H. Gonzalez-Serratos, L. Hill and R. Valle-Aguilera, Effects of Catecholamines and Cyclic Amp on Excitation-Contraction Coupling in Isolated Skeletal Muscle Fibres of the Frog, *J. Physiol.* **315,** 267-282 (1981).
35. M. Endo and Y. Nakajima, release of clcium induced by depolarization of the sarcoplasmic reticulum membrane, *Nature NewBiology* **246,** 216-218 (1973).
36. J. R. Lepock, A. M. Rodahl, C. Zhang, M. Heynen, B. Waters and K.-H. Cheng, Thermal denaturation of the Ca^{2+} ATPase of sarcoplasmic reticulum reveals two thermodynamically independent domains, *Biochemistry* **29,** 681-689 (1990).
37. M. Chiesi and A. Martonosi, Clcium transport in sarcoplasmic reticulum vesicles isolated from rabbit skeletal muscle, *Membrane Biochemistry,* edited by E. Carafoli and G. Semenza (Springer-Verlag, N.Y., 1989) pp. 51-61.

DISCUSSION

Pollack: Could the restoring force arise from the tendency of the shortened thick filament to restore its natural length? There is ample evidence in the literature to suggest that thick filament shorten, particularly at short sarcomere length.

Gonzalez-Serratos: We published two papers addressing that there is no change in the myosin filament length.

Huxley: Is it possible that extension force in relaxing fibrils comes from overlapped actin filaments rebelling each other? Perhaps it could be checked by looking at fibers from muscles in which I-filaments are longer than in frog, e.g. rabbit, human.

Gonzalez-Serratos: It could be, but I can not answer that with precision, as experiments have not been done in fibers with different actin filament lengths. It is clear that indeed some sort of force must exist, that moves back the actin filaments to always the point, at which they just meet in front of each other. We could not figure out what was that force.

Winegrad: Tom Robinson and I looked at the length of thin filaments in frog and rat hearts. We examined serial transverse sections and found that as double overlap of thin filaments occurred, the thin filaments assumed a different arrangement around thick filaments. Using isolated myofibers and myofibrils, we found that 3-5% of maximum Ca activated force was necessary to overcome the resistance to rearrangement of thin filaments. Double overlap in rat heart occurs at approximately 1.95μ sarcomere length.

Gonzalez-Serratos: This agrees well with our finding. But your observations did not provide an explanation of what would be the restoring force to slide back the filaments to 1.95μm sarcomere length.

V. CARDIAC AND SMOOTH MUSCLE

CARDIAC MYOSIN BINDING PROTEIN C: MODULATOR OF CONTRACTILITY

Saul Winegrad[1]

1. INTRODUCTION

One of the basic mechanisms for movement in eukaryotic cells is the interaction of a molecular motor with a polymer. The properties of different molecular motors and polymers have evolved to allow diversity. The breadth of function is also achieved by the presence of additional proteins that are crucial for the three dimensional organization of the contractile proteins and the modulation of their interaction. In striated muscle, force is generated by an interaction between the molecular motor myosin and filamentous actin, the major proteins of, respectively, thick and thin filaments. Although they generate force, actin and myosin alone cannot reproduce all of the fundamental properties of the contractile system of striated muscle. A pure actin-myosin system lacks the calcium dependent switch that controls the transition between the resting and force generating states. Regulatory proteins in the thin filaments, tropomyosin and the three subunits of troponin, provide this control.

In addition to myosin, the thick filament contains another protein that has been identified as myosin binding protein-C (MyBP-C; Fig 1)(Offer et al 1973). There are other myosin binding proteins, but MyBP-C is present in the largest amount, comprising 1-2 % of the myofibrillar mass. It is located along stripes in the region of the A band referred to as the C zone, each stripe separated by 43 nm from the adjacent ones (Craig and Offer 1976). With this protein present in addition to myosin, it is possible to synthesize thick filaments with

[1] Saul Winegrad, Department of Physiology, University of Pennsylvania, Philadelphia, Pa

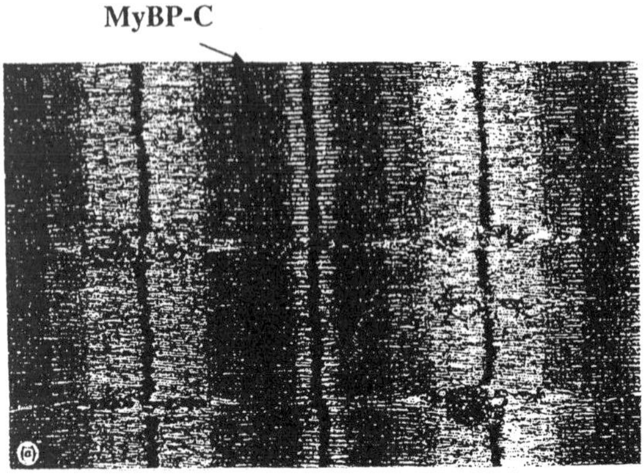

Fig 1. Electron micrograph showing location of MyBP-C in C zone of skeletal muscle.

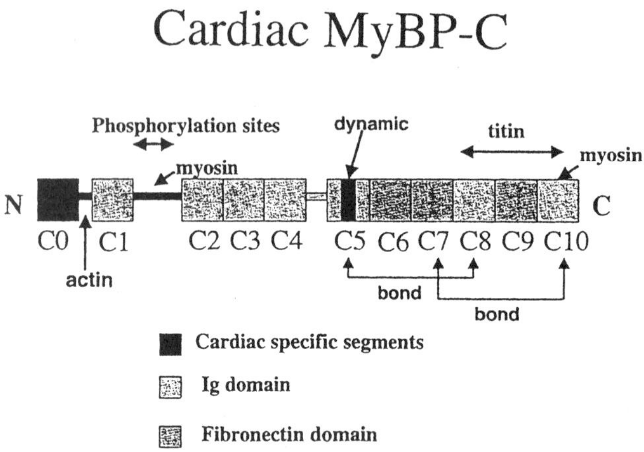

Fig 2. Schematic drawing of cardiac MyBP-C molecule showing 11 domains, cardiac specific regions and binding sites (upper), and drawing showing location on myosin of the binding sites for MyBP-C (lower).

central bare zones, tapered ends and periodically distributed cross bridges arranged helically around the circumference of the filaments. Without this protein, the filaments are thicker, their lengths and thicknesses are very heterogeneous, clear central bare zones are not present and myosin cross bridges may not be clearly discernible or ordered.

MyBP-C is present only in striated muscle, existing in 3 isoforms: two of the three isoforms are found only in skeletal muscle and the third only in the heart. Each isoform is encoded in a separate gene. The skeletal MyBP-C isoforms contain 10 domains, consisting of 7 immunoglobulin C2 motifs and 3 fibronectin type III motifs arranged in the same order (identified as modules 1 to 10, N to C terminus) (Fig. 2) (Gautel et al 1995). The cardiac isoform differs from the skeletal isoforms in three important ways: 1) cardiac MyBP-C contains an additional Ig module with 101 residues at the N terminus (module 0 or C0); 2) the 105 residue linker between the CI and C2 Ig domains (MyBP-C motif) contains 9 additional residues and 3 phosphorylation sites that are unique to the cardiac isoform; and 3) an additional 28 residues rich in proline are present in the C5 Ig domain. The skeletal isoforms are about 3 nm in diameter and 32 nm long, while the cardiac isoform is about 40-44 nm long because it contains the MyBP-C motif and an additional Ig domain (C0) at the N terminus.

MyBP-C has attracted greater attention since the demonstration that mutations in the gene encoding the cardiac isoformproduce about 40% of familial hypertrophic cardiomyopathy (FHC) (Watkins et al 1995, Bonne et al 1995, Carrier et al 1997), and mutations in the region of subfragment 2 that binds MyBP-C can cause FHC. The majority of mutations in MyBP-C produce at least one abnormal transcript and lead to proteins lacking the LMM binding site, in some cases the titin binding site and even the phosphorylation sites. No skeletal myopathy associated with cardiomyopathy due to mutations of MyBP-C has been described. Because the two skeletal isoforms can be coexpressed in skeletal muscle, but only the cardiac isoform is expressed in cardiac muscle, cardiac muscle should be more vulnerable to gene mutation in MyBP-C than skeletal muscle.

2. EFFECT ON CONTRACTILE FUNCTION

In order to understand the effect of MyBP-C on the contractile properties of heart muscle, it is very important to consider both the stoichiometry between MyBP-C and myosin and the localization of MyBP-C (Winegrad 1999). There are approximately 7 myosin molecules for every molecule of MyBP-C, and MyBP-C is restricted to the C zone, which includes only about 40% of the A band. As discussed below, changes in MyBP-C can cause changes in force generation as large as 50% and even greater. In view of the stoichiometry and localization of MyBP-C, it is clear that its effect on contractility cannot be due to a reaction between individual myosin and MyBP-C molecules. This conclusion suggests that some structural or cooperative effect is involved.

In solution, MyBP-C can change actomyosin ATPase activity, but only if actin and the entire myosin heavy chain, including the rod portion that forms the backbone of the normal thick filament and both light chains, are present. However actin and myosin in solution do not generate force, and therefore may not be an adequate model for studying the function of MyBP-C. For this reason skinned cardiac trabeculae or myocytes, in which the membrane has been removed chemically but the contractile filaments remain in the normal lattice, are used. They contract in response to calcium ions. Partial extraction of MyBP-C for 1 hour from isolated cardiac myocytes increases force at submaximal activation by Ca without changing maximum force at optimal Ca (Fmax) (Hofmann et al 1991). Restoration of MyBP-C reverses these changes. When MyBP-C is extracted from skinned cardiac trabeculae over a longer period, Fmax is reduced even though the total amount of MyBP-C extracted may be less than the amount extracted from isolated cardiac myocytes (Kulikovskaya et al 2003). The change in Fmax is completely reversed following restoration of MyBP-C. Apparently there is either a significant difference of MyBP-C in isolated skinned myocytes from that in skinned trabeculae, or there is some time-dependent change in thick filament function (structure) that is not clearly detectable in 1 hour and causes a decrease in Fmax.

The structure of the thick filament changes in parallel with the extraction and replacement of MyBP-C (Kulikovskaya et al 2003). Before extraction the myosin heads in a majority of thick filaments were well ordered. After extraction for 4 hours the majority of thick filaments had disordered myosin heads. Order is restored and Fmax raised following incubation with MyBP-C. These correlations suggest Fmax is reduced with extraction of MyBP-C by a change in the structure of the thick filament that affects the steric relationship of the myosin heads to the backbone of the thick filament and actin in the thin filament.

Three different approaches have been used to produce preparations that either do not contain cardiac MyBP-C or MyBP-C is primarily abnormal. Harris et al (2002) and Carrier et al (personal communication) have generated knockouts, in which no cardiac MyBP-C is present in heart muscle and there has been no expression of the skeletal isoforms in compensation. McConnell et al (1999) et al have prepared mouse hearts in which the LMM binding site of MyBP-C has been deleted, resulting in proteolysis and an almost complete absence of the protein. Yang et al have used transgeneic techniques to produce hearts in which the cardiac MyBP-C is primarily mutant. In all of these preparations the hearts are abnormal and contractility is compromised. The sarcomere pattern is essentially normal in the knockouts, but there are areas with abnormal cells and fibrosis. The hearts with mutant MyBP-C have areas of fibrosis and structural disorder. Maximum velocity of loaded shortening and redevelopment of tension are increased in the knockout of Harris et al. (Korte et al 2003).

3. MYOSIN-MYBP-C INTERACTIONS

MyBP-C has myosin binding domains in the C terminus module and in the C1C2 fragment at the N terminus (Fig 2; 3). Two MyBP-C binding domains exist in myosin, one in subfragment 2 (S2) near the junction of the S2 with the myosin rod (LMM) in the hinge region of the myosin molecule and the other in the myosin rod (Craig and Offer 1976). MyBP-C does not bind to the head of the myosin or the cross bridge region. The C terminal module (module 10 or C10) alone is insufficient for maximal binding to myosin. The affinity of MyBP-C for myosin increases progressively as modules 8-9 are added at the C terminal. In addition to binding to myosin, 3 domains in the C terminus of MyBP-C, C8-C10, bind to a specific subset of immunoglobulin and fibronectin domains in titin (Fischman) . From the

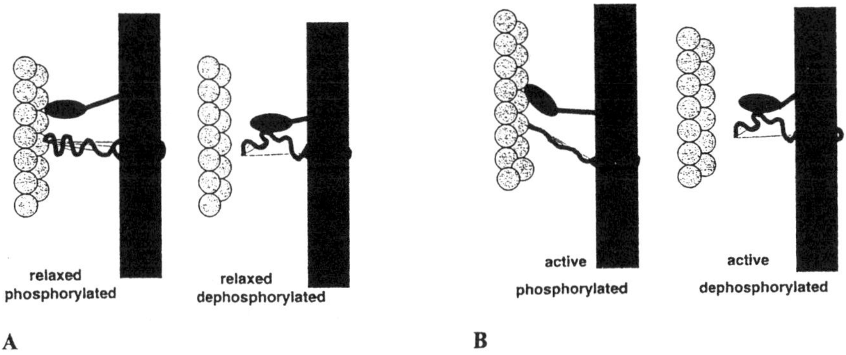

Fig 3. Cartoon of cross bridge showing the effect of phosphorylation of MyBP-C on cross bridge formation and cycling. A:Relaxed; B:active. Phosphorylated: left; dephosphorylated: right.

locations of the binding one would predict that MyBP-C could have significant effects on cross bridge movement and myosin packing.

The 43 nm periodicity of MyBP-C corresponds very closely to that of the helically arranged myosin heads in the thick filament. MyBP-C binding is restricted to the C zone probably because its binding is sensitive to the packing of myosin in the thick filament, and this packing is different near the bare zone and the tapered ends of the filament or because titin binds to myosin and to MyBP-C in the C zone. Because MyBP-C is 32-44 nm long but antibody staining of the thick filament produced a band only about 7 nm, the molecule is probably oriented perpendicular to the long axis of the filament, but X-ray diffraction studies have failed to produce a reflection that would be expected with a completely transverse orientation. Based on the stoichiometry of MyBP-C to myosin each stripe most likely contains 3 molecules of MyBP-C per thick filament (Craig and Offer, 1976). Its accessibility to antibodies suggests that at least part of MyBP-C lies on the surface of the thick filament (Craig and Offer, 1976). The 43 rather than 14 nm repeat (every 3 rather than every 1 cross bridge) suggests that all myosin molecules within the filament may not be equivalent.

4. PHOSPHORYLATION OF CARDIAC MYBP-C

Based on amino acid sequence of the human cardiac isoform, four phosphorylation sites should exist within residues 157-259 in the MyBP-C motif (Gautel et al 1995, Gruen et al 1999). Only three sites can actually be phosphorylated by PKA or Ca-calmodulin regulated kinase (CAMK) presumably because the fourth site is inaccessible. In the isolated protein, PKA can phosphorylate the three sites although in the intact cell, it appears that one of the three sites is preferentially phosphorylated by CAMK, a kinase normally present in thick filaments (McClellan et al 2001). Besides their relative sensitivity to CAMK, phosphorylation the sites are not equivalent in other ways. If a specific one of the three sites is mutated to prevent phosphorylation, the ability of PKA to phosphorylate the other sites is markedly reduced (Gautel et al 1995). In skinned trabeculae the site that must be preferentially phosphorylated first before the other two can be phosphorylated is the CAMK phosphorylated site. The order of phosphorylation means that intracellular Ca concentration has an important effect on the phosphorylation state of MyBP-C. Soaking intact, quiescent heart muscle in a solution with a high concentration of Ca or stimulating the cells to contract raises the level of phosphorylation. Changes in phosphorylation from changes in intracellular Ca occur within a few minutes (McClellan et al 2001).

Five major or potentially major regulators of contractility, alpha and beta adrenergic activation, cholinergic stimulation, Ca and endothelin, lead to changes in phosphorylation of MyBP-C in the same direction as their effect on contractility (Winegrad 1999): cholinergic agents lower phosphorylation and contractility; the other four increase both. Isoproterenol and carbachol each induce changes in relaxation time that correlate well with the degree and time course of phosphorylation of MyBP-C (Hartzell 1984). In preparations with normal intact filament lattice, alpha and beta adrenergic stimulation singly and in combination produce changes in the rate of ATP hydrolysis by actomyosin that correlate well with the degree of phosphorylation of MyBP-C (and not with the degree of phosphorylation of either TNI or RLC. McClellan et al 1994).

Demonstration of an effect of MyBP-C phosphorylation on contractility requires the structural integrity of the myofilament lattice. In a reconstituted actomyosin system, removing unphosphorylated MyBP-C and replacing it with phosphorylated MyBP-C does not change the Ca activation of Mg ATPase activity, but this system does not develop force. A change in cross bridge kinetics that is either sensitive to stress or force on the cross bridge or involves a specific structural relationship between cross bridge and thin filament as suggested above may not be detected.

The effect of fragments prepared from cardiac MyBP-C on the contractile properties of skeletal muscle, where phosphorylation of endogenous MyBP-C does not occur, have been examined (Kunst et al 2000). Of particular interest were the results with fragments Cl to C2 (ClC2), which contained the phosphorylation sites in the MyBP-C motif, and C0Cl, which did not contain the phosphorylation sites. Unphosphorylated ClC2 was bound to the skinned slow skeletal fibers (in which the isoform of myosin is very similar to that in cardiac muscle, but the isoform of MyBP-C is the skeletal variety) and reduced Fmax by about 50%. The binding changed some of the physical properties of fibers such as stiffness of cross bridges. Phosphorylated C1C2 did not bind to myosin. From these biochemical data and others, Kunst et al (2000) proposed that the properties of the cross bridges and the force that they generate can be modified by binding of ClC2 in MyBP-C to myosin near its hinge region.

5. ACTIN MYBP-C INTERACTIONS

MyBP-C can bind to actin in the micromolar range (Moos et al 1978,Kulikovskaya et al 2003a). By forming stable actin-myosin aggregates in solution, MyBP-C binding to actin stimulates actin activated but not Ca, Mg or EDTA activated myosin ATPase activity. MyBP-C binds strongly to I bands and, in transfected cardiac myocytes, to a region within the A bands. Direct binding studies with cardiac proteins indicate that the binding site is in C0, and x-ray diffraction of cardiac muscle supports that conclusion (Squire et al 2003). The affinity of the fragment for actin however is increased if the fragment contains C1 and C2 (Kulikovskaya et al 2003a).

Although phosphorylation of fragments of MyBP-C does not alter the binding to pure actin, phosphorylation does influence the binding of MyBP-C to actin in an intact filament lattice (Kulikovskaya et al 2003a). The affinities of the fragments for pure actin and actin in myofibrils are similar, but in the intact filament lattice there is competition between actin and the S2 region of myosin for the N terminus of MyBP-C (Kulikovskaya et al 2003a). Exogenous C0 at a low concentration does not bind to trabeculae when the MyBP-C is phosphorylated and endogenous C0 in the MyBP-C within the trabeculae is presumably bound to actin, but exogenous C0 does bind to actin when MyBP-C is not phosphorylated and is bound to myosin through its N terminus. The converse with C1C2 is also true. Exogenous C1C2 binds when MyBP-C is phosphorylated and endogenous C1C2 is not bound to myosin, but exogenous C1C2 does not bind when endogenous MyBP-C is not phosphorylated. Therefore it is reasonable to conclude that C0 added to skinned cardiac trabeculae binds at the same site as endogenous C0. This leads to the important conclusion that the N-terminus of endogenous MyBP-C binds to actin or to myosin depending on the phosphorylation state of endogenous MyBP-C (Fig 3).

In higher concentrations, exogenous C0 decreases Fmax but only when MyBP-C is phosphorylated and is bound not to S2 but presumably to actin. Since C0 added in sufficient concentration appears to disrupt the interaction between endogenous C0 and actin and decrease development of force, C0 binding to actin enhances contractility only when it is part of the intact MyBP-C with its C terminus bound to LMM. These results suggest that phosphorylation determines where the N terminus of MyBP-C binds and, in turn, the location of the N terminus modulates contractility through an influence on formation of weak bonds between actin and myosin (Fig 4; see below and 10).

If phosphorylated MyBP-C binds to actin, it will act as a load on the contraction. At first thought this should have an undesirable effect on the development of force. However, in the case of cardiac muscle (skeletal isoforms of MyBP-C do not contain C0 or phosphorylation sites) such a load could serve a useful function if cardiac MyBP-C contains a compliant region (Kulikovskaya et al 2003a). In the presence of such a compliance some of the energy generated during systole would be stored in the compliance to assist in the filling of the heart during diastole. There are at least six independent observations in the literature that support such a mechanism. Partial removal of MyBP-C from isolated cardiac cells increased maximum velocity of unloaded shortening (Vmax), an effect they attributed to removal of an internal load (Hofmann et al 1991) and velocity of shortening in cardiac

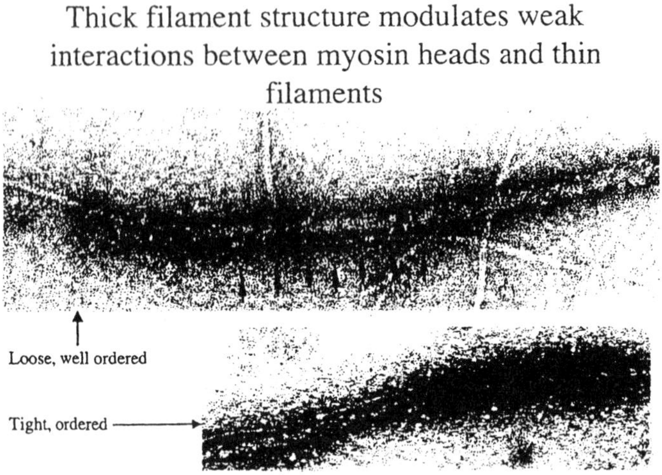

Fig 4. Electron micrographs of thick and adjacent thin filaments showing the effect of phosphorylation of MyBP-C on the formation of weak bonds between myosin and actin.

muscle is higher in the MyBP-C knockout than in the wild type (Korte et al 2003). Harris et al (2002) have observed impairment of diastole in their MyBP-C knockout preparations of heart. The rate of relaxation of isolated heart is also enhanced by phosphorylation of cardiac MyBP-C (Hartzell1984). Excess S2, which would be expected to bind strongly to endogenous MyBP-C, slows relaxation Calaghan et al (2000). The reflection associated with MyBP-C decreases in intensity or disappears during contraction, consistent with a broader distribution of its mass during contraction (see model on page). Striated muscle contains a parallel compliance that increases as the muscle shortens during contraction ().

MyBP-C probably contains compliant regions (Kulikovskaya et al 2003a). A comparison of the amino acid sequence of the proline-rich region of C0 and C1 with the sequence of the compliant PEVK region of titin reveals a high level of homology and MyBP-C also contains tandem Ig domains, which provide extensibility to titin (a large compliant molecule that acts as a compliance in parallel with the contractile filaments). In the presence of phosphorylation of MyBP-C, C0 should bind to actin with a compliant region of MyBP-C in parallel with the force generators. Energy stored in the compliance during systole would be released during diastole to assist filling. The rate of diastolic filling of the ventricles becomes progressively more important as the heart rate increases, and beta adrenergic stimulation, which is one of the major mechanisms for increasing heart rate, also increases phosphorylation of MyBP-C. In the absence of phosphorylation of the motif between C1 and C2, C1C2 would bind to myosin, and the compliance of C0 would not be in parallel with the force generators.

6. ROLE IN FILAMENT AND SARCOMERE FORMATION

Although MyBP-C is not necessary for the formation of myosin filaments *in vitro*, the dimensions of synthetic myosin filaments are modified by the presence MyBP-C, resulting in an increase in length, a decrease in the diameter, and increased uniformity of the diameters of the filaments apparently from greater compactness of the filament (Koretz 1979). If the myosin filaments are produced in the presence of MyBP-C in the physiological molar ratio, the range of diameters is much narrower. MyBP-C reduces the critical concentration of myosin required for polymerization. There is decreased fraying suggesting the lateral myosin-myosin interactions are strengthened, and only in the presence of the normal content of MyBP-C do synthetic thick filaments resemble native thick filaments in their thickness, length, bare zone and distribution of myosin heads. Skeletal myosin expressed in transfected COS cells forms thicker filaments than those in skeletal muscle, but when myosin and MyBP-C are co-expressed, the dimensions of the thick filament are more physiological: the length of the myosin filaments is increased and the diameter decreased.

In spite of observations suggesting that the formation of normal thick filaments *in vitro* and the development of normal striated muscle in the embryo require MyBP-C, normal appearing sarcomeres are produced in cardiac muscle of a cardiac MyBP-C knockout (Harris et al 2002). In low magnification transmission electron micrographs of these hearts, the thick filaments do not show any obvious abnormalities, and the only major difference in myofibrillar structure is absence of the transverse alignment of A and I bands, The difference in the requirement of MyBP-C for formation of thick filaments *in vitro* and in developing cardiac muscle remains unresolved. The most likely explanation is the occurance of some sort of compensation in the knockout, presumably from some other myosin binding protein such as myomesin, M line protein, MyBP-H or even titin. Up-regulation of the normally unexpressed skeletal isoforms is unlikely as there is no clear evidence fo their presence in the hearts of the knockout mice.

7. INTERACTIONS BETWEEN MOLECULES OF MYBP-C

Most mutations of cardiac MyBP-C produce truncations that eliminate the C terminus of the protein, with its binding site for LMM. The result is that little MyBP-C is bound to the filament lattice. However with either of two mutations in C5, the LMM binding site is present in the abnormal protein, but the heart still becomes myopathic. This suggests that C5 *per se* may have an important function in cardiac muscle. The role in contractility may be related to the proline-rich 28 amino acid sequence not found in either of the skeletal isoforms of MyBP-C and its effect on the structure of cardiac MyBP-C *in situ*.

Moolman-Smook et al (2002) have recently shown that C5 binds to C8 and that the strength of binding is reduced by two missense mutations of C5 associated with familial hypertrophic cardiomyopathy without truncation of MyBP-C. They also have produced evidence that C7 and C10 interact. Based on these observations they have proposed a model with 3 molecules of MyBP-C forming a collar around a thick filament every 43 nm in the C zone of the sarcomere. The 3 molecules overlap in a parallel arrangement.

The role of C5 in the contractile process has been examined by adding fragments containing C5 to the solution bathing skinned cardiac trabecula to disrupt the existing interactions of C5 in the endogenous MyBP-C. C2C5 fragments produced a marked reduction in force generated in sub-optimal concentrations of Ca. The change in force with C2C5 completely reverses with removal of the fragment. C2C4 fragments at the same or 3 times higher concentration did not change force or Ca sensitivity. The effects of C5-containing fragments on force indicate that disruption of the collar around the thick filament will alter the function of myosin heads and their formation of cross bridges. MyBP-C maintains myosin within the thick filament in a specific position (orientation) that helps to define the contractile state.

From these results, it appears that the collar formed by MyBP-C around the thick filaments stabilizes the structure of the thick filament. C5 is in a strategic location to influence the structure of the collar and the relation of the collar to the N terminal end of MyBP-C. It is not however directly phosphorylated, and does not appear to be directly involved in a physiological regulatory mechanism.

8. STRUCTURE-PHOSPHORYLATION

Thick filaments isolated from rat ventricle trabeculae can exist with one of at least three different structures as defined by the thickness of the filament and the position and the degree of order of the myosin heads: tightly packed, disordered, and loosely packed (Weisberg and Winegrad, 1996). In the "tight" structure, thick filaments have an average diameter outside the bare zone of 29.8 ± 0.12 nm. A significant degree of order of myosin heads can be seen in electron micrographs of individual filaments and their optical diffraction patterns. In the "disordered" structure myosin heads are extended from the backbone of thick filaments without any consistent pattern. The diffraction pattern is weak or diffuse as a consequence of the lack of uniformity of position of myosin heads. In the "loose" structure, the filaments are thicker (average diameter of 34.2±0.8 nm). The backbone of the filament appears to be less tightly packed than in either of the other structures, and myosin heads are well ordered without being extended. In filaments from resting bundles soaked in normal Krebs' solution, the distribution of structures was approximately 12% disordered, 30% loose and 52% tight.

The relative frequency of each of these structures varies according to the degree of phosphorylation of MyBP-C (Levine et al 2001). Filaments with a high degree of phosphorylation have ordered loose structure. Thick filaments with a low level of phosphorylation primarily have disordered structure. A transition among the three structures can be produced in a matter of minutes by modifying the degree of phosphorylation of MyBP-C.

9. THICK-THIN FILAMENT INTERACTION

The effect of phosphorylation of MyBP-C on the movement of myosin heads from the thick filament can be seen with individual filaments (Fig 4; (Levine et al 2001). Among phosphorylated thick filaments, 92% had cross bridges extending to the surface of the adjacent thin filaments with 43 nm periodicity (Fig. 4). Only 47% of moderately phosphorylated thick filaments had similarly extended cross bridges. A similar correlation exists between thick filament structure and the presence of extended myosin heads forming cross bridges. Twenty-nine percent of the adjacent thick and thin filaments with tight structure had myosin heads periodically extended to the thin filaments, 89% of the loose structures had similar interactions and only 16% of unphosphorylated, disordered thick filaments had interactions.

At the time of their staining in these studies thick and thin filaments were incubated in a solution containing 2.5 mM MgATP, 0 Ca and 1.0 mM EGTA to prevent the formation of rigor links or actively cycling cross bridges. The ionic strength was kept near physiological to permit the formation of weak attachments of cross bridges to thin filaments. This weakly bound state is non-force producing and exists before cross bridges with bound ADP and Pi release Pi to form a strongly bound state that leads to force and movement. Therefore it is likely that the interactions that were present in adjacent thick and thin filaments were weak, non-force generating bonds.

Optical diffraction patterns of paired thick and thin filaments with periodic connecting structures contained reflections along a 36 nm layer line, which is a characteristic period for cross bridges bound to thin filaments. Separate thick and thin filaments did not produce this reflection. The reflection, present only when there were periodic structures that appear to be cross bridges, indicates that cross bridges were probably attached to actin in the thin filament in a site-specific manner. X-ray diffraction studies in rhythmically contracting heart muscle have shown that the myosin heads do not move away from the thin filament during diastole (Matsubara). They remain close or bound to the thin filament without generating force, in, presumably, the weakly bound state.

If formation of weak bonds is rate limiting in cycling of cross bridges in cardiac muscle, an increase in the concentration of weakly bound cross bridges would increase the number of force generating cross bridges when the muscle is activated by Ca. Such a mechanism would explain the relationship between phosphorylation of MyBP-C and contractility when contractility is changed by stimulation with alpha or beta adrenergic agonists, endothelin, cholinergic stimulation or Ca. In reconstituted systems of actin and myosin lacking the steric constraints of an intact filament lattice, the weakly bound state should form more easily and be less dependent on phosphorylation of MyBP-C. In intact cardiac muscle, enhancement of Ca activated force from increased formation of weak bonds between actin and myosin would be expected with phosphorylation of MyBP-C.

10. A HYPOTHETICAL MECHANISM (FIG 3)

Phosphorylation of cardiac MyBP-C modulates the binding of the N terminal of MyBP-C between two sites one on myosin and the other on actin. Binding at the actin site produces a structural change in the thick filament that favors formation of the weakly bound state between actin and myosin and entry into the force generating cycle when Ca concentration rises above the threshold for activation. Binding at the myosin site produces as structural change that decreases the probability of formation of the weakly bound state and decreases the probability of myosin heads cycling as force generators. Phosphorylation of MyBP-C modulates the fraction of myosin heads that will respond to Ca by cycling as force generators, and the concentration of activating Ca determines the percentage of these potential force generators that actually generate force during a given contraction.

This model of the effect on cross bridge formation and the production of a parallel compliance, both under the control of phosphorylation of MyBP-C, predicts that the degree of phosphorylation of MyBP-C modulates the distribution of of cardiac force generators between a high level of activity with a parallel compliance assisting in relaxation (and diastolic filling) and a lower level of contractility without a parallel compliance to assist in relaxation. Thus, with catecholamine stimulation, phosphorylation increases, heart rate increases decreasing filling time, contractile force increases, and the rate of cardiac filling is enhanced by a restoring force. Without catecholamine stimulation, heart rate is slower, contractility and cardiac output are reduce and the need for rapid relaxation is reduced.

REFERENCES

Bonne, G., L. Carrier, J. Bercovici, C. Cruard, P. Richard, B. Hainque, M. Gautel, S. Labiet, M. James, J. Beckman, J. Weissenbach, H.P. Vosberg, M. Fiszman, M. Komajda and Schwartz, K. Cardiac myosin binding protein C gene splice acceptor site mutation is associated with familial hypertrophic cardiomyopathy. Nature Genet. 11: 438-440,1995.

Calaghan SC. J. Trinick, PJ Knight. E. White A role for C-protein in the regulation of contraction and intracellular Ca2+ in intact rat ventricular myocytes.. J. Physiol (Lond.). 528:151-6. 2000

Carrier L., G.Bonne, E.Bahrend, B.Yu, P.Richard, F.Niel, B. Hainque, C. Cruaud, F. Gary, S.Labeit, JB. Bouhour, O. Dubourg, M. Desnos, AA. Hagege, RJ. Trent,M. Komajda, M. Fiszman and K. Schwartz , Organization and sequence of human cardiac myosin binding protein C gene (MYBPC3) and identification of mutations predicted to produce truncated proteins in familial hypertrophic cardiomyopathy. Circ. Res. 80:427-34. 1997

Craig, R. and G. Offer. The location of C-protein in rabbit skeletal muscle. Proc. R. Soc. (Lond) 192:451-461,1976

Gautel,M., O. Zuffardi, A. Freiberg, and S. Labeit. Phosphorylation switches specific for the cardiac isoforms of myosin binding protein C: a modulator of cardiac contraction. EMBO J. 14: 1952-1960, 1995

Gruen, M,,H. Prinz and M.Gautel. , cAPK-phosphorylation controls the interaction of the regulatory domain of cardiac myosin binding protein C with myosin-S2 in an on-off fashion. FEBS Letters. 453(3):254-9. 1999

Hofmann, PA,HC Hartzell, and RL Moss, Alterations in Ca sensitive tension due to partial extraction of C-protein from rat skinned cardiac myocytes and rabbit skeletal muscle fibers. J. Gen. Physiol. 97: 1141-1163,1991

Harris, SP., CR. Bartley, T.A, Hacker, K.S. McDonald, PS, Douglas, M.L, Greaser, P.A. Powers and RL Moss,; Hypertrophic cardiomyopathy in cardiac myosin binding protein-C knockout mice. Circulation Research. 90:594-601, 2002

Hartzell H.C. Phosphorylation of C-protein in intact amphibian cardiac muscle. Correlation between ^{32}P incorporation and twitch relaxation. J. Gen. Physiol. 83: 563-88, 1984

Hofmann, P.A, H.C Hartzell and RL. Moss, Alterations in Ca sensitive tension due to partial extraction of C-protein from rat skinned cardiac myocytes and rabbit skeletal muscle fibers. J. Gen. Physiol. 97: 1141-1163, 1991

Koretz, J. F.. Effects of C-protein on synthetic myosin filament structure. Biophys. J. 27: 433-446. 1979

Korte, FS., KS. McDonald, SP Harris, and RL Moss Loaded shortening, power output, and rate of force redevelopment are increased with knockout of cardiac myosin binding protein. Circ. Res. 93: 752-758, 2003.

Kulikovskaya, I. G. McClellan, R. Levine and ,S. Winegrad. Effect of extraction of myosin binding protein C on contractility of rat heart. Amer. J. Physiol. 285: H857-H865, 2003

Kulikovskaya, I. G. McClellan, J. Flavigny, L. Carrier and S. Winegrad. Effect of MyBP-C Binding to Actin on Contractility in Heart Muscle. J.Gen. Physiol.122: 761-774, 2003a

Kunst, G. , K.R. Kress, M, Gruen, D. Uttenweiler, M. Gautel and R.H.A. Fink. MyBP-C (C-Protein)- a phosphorylation dependent force regulator in muscle that controls the attachment of myosin heads by its interaction with myosin – S2. Circ. Res 86: 51-58, 2000.

Levine, R.J.C., A. Weisberg, I. Kulikovskaya, G. McClellan and S. Winegrad. Multiple structures of thick filaments in resting cardiac muscle and their influence on cross bridge interactions. Biophys. J. 81: 1070-1082, 2001.

McClellan,G., A. Weisberg, and S.Winegrad, , CAMP can raise or lower cardiac actomyosin ATPase activity depending on alpha adrenergic activity. Amer. J. Physiol. 267: H431-442,1994

McClellan,G., I. Kulikovskaya and S. Winegrad. Structural and functional responses of the contractile proteins to changes in calcium concentration in the heart. Biophys. J. 81: 1083-1092, 2001.

McConnell BK. KA. Jones D. Fatkin LH. Arroyo RT. Lee O. Aristizabal DH. Turnbull D. Georgakopoulos D. Kass M. Bond H. Niimura FJ. Schoen D. Conner DA. Fischman CE. Seidman JG. Seidman Fischman DH. Dilated cardiomyopathy in homozygous myosin-binding protein-C mutant mice. J. Clin. Invest. 104:1235-44,1999

Moolman-Smook, J., E. Flashman, W. de Lange, Z. Li, C. Redwood, V. Corfield, and H. Watkins. Identification of novel interactions between domains of myosin-binding protein C that are modulated by hypertrophic cardiomyopathy missense mutations. Circ. Res. 91: 704-711, 2002.

Moos, C., C.M. Mason, J.M. Besterman, M. Feng, J.H. Dubin. The binding of skeletal muscle C-protein to F-actin, and its relation to the interaction of actin with myosin subfragment-1 J. Mol. Biol. 124: 571-586, 1978.

Offer, G., C.,Moos,and R. Starr. A new protein of the thick filaments of vertebrate skeletal myofibrils. Extraction. purification and characterization. J. Mol. Biol.; 74: 653-676, 1973.

Okagaki,T., FE Weber, DA Fischman,.KT Vaughan, T., Mikawa, and FC Reinach, , The major myosin binding domain of skeletal muscle MyBP-C (C protein) resides in the COOH-terminal, immunoglobulin C2 repeat. J. Cell Biol. 123: 619-626, 1993

Squire J.M., P.K. Luther and C. Knupp. Structural evidence for the interaction of C-protein(MyBP-C) with actin and sequence identification of a possible actin-binding domain. J. Molec. Biol. 331: 713-724, 2003

Watkins, H, D Conner, L Thierfelder, JA Jarcho, C MacCrea, WJ McKenna, BJ Maron, JG Seidman,and CE Seidman, Mutations in the cardiac myosin binding protein C on chromosome 11 cause familial hypertrophic cardiomyopathy. Nat. Genetics **1:** 433-438. 1995

Weisberg,A. and S. Winegrad. Alteration in myosin cross bridges by phosphorylation of myosin-binding protein C in cardiac muscle. Proc. Nat. Acad. Sc (USA) 93: 8999-9003, 1996.

Yang Q, AJ Sanbe, H Osinska T.E. Hewett, R Klevitsky and J Robbins, A mouse model ofcardiac myosin binding protein C human familial hypertrophic cardiomyopathy J. Clin. Investig.102: 1292- 1300, 1998

DISCUSSION

Maughan: Fascinating work, Saul. As you know, there are also cardiac C-protein nulls in mice models, one is a true null generated by Rick Moss, and another is an effective null generated by Tom Seidman. Am I correct in assuming from your talk that, because these mice live and therefore have nearly fully functional myocardium, it is in the absence of C-protein and its stabilizing role in the thick filament but in the presence of certain C-protein fragment, which would otherwise not be present if they were not broken down in the nulls.

Winegrad: C-Protein is not essential for development, not essential for life, but life is better with C-Protein. There are three null preparations: two knockout and one transgenic. These three preparations show normal sarcomeres and normal-appearing thick filaments, but by three months, they show cardiac hypertrophy and decreased response to stress. Therefore lack of MYBP-C does not appear to prevent formation of thick filaments as formerly thought, but in some way impairs the performance of heart.

HUMAN ATRIAL MYOSIN LIGHT CHAIN 1 EXPRESSION ATTENUATES HEART FAILURE

Ahmed Ihab Abdelaziz,* Ines Pagel,* Wolfgang-Peter Schlegel,* Monika Kott,* Jan Monti,* Hannelore Haase* and Ingo Morano*#

ABSTRACT

Most patients with hypertrophic cardiomyopathy and congenital heart diseases express the atrial essential myosin light chains (ALC-1) in their ventricles, replacing the ventricular essential light chains (VLC-1). VLC-1/ALC-1 isoform shift is correlated with increases in cardiac contractile parameters of a transgenic rat model overexpressing hALC-1 in the heart (TGR/hALC-1) compared to normal WKY rats. To investigate, whether the benefical effects of the hALC-1 on cardiac contractility could attenuate contractile failure of the overloaded heart, aortocaval shunt operations of 9-10 weeks old WKY and TGR/hALC-1 were performed. 5 weeks later, both animals groups were sacrificed for analysis of cardiac contraction and transgene expression. Control animals were operated but remained normal body and heart weights. The whole heart contractility parameters were evaluated using the Langendorff heart preparation. Shunt-operated TGR/hALC-1 and WKY rats developed comparable levels of cardiac hypertrophy which was associated with significant reduction of contractile parameters of the Langendorff hearts. However, the decline of cardiac contractility was less pronounced in shunt-operated TGR/hALC-1 compared to shunt-operated WKY. In fact, developed left ventricular pressure as well as maximal velocity of pressure development and relaxation were significantly higher in shunt-operated TGR/hALC-1 as compared to shunt-operated WKY. Expression of hALC-1 was 17μg/mg whole SDS-protein in control (sham-

* *Max-Delbrück-Center for Molecular Medicine, Robert-Rössle-Str. 10, 13125 Berlin, Germany;*
# *Humboldt-University, Charité, Johannes-Müller-Institute for Physiology, Tucholskystr. 2, 10115 Berlin, Germany*

operated) controls and declined significantly to 14 µg/mg whole SDS-protein in hypertrophied TGR/hALC-1. These results demonstrate that the expression of hALC-1 could have a beneficial effect on the overloaded hypertrophied heart.

1. INTRODUCTION

Myosin is the motor protein of all muscle types. It cyclically interacts as the "cross-bridge" with the thin (actin) filament thus producing force and shortening while consuming ATP (Huxley,1957). The molecular substructure of myosin has been determined since the early 1930 by using reagents which break secondary linkages but do not break peptide bonds. H.H. Weber and Stöver (1933) observed a decreased viscosity and depolymerizationof a (still impure) myosin preparation with 526 kDa into smaller subunits with around 100.000 kDa if it was denatured with 45% urea. This effect of urea was than studied in more detail by Tsao (1953): he recovered small and large molecular weight components in experiments in which myosin was stored in 6.5 M urea for 12-16 months, with the small units of molecular weights around 16.000 and the larger units with around 170.000 dalton. Furthermore, treatment of myosin with alkali (Kominz et al., 1959), guanidine hydrochloride (Dreizen et al., 1967), or by acylation of the _-amino groups of lysine with acetic anhydride (Locker and Hagyard, 1967), or denaturation with heat (Locker, 1956), all of which resulted in the loss of ATPase activity, dissociated light chains from myosin. Reaction of thiol groups with 5,5′-dithiobis-(2-nitrobenzoic acid) (DTNB) liberates a single class of light chains without significant loss of ATPase activity (Gazith et al., 1970, Weeds and Lowey, 1971). This light chain is termed the DTNB or NBs2 light chains. In molluscan muscles, selective removal of this type of light chain from myosin with EDTA leads to a loss of calcium sensitivity of myosin (Szent-Györgyi and Szentkiralyi, 1973): The EDTA-light chain inhibited actin-activated ATPase activity of myosin in the absence of $Ca2^{+}$, while the inhibitory effect is removed in the presence of Ca2+. Perrie et al., 1972 observed that the DTNB light chain could exist in phosphorylated and non-phosphorylated forms in muscle.

The other light chains were called the Alkali light chains. They contain one thiol, the DTNB two thiol groups in skeletal muscle (Weeds, 1969). The stoichiometry of the light chains, determined by radioisotope dilution using thiol peptides as markers, indicates that skeletal myosin contains two moles of DTNB and two moles of alkali light chains per mole (Weeds and Lowey, 1971).

SDS-PAGE of fast skeletal muscle myosin in the presence of sodium dodecylsulfate (SDS-PAGE) showed the presence of three light chains with apparent molecular weights of 25.000, 18.000, and 16.000 (Weeds and Lowey, 1971). Both the 25.000 (A1) and the 16.000 daltons (A2) MLC turned out to be generated by alternative splicing of the same gene, A2 lacking the N-terminal 41 amino acids (Frank and Weeds, 1974; Nabeshima et al., 1984; Periasamy et al., 1984). SDS-PAGE from smooth muscle cells revealed light chains with 20.000 daltons and 17.000 daltons which correspond to the regulatory and Alkali light chains, respectivley (Burridge, 1974). The 20 kDa light chain can be phosphorylated, a process necessary for actin-activation of the myosin molecule and smooth muscle contraction (Sobieszek, 1977, Aksoy et al., 1983).

There are only two light chains only with 27.000 and 18.000 of slow skeletal and cardiac myosin (Lowey and Risby, 1971, Sarkar et al., 1971). The 18.000 dalton component corresponds to the DTNB, the 27 kDa to the Alkali light chains (Weeds and

Lowey, 1971; Lowey and Risby, 1971). According to their electrophoretic mobility, Alkali and DTNB light chains were also desingated as MLC-1 and MLC-2, respectively. Densitometry measurements of SDS-PAGE (Lowey and Risby, 1971), and radioisotope dilution (Weeds and Frank, 1973) revealed a stoichiometry of 2 mol DTNB and 2 mole Alkali light chain per mole cardiac or slow skeletal myosin. The genes of the different light chains of muscular tissues were found to be distributed all over the genome (Barton and Buckingham, 1985).

The three-dimensional structure of the subfragment-1 of myosin of chicken skeletal muscle with the precise binding domains for the MLC was elucidated by X-ray crystallographic analysis in the early 1990s (Rayment et al., 1993). It consists of an alpha-helical heavy chain (MHC), which folds at the N-terminus into an asymmetric globular head domain. This head has a length of 16.5nm, a width of 6.5nm and a thickness of 4nm, Amino acids (aa) 771-843 at the C-terminus form a 8.5nm α helical structure -the neck region - which binds the two types of light chains. The essential myosin light chain binds between aa 783 and 806 and the regulatory light chains downstream between aa 808 and 842.

MLC-1 binds not only to the neck domain of the MHC (Rayment et al., 1993), but also with its N-terminus to the C-terminus domain of actin (Sutoh, 1982; Trayer et al., 1987; Stepkowski, 1995). Thus, MLC-1 may tether the MHC to the actin filament. Binding of MLC-1 to actin cannot be predicted from the crystal structure of myosin S1, due to the limited resolution of the three-dimensional structure of the N-terminus of MLC-1 (Rayment et al., 1993). In fact, around 40 N-terminal amino acid residues of MLC-1 are not seen in the crystal structure. This part of MLC-1 containes several Pro and Ala residues (Fodor et al., 1989, Seharaseyon et al., 1990), which could form an antenna-like structure long enough to bridge the gap to the actin filament. Experimental evidence for the functional importance of the MHC/actin tether could be obtained by weakening the thether on the MLC-1/actin interface and/or MLC-1/MHC interface and simultaneous registration of cross-bridge function. Inhibition of the MLC-1/actin interaction by peptide competition using synthetic N-terminal MLC-1 peptides increases force production and shortening velocity of both demembranated (skinned) and of intact electrically driven human ventricular fibers (Morano et al., 1995) as well as myofibrillar ATPase activity (Rarick et al., 1996).

The Alkali light chains appeared initially to be essential for ATPase activity of myosin, since they were removed under mild conditions with 4 M LiCl and recombination of the separated light and heavy chains results in a partial recovery of ATPase activity (Stracher, 1969; Dreizen and Gershman, 1970). Alkali light chains are thus synonymous with "essential" myosin light chains. However, a decade later it could be clearly demonstrated, that the pure heavy chain, i.e. without any Alkali light chains revealed both normal $Ca2^{+}$ or K+/EDTA activated as well as actin-activated myosin ATPase activities from skeletal (Wagner and Giniger, 1981; Sivaramakrishnan and Burke, 1982) and cardiac muscle (Mathern and Burke, 1986).

Although Alkali MLC are not essential for myosin heavy chain enzymatic activity, they may play significant roles during the chemomechanical transduction process. Thus, in vitro motility assays demonstrated significantly higher actin transport velocities of myosin heavy chains associated with the Alkali light chains than with myosin heavy chains alone (Lowey et al., 1993). Furthermore, different isoforms of the alkali light chains (A1, A2) modulate actin-activated ATPase activity. at low ionic strength, the actin concentration required for half-maximal ATPase activity of myosin subfragment 1(S-1) is considerably lower for S-1(A1) than for S-1(A2) (Weeds and Taylor ,1975; Chalovich et al., 1984). In addition, the maximum turnover rate in the presence of actin for S-1(A2) is higher than that for S-1(A1 (Weeds and Taylor, 1975). These differences disappear, however at high ionic strength (Chalovich et al., 1984, Wagner et al., 1979). Myosin ATPase activities of these two species are identical in the absence of actin (Weeds and Taylor, 1975). In vitro motility experiments demonstrated that A2 myosin translocated actin filaments with a higher velocity than A1 myosin (Lowey et al., 1993). A significant role of Alkali light chain isoforms during the conversion of chemical energy into mechanics could also be observed in chemically skinned human heart preparations: there was a significant improvement of cross-bridge function upon partial replacement of VLC-1 by ALC-1 (Morano et al., 1996; Morano et al., 1997). These results have been supported in a transgenic mouse model overexpressing the mouse ALC-1 in the ventricle (Fewell et al., 1998): 95% replacement of the mouse VLC-1 by the mouse ALC-1 was accompanied by a 1.78 fold increase in Vmax skinned heart preparations (Fewell et al., 1998). However in skinned human heart preparations, a 20% replacement of the human VLC-1 by the human ALC-1 leads to 1.88 fold increase in Vmax of shortening (Morano et al., 1996). The higher effectiveness of the human ALC-1 compared to the mouse ALC-1 could recently be verified by a transgenic rat model expressing hALC-1 in the heart (Abdelaziz et al., 2004) which may be caused by the considerable differences of the primary sequence of the N-terminal domains (Fodor et al., 1989; Seharaseyon et al., 1990).

To demonstrate that cardiac hALC-1 expression could also induce beneficial effects on the failing heart, we induced cardiac failure by an aortocaval shunt operation of normal WKY rats and TGR/hALC-1. This operation procedure has the capacity to induce a high degree of cardiac hypertrophy, ANP/BNP expression, as well as significant functional deteriorations of the heart (Langenickel et al., 2000). Expression of the hALC-1 could indeed significantly attenuate the functional decline of the hypertrophied hearts of TGR/hALC-1 compared to the hypertrophied hearts of control (WKY) animals.

2. MATERIAL AND METHODS

2.1. Animal model and aortocaval shunt operation

We investigated male transgenic rats which overexpressed the human ALC-1 under the control of the rat alpha-myosin heavy chain promoter in the heart (TGR/hALC-1) and WKY rats (cf. Abdelaziz et al., 2004). 12 weeks old male WKY and TGR/hALC-1 were anaesthetized with ether. A laparotomy was performed, and the aorta was punctured distal to the renal arteries with a 1.8mm disposable needle. The needle was advanced in the adjacent inferior vena cava. After the aorta had been temporarily clamped, the needle was withdrawn , and the punctured site was sealed with cyanoacrylate glue. 5 weeks later animals were sacrified for functional analysis. Corresponding control animals were operated in the same manner but remained normal body and heart weights.

2. 2. Isolated perfused hearts (Langendorff heart)

Hearts from anaesthetized (30 mg/kg sodium Phenobarbital) and heparinized (500 U/kg) male transgenic and WKY rats (200–400 g body weight) were excised, weighted, and cannulated for retrograde aortic perfusion with a modified Krebs-Henseleit solution containing (mM) NaCl (118), KCl (4.7), CaCl2 (1.5), MgSO4 (1.2), NaHCO3 (25), Na2EDTA (0.05), KH2PO4 (0.23), 2.5% albumin and glucose (11.1). The solution was saturated with 95% O2/5% CO2 pH 7.4. The perfusion apparatus was From Hugo-Sachs Electronic (Germany) and the corresponding software was MEM Notocord (Germany). Systolic pressure was measured with a latex balloon, filled with ethanol/H2O, which was inserted into the left ventricle through the left atrium via a catheter to an (Isotec) transducer. The pressure in the balloon was set from 14–18 mmHg. The stimulation frequency was fixed at 340 beats/min. The perfusion was carried out at 37°C with a constant aortic pressure of 70 mm Hg. The recorded functional parameters were coronary flow, perfusion pressure, developed left ventricular pressure (LVP), maximal rate of pressure development (+dp/dtmax) and maximal relaxation rate (-dp/dtmax).

2.3. Quantitative estimation of expresed hALC-1

Ventricular tissues were homogenized in 5% SDS, 50mM Tris/HCL Buffer (PH 7.5) 0.25M sucrose, 75mM urea, 0.2M DTT (Dithiotreitol) and put in the thermal shaker at 95°C for 3 minutes. 80 μg protein was applied to a SDS-polyacrylamide gel (12%) and electrophoresis was performed for 2 hours at 120 V (constant). Proteins were transferred from SDS gels to a nitrocellulose membrane (60 minutes 3mA/cm_). The membrane was subsequently overnight incubated the primary antibody (anti-human ALC-1 antibody; cf. Abdelaziz et al. 2004) at 4 °C and then with the second antibody (anti-rabbit) for 1 hour at room temperature. Proteins were visualized with the enhanced chemiluminescence's reaction Kit (ECL, Amersham) and X ray film (X-Omat, Kodak, Rochester, NY). The signals were scanned densitometrically using BIO–RAD GS-710 (calibrated imaging densitometry, California, USA). The amount of expressed hALC-1 transgene in the rat heart tissue was evaluated by Western blot analysis. Defined amounts of expressed recombinant hALC-1 (hALC-1HIS), which revealed the same response to the hALC-1 antibody as the native hALC-1 (cf. Abdelaziz et al. 2004) were used as protein standard.

2.4. Statistical analysis

Values are expressed as mean ± SEM. Significant analysis is performed with student's t test.

3. RESULTS

3.1. Functionals analysis of the isolated perfused heart contraction

As compared to control (sham-operated) WKY rats and TGR/hALC-1, which revealed normal heart weight/body weight (H/B) ratios of 4.37±0.12 mg/g (n=6) and 4.48±0.27 mg/g (n=6). Shunt-operated rats developed significant levels of cardiac hypertrophy with H/B ratios of 7,55±0.72 mg/g (WKY; n=6) and 7.17±0.48 mg/g (TGR/hALC-1 n=6). The hypertrophy levels was not significantly different between WKY and TGR/hALC-1.

Control TGR/hALC-1 revealed statistically significantly higher contractility parameters of the Langendorff hearts than the corresponding WKY group (Fig. 1).

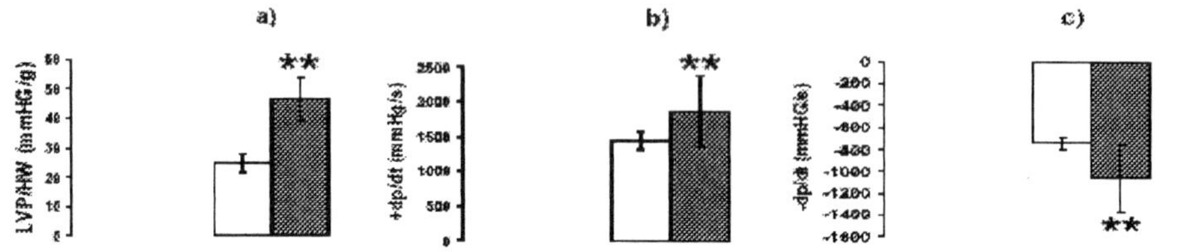

Figure 1. Contractile parameters of control WKY rats (open bars) and TGR/hALC-1 (closed bars). a) developed pressure per g heart weight. b) maximal velocity of pressure development (+dp/dt; mmHg/s); c) maximal relaxation rate (-dP/dt; mmHg/s)

The hypertrophied hearts revealed significantly lower contractile parameters than the corresponding control group (cf. Fig. 1 vs Fig. 2). However, the contractile parameters of hypertrophied hearts obtained from TGR/hALC-1 remained significantly higher ($p<0.01$) if compared to the hypertrophied WKY rat hearts (Figure 2).In fact, there was a negative linear correlation between the degree of cardiac hypertrophy and developed left ventricular pressure in both WKY and TGR/hALC-1. However, the decline of contraction with increasing level of cardiac hypertrophy was less pronounced in TGR/hALC-1 compared to WKY.

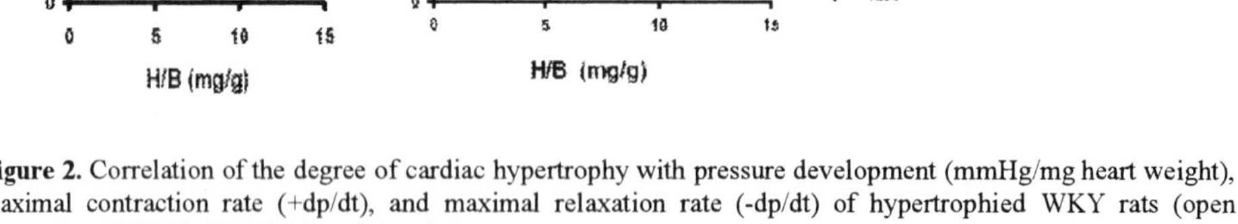

Figure 2. Correlation of the degree of cardiac hypertrophy with pressure development (mmHg/mg heart weight), maximal contraction rate (+dp/dt), and maximal relaxation rate (-dp/dt) of hypertrophied WKY rats (open squares) and TGR/hALC-1 (closed rhombs).

3.2. Quantification of the expressed hALC-1

We used hALC-1_{HIS} as a standard to quantify the amount of expressed hALC-1 by Western blot analysis (Figure 3a). Please note, that there was a linear ratio between the amount of recombinant hALC-1 and the optical signal between 1 and 5 µg protein. TGR-hALC-1, but not WKY expressed the hALC-1 transgene (Figure 3b).

The average hALC-1 expression of control TGR/hALC-1 decreased significantly from 17±4 µg/mg whole SDS-protein to 14±2 µg/mg whole SDS-protein, in hypertrophied hearts ($p<0.05$).

Assuming around 45 % of the whole protein is myosin, and around 10% of the whole myosin represents MLC-1 the hALC-1 represents about 38 % and 31 % of whole MLC-1 in control and hypertrophied hearts, respectively of TGR/hALC-1.

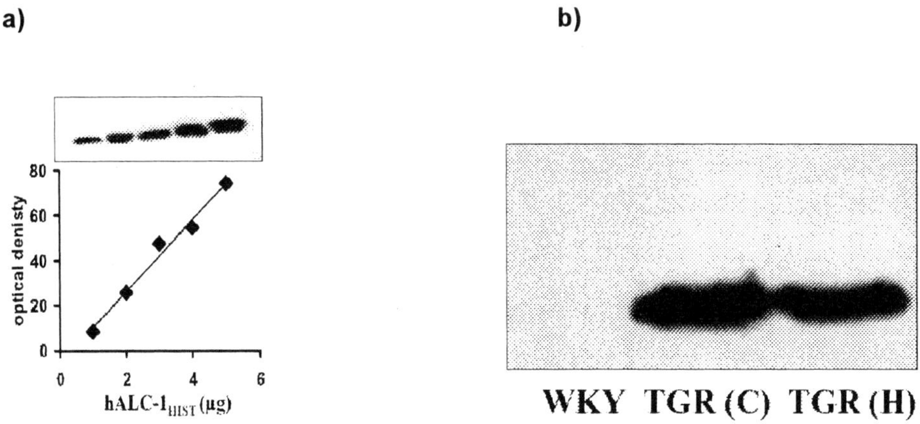

Fig. 3. Quantification of hALC-1 transgene expression. a) Western-blot of hALC-1_{HIS} as a protein standard. Insert: original ECL signals of 1-5µg hALC-1_{HIS}. Bottom: evaluation of hALC-1_{His} ECL signals. b) Representative Western-blot of cardiac SDS-extracts of hearts obtained from a WKYrat, as well as control (C) and hypertrophied (H) TGR/hALC-1.

4. DISCUSSION

Previous research work identified the beneficial functional effects of the hALC-1 on cross-bridge function of *in vitro* cardiac preparations (human ventricular chemically skinned fibres) (Morano et al., 1996; Morano et al., 1997). Cardiac-specific overexpression of the mouse ALC-1 in transgenic mice (Fewell et al., 1998) as well as human ALC-1 in transgenic rats (Abdelaziz et al., 2004) verified the improvements of cardiac contractile function by ALC-1 expression at the isolated perfused mouse heart and skinned fibre levels. The human ALC-1, however, turned out to be more effective than the mouse ALC-1: Around the same relative improvements of cardiac contractile functions could be obtained with only a fraction of expressed human ALC-1 compared to mouse ALC-1 (Abdelaziz et al., 2004). These results directed our future prospects towards patients with heart failure as target for therapeutic up-regulation of ALC-1 expression by gene therapy. To demonstrate, that cardiac hALC-1 expression could indeed induce beneficial effects on the functionally impaired heart, we induced cardiac hypertrophy and impaired function by an aortocaval shunt operation of normal WKY rats and TGR/hALC-1. This operation procedure has the capacity to induce a high degree of cardiac hypertrophy, ANP/BNP expression, as well as significant functional deteriorations of the heart (Langenickel et al., 2000). Expression of hALC-1 declined in the hypertrophied hearts compared to the hearts of control TGR/hALC-1. Please note that we used the alpha-myosin heavy chain promoter to control cardiac-specific hALC-1 transgene expression. The activity of this promoter is reduced with age and during cardiac hypertrophy (Schiaffino and Reggiani, 1996), which could explain the reduced hALC-1 expression in the hypertrophied hearts of TGR/hALC-1.

There was a significant negative correlation between cardiac contractile parameters of isolated perfused hearts and the level of cardiac hypertrophy in both WKY and TGR/hALC-1. However, this decline of heart function with the hypertrophied state was much less pronounced in TGR/hALC-1 as compared to WKY, demonstrating the beneficial effects of hALC-1 on the function of the overloaded heart. This positive effect of hALC-1 is even underestimated, since the level of transgene expression was reduced in the hypertrophied hearts of TGR/hALC-1 as compared to control TGR/hALC-1.

In summary, expression of the hALC-1 could significantly attenuate the functional decline of the functionally impaired hypertrophied heart in a transgenic rat model. Therapeutic upregulation of hALC-1, therefore may represent a valuable gene therapeutic tool for the treatment of cardiac failure.

ACKNOWLEDGEMENTS

This work was supported by fellowships from Vacsera and Graduiertenkolleg 754 for IAA.

5. REFERENCES

Abdelaziz, I., Segaric, J., Batsch, H., Petzhold, D., Schlegel, W.P., Kott, M., Seefeldt, I., Klose, J., Bader, M., Haase, H., Morano, I., 2004; Functional characterization of the human arial essential myosin light chain (hALC-1) in a transgenic rat model. *J. Mol. Med.*, **82**: 265 – 274

Aksoy, M.O., Mras, S., Kamm, K.E., Murphy, R.A., 1983; $Ca2^+$, cAMP, and changes in myosin phosphorylation during contraction of smooth muscle. Am. J. Physiol. 245: C255-C270

Barton, P.J.R., Buckingham, M.E., 1985; The myosin alkali light chain proteins and their genes. *Biochem J* **231**:249-261

Burridge, K., 1974; A comparisn of fibroblast and smooth muscle myosins. *FEBS Lett.* **45**: 14-17

Chalovich, J.M., Stein, L.A., Greene, L.E., Eisenberg, E., 1984; Interaction of isozymes of myosin subfragment 1 with actin: effect of ionic strength and nucleotide. *Biochem.* **23**: 4885-4889

Dreizen, P., Gershman, L.C., 1970; Relation of structure to function in myosin. II. Salt denaturation and recombination experiments. *Biochem.* **9**: 1688-1693

Dreizen, P., Gershman, L.C., Trotta, P.P., Stracher, A., 1967; Myosin Subunits and their interactions. *J. Gen Physiol.* **50**: 85-118

Fewell, J.G., Hewett, Sanbe, A., Klevitsky, R., Hayes, E., Warshaw, D., Maughan, J., Robbins, J., 1998; Functional significance od cardiac essential light chain isoform switching in transgenic mice. *J Clin Invest* **101**: 2630-2639

Fodor, W.L., Darras, B., Seharaseyon, J., Falkenthal, S., Francke, U., Vanin, E.F., 1989; Human ventricular/slow twitch myosin alkali light chain gene characterization, sequence, and chromosomal location. *J Biol Chem* **264**: 2143-2149

Frank, G., Weeds, A.G., 1974; The amino acid sequence of the alkali light chains of rabbit skeletal muscle myosin. *Eur. J. Biochem.* **44**: 317-334

Gazith, J., Himmelfarb, S., Harrington, W.F., 1970; Studies on the subunit strucutre of myosin. *J. Biol. Chem.* **245**: 15-22

Huxley, A.F., 1957; Muscle structure and theories of contraction. *Prog. Biophys.Biophys. Chem.* **7**: 255-318

Kominz, D.R., Carroll, W.R., Smith, E.N., Mitchell, E.R., 1959; A subunit of myosin. *Arch. Biochem. Biophys.* **79**: 191-196

Langenickel, T., Pagel, I., Höhnel, K., Dietz, R., Willenbrock, R. 2000; Differential regulation of cardiac ANP and BNP mRNA in different stages of experimental heart failure. *Am. J. Physiol. Heart Circ. Physiol.* **258**: H1500-H1506

Locker, R.H., 1956; The dissociation of myosin by heat coagulation. *Biochim. Biophys. Acta* **20**: 514

Locker, R.H., Hagyard, C.J., 1967; Small subunits in myosin. *Arch. Biochem. Biophys.* **120**: 454-461

Lowey, S., Risby, D. 1971; Light chains from fast and slow muscle myosins. *Nature* **234**: 81-85

Lowey, S., Waller, G.S., Trybus, K.M. 1993; Skeletal muscle myosin light chains are essential for physiological speeds of shortening. *Nature* **365**: 454-456

Lowey, S., Waller, G., Trybus, K.M. 1993; Function of skeletal muscle myosin heavy and light chain isoforms by an in vitro motility assay. *J. Biol. Chem.* **268**: 20414-20418

Mathern, B.E., Burke, M. 1986; Stability and substucture of cardiac myosin subfragment 1 and isolation and properties of its heavy-chain subunit. *Biochem.* **25**: 884-889

Morano, I., Hädicke, K., Haase, H., Bohm, M., Erdmann, E., Schaub, M.C. 1997; Changes in essential myosin light chain isoform expression provide a molecular basis for isometric force regulation in the failing human heart. *J Mol Cell Cardiol* **29**: 1117-1187

Morano, I., Ritter, O., Bonz, A., Timek, T., Vahl, C.F., Michel, G. 1995; Myosin light chain-actin interaction regulates cardiac contractility. *Circ. Res.* **76**: 720-725

Morano, M., Zacharzowski, U., Maier, M., Lange, P.E., Alexi-meskishvili, V., Haase, H., Morano, I. 1996; Regulation of human heart by essential myosin light chain isoforms. *J Clin Invest* **98**: 467-473

Nabeshima, Y., Fujii-Kuriyama, Y., Muramatsu, M., Ogata, K. 1984; Alternative transcription and two modes of splicing results in two myosin light chains from one gene. *Nature* **308**: 333-338

Periasamy, M., Strehler, E.E., Garfinkel, L.I., Gubits, R.M., Ruiz-Opazo, N., Nadal-Ginard, B. 1984; Fast skeletal muscle myosin light chains 1and 3 are produced from a single gene by a combined process of differential RNA transcription and splicing. *J. Biol. Chem.* **259**: 13595-13604

Perrie, W.T., Smillie, L.B., Perry, S.V. 1972; A phosphorylated light chain component of myosin. *Biochem. J.* **128**; 105P-106P

Rarick, H.M., Opgenorth, T.J., von Geldern, T.W., Wu-Wong, J.R., Solaro, R.J. 1996; An essential myosin light chain peptide induces supramaximal stimulation of cardiac myofivrillar ATPase activity. *J. Biol. Chem.* **271**: 27039-27043

Rayment, I., Rypniewski, W.R., Schmidt-Base, K., Smith, R., Tomchick, D.R., Benning, M.M., Winkelmann, D.A,. Wesenberg, G., Holden, H.M. 1993; Three-dimensional structure of myosin subfragment-1: a molecular motor. *Science* **261**: 50-58

Sarkar, S.F., Sreter, F.A., Gergely, J. 1971; Light chains of myosin from white, red , and cardiac muscles. *Proc. Nat. Acad. Sci.* **68**: 946-950

Schiaffino, S., Reggiani, C. 1996; Molecular diversity of myofibrillar proteins: gene regulation and functional significance. *Physiol. Rev.* **76**: 371-423

Seharaseyon, J., Bober, E., Hsieh, C.L., Fodor, W.L., Francke, U., Arnold, H.I.H., Vanin, E.F. 1990; Human embryonic/atrial myosin alkali light chain gene: characterization, sequence, and chromosomal location. *Genomics* 7:289-293

Sivaramakrishnan, M., Burke, M. 1982; The free heavy chain of vertebrate skeletal myosin subfragment 1 shows full enzymatic activity. *J. Biol. Chem.* **257**: 1102-1105

Sobieszek, A. 1977; Ca2+ -linked phosphorylation of a light chain of vertebrate smooth muscle myosin. *Eur. J. Biochem.* **73**: 477-483

Stepkowski, D., 1995; The role of the skeletal muscle myosin light chains N-terminal fragments. *FEBS Lett* **374**: 6-11

Stracher, A. 1969; Evidence for the involvement of licht chains in the biological functioning of myosin. *Biochem. Biophys. Res. Commun.* **35**: 519-820

Sutoh, K. 1982; Identification of myosin-binding sites on the actin sequence. *Biochemistry* **21**: 3654-3661

Szent-Györgyi, A.G., Szentkiralyi, 1973; The light chains of scallop myosin as regulatory subunits. *J. Mol. Biol.* **74**: 179-203

Trayer, I.P., Trayer, H.P., Levine, B.A. 1987; Evidence that the N-terminal region of A1-Light chain of Myosin interacts directly with the C-terminal region of actin. A proton magnetic resonance study. *Eur J Biochem.* **164**: 259-266

Tsao, T.C. 1953; Fragmentation of the myosin molecule. *Biochim Biophys. Acta* **11**: 368-3823.

Wagner, P.D., Giniger, E. 1981; Hydrolysis of ATP and reversible binding to F-actin by myosin heavy chains free of all light chains. *Nature (London)* **292**: 560-562

Wagner, P.D., Slater, C.S., Pope, B., Weeds, A.G. 1979; Studies on the actin activation of myosin subfragment-1 isoenzymes and the role of myosin light chains. *Eur. J. Biochem.* **99**: 385-394

Weber, H.H., Stöver, R. 1933; Das kolloidale Verhalten der Muskeleiweißkörper. *Biochem. Z.* **259**: 269-284
Weeds, A.G. 1969; Light chains of myosin. *Nature* **223**: 1362-1364
Weeds, A.G., Frank. G. 1973; Structural studies on the light chains of myosin. *In: The Mechanism of Muscle Contraction. Cold spring Harbor Symposia on quantitative Biology,* Vol. **XXXVII**, pp 9-14
Weeds, A.G., Lowey, S. 1971; Substructure of the myosin molecule. II. The light chains of myosin. *J. Mol. Biol.* **61**: 701-725
Weeds, A.G., Taylor, R.S. 1975; Separation of subfragment-1 isoenzymes from rabbit skeletal muscle myosin. *Nature (London)* **257**: 54-56

DISCUSSION

Suga: Have you measured O_2 consumption in these contractility-improved heart?

Morano: We measured the mechanical parameters which would be obtained with Langendorf preparations, but did not determine O_2 consumption.

Cecchi: Is the effect of increased performance in transgenic animals permanent?

Morano: There is a decline of transgene expression with age and, at the same time, decline of the improved contractile parameter to control values.

CYTOPLASMIC FREE CONCENTRATIONS OF Ca^{2+} IN SKELETAL MUSCLE CELLS

Masato Konishi *

1. INTRODUCTION

Because cytoplasmic free Ca^{2+} regulates muscle contraction (Ebashi and Endo, 1968), it is essential to obtain accurate information on the levels of cytoplasmic free Ca^{2+} concentration ($[Ca^{2+}]_i$), as well as the time course and amplitude of $[Ca^{2+}]_i$ changes (Ca^{2+} transients), to study excitation-contraction coupling of muscle cells. Ca^{2+} transients in skeletal muscles have been measured with many different methodologies, such as optical indicators, bioluminescent proteins and Ca^{2+}-selective microelectrodes. Since each method has various advantages and disadvantages in the measurement of Ca^{2+} transients, significant controversy still remains concerning the quantities reported by different techniques. In this chapter, I will review developments of $[Ca^{2+}]_i$ measurements in skeletal muscle cells, and also briefly describe our experiments carried out to estimate $[Ca^{2+}]_i$ at rest and during steady state contraction in frog skeletal muscle fibers.

The first measurements of Ca^{2+} transients in skeletal muscle were reported by Jöbsis and O'Connor (1966) who introduced murexide into toads by intraperitoneal injection and observed changes in murexide absorbance in the sartorius muscle in response to electrical stimulation. Because murexide is a dye that undergoes absorbance changes upon Ca^{2+} binding, the authors attributed the optical signals to Ca^{2+} transients. However, their "Ca^{2+} transients" have been questioned, because of their very slow time course (time to peak 50-100 ms after stimulation) For example, Maylie et al. (1987b) used tetramethylmurexide, a dye structurally related to murexide, and showed very rapid changes in its absorbance with a time to peak of about 5 ms after stimulation.

* Department of Physiology, Tokyo Medical University, Tokyo 160-8402, Japan

2. MEASUREMENTS OF Ca^{2+} TRANSIENTS

2.1. Bioluminescent proteins

Shimomura et al. (1962) extracted aequorin, a Ca^{2+} sensitive bioluminescent protein, from a jelly fish *Aequorea aequorea* and clarified fundamental properties of its luminescence. Ashley and Ridgway (1968) injected aequorin into striated muscles of a barnacle and simultaneously measured the membrane potential, Ca^{2+} transients and tension. This work clearly demonstrated the central role of cytoplasmic Ca^{2+} in excitation-contraction coupling. Properties of aequorin were further studied extensively by Blinks and his collaborators (1976, 1982), and aequorin was applied to frog skeletal muscle fibers [Rüdel and Taylor (1973), Blinks et al. (1978)] and also to cardiac and smooth muscles [Allen and Blinks (1978), Allen and Kurihara (1982), Morgan and Morgan (1982)]. In general, aequorin is a good choice to simultaneously measure Ca^{2+} transients and mechanical activities in skeletal, cardiac and smooth muscles. Recently, recombinant versions of aequorin can be inserted and expressed in target organelles (e.g., mitochondria) of cultured cells as a useful tool to measure Ca^{2+} concentrations in organelles.

2.2 Metallochromic dyes

Metallochromic dyes undergo absorbance changes upon Ca^{2+} binding. Among this family of dyes, arsenazao III and antipyrylazo III have been widely used. Miledi et al. (1977) injected arsenazao III into frog skeletal muscle fibers, and measured Ca^{2+} transients under voltage-clamp or following action potentials. Disadvantages of metallochromic dyes include complex stoichiometry of Ca^{2+}-dye binding; for example, arsenazao III binds Ca^{2+} with 1:1, 1:2, 2:1 and 2:2 stoichiometries (Dorogi and Neumann, 1981; Rios and Schneider 1981; Palade and Vergara, 1983). This makes it very difficult to calibrate the dye signals in terms of $[Ca^{2+}]_i$. For antipyrylazo III, the 1:2 Ca^{2+}:indicator complex appears to be predominantly formed at the concentrations of antipyrylazo III and Ca^{2+} normally encountered in the cytoplasm (Rios and Schneider, 1981; Hollingworth et al., 1987). The absorbance signal from the 1:2 Ca^{2+}:antipyrylazo III complex is considered to follow a rapid change in $[Ca^{2+}]_i$ during twitch response with a relatively small (1-2 ms) kinetic delay in frog skeletal muscle fibers (Baylor et al., 1985; Maylie et al., 1987a). On the other hand, the peak amplitude of the cytoplasmic Ca^{2+} transient calibrated from antipyrylazo III is 2-3 μM (Maylie et al., 1987a; Baylor and Hollingworth, 1988), which may be erroneously small, probably because ~75% of the cytoplasmic antipyrylazo III molecules appear to be bound to intracellular constituents, and the properties of the bound indicator molecules are likely to be different from those in salt solutions (Baylor et al., 1986). The calibration difficulties arising from the binding of an indicator to cellular constituents applies not only to metallochromic indicators, but, as discussed below, are general problems shared among all optical indicators.

2.3. High affinity fluorescent indicators

Tsien and his collaborators synthesized a series of fluorescent Ca^{2+} indicators that have useful properties for biological applications (Grynkiewicz et al., 1985; Minta et al.

1989). In particular, fura-2 and indo-1 have been widely used in many types of cells. Fura-2 and indo-1 have dissociation constant values in the sub-micromolar range ("high affinity" indicators), which makes these indicators useful to measure low $[Ca^{2+}]_i$ near resting levels. During the large Ca^{2+} transients of skeletal muscles, on the other hand, these indicators are likely saturated, particularly in the localized space near the terminal cistern of the sarcoplasmic reticulum, the site of Ca^{2+} release. It has been also reported that fura-2 cannot track fast Ca^{2+} transients of skeletal muscle fibers because of slow kinetics due to its small dissociation rate constant (Baylor and Hollingworth, 1988). Therefore, these high affinity indicators are advantageous for measurements of resting $[Ca^{2+}]_i$, but not suitable for tracking large and fast Ca^{2+} transients of skeletal muscles. However, the high affinity indicators appear to be useful for cell types other than skeletal muscles, such as cardiac and smooth muscle cells as well as non-muscle cells. Fluo-3 has been usefully applied in combination with con-focal imaging to detect localized increase in $[Ca^{2+}]$, e.g., Ca^{2+} sparks (Cheng et al., 1993).

2.4. Low affinity indicators

Chandler and his colleagues first suggested that Ca^{2+} indicators that have a low affinity for Ca^{2+} ("low affinity" indicators) may be useful for measurements of Ca^{2+} transients in skeletal muscle fibers (Maylie et al., 1987b). Southwick and Waggoner (1989) synthesized two purpurate indicators: purpurate-3,3'-diacetic acid (PDAA) and 1,1'-dimethylpurpurate-3,3'diacetic acid (DMPDAA), analogues of murexide and tetramethylmurexide, respectively. These indicators have a low affinity for Ca^{2+} (K_D 0.8-1.0 mM range), and PDAA is highly selective for Ca^{2+} over Mg^{2+} and H^+ (Hirota et al., 1989). When applied to frog cut muscle fibers, both PDAA and DMPDAA are thought to track cytoplasmic Ca^{2+} transients without kinetic delay. The time course of the Ca^{2+} transients was found to be surprisingly fast; time to peak ~5 ms, halfwidth 7-9 ms (16°C). Similar results were obtained in intact frog skeletal muscle fibers injected with PDAA (Konishi and Baylor, 1991). Interestingly, these new indicators were also reported to have a large amplitude for the cytoplasmic Ca^{2+} transients, as large as 20 μM, stimulated by an action potential. Another important advantage of these new purpurate indicators is that the percentage of the indicator that appeared to be bound to intracellular constituents is relatively small (~20% in cut fibers, Hirota et al., 1989; 24-43% in intact fibers, Konishi and Baylor, 1991), in comparison with the large bound fraction (~60-90%) detected with the other indicators. Because an indicator with a smaller bound fraction is likely to yield a more reliable estimate of the amplitude of the cytoplasmic Ca^{2+} transients, the amplitude of Ca^{2+} transients is considered to be more reliable when estimated with PDAA (or DMPDAA) than other indicators.

Recently, many low-affinity fluorescent Ca^{2+} indicators have become available. Furaptra is one of the low-affinity tricarboxylate indicators; although originally designed as a Mg^{2+} indicator (Raju et al., 1990), it has often been used in muscle fibers as a Ca^{2+} indicator with rapid kinetics (Konishi et al., 1991; Claflin et al., 1994; Hollingworth et al., 1996). Although this indicator is commercially available under the name 'mag-fura-2' (Molecular Probes, Inc., Eugene, OR, USA), it has 1.5-2 orders higher affinity for Ca^{2+}

(K_D ~50 μM at 16-37°C) than for Mg^{2+} (K_D = 5.3 mM and 1.5 mM at 16°C and 37°C, respectively)(Raju et al., 1990; Konishi et al., 1991). As expected from these K_D values, furaptra was found to be a useful indicator for measuring the cytoplasmic [Mg^{2+}] in resting muscle fibers (Konishi et al., 1993) and as a Ca^{2+} indicator in stimulated fibers (Konishi et al., 1991). In frog muscle fibers, the brief time course of the cytoplasmic Ca^{2+} transient reported with furaptra is very similar to that reported with PDAA. Thus, furaptra seems to be a useful tool to track the cytoplasmic Ca^{2+} transient without kinetic delay. As in frog fibers, furaptra appears to be useful for tracking the brief time course of Ca^{2+} transients in mammalian skeletal muscle fibers, which generally have diameters about half as large as frog fibers. The cytoplasmic Ca^{2+} transients measured in mouse extensor digitorum longus muscle appear to have a very brief time course, even briefer than those in frog fibers (Hollingworth et al., 1996); after single action potential stimulation, the halfwidth is 4.6 ms at 16°C, and 2.0 ms at 28°C. The results further necessitate the use of low-affinity indicators to faithfully monitor the time course of the cytoplasmic Ca^{2+} transients in skeletal muscle fibers. Thus, our current understanding is that Ca^{2+} transients in frog and mammalian skeletal muscle fibers are large in amplitude (~10-20 μM) and very brief in time course, with a time to peak of ~5 ms and a halfwidth of <10 ms.

3. MEASUREMENTS OF RESTING $[Ca^{2+}]_I$

Two different types of methods have been applied to estimate the resting levels of $[Ca^{2+}]_i$: Ca-ISEs and optical indicators (for the list of estimated resting $[Ca^{2+}]_i$, see Table 1 of Konishi, 1998). Ca^{2+}-selective microelectrodes (Ca-ISEs) are prepared by filling the tip of glass microelectrodes with Ca^{2+} exchange resins. Among the Ca^{2+} exchange resins introduced, the neutral carrier, ETH-1001, has been used most successfully. Tsien and Rink (1980) applied Ca-ISEs with the neutral carrier to frog muscle fibers, and showed that the Ca-ISEs can be used to measure $[Ca^{2+}]_i$. The output of the electrodes should be directly related to the [Ca^{2+}] at the tip; ideally ~30 mV change for a 10-fold change in Ca^{2+} activity. However, at very low Ca^{2+} levels around resting $[Ca^{2+}]_i$ and in the presence of ions which interfere with the electrodes (most importantly K^+, Na^+ and Mg^{2+}), the response of actual electrodes shows much shallower slopes than that expected from Nernstian behavior (for review, see Blinks et al., 1982). Thus, the small errors in the estimation of the cytoplasmic concentration of interfering ions and in membrane potential subtraction lead to a non-negligible difference in the estimation of Ca^{2+} concentration. Another potential source of error of the Ca-ISEs is the leakage of Ca^{2+} around the sites of mechanical impalement, which tends to elevate the local [Ca^{2+}]. Some of the estimates obtained with Ca-ISEs may be erroneously high as a consequence of local membrane damage, as pointed out by Weingart and Hess (1984).

Aequorin is weakly luminescent even in the virtual absence of Ca^{2+} (Ca^{2+}-independent luminescence), but emits large additional luminescence upon Ca^{2+} binding (Allen et al., 1977). The difficulty in the use of aequorin for determination of resting $[Ca^{2+}]_i$ is that the fractional light level obtained from resting muscle is very close to the Ca^{2+}-independent luminescence; even a small error in the in vitro calibration curve greatly influences estimates of $[Ca^{2+}]_i$. Blatter and Blinks (1991) also found that the fractional luminescence levels of most resting fibers are below the Ca^{2+}-independent luminescence of the calibration curve, giving apparently 'negative' values for $[Ca^{2+}]_i$. A possible reason for the 'negative' estimates for $[Ca^{2+}]_i$ is that myoplasm contains a diffusible macromolecule that interacts with aequorin to reduce Ca^{2+}-independent light emission (Blatter and Blinks, 1991). Thus, to date only upper limit estimates of resting $[Ca^{2+}]_i$ have been obtained from aequorin under experimental conditions in which $[Ca^{2+}]_i$ was raised (and L/L_{max} was increased well above the Ca^{2+}-independent level) either by partial depolarization (Blatter and Blinks, 1991) or application of caffeine (Konishi and Kurihara, 1987).

Fura-2 and indo-1 are the high affinity Ca^{2+} indicators. The high Ca^{2+}-affinity of these indicators is advantageous for monitoring $[Ca^{2+}]_i$ around resting levels. As the vast majority of measurements of $[Ca^{2+}]_i$ are made with fluorescent high affinity dyes, only a few studies have provided useful information on the precise $[Ca^{2+}]_i$ values in muscle fibers. This is primarily because there are difficulties related to calibration of indicator signals in terms of $[Ca^{2+}]_i$. In general, Ca^{2+} indicator molecules are heavily bound to cellular constituents (possibly proteins) within the cytoplasm (Konishi et al., 1988). This binding alters both spectral and binding properties of the indicators; in vitro studies show that the binding to soluble proteins changes both fluorescence and Ca^{2+} binding of fura-2 (Konishi et al., 1988), indo-1 (Baker et al., 1994), fluo-3 (Harkins et al, 1993) and fura-red (Kurebayashi et al, 1993). In frog skeletal muscle fibers, the fluorescence spectra of fura-2 is shifted to longer wavelengths (Konishi and Watanabe, 1995), and as a consequence, a simple comparison of the fluorescence ratio of fura-2 in the cytoplasm and in salt solutions lacking proteins indicates 'negative' $[Ca^{2+}]_i$ (Suda and Kurihara, 1991). Thus, the simple assumption that the indicators behave in the same way in the intracellular environment as in salt solutions is probably not valid, necessitating calibration in the 'intracellular' environment.

Technical difficulties in calibrating optical signals probably contribute to the large discrepancies between recent estimates of $[Ca^{2+}]_i$ in frog muscle at rest (range between <50 nM (Blatter and Blinks, 1991) and 300 nM (Kurebayashi et al., 1993). In an attempt to more accurately estimate resting $[Ca^{2+}]_i$, Konishi and Watanabe (1995) injected fura-2 conjugated to high molecular weight dextran (fura dextran, MW ~10,000) into frog skeletal muscle fibers. To calibrate the indicator's fluorescence in terms of $[Ca^{2+}]_i$, we applied β-escin to permeabilize the cell membrane. After β-escin treatment, small molecules (e.g., Ca^{2+}, ATP etc.) could quickly permeate the cell membrane, whereas 10-kD fura dextran only slowly leaked out from the fiber. It was thus possible to calibrate the indicator fluorescence within the fibers by adjusting the bath solution $[Ca^{2+}]$ to various levels. From the calibration obtained in the β-escin treated fibers, resting $[Ca^{2+}]_i$ in frog skeletal muscle fiber was, on average, 60 nM. Thus it seems reasonable to assume that resting $[Ca^{2+}]_i$ in frog skeletal muscle fibers is within 50-100 nM range.

4. RELATION BETWEEN $[Ca^{2+}]_i$ AND FORCE

The steady-state relation between $[Ca^{2+}]$ and force has been studied primarily with skinned fibers in which the cell membrane is removed to easily establish the known levels of $[Ca^{2+}]$ in the myofibrillar space (Natori, 1954; Endo and Iino, 1980). In spite of the usefulness of skinned fibers, it is desirable to obtain a quantitative relation between $[Ca^{2+}]_i$ and force in intact muscle fibers, because removal of the cell membrane should produce a large alteration in the environment surrounding the myofibrils and may well influence the $[Ca^{2+}]$-force relation.

In intact muscle fibers stimulated by either a single action potential or a train of action potentials, force generation follows a rapid rise and fall of $[Ca^{2+}]_i$ with a significant delay, forming a non-equilibrium relation between $[Ca^{2+}]_i$ and force. However, under experimental conditions that significantly slow the change in $[Ca^{2+}]_i$ to reach equilibrium with force generation, simultaneous measurements of $[Ca^{2+}]_i$ and force should provide information on the steady-state relation between $[Ca^{2+}]_i$ and force. Konishi and Watanabe (1998) measured $[Ca^{2+}]_i$ with fura dextran (see above) in frog skeletal muscle fibers, and plotted $[Ca^{2+}]_i$ versus force during high K^+ contractures and during the slow relaxation phase after tetani in the presence of an inhibitor of the SR Ca^{2+} pump. The $[Ca^{2+}]_i$-force data obtained from the two types of measurements agreed fairly well, consistent with a Hill coefficient of 3.2-3.9 and a Ca_{50} of 1.5-1.7 μM (sarcomere length 2.6-2.8 μm, 16-18°C). We concluded that the steady state force was a 3rd to 4th power function of $[Ca^{2+}]_i$, and half maximal force was achieved in the low micromolar range in intact frog skeletal muscle fibers. However, there are still significant contradictions and uncertainties concerning the steady state relation between $[Ca^{2+}]_i$ and force in intact skeletal muscles. Clearly, further improvements for measurements of both $[Ca^{2+}]_i$ and force are necessary to resolve this important issue.

5. INHOMOGENEOUS ACTIVATION WITHIN SARCOMERES

In the previous sections, $[Ca^{2+}]_i$ refer to the spatially averaged $[Ca^{2+}]$ in the cytoplasm. In the steady state, $[Ca^{2+}]$ levels in the sarcomeres are thought to be uniform. However, when Ca^{2+} is released from the terminal cistern of the SR, the concentration gradient of cytoplasmic free Ca^{2+} should develop. When the SR Ca^{2+} release is turned off, cytoplasmic Ca^{2+} diffusion causes the gradient to dissipate. During the Ca^{2+} transient, therefore, the local level of $[Ca^{2+}]$ within the sarcomeres is higher or lower than $[Ca^{2+}]_i$, in accordance with the distance from the Ca^{2+} release channels of the SR and the time after initiation of the release. Hollingworth et al. (2000) experimentally confirmed the sarcomeric Ca^{2+} gradient during twitch activation of frog skeletal muscle fibers imaged with two-photon microscopy. They found that the time course of fluo-3 fluorescence changes was very different at the z- and m-lines, with a significantly delay (about 1 ms) between the rise of fluorescence at the z-line (near the termincal cistern) and the m-line (the middle of sarcomeres). These localized changes in $[Ca^{2+}]_i$ which would lead to non-uniform activation of myofilaments are complex functions of intrasarcomeric geometry and time, and their significance requires clarification by future studies.

Acknowledgements
We thank Prof. J. Patrick Barron of the International Medical Communications Center of Tokyo Medical University for reading the manuscript.

6. REFERENCES

Allen, D.G. and Blinks, J.R. and Prendergast, F.G. (1977) Aequorin luminescence: Relation of light emission to calcium concentration - A calcium-independent component. *Science* 195, 996-998.

Allen, D.G. and Blinks, J.R. (1978) Calcium transients in aequorin-injected frog cardiac muscle. *Nature* 273, 509-513.

Allen, D.G. and Kurihara, S. (1982) The effects of muscle length on intracellular calcium transients in mammalian cardiac muscle. *J. Physiol. (Lond.)* 327, 79-94.

Ashley, C.C. and Ridgway, E.B. (1968) Simultaneous recording of membrane potential, calcium transient and tension in single barnacle muscle fibres. *Nature* 219, 1168-1169.

Baker, A.J., Brandes, R., Schreur, J.H.M., Camacho, S.A. and Wiener M.W. (1994) Protein, and acidosis alter calcium-binding, and fluorescence spectra of the calcium indicator indo-1. *Biophys. J.* 67, 1646-1654.

Baylor, S.M. and Hollingworth, S. (1988) Fura-2 calcium transients in frog skeletal muscle fibres. *J. Physiol. (Lond.)* 405, 151-192.

Baylor, S.M., Hollingworth, S., Hui, C.S. and Quinta-Ferreira, M.E. (1985) Calcium transients from intact frog skeletal muscle fibres simultaneously injected with Antipyryrazo III and Azo 1. *J. Physiol. (Lond.)* 365, 70P.

Baylor, S.M., Hollingworth, S., Hui, C.S. and Quinta-Ferreira, M.E. (1986) Properties of the metallochromic dyes arsenazo III, antipyrylazo III and azo 1 in frog skeletal muscle fibres at rest. *J. Physiol. (Lond.)* 377, 89-141.

Blatter, L.A. and Blinks, J.R. (1991) Simultaneous measurement of Ca^{2+} in muscle with Ca electrodes and aequorin. Diffusible cytoplasmic constituent reduces Ca^{2+}-independent luminescence of aequorin. *J. Gen. Physiol.* 98, 1141-1160.

Blinks, J.R., Prendergast F.G. and Allen, D.G. (1976) Photoproteins as biological calcium indicators. *Pharmacol. Rev.* 28, 1-93.

Blinks, J.R., Rudel, R. and Taylor, S.R. (1978) Calcium transients in isolated amphibian skeletal muscle fibres: Detection with aequorin. *J. Physiol.* 277, 291-323.

Blinks, J.R., Wier, W.G., Hess, P. and Prendergast F.G. (1982) Measurement of Ca^{2+} concentrations in living cells. *Prog. Biophys. Molec. Biol.* 40, 1-114.

Cheng, H., Ledeler, W.J. and Cannell, M.B. (1993) Calcium sparks: elementary events underlying excitation-contraction coupling in heart muscle. *Science* 262, 740-744.

Claflin, D.R., Morgan, D.L., Stephenson, D.G. and Julian, F.J. (1994) The intracellular Ca^{2+} transient and tension in frog skeletal muscle fibres measured with high temporal resolution. *J. Physiol. (Lond.)* 475, 319-325.

Dorogi, P.L. and Newmann, E. (1981) Spectrophotometric determination of reaction stoichiometry and equilibrium constants of metallochromic indicators. II. The Ca^{2+}-arsenazo III complexes. *Biophys. Chem.* 13, 125-131.

Ebashi, S. and Endo, M. (1968) Calcium ion and muscle contraction. *Prog. Biophys. Molec. Biol.* 18, 123-183.

Endo, M. and Iino, M. (1980) Specific perforation of muscle cell membranes with preserved SR functions by saponin treatment. *J. Muscle Res. Cell Motil.* 1, 89-100.

Godt, R. E. and Lindley, B. D. (1982) Influence of temperature upon contractile activation and isometric force production in mechanically skinned muscle fibers of the frog. *J. Gen. Physiol.* 80, 279-297.

Grynkiewicz, G., Poenie, M. and Tsien, R. (1985) A new generation of Ca^{2+} indicators with greatly improved fluorescence properties. *J. Biol. Chem.* 260, 3440-3450.

Harkins, A.B., Kurebayashi, N. and Baylor, S.M. (1993) Resting myoplasmic free calcium in frog skeletal muscle fibers estimated with fluo-3. *Biophys. J.* 65, 865-881.

Hirota, A., Chandler, W.K., Southwick, P.L. and Waggoner, A.S. (1989) Calcium signals recorded from two new purpurate indicators inside frog cut twitch fibers. *J. Gen. Physiol.* 94, 597-631.

Hollingworth, S., Aldrich, R.W. and Baylor S.M. (1987) In vitro calibration of the equilibrium reactions of the metallochromic indicator dye antipyrylazo III with calcium. *Biophys. J.* 51, 383-393.

Hollingworth, S., Soeller C., Baylor, S. M. and Cannell, M. B. (2000) Sarcomeric Ca^{2+} gradients during activation of frog skeletal muscle fibers imaged with confocal and two-photon microscopy. *J. Physiol. (Lond.)* 526, 551-560.

Hollingworth, S., Zhao, M. and Baylor S.M. (1996) The amplitude and time course of the myoplasmic free

[Ca^{2+}] transient in fast-twitch fibers of mouse muscle. *J. Gen. Physiol.* 108, 455-469.

Jobsis, F.F. and O'Connor, M.J. (1966) Calcium release and reabsorption in the sartorius muscle of the toad. *Biochem. Biophys. Res. Commun.* 25, 246-252.

Konishi, M. (1998) Cytoplasmic free concentrations of Ca2+ and Mg2+ in skeletal muscle fibers at rest and during contraction. *Jpn. J. Physiol.* 48, 421-438.

Konishi, M. and Baylor, S.M. (1991) Myoplasmic calcium transients monitored with purpurate indicator dyes injected into intact frog skeletal muscle fibers. *J. Gen. Physiol.* 97, 245-270.

Konishi, M., Hollingworth, S., Harkins, A.B. and Baylor, S.M. (1991) Myoplasmic calcium transients in intact frog skeletal muscle fibers monitored with the fluorescent indicator furaptra. *J. Gen. Physiol.* 97, 271-301.

Konishi, M. and Kurihara, S. (1987) Effects of caffeine on intracellular calcium concentration in frog skeletal muscle fibres. *J. Physiol. (Lond.)* 383, 269-283.

Konishi, M., Olson, A., Hollingworth, S. and Baylor, S.M. (1988) Myoplasmic binding of fura-2 investigated by steady-state fluorescence and absorbance measurements. *Biophys. J.* 54, 1089-1104.

Konishi, M., Suda, N. and Kurihara, S. (1993) Fluorescence signals from the Mg^{2+}/Ca^{2+} indicator furaptra in frog skeletal muscle fibers. *Biophys. J.* 64, 223-239.

Konishi, M. and Watanabe, M. (1995) Resting cytoplasmic free Ca^{2+} concentration in frog skeletal muscle measured with fura-2 conjugated to high molecular weight dextran. *J. Gen. Physiol.* 106, 1125-1150.

Konishi, M. and Watanabe, M. (1998) Steady state relation between cytoplasmic free Ca^{2+} concentration and force in intact frog skeletal muscle fibers. *J. Gen. Physiol.* 111, 505-519.

Kurebayashi, N., Harkins, A.B. and Baylor, S.M. (1993) Use of fura red as an intracellular indicator in frog skeletal muscle fibers. *Biophys. J.* 64, 1934-1960.

Maylie, J., Irving, M., Sizto, N.L. and Chandler, W.K. (1987a) Calcium signals recorded from cut frog twitch fibers containing antipyrylazo III. *J. Gen. Physiol.* 89, 83-143.

Maylie, J., Irving, M., Sizto, N.L. and Chandler, W.K. (1987b) Calcium signals recorded from cut frog twitch fibers containing tetramethylmurexide. *J. Gen. Physiol.* 89, 145-176.

Miledi, R., Parker, I. and Schalow, G. (1977) Measurement of calcium transients in frog muscle by the use of arsenazo III. *Proc. R. Soc. Lond.* B198, 201-210.

Minta, A., Kao, J.P.Y. and Tsien, R.Y. (1989) Fluorescent indicators for cytosolic calcium based on rhodamine and fluorescein chromophores. *J. Biol. Chem.* 264, 8171-8178.

Morgan, J.P. and Morgan, K.G. (1982) Vascular smooth muscle: the first recorded Ca^{2+} transients. *Pflugers Arch.* 395,75-77.

Natori, R. (1954) The role of myofibrils, sarcoplasma and sarcolemma in muscle-contraction. *Jikeikai Med. J.* 1, 18-28.

Palade, P. and Vergara, J. (1983) Stoichiometries of arsenazo III-Ca complexes. *Biophys. J.* 43, 355-369.

Raju, B., Murphy, E., Levy, L.A., Hall, R.D. and London, R.E. (1989) A fluorescent indicator for measuring cytosolic free magnesium . *Am. J. Physiol.* 256, C540-C548.

Rios, E. and Schneider, M.F. (1981) Stoichiometry of the reactions of calcium with the metallochromic indicator dyes antipyrylazo III and arsenazo III. *Biophys. J.* 36, 607-621.

Rüdel, R and Taylor, S. R. (1978) Aequorin luminescence during contraction of amphibian skeletal muscle. *J. Physiol. (Lond.)* 233, 5-6P.

Shimomura, O., Johnson, F.F. and Saiga, Y. (1962) Extraction, purification, and properties of aequorin, a bioluminescent protein from the luminous hydromedusan, Aequorea. *J. Cell. Comp. Physiol.* 59, 223-239.

Southwick, P.L. and Waggoner, A.S. (1989) Synthesis of purpurate-1,1'-diacetic acid PDAA tripotassium salt: a new calcium indicator for biological applications. *Org. Prep. Proc. Int.* 21, 493-500.

Suda, N. and Kurihara, S (1991) Intracellular calcium signals measured with fura-2 and aequorin in frog skeletal muscle fibers. *Jpn. J. Physiol.* 41, 277-295.

Tsien, R.Y. and Rink, T.J. (1982) Neutral carrier ion-selective microelectrodes for measurement of intracellular free calcium. *Biochim. Biophys. Acta* 599, 623-638, 1980.

Weingart, R. and Hess, P. (1984) Free calcium in sheep cardiac tissue and frog skeletal muscle measured with Ca^{2+}-selective microelectrodes. *Pflügers Arch.* 402, 1-9.

DISCUSSION

Gonzalez-Serratos: 1) We have reported that, during activation of isolated frog muscle fibers, Ca^{2+} is released as spots that appear and disappear continually during all the period of titanic activation. 2) You showed that, with 30 mM [K^+], the force is nearly maximum. However, it has been shown by several researchers including me, that 30 mM [K^+] produce only around 1 /3 of the maximum force. Do you think that the calcium probe you are using may change contractility?

Konishi: I have seen your abstract in the Biophysical Society, although I did not know if it was published as a full paper. Probably I should have referred to your work. However, what I liked to emphasize in this talk was non-uniformity within a sarcomere. Non-uniformity within a fiber that you pointed out is another important factor to modify fiber activity. We observed some fiber-to-fiber variation of the [K^+]-force relation, but you may be right that we have in general slightly more force generation at 30mM [K^+] than that reported in the literature. We do not think that the relation was modified in the presence of fura-dextran, because we obtained very similar results in the fibers without indicator infection or the fibers injected with aequorin. I don't know what makes the difference, and I can only say that this is what we observed under our experimental conditions. We prepared high K^+ solutions of constant [K^+]x[Cl^-] product by equimolar substitution of NaCl with K-methanesulfonate.

Maughan: How large was the dextran probe?

Konishi: 10 K mol wt.

Maughan: A comment: you mentioned calcium spatial heterogenity reflected by Ca^{2+} sparks. Another source of calcium spatial heterogenity is the gradient set up by the presence of highly negatively charged myofilaments, where the Ca^{2+} concentration is much greater in the vicinity of the thin filaments and troponin C, and when the calcium probe Fula-2, a highly charged anion is excluded to some extent from the vicinity of the troponin. Have you done calculations to assume the degree of what your in vivo force P-Ca curves are modified to reflect external conditions at the troponins?

Konishi: Simple calculation of the Donnan equilibrium predicts that the intramyofibrillar Ca^{2+} concentration would be somewhat higher than the extramyofibrillar concentration. However, the myofilament charges might not fully influence the Ca^{2+} indicator molecules between the myofilament separated by about 40nm, which is a distance much greater than the Debye length of about 1nm. This is a very difficult situation to model.

Hexley: What is the dissociation of troponin for Ca^{2+}? Does binding constant of troponin (in these works for calcium agree with free Ca^{2+} level needed for activation?

Konishi: A dissociation constant of troponin is generally assumed to be about 2 μM, which is consistent with the findings that Ca^{2+} transients are large in amplitude. However, attachment of myosin head significantly lowers the troponin dissociation constant, so the situation is not simple and not totally clear.

MYSTERIOUS BEAUTY OF BEATING HEART: Cardiac mechano-energetico-informatics

Hiroyuki Suga

1. INTRODUCTION

Since 1966, I have been characterizing the contractile and pump function of the heart as a beautifully integrated organ. In those days, three major frameworks of cardiac function existed. They are: i) Frank's pressure-volume (P-V) or work diagram[1] found at the ventricle level, ii) Starling's cardiac output curve[2] (law of the heart[2]) and Sarnoff's ventricular function curve[3] found at the heart level, and iii) Sonnenblick's force-velocity relation[4] and unloaded maximal shortening velocity[4] (Vmax) found at the myocardium level in the historical order.

Of these, the cardiac output and ventricular function curves were recognized to be the most reliable measures of the cardiac pump function, and Vmax was recognized to be the most reliable index of the myocardial contractility in those days. Although there were several other indexes of cardiac function such as EF (ejection fraction), maximum ±dP/dt (time derivative of pressure rise and fall), Vcf (circumferential shortening velocity), etc, these practical indexes were not exactly related to the specific conceptual frameworks. The 1957 Huxley sliding filament theory[5] had not yet explicitly been related to myocardial contraction.

When I learned all these concepts and indexes in those days, none of them appeared ideal for characterization of cardiac contractile and pump functions. I hoped that there must be a new measure or framework to characterize ventricular contractility and cardiac performance of a beating heart in a better manner on a sound physiological and engineering basis. Then, I started my canine heart research in Tokyo.[6]

National Cardiovascular Center Research Institute, 5-7-1 Fijishirodai, Suita, Osaka, 565-8565, JAPAN
E-mail: hsuga@ri.ncvc.go.jp, Fax +81-6-6833-1421, Tel +81-6-6833-5211

2. MECHANICS: TIME-VARYING ELASTANCE AND Emax

I became interested in the instantaneous changes in the left ventricular pressure curves during ejection under various degrees of an intra-beat transient aortic occlusion manually performed in anesthetized, open-chest dogs.[6] With the transient aortic occlusion and its immediate release, the aortic flow measured with an electromagnetic flowmeter dipped and then returned, respectively, and simultaneously ventricular systolic pressure suddenly humped and returned. None of Frank's, Starling's, and Sarnoff's frameworks[1-3] could quantitatively account for these results. Even Sonnenblick's framework[4] could not directly handle ventricular responses under such anisotonic, complex afterload changes. Although Brutsaert's modification[7] of Sonnenblick's framework,[4] though not yet proposed in those days, could deal with such physiological afterloading conditions, the ventricular pressure and aortic flow changes on and off the aortic clamping seem too complicated to be fully accounted for even by this modification.[7, 8]

One day, I experienced my first serendipity that a time-varying elastance model could reasonably account for my observation.[6] Although some circulatory modelers had intuitively used similar models,[9] they had no firm physiological evidence explicitly supportive of their models. To validate the time-varying elastance model physiologically, I designed new canine experiments to measure left ventricular instantaneous pressure (P) and volume (V) under various pre- and afterloads and chronotropic and inotropic backgrounds.[8]

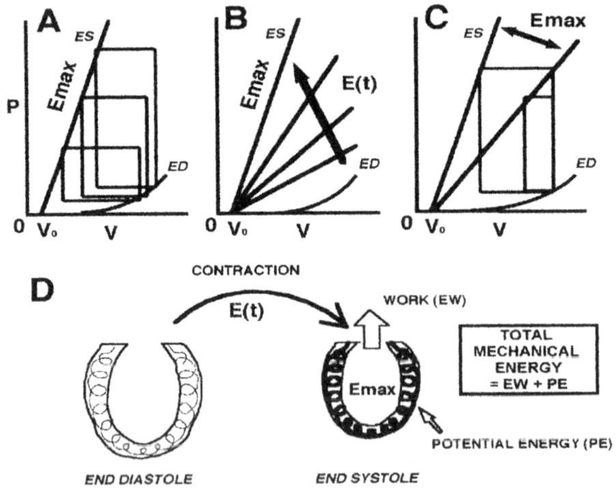

Figure 1. Ventricular pressure (P)-volume (V) loops and end-systolic P-V relation whose slope is Emax (Panel A), time-varying elastance E(t) (Panel B), and contractility index (Emax) (Panel C). A ventricular contraction produces external work (EW) and potential energy (PE) in the time-varying elastance (E(t)) model of the ventricle (Panel D). The sum of EW and PE equals the total mechanical energy generated by a ventricular contraction. V_0 = unstressed ventricular volume. ES: end systole. ED: end diastole.

After analyzing these P-V data, I concluded that a contracting and relaxing left ventricle has a time-varying elastance E(t) that increases from the end-diastolic minimum to the end-systolic maximum load-independently[6] (Fig. 1A and B). The maximum elastance (Emax) corresponds to the slope of the line connecting a dead volume (V_0) on the volume axis, at and below which the peak isovolumic pressure was zero and negative, respectively. I also found that Emax increased and decreased sensitively with positive and negative, respectively, inotropic interventions[6] (Fig. 1C). Then, I proposed it as a novel index of ventricular contractility.[6] This work[6] became my PhD thesis in 1970.

When I joined Prof. Kiichi Sagawa at the Johns Hopkins University in 1971, he encouraged me to continue my Emax study further toward its establishment. Our Emax studies in 1971-1973 were published in Circulation Research,[10, 11] later summarized in our 1988 Oxford University book[12] combined with my further research output in Japan, and crystallized in my solicited reviews and academic honors[13-18] after I combined the Emax concept with my novel cardiac energetics concept PVA (systolic P-V area, see below) as the deductive extrapolation of Emax. Fortunately, our Emax concept has been attracting cardiac physiologists and cardiologists and cited recently as a core concept on cardiac pump function because no better concept has been proposed to exceed Emax.[19, 20]

3. ENERGETICS: TOTAL MECHANICAL ENERGY AND PVA

While continuing studies in mid 1970s to expand the generality of the Emax concept,[10, 11] I searched for a novel concept theoretically deducible from the time-varying elastance E(t) of the ventricle. Then, I experienced my second serendipity. Namely, a time-varying elastance can provide a new measure of the total mechanical energy of cardiac contraction[12-18] (Fig. 1D). This energy consists of the mechanical potential energy generated within the ventricular wall (PE) and the external mechanical work (EW) performed on the stroke volume ejected against the arterial afterload. This is an analogy to the mechanical potential energy stored in a stretched spring and the mechanical work performed by its shortening

The total mechanical energy thus defined corresponds to the systolic P-V area (PVA) in the ventricular P-V diagram[12-18] (Fig. 2A). The PVA is the sum of the two areas, one for the external work (EW) within the P-V loop and the other for the elastic potential energy (PE) under the Emax line on the origin side of the P-V loop.

When I proposed this novel cardiac energetic concept, I was criticized in that the Emax and PVA concepts should be invalid because of their similarity to the classic cocked spring model of the skeletal muscle and hence the PVA concept should contradict the Fenn effect.[21] However, there are essential differences between the cardiac and skeletal muscle Fenn effects as well as between the cardiac time-varying elastance and skeletal muscle cocked spring models.[14, 22] The skeletal muscle Fenn effect is specifically different from the cardiac one in that the former has a considerably greater energy consumption for a shortening contraction than the latter.[14, 22] The key difference in the model exists in the increasing rate of the elastance, i.e., gradual in cardiac muscle versus instant in skeletal muscle. Furthermore, the following evidence has eventually conquered the criticism to the PVA.

We discovered in canine heart experiments that PVA surprisingly well correlated with Vo_2 in a linear manner (Fig. 2B) and that increasing and decreasing Emax sensitively shifted up and down the PVA-Vo_2 relation without changing its slope (Fig. 2C).[14, 23] The linearity of the PVA-Vo_2 relation does not require the linearity of the end-systolic P-V relation (ESPVR) since the former is linear even when the latter is curved in puppy and rat left ventricles.[24, 25]

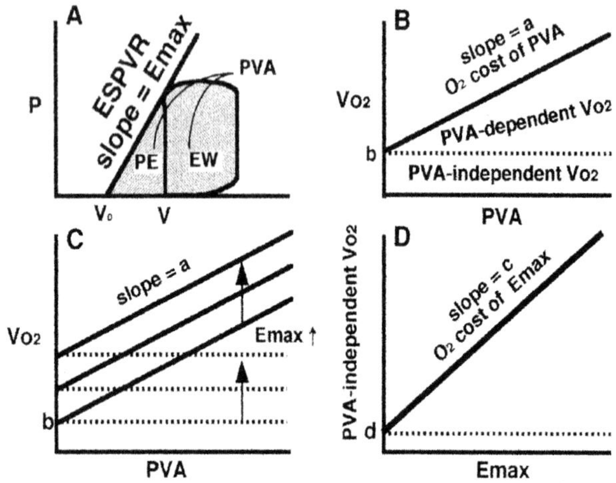

Figure 2. Cardiac mechano-energetic framework based on contractility index Emax and total mechanical energy PVA = external work (EW) + potential energy (PE) (Panel A). Linear relationship between cardiac O_2 consumption (Vo_2) and PVA (Panel B). Its parallel elevation with increasing Emax (Panel C) and a linear relation between PVA-independent Vo_2 and Emax (Panel D). Vo_2 consists of PVA-dependent Vo_2 primarily for cross-bridge cycling and PVA-independent Vo_2 (b) primarily for Ca^{2+} handling plus basal metabolism (d). Slope (a) in Panel B represents O_2 cost of PVA. Slope (c) in Panel D represents O_2 cost of Emax.

The PVA has overcome the serious problems of the conventional determinants of Vo_2, such as EW, total (= external + internal) mechanical work, contractile element work, pressure-rate (or double) product, tension-time integral, Vmax, (dP/dt)max, etc.[14] None of these conventional ones could consistently predict Vo_2 over the entire physiologically working range under a variety of pre- and afterloading conditions.[14] We formulated the observed mechanoenergetic relation as Vo_2 = a PVA + b = aPVA + cEmax + basal metabolism. Here, a = O_2 cost of PVA or O_2 cost of contractile energy apparently generated by cross-bridge cycling (Fig. 2B), b = cEmax + basal metabolism, and c = O_2 cost of Emax or O_2 cost of contractility apparently achieved by calcium handling (Fig. 2D).[14-18]

We have confirmed that the above PVA-Vo_2 relation and the resultant O_2 cost of PVA are virtually load-independent in that Vo_2 remains unchanged even when the left ventricular end-diastolic volume, stroke volume, ejection pressure, and end-systolic pressure are changed as long as PVA remains the same (Fig. 3).[14] Moreover, Vo_2 remains the same whether PE is either remaining unchanged isovolumically, converted to EW during the relaxation, or decreased by a negative EW within the same PVA (Fig. 3).[14] No other so-called Vo_2 determinant can predict Vo_2 reliably to the same extent as PVA.[14]

The O_2 cost of PVA can be expressed dimensionless since both Vo_2 and PVA can have the same dimensions of energy. Namely, 1 ml O_2 of Vo_2 is equivalent approximately to 20 J under aerobic metabolism, and 1 mmHg ml of PVA is convertible to 1.333×10^{-4} J.[14] We empirically found the dimensionless value for the O_2 cost of PVA to be 2.5-3.0 and its reciprocal to be 0.3-0.4 (also dimensionless).[14] This reciprocal value indicates that the contractile efficiency of cross-bridge cycling to generate mechanical energy from Vo_2 is 30~40% (~35% on average).[14] Since the phosphorylation efficiency to reproduce ATP from ADP is ~60% in myocardium,[26] the chemomechanical efficiency of cross-bridge cycling is calculated to be ~60% (~0.35/~0.6) which is surprisingly high as a mechanical pump.[14, 26]

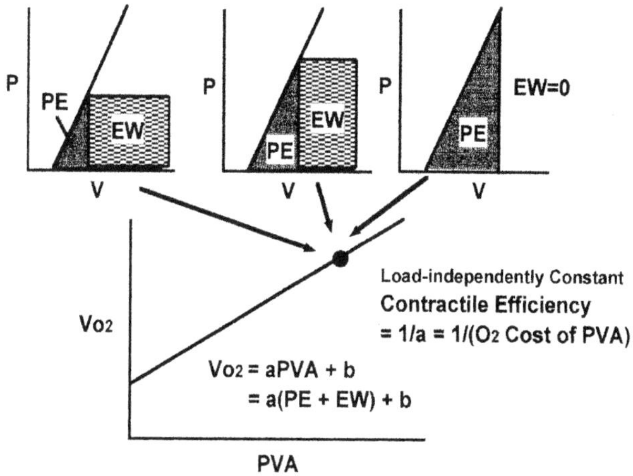

Figure 3. Independence of the PVA-Vo_2 relation (bottom) from the mode of contraction, pre- and afterloading conditions, and the relative magnitudes of external work (EW) and potential energy (PE) (top). a = O_2 cost of PVA. 1/a = contractile efficiency, which was found to be ~35%, independent of loading conditions and relatively constant among hearts. b = PVA-independent Vo_2.

We have studied how pathophysiological conditions of the heart would affect the O_2 cost of PVA. Most positive and negative inotropic agents such as Ca^{2+}, Ca^{2+} antagonists, Ca^{2+} sensitizers, catecholamines, β blockers, phosphodiesterase inhibitors, etc did not affect the cost of PVA and the contractile efficiency in canine hearts.[12-18] Therefore, our results surprisingly suggest that the overall efficiency of cross-bridges as a whole remains high and constant independent of ventricular loading, heart rate, and inotropic conditions. This is not expectable from the skeletal muscle Fenn effect.[22] We will return to this point later because this aspect is related to the mystery of a beating heart in this review.

The O_2 cost of Emax has the unit of energy per elastance.[14] This unit is not directly related to efficiency. This unit rather means the reciprocal of the economy of contractility. We have found that acute positive and negative inotropic agents with different pharmacological mechanisms do not affect the O_2 cost of Emax in normal canine hearts (Fig. 2D).[14] However, we have found that this cost increases in failing hearts under such conditions as stunning, ryanodine treatment, and acidosis.[12-18] Other investigators have reported that some Ca^{2+} sensitizers advantageously decrease the increased cost of failing hearts toward normal. Hypothermia[14] and alkalosis[27] also decrease the O_2 cost of Emax, whereas hyperthermia increases the cost.[28] We will come back to this aspect later because this is related to the total calcium handling for the excitation-contraction coupling.

4. INFORMATICS 1: CROSSBRIDGES

In relation to the Emax, I had a lucky chance in 1975 with Dr. Ichiro Matsubara to study the cross-bridge mass behavior in the right ventricular papillary muscle in either saline- or blood-perfused canine excised heart preparation by a narrow-angle equatorial X-ray diffraction analysis.[29, 30] We found that the intensity ratio of the 1, 0 and the 1, 1 reflections decreased during contraction and the calculated electron density of the thick (myosin) filaments decreased and that of the thin (actin) filaments increased during contraction and all these systolic changes returned to the diastolic levels during relaxation.[29, 30]

Recently, Dr. Naoto Yagi (Dr. Matsubara's ex-fellow), Dr. Hiroshi Okuyama, and I have restarted similar X-ray diffraction studies on both rat beating papillary muscles and hearts at the Japan Synchrotron Radiation Research Institute (JASRI) SPring 8 with several coworkers.[31, 32] Thanks to the much stronger X-ray beam, these recent results not only support our previous results on the (1, 0)/(1, 1) intensity ratio,[29, 30] but also its much finer time resolution has newly indicated that the ratio change advances the force development during the contraction phase and lags behind the force reduction during the relaxation phase.[31] This means that the above ratio may not proportionally reflect the number of the force generating cross-bridges in one cardiac cycle. The relation of this ratio with the myocardial or ventricular time-varying elastance E(t) also remains unknown. The relation between the (1, 0)/(1, 1) intensity ratio and the cross-bridge cycling remains unknown, either.

Apart from the X-ray diffraction study, we have calculated the number of cross-bridge cycling from the PVA-Vo_2 relation, assuming the above-mentioned high and load-independently constant ~60% contractile efficiency of cross-bridge cycling from ATP.[14] Based on this efficiency and the free energy of ATP of 57 kJ/mol[26] described above, we calculated that a representatively maximal PVA of 4000 mmHg ml/beat per 100 g myocardium, equivalent to 0.53 J, could be generated by consuming ~15 μmol of ATP in one cardiac cycle.[33] This amount of ATP is comparable to the concentration of cross-bridges (two per myosin) known to be ~15 μmol/100 g myocardium.[33] We could then speculate that, if each of the ~15 μmol cross-bridges per 100 g myocardium cycles at least once or a little more (e.g., ~1-1.2 times) during each heart beat, all the cross-bridges could develop such a maximal PVA in one beat.[33] Thus, the PVA-Vo_2 relation integratively comprehends the cross-bridge mass behavior in a beating heart.[33]

This ~1-1.2 cycle per attached cross-bridge per ATP must hold theoretically at any PVA because the PVA-dependent Vo_2 and hence ATP consumption must be proportional to PVA under the constant O_2 cost of PVA (Fig. 3). Moreover, the same must hold regardless of the shape of the PVA, the ratio of EW and PE, or ejection fraction under the load-independently constant O_2 cost of PVA (Fig. 3). Furthermore, the constant O_2 cost of PVA (a) and hence contractile efficiency (1/a) at any Emax mean that the ~1-1.2 cycle per attached cross-bridge per ATP must hold regardless of ventricular inotropic backgrounds.

We have other pieces of evidence supportive of this ~1-1.2 cycle per attached cross-bridge in one cardiac cycle. The first one is the result of our hybrid logistic curve fitting to both left ventricular isovolumic pressure and myocardial isometric force curves (Fig. 4).[34-36] The hybrid logistic curve that we proposed always fitted to the observed curves of both ventricular isovolumic pressure and myocardial isometric force more precisely than any other conventional curves.[34-36] Moreover, we could theoretically consider the increasing and decreasing components of the logistic curves to represent the increasing numbers of the attached and detached, respectively, cross-bridges and their difference to represent the first increasing and then decreasing number of the force-generating cross-bridges during the contraction and relaxation periods.[34]

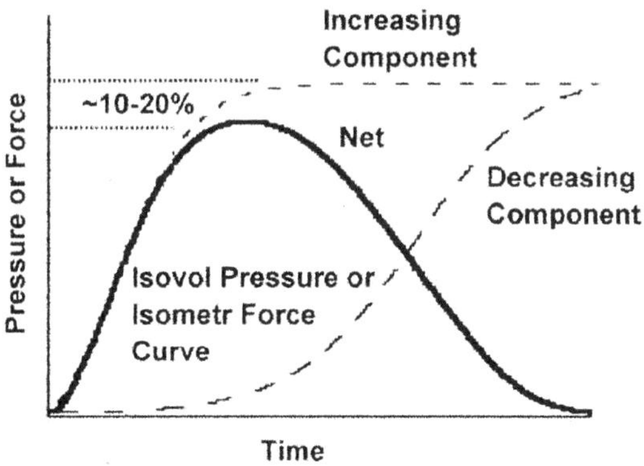

Figure 4. Schematic drawing of a hybrid logistic curve fitted to either ventricular isovolumic (isovol) pressure or myocardial isometric (isometr) force curve. Thick curve is a representative hybrid logistic curve as the net of the advanced logistic curve (dashed, increasing component) minus the lagged logistic curve (dashed, decreasing component). Fitting correlation coefficients were always almost unity (> 0.999).

The fitting results have shown that, at the peak of either isovolumic or isometric contraction, the logistic curve for the attached cross-bridges (increasing component) only slightly (~10-20%) exceeds the peak pressure or force, respectively, curve and has almost reached the plateau level (Fig. 4).[34-36] Moreover, the logistic curve for the detached cross-bridges (decreasing component) slowly started to rise before the peak of contraction, rose rapidly thereafter, and reached the plateau at the end of relaxation. These results suggest that the cross-bridge attachment-detachment cycle is slightly more than once per attached cross-bridge in one beat.[34] This is compatible with the cardiac energetics that Vo_2 is saved only by ~10-20% when ventricular contraction is quickly aborted by volume unloading at the peak of contraction.[37-39]

Similar fitting results were also obtained by the hybrid Weibull function curve,[40] where the Weibull function characterizes the cumulative loss of performance (e.g., cross-bridge attachment and detachment) of a complex system in reliability engineering.[40] The goodness of the Weibull fitting also supports that every once attached cross-bridge cycles ~1-1.2 times on average within a beat.[40] If every cross-bridge cycles more frequently (e.g., 2~3 times) within a beat, the increasing logistic or Weibull curve would exceed the peak pressure and force curves by much more (e.g., 2~3 times) than ~1-1.2 times, and the decreasing curve would rise rapidly before the peak of contraction.

More recently, Prof. Seiryo Sugiura has revealed together with Prof. Haruo Sugi that a myocardial cross-bridge either develops a unitary force of ~1.5 pN with no sliding or a unitary sliding step of ~10 nm (twice actin molecule size) with no loading per ATP at its low concentration in an in-vitro motility assay.[41, 42] These unitary force and step size of a myocardial cross-bridge are comparable to the smallest value for unitary force of ~1.5 (average)-6 pN (peak) and twice the smallest value for unitary step size or sliding distance of ~5 (actin molecule size)-30 nm, respectively, per ATP of a skeletal muscle cross-bridge.[43, 44] They showed the possibility that the ~10-30 nm sliding consists of multiple stepping per ATP.[43, 44] Nevertheless, the myocardial cross-bridge behavior at the molecular level in an in-vitro motility assay suggests that its unitary cycle with ~1.5 pN force and ~10 nm step per ATP could generate a mechanical energy of ~1.5 x 10^{-20} J at most. Its half seems more likely to be the maximum energy since the developed force would decrease with sliding as in the 1957 Huxley model.[5]

From the above mechanical energy of a myocardial cross-bridge combined with the nominal free energy of ~10 x 10^{-20} J per ATP, the efficiency from one ATP to the unitary cross-bridge cycle would be ~7.5-15%. This efficiency is only ~1/8-1/4 of the ~60% ATP to PVA efficiency at the heart level. This calculation suggests that each cross-bridge should repeat the unitary force and step cycle multiple times, e.g., ~4-8 times, per ATP to reach the high constant contractile efficiency of ~60% in each cardiac contraction.

To solve these ~4-8 times higher efficiency at the organ level than at the in-vitro molecular level, it seems quite reasonable to speculate that each attached cross-bridge cycles ~4-8 times per ATP, but not merely once, to convert the chemical energy of ATP to the total mechanical energy at the high overall efficiency of ~60%.

We further speculated the numbers of unitary steps of cross-bridge sliding per ATP under different loading conditions. First, in an isometric contraction, myocardial sarcomeres usually shorten by stretching the series elasticity that is several times more compliant than that of skeletal muscles.[22] Moreover, myocardium normally shortens more or less in a somewhat asynchronously contracting ventricular wall, even without ejection in an isovolumic contraction. These mechanical properties of the heart and myocardium would require each attached cross-bridge to make multiple steps per ATP even when a myocardium contracts isometrically and the ventricle contracts isovolumically.

More quantitatively, when the series elasticity is stretched by ~7% of the muscle length at the peak of ventricular contraction, it allows a half sarcomere of ~1 μm length to shorten at least by ~70 nm, which is 7 times the unitary cross-bridge step of ~10 nm. This suggests that each attached cross-bridge would slide over ~7 unitary steps on average per ATP to develop a high pressure during an isovolumic contraction (Fig. 5). If each attached cross-bridge hydrolyzed ATP in each unitary step, ~7 times more ATP would be consumed and the ~60% contractile efficiency could not be maintained. The contractile efficiency would then be ~8%, unrealistically far smaller than the experimentally obtained ~60%.[14]

Next, during ejection with e.g., a typical normal ejection fraction of ~70%, myocardium and hence a half sarcomere of ~1 μm length would shorten by ~15% on average, though varying among different layers of the ventricular wall. This shortening corresponds to ~150 nm and hence ~15 times the unitary cross-bridge step (Fig. 5). If each attached cross-bridge hydrolyzed ATP in each unitary step, ~15 times more ATP would be consumed and the ~60% contractile efficiency could not be maintained here, either.

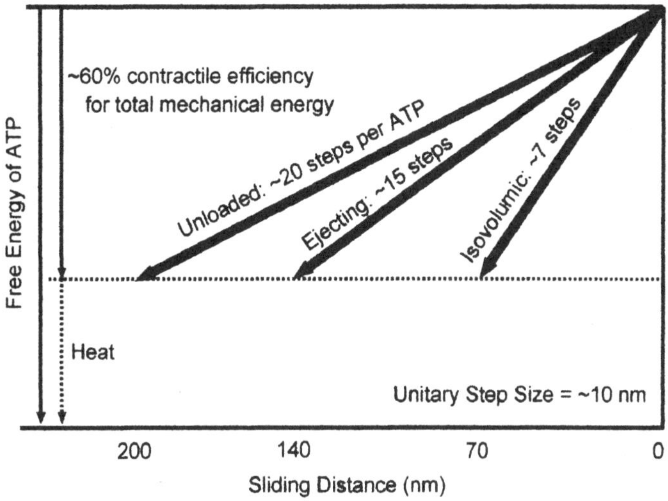

Figure 5. Variable sliding distances and hence variable numbers (~7-20) of unitary (~10 nm) steps of an attached cross-bridge per ATP on average to achieve a high constant contractile efficiency of ~60% from ATP in a beating ventricle under three different afterload conditions (from right to left: isovolumic, moderately afterloaded ejecting, and almost fully unloaded) from a given high preload.

Last, as an extremely unloaded case, the ventricle can eject its entire volume against a small afterload with almost no extra Vo_2 for a small PVA.[45] In this case, we would speculate that the half sarcomere shortens from ~1 μm to ~0.8 μm by ~200 nm corresponding to ~20 unitary steps per ATP (Fig. 5). If each attached cross-bridge hydrolyzed ATP in each unitary step, ~20 times more ATP would be consumed and the ~60% contractile efficiency could not be maintained here, either.

When the force generated by each sliding cross-bridge decreases with its increasing sliding distance or shortening velocity as in the 1957 Huxley model,[5] the overall ~60% efficiency also suggests that the number of unitary steps of an attached cross-bridge per ATP should increase reciprocally with the decreasing unitary force despite the constant ~10 nm unitary step of a cross-bridge (Figs. 5 and 6). This should hold because even the ejecting contraction requires Vo_2 proportional to PVA without requiring extra Vo_2 for ejection and muscle shortening[8, 13] in contrast to skeletal muscles.[4] This is the essential difference of the cardiac Fenn effect from the skeletal one.[21, 22] This is also the most mysterious aspect of the cross-bridge mechanoenergetics in that the constant ~10 nm unitary step of an attached cross-bridge per ATP observed at the molecular level does not seem to limit the cross-bridge behavior at the integrated myocardium and organ levels.

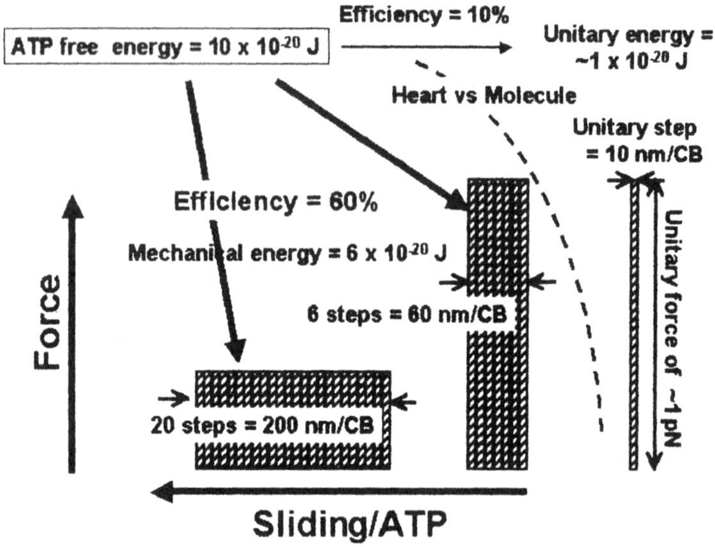

Figure 6. Reciprocal changes in force and number of multiple unitary steps of an attached cross-bridge per ATP on average to achieve a high constant contractile efficiency of ~60% from ATP or ~35% from O_2 consumption under different afterload conditions in a beating ventricle. Two shaded areas on the left represent the same total mechanical energy generated by an attached cross-bridge with variable numbers of unitary steps under two different sets of force and shortening. The right-most shaded area represents the unitary force and step per ATP in an in-vitro motility assay.

These ~4 to ~20 multiple unitary steps of each cross-bridge per ATP[46] (Figs. 5 and 6) may appear contradictory to the above described calculation that each attached cross-bridge hydrolyzes ATP ~1-1.2 times during an either isovolumic or ejecting contraction of the heart.[14, 33] This ~1-1.2 ATP per cross-bridge was, as already mentioned above, calculated from a reasonably maximum PVA of 4000 mmHg ml = ~0.53 J per 100 g left ventriclc (LV) generated by all the 15 μmol cross-bridges per 100 g LV.[14, 33] This mechanical energy corresponds to ~6 x 10^{-20} J generated by each cross-bridge. Since each ATP has ~10 x 10^{-20} J as mentioned above, the 60% efficiency of the above mechanical energy per ATP leads to ~6 x 10^{-20} J, which is comparable to the above-mentioned total mechanical energy of ~6 x 10^{-20} J per cross-bridge. This coincidence implies that each once attached cross-bridge hydrolyzes ATP 1~1.2 times regardless of the number of its multiple unitary steps during any cardiac contraction under a variety of loading conditions.[46]

We have previously shown that the 1957 Huxley model is compatible with the PVA-Vo_2 relationship and hence the cardiac Fenn effect if the number of cross-bridge unitary steps per ATP is allowed to increase under lower afterloaded contractions.[47] However, the energetically reasonable numbers of multiple steps such as above had not been estimated. Therefore, more direct evidence remains to be obtained for the load-dependent multiple unitary steps of cross-bridges in the heart.

Multiple numbers of unitary steps of a skeletal muscle cross-bridge per ATP has recently been shown to exist not only under unloaded sliding but also under finite loading conditions in in-vitro motility assays.[43, 44] However, it remains to be elucidated whether such a result of motility assay at the molecular level matches the skeletal muscle Fenn effect at the integrative level.

5. INFORMATICS 2: CALCIUM

In cardiac muscle, the number of attached cross-bridges and hence the magnitude of PVA both depend on the amount of calcium recruited in the excitation-contraction coupling. Although a calcium chelator, such as aequorin and fura-2, can measure the concentration of sarcoplasmic free calcium, the total amount of calcium recruited in the excitation-contraction coupling is still impossible to measure directly in a beating heart. The free calcium is like an peak of a floating iceberg in the sea relative to the total amount of the recruited calcium.

However, not only the PVA generation by cross-bridges but also the calcium handling for the excitation-contraction coupling demand a considerable fraction of total Vo_2 and are altered by the myocardial inotropism. Moreover, the O_2 cost of PVA remains reasonably constant but the O_2 cost of Emax varies with cardiac conditions.[14] The constancy of the former has been accounted for by the variable multiple unitary steps of a cross-bridge as reviewed above.

However, the variable O_2 cost of Emax means the variability of the economy of contractility. To better understand the variable O_2 cost of Emax, we have recently developed an integrative analysis method to assess the total amount of Ca^{2+} handled (i.e., released and removed) for contraction.[16-18, 48-50] We have developed the following equation to calculate the total Ca^{2+} handling per beat in μmol/kg wet myocardium from the total Ca^{2+} handling Vo_2 (ml O_2/100 g per beat) and the internal Ca^{2+} recirculation fraction (RF, dimensionless).[16-18, 48-50]

$$\text{Total } Ca^{2+} \text{ handling} = 6 \times 10^7 \, [(\text{total } Ca^{2+} \text{ handling } Vo_2)/22400] \, /[RF/2 + (1 - RF) + N\,RF/2]$$

Here, 6 stands for the $6P{:}O_2$ molecular ratio, i.e., two times the nominal 3P:O atomic ratio, in the oxidative phosphorylation; 10^7 is the conversion factor of Vo_2 from μmol/kg to mol/100 g; 24000 is the gas constant (ml/mol); denominator 2 dividing RF stands for the nominal $2Ca^{2+}$:ATP stoichiometry of the sarcoplasmic Ca^{2+} pump relative to the nominal $1Ca^{2+}$:ATP stoichiometry of the sarcolemmal Na^+/Ca^{2+} exchange coupled with the sarcolemmal Na^+/K^+ pump. (1 - RF) is the extrusion fraction of total Ca^{2+} removed transsarcolemmally with the $1Ca^{2+}$:ATP stoichiometry.

In the above equation, the total Ca^{2+} handling Vo_2 is measured as Vo_2 for myocardial activation, namely, Vo_2 for zero PVA minus basal metabolism. RF is obtained from the decay speed of an experimentally produced postextrasystolic potentiation.[48-50]

By this method, we successfully assessed the total Ca^{2+} recruited for excitation-contraction coupling in the range of 40-110 μmol/kg.[48-50] For further details, please see these references.[16-18, 48-50] These variable amounts of calcium are recruited for the excitation-contraction coupling, depending on the inotropic backgrounds and pathological conditions.

6. DISCUSSION

I have briefly reviewed the cardiac mechano-energetico-informatics that I have developed over these nearly 40 years with my coworkers using the excised, cross-circulated canine heart preparation. The Emax and PVA concepts originated from my intuitive analogy of the heart to a time-varying elastance first in mechanics and then in energetics. This successful analogy contrasts with the failure of the same analogy in the skeletal muscle based on the Fenn effect[21] found in 1924. The Fenn effect is the symbolic mechanoenergetic difference between the heart and the skeletal muscle. The Emax and PVA concepts match the cardiac Fenn effect, but do not match the skeletal muscle Fenn effect.[14, 22]

The cardiac and skeletal muscles are destined to perform differently through the evolution. The heart is designed to beat and eject blood continuously but periodically without much inter-beat rest through the individual's life under autonomic control and respond to the circulatory demand under daily variable physical loads. In contrast, the skeletal muscle is designed to perform exercise and rest at the individual's will. This difference may be related to the different Fenn effects and further to the constant O_2 cost of PVA and hence the variably multiple unitary steps of a cross-bridge per ATP in the heart.

This mechanoenergetic characteristic of the heart is advantageous because the normal heart is ejecting against an intermediate afterload in every beat and its economy of contraction for ejection and myocardial shortening can be much higher than that of the skeletal one. Behind this advantageous mechanoenergetics of the heart, I would suspect that the multiple unitary steps of a cross-bridge per ATP are designed at the integrative level in the heart and optimized by evolution.

The heart is required to eject ~3 times different stroke volumes at ~3 times different heart rates by inotropic and chronotropic controls during the daily life ranging from rest to exercise. I would also suspect that the multiple unitary steps of a cross-bridge per ATP are designed in the heart to keep always the advantageous mechanoenergetics in terms of the high constant contractile efficiency throughout the variable daily life over ~70-80 years.

Taken together, the multiple unitary steps of a cross-bridge per ATP in the heart as estimated from cardiac mechanoenergetics combined with cardiac informatics is still considerably mysterious in the cardiac contractile mechanism.[46] This beautifully integrated but still mysterious mechanism should be elucidated by the intimate collaboration between the elemental and integrative, or molecular and organ physiology for better physiomic understanding of the heart.

Acknowledgments

I thank all my coworkers for enthusiastic collaborations with me at any time in my nearly 40-year research on cardiac function. I also appreciate all the past and present financial supports of my research, particularly the ongoing Cardiovascular Diseases Research Grant (14A-1) from the Ministry of Health, Labour and Welfare of Japan.

7. REFERENCES

1) K. Sagawa, R.K. Lie, and J. Schaefer. Translation of Otto Frank's paper "Die Grundform des Arteriellen Pulses" in Zeitschrift fur Biologie. 37: 483-526 (1899). *J. Mol. Cell Cardiol.* 22: 253-277 (1990).
2) S.W. Peterson and E.H. Starling. On the mechanical factors which determine the output of the ventricles. *J. Physiol.* 48: 357-379 (1914).
3) S.J. Sarnoff and E. Berglund. Ventricular function. I. Starling's law of the heart studied by means of simultaneous right and left ventricular function curves in the dog. *Circulation.* 9: 706-718 (1954).
4) E.H. Sonnenblick. Force-velocity relations in mammalian heart muscle. *Am. J. Physiol.* 202: 931-939 (1962).
5) A.F. Huxley. Muscle structure and theories of contraction. *J. Prog. Biophys. Biophys. Chem.* 7: 255-318 (1957).
6) H. Suga. Analysis of left ventricular pumping by its pressure-volume coefficient. (in Japanese with English abstract and legends). *Jpn. J. Med. Electr. Biol. Eng.* 7: 406-415 (1969).
7) D.L. Brutsaert and E.H. Sonnenblick. Cardiac muscle mechanics in the evaluation of myocardial contractility and pump function: problems, concepts, and directions. *Prog. Cardiovasc. Dis.* 16: 337-361 (1973).
8) E. Braunwald, J. Ross, and E.H. Sonnenblick. Mechanisms of Contraction of the Normal and Failing Heart. Little, Brown & Co., Boston, USA (1976).
9) J.E.W. Beneken. Electronic analog computer model of the human blood circulation. in: *Pulsatile Blood Flow.* edited by E.O. Attinger. (McGraw Hill, New York, 1964) pp. 423-432.
10) H. Suga, K. Sagawa, and A.A. Shoukas. Load independence of the instantaneous pressure-volume ratio of the canine left ventricle and effects of epinephrine and heart rate on the ratio. *Circ. Res.* 32: 314-322 (1973).
11) H. Suga H and K. Sagawa. Instantaneous pressure-volume relationships and their ratio in the excised, supported canine left ventricle. *Circ. Res.* 35: 117-126 (1974).
12) K. Sagawa, L. Maughan, H. Suga, and K. Sunagawa. *Cardiac Contraction and the Pressure-Volume Relationship.* (Oxford Univ Press, New York, 1988).
13) H. Suga. Cardiac mechanics and energetics - from Emax to PVA. *Frontiers Med. Biol. Engng.* 2: 3-22 (1990).
14) H. Suga. Ventricular energetics. *Physiol. Rev.* 70: 247-277, 1990.
15) H. Suga. Paul White Dudley Lecture: Cardiac performance as viewed through the pressure-volume window. *Jpn. Heart J.* 35: 263-280 (1994).
16) H. Suga. Cardiac function. Chapter 5. in: *Pediatric Cardiovascular Medicine.* edited by J.H. Moller, J.I.E. Hoffman. (Churchill Livingstone, New York, 2000) pp. 65-77.
17) H. Suga. Global cardiac function: mechano-energetico-informatics. *J. Biomech.* 36:713-720 (2003).
18) H. Suga. Cardiac energetics: from Emax to pressure-volume area. *Clin. Exp. Pharmacol. Physiol.* 30: 580-588 (2003).
19) E. Braunwald, D.P. Zipes, and W.B. Libby (eds). *Heart Disease. A Textbook of Cardiovascular Medicine.* 6th ed. (Saunders Co, Philadelphia, 2001).
20) W.F. Ganong. *Review of Medical Physiology.* 21st ed. (McGraw Hill, New York, 2003).
21) W.O. Fenn. The relation between the work performed and the energy liberated in muscular contraction. *J. Physiol (London)* 58: 373-395 (1924).
22) H. Suga. Energetics of the time-varying elastance model, a viscoelastic model, matches Mommaerts' unifying concept of the Fenn effect of muscle. *Jpn. Heart. J.* 31: 341-353 (1990).
23) T. Nozawa, Y. Yasumura, S. Futaki, N. Tanaka, and H. Suga. The linear relation between oxygen consumption and pressure-volume area can be reconciled with the Fenn effect in dog left ventricle. *Circ. Res.* 65: 1380-1389 (1989).
24) H. Suga, O. Yamada, Y. Goto, Y. Igarashi, Y. Yasumura, T. Nozawa, and S. Futaki. Left ventricular O_2 consumption and pressure-volume area in puppies. *Am. J. Physiol.* 253: H770-776 (1987).
25) Y. Hata, T. Sakamoto, S. Hosogi, T. Ohe, H. Suga, and M. Takaki. Linear O_2 use-pressure-volume area relation from curved end-systolic pressure-volume relation of the blood-perfused rat left ventricle. *Jpn. J. Physiol.* 48: 197-204 (1998).
26) C.L. Gibbs and J.B. Chapman. Cardiac mechanics and energetics: chemomechanical transduction in cardiac muscle. *Am. J. Physiol.* 249: H199-206 (1985).
27) K. Onishi, K. Sekioka, R. Ishisu, H. Tanaka, M. Nakamura, Y. Ueda, and T. Nakano. Decrease in oxygen cost of contractility during hypocapnic alkalosis in canine hearts. *Am. J. Physiol.* 270: H1905-1913 (1996).
28) T. Mikane, J. Araki, S. Suzuki, J. Mizuno, J. Shimizu, S. Mohri, H. Matsubara, M. Hirakawa, T. Ohe, and H. Suga. O_2 cost of contractility but not of mechanical energy increases with temperature in canine left ventricle. *Am. J. Physiol.* 277: H65-73 (1999).

29) I. Matsubara, A. Kamiyama, and H. Suga. X-ray diffraction study of contracting heart muscle. *J. Mol. Biol.* 111: 121-128 (1977).
30) I. Matsubara, H. Suga, and N. Yagi: An X-ray diffraction study of the cross-circulated canine heart. *J. Physiol. (London)* 270: 311-320 (1977).
31) N. Yagi, H. Okuyama, H. Toyota H, J. Araki, J. Shimizu, G. Iribe, K. Nakamura, S. Mohri, K. Tsujioka, H. Suga, and F. Kajiya. Sarcomere-length dependence of lattice volume and radial mass transfer of myosin cross-bridges in rat papillary muscle. *Pfluegers. Arch.* (in press).
32) N. Yagi, J. Shimizu, S. Mohri, J. Araki, K. Nakamura, H. Okuyama, H. Toyota, T. Morimoto, Y. Morizane, M. Kurusu, T. Miura, K. Hashimoto, K. Tsujioka, H. Suga, and F. Kajiya. X-ray diffraction from a left ventricular wall of rat heart. *Biophys. J.* (in press).
33) H. Suga, Y. Goto, O. Kawaguchi, K. Hata, T. Takasago, A. Saeki, and T.W. Taylor. Ventricular perspective on efficiency. *Basic Res. Cardiol.* 88 (suppl 2): 43-65 (1993).
34) H. Matsubara, J. Araki, M. Takaki, S.T. Nakagawa, and H. Suga. Logistic characterization of left ventricular isovolumic pressure-time curve. *Jpn. J. Physiol.* 45: 535-552 (1995).
35) T. Sakamoto, M. Takaki, Y. Hata, H. Matsubara, J. Araki, and H. Suga. Hybrid logistic characterization of isometric twitch force-time curve of intact blood-perfused canine right ventricular papillary muscle. *Jpn. J. Physiol.* 47: 283-289 (1997).
36) J. Mizuno, T. Mikane, J. Araki, M. Hatashima, T. Moritan,T. Ishikawa, K. Komukai, S. Kurihara, M. Hirakawa, and H. Suga. Hybrid logistic characterization of isometric twitch force curve of isolated ferret right ventricular papillary muscle. *Jpn. J. Physiol.* 49: 145-158 (1999).
37) R.G. Monroe. Myocardial oxygen consumption during ventricular contraction and relaxation. *Circ. Res.* 14: 294-300 (1964).
38) Y. Yasumura, T. Nozawa, S. Futaki, N. Tanaka, and H. Suga. Time-invariant oxygen cost of mechanical energy in dog left ventricle: consistency and inconsistency of time-varying elastance model with myocardial energetics. *Circ. Res.* 64: 764-778 (1989).
39) C.L. Gibbs, I.R. Wendt, G. Kotsanas, and I.R. Young. The energy cost of relaxation in control and hypertrophic rabbit papillary muscles. *Heart Vessels.* 5: 198-205 (1990).
40) J. Araki, H. Matsubara, J. Shimizu, T. Mikane, S. Mohri, J. Mizuno, M. Takaki, T. Ohe, M. Hirakawa, and H. Suga. Weibull distribution function for cardiac contraction: integrative analysis. *Am. J. Physiol.* 277: H1940-1945 (1999).
41) S. Sugiura, N. Kobayakawa, H. Fujita, H. Yamashita, S. Momomura, S. Chaen, M. Omata, and H. Sugi. Comparison of unitary displacements and forces between 2 cardiac myosin isoforms by the optical trap technique: molecular basis for cardiac adaptation. *Circ. Res.* 82: 1029-1034 (1998).
42) S. Sugiura. Actin-myosin interaction. *Cardiovasc. Res.* 44: 266-273 (1999).
43) T. Yanagida, S. Esaki, A.H. Iwane, Y. Inoue, A. Ishijima, K. Kitamura, H. Tanaka, and M. Tokunaga. Single-motor mechanics and models of the myosin motor. *Philos. Trans. R. Soc. Lond. B. Biol. Sci.* 355: 441-447 (2000).
44) K. Kitamura and T. Yanagida. Stochastic properties of actomyosin motor. *BiosSystems* 71: 101-110 (2003).
45) H. Suga, O. Yamada, Y. Goto, and Y. Igarashi. Oxygen consumption and pressure-volume area of abnormal contractions in canine heart. *Am. J. Physiol.* 246: H154-H160 (1984).
46) H Suga. Mechanoenergetic estimation of multiple crossbridge steps per ATP in beating heart. *Jpn. J. Physiol.* (in press).
47) T.W. Taylor, Y. Goto, and H. Suga. Variable cross-bridge cycling-ATP coupling accounts for cardiac mechanoenergetics. *Am. J. Physiol.* 264: H994-1004 (1993).
48) J. Shimizu, J. Araki, J. Mizuno, S. Lee, Y. Syuu, S. Hosogi, S. Mohri, T. Mikane, M. Takaki, T.W. Taylor, and H. Suga. A new integrative method to quantify total Ca^{2+} handling and futile Ca^{2+} cycling in failing hearts. *Am. J. Physiol.* 275: H2325-2333 (1998).
49) J. Araki, S. Mohri, G. Iribe, J. Shimizu, and H. Suga. Total Ca^{2+} handling for E-C coupling in the whole heart: an integrative analysis. *Can. J. Physiol. Pharmacol.* 79: 87-92 (2001).
50) J. Araki, J. Shimizu, G. Iribe, S. Mohri, T. Kiyooka, Y. Oshima, W. Fujinaka, Y. Doi, and H. Suga. Assessment of total Ca^{2+} handling for excitation-contraction coupling in beating left ventricle. *J. Mech. Med. Biol.* 1: 123-138 (2001).

DISCUSSION

Huxley: Tension per cross-bridge is likely to be much higher than that you assume. Recent single molecule measurements under more isometric conditions indicate much higher values, about 10pN or more; also, there are indications that only 10-20% of cross-bridges may be attached to actin in isometric contraction at any one time, so average tension per cross-bridge is likely to be much higher than the 1-2pN value.

Suga: I used unitary force of ~1.5 pN per cross-bridge referring to Prof. Sugiura's paper (Sugiura, Kobayashi, Fujita, Yamashita, Momomura, Chaen, Omata & Sugi, Circ. Res. 82: 1029-1034, 1998). If the unitary force is 10 pN, as you suggest, my results will be different.

IS MYOSIN PHOSPHORYLATION SUFFICIENT TO REGULATE SMOOTH MUSCLE CONTRACTION?

Gabriele Pfitzer[1*], Mechthild Schroeter[1], Veronika Hasse[1], Jun Ma[1], Klaus-Henning Rösgen[1], Silvia Rösgen[1], Neil Smyth[2]

1. INTRODUCTION

Since the early work of Rüegg and coworkers in 1965 (Filo et al., 1965) it is known that in smooth muscle as in striated muscle, contraction is activated by an increase in cytosolic Ca^{2+}. The fundamental mechanism of contraction in both striated and smooth muscle is believed to be the sliding of the thin and thick filaments passed each other driven by the cyclic interaction between myosin crossbridges and actin (Arner and Pfitzer, 1999). However, the mechanism by which changes in cytosolic Ca^{2+} are sensed and translated to result in contraction or relaxation of smooth muscle and the crossbridge cycle itself are quite distinct from its striated muscle counterpart.

Originally it was thought that a troponin-like mechanism accounted for the regulation of smooth muscle contraction (Ebashi et al., 1977; Hartshorne and Mrwa, 1982, and Marston 1982 for reviews of the early work). However, while with the skeletal system pure actin and pure myosin exhibit a Mg-ATPase activity that is close to maximal, the smooth muscle system loses activity upon purification (Schirmer et al., 1965). Thus, a Ca^{2+}-dependent activating mechanism of the dormant smooth actomyosin is required. In 1977, Sobieszek discovered that phosphorylation of the regulatory light chains of myosin (RLC) allowed smooth muscle myosin to be activated by actin leading finally to the formulation of the phosphorylation theory of smooth muscle regulation (Kamm and Stull 1985 for review).

The enzyme responsible for phosphorylation of RLC has been identified as the Ca^{2+}-dependent myosin light chain kinase, MLCK, consisting of two subunits, a larger subunit containing the catalytic core and a 17 kDa subunit that was identified as the Ca^{2+}-sensor calmodulin (Dabrowska et al., 1978). At the same time it was recognized that MLCK is phosphorylated by the cAMP dependent protein kinase resulting in a decrease in its affinity for the Ca^{2+}-calmodulin complex (Adelstein et al., 1978) and a decrease in the

[1] Department of Vegetative Physiology and [2] Institute of Biochemistry II, University of Cologne, 50931 Koeln, Germany, [*]corresponding author

Ca^{2+}-sensitivity of contraction in triton skinned smooth muscle (Pfitzer et al., 1985). This was the first example that the dependence of force on Ca^{2+} is not unique but rather may be modulated (Pfitzer 2001; Somlyo and Somlyo, 2003).

The enzyme that catalyzes dephosphorylation of RLC was identified about a decade after the discovery of MLCK as a type 1 phosphatase (MLCP) which consists of three subunits, a 130 kDa regulatory subunit that targets MLCP to myosin and confers specificity, a catalytic subunit and a 20 kDa subunit of unknown function (Hartshorne, 1998). Originally it was thought for lack of evidence to the contrary that MLCP was a non-regulated, permanently active enzyme. We now know that MLCP is a highly regulated and its activity may be up- and down-regulated by several intracellular signaling cascades (Somlyo and Somlyo, 2003 for review). As the degree of phosphorylation of RLC depends on the ratio of the activities of MLCK and MLCP, force in smooth muscle is thus not primarily determined by the absolute Ca^{2+} concentration but rather by a "signaling network" created by the different interacting signaling cascades in the smooth muscle cell, activated or inhibited in response to extracellular messages.

Matters are, however, more complex since it was shown early on that the relation between RLC phosphorylation and force is not unique. Rather in the course of a contraction RLC phosphorylation declines from its initial high levels to significantly lower yet suprabasal levels while tension is sustained. This is associated with a decline in ATP consumption and maximal shortening velocity. This highly economical contraction was denoted as "latch" (Dillon et al., 1981). During latch, force is assumed to be maintained by dephosphorylated crossbridges that cycle slowly or not at all. From this the important conclusion emerged that the crossbridge cycle in smooth muscle is regulated on a short-term basis.

Force measurements using single smooth muscle S1 molecules (Veigel et al., 2003), as well as in biochemical (Cremo and Geeves, 1998) and structural studies (Gollub et al., 1999) revealed that smooth muscle myosin II differs from its skeletal counterpart by its high affinity for MgADP and that it produces its power stroke in two steps (Veigel et al., 2003). According to these studies a so called intermediate, ADP-bound state precedes the final state from which the crossbridges detach (Gollub et al., 1999) which would slow down the enzymatic cycle and lower the rate of energy consumption. The occupation of this intermediate state depends on the concentration of MgADP as well as on the phosphorylation of RLC and the strain in such a manner that this state could contribute to the ability of smooth muscle to maintain force at low levels of RLC phosphorylation (Khromov et al., 1995). Hence, the properties of smooth muscle myosin II could account for the latch state (cf. also J. C. Rüegg, this volume).

Different models have been suggested to explain, how attached, dephosphorylated crossbridges may be generated (for review Arner and Pfitzer, 1999). The Hai and Murphy model (1989) assumes that "latch" bridges are generated by dephosphorylation of attached phosphorylated crossbridges. To allow a significant flux from attached phosphorylated crossbridges to dephosphorylated crossbridges the activity of MLCP has to be high. This model implies that while the degree of RLC phosphorylation depends on the activity ratio of MLCK and MLCP, the coupling between force and phosphorylation depends on the absolute activities of MLCK and in particular of MLCP.

Alternative models suggest that a small number of phosphorylated, attached crossbridges allows the cooperative attachment of dephosphorylated crossbridges (Somlyo et al., 1988; Butler and Siegman, 1998).

Further there is longstanding biochemical evidence of one or more thin filament based regulatory systems (Ebashi, 1977; Marston 1982). Originally it was suggested that latch state is regulated by such an additional regulatory mechanism. From the above reviewed evidence it is clear that this is not necessary. However, we (Katoch et al., 1997) and others (Barany and Barany 1993; Tansey et al., 1992) have shown, that relaxation may occur at high levels of RLC phosphorylation which violates the phosphorylation theory that requires that relaxation is preceded by dephosphorylation of RLC. Furthermore, relaxation may be accelerated under conditions when RLC is basal (Fischer and Pfitzer, 1998). Therefore, we suggest that one or more alternative inhibitory mechanisms are required to switch off smooth muscle activation independent of RLC dephosphorylation. We have shown that the actin and myosin binding protein, caldesmon inhibits contraction at elevated RLC phosphorylation (Pfitzer et al., 1993). Extraction of calponin, another thin filament linked protein, induced a contraction without an increase in RLC phosphorylation (Malmqvist et al., 1997). Thus, these proteins could inhibit attachment of dephosphorylated crossbridges.

In this study we will present evidence that maximal shortening velocity depends on the degree of RLC phosphorylation and but not on the absolute activities of MLCK or MLCP favouring a model in which the latch state is generated by cooperative attachment of dephosphorylated crossbridges. We will further present evidence that caldesmon can switch off attachment of dephosphorylated crossbridges.

2. METHODS

The details for determination of isometric tension in intact and triton-X 100 skinned chicken gizzard fibers and the solutions used in the experiments have previously been published (Fischer and Pfitzer, 1989, Pfitzer et al., 1993).

Maximum unloaded shortening velocity (v_{us}) was determined as in (Bialojan et al., 1987). In brief, thin fiber bundles (100-150 µm in diameter) were glued between an AME 801 force transducer and a length step generator. After attaining steady state force, the fibers were subjected to a series of quick releases ranging between 8 and 15% L_o which caused force to drop to zero. A plot of the amount of release (muscle shortening) versus the time elapsing until the onset of redevelopment of isometric force (i.e. slack time at zero force) yields a approximately straight line extrapolating to an ordinate intercept of about 8%, the gradient being a measure of v_{us}.

Caldesmon was extracted from skinned fibers in rigor/BDM solution containing (in mM) imidazole 25 (pH 7.0 at 4°C), KCl 50, $MgCl_2$ 50, EGTA 4, BDM 15, DTE 2, and 10 µM leupeptin for 8 hrs which resulted in about 80% extraction of caldesmon. In such extracted fibers the force-pCa relation was determined with and without prior reconstitution with 1 mg/ml caldesmon for 2.5 hrs contained in rigor/BDM solution.

For thiophosphorylation of RLC, fibers depleted of endogenous ATP by incubation in rigor solution with 400 U/ml hexokinase and 10 mM glucose were incubated for 15 min in rigor solution with different [Ca^{2+}] containing 1 mM ATPγS to induce thiophosphorylation of RLC. This was followed by washing in rigor solution to remove Ca^{2+} and ATPγS and incubation with caldesmon buffer or caldesmon (1mg/ml) for 3 hrs. A contraction was induced in ATP containing relaxing solution whereby the amplitude of the contraction depended on the pCa used to induce thiophosphorylation of the RLC.

RLC phosphorylation and thiophosphorylation was determined by 2D-PAGE as described previously (Fischer and Pfitzer, 1989; Wirth et al., 2003).

To ablate exon2 of the caldesmon gene, a targeting vector was constructed which replaced exon2 and part of intron 2 by a neomycin cassette. This vector, which was used for homologous recombination in embryonic stem cells, allows in frame splicing from exon1 to exon3a. Correct integration of the vector was verified with southern blots. Western blots from bladder and ileum of the mice show a band which is immunoreactive with caldesmon antibodies and of lower relative mobility indicating the expression of a truncated caldesmon. Longitudinal smooth muscle strips from the mouse ileum and circular muscle strips from the gastric fundus were permeabilized with triton X-100, mounted in a myograph and incubated in solutions as in Wirth et al., (2003).

Statistics. The data are expressed as the mean ±SEM. Student´s t-test was used to determine significant differences between groups and a P-value of <0.05 was considered to be statistically significant.

3. RESULTS AND DISCUSSION

Intact chicken gizzard responds to electrical field stimulation with a very rapid and transient rise in tension which is typical for a phasic smooth muscle. In line with the phosphorylation theory the rise and fall in tension is preceded by a respective increase and decrease in RLC phosphorylation (Fig. 1A). The rate of dephosphorylation is faster ($0.7\pm0.03\ s^{-1}$) than the rate of relaxation ($0.26\pm0.02 s^{-1}$). A similar phasic contraction was seen in response to stimulation with the muscarinic agonist, carbachol (Fischer and Pfitzer, 1989).

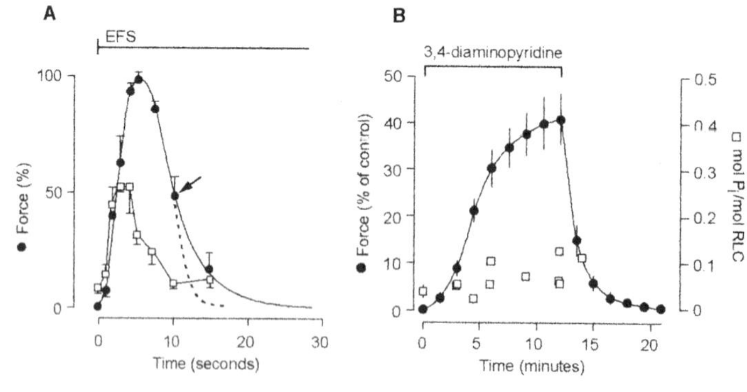

Figure 1. Activation of intact fiber bundles from chicken gizzard with electrical field stimulation (EFS, A) or 3,4-diaminopyridine (B). The tissues were shock frozen for determination of RLC phosphorylation at the indicated time points. Symbols with error bars are mean ±SEM, n=4. Phosphorylation values in B are single values.

A slowly developing, tonic contraction at apparently basal levels of RLC phosphorylation can be induced by the potassium channel blocker, 3,4 diaminopyridine (DAP, Fig. 1B). We cannot exclude the possibility that there is a small but not significant increase in RLC phosphorylation but this does not precede contraction. The active state determined by quick release recovery was much less than in the electrical field induced contraction suggesting that the crossbridge cycle rate is slow (data not shown).

The contraction elicited by DAP in chicken gizzard fibers is similar to the latch state seen during the sustained phase of a contraction in a tonic type of smooth muscle. This is interesting because it is believed that in mammalian smooth muscle, latch is observed in tonic but not in phasic smooth muscle (Khromov et al., 1995).

The regulation of a tonic contraction in mammalian smooth muscle is complex, because after an initial rise cytosolic Ca^{2+} typically declines to lower values during the sustained phase suggesting a low activity of MLCK. At the same time MLCP activity may also be low because of inhibition by the Rho/ Rho kinase pathway and/or PKC (Pfitzer, 2001 for review). Thus not only the ratio of the activity of MLCK and MLCP but also the absolute activities change in the course of a contraction. According to the "latch-model" (Hai and Murphy, 1989), the absolute rates of phosphorylation and in particular dephosphorylation may determine the number of attached, dephosphorylated "latch-bridges". Therefore, it is important to investigate the effects of Ca^{2+} and the inhibition of phosphatase separately. As it is possible to induce a latch like state in the triton skinned chicken gizzard fiber (Wagner and Rüegg, 1986) we used this preparation to study these pathways separately by either activating MLCK with Ca^{2+} at constant MLCP activity or by inhibiting MLCP with okadaic acid at a given pCa.

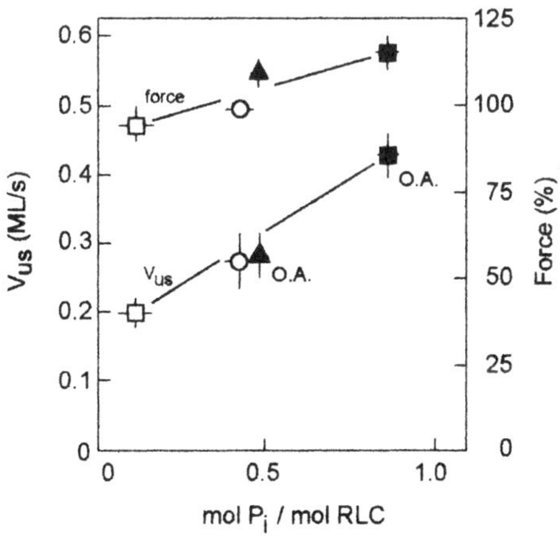

Figure 2. Dependence of unloaded shortening velocity (v_{us}) and isometric tension on RLC phosphorylation in triton skinned chicken gizzard fibers. Open symbols, activation with Ca^{2+}-calmodulin (CaM) only, closed symbols activation in the presence of the MLCP inhibitor, okadaic acid (O.A., 10 μM). Squares: 1.7 μM Ca^{2+}, 0.15 μM CaM; circles: 15 μM Ca^{2+}, 1 μM CaM; triangles: 1.7 μM Ca^{2+} without exogenous CaM which without O.A. would not induce a contraction

The conditions chosen were such (Fig. 2) that isometric force changed only little (between 95% and 117% of nominal maximal force elicited with 15 μM Ca^{2+} and 1 μM calmodulin) despite the large change in RLC phosphorylation (between 11% and 84%). In contrast to isometric force, unloaded shortening velocity (v_{us}) linearly depended on the degree of RLC phosphorylation and increased about two fold (Fig. 2). The dependence of v_{us} on RLC phosphorylation appears to be independent of the absolute activities of either MLCK or MLCP. This is particularly obvious when a similar degree of RLC phosphorylation was obtained by activation with 15 μM Ca^{2+} in the presence of 1 μM calmodulin (high MLCP activity) or by inhibition of MLCP with okadaic acid at 1.7 μM Ca^{2+} without exogenously added calmodulin (low MLCP activity). Thus, we found that under conditions where force was maximal, v_{us} a measure of crossbridge turnover at isotonic conditions depends only on the degree of RLC phosphorylation, results contrasting with other findings (Malmqvist & Arner, 1996).

Our results clearly indicate that no additional Ca^{2+}-regulatory mechanism besides RLC phosphorylation is necessary to regulate maximum shortening velocity in support of models that suggest that sustained force at low levels of RLC phosphorylation is due a to strongly bound and long lived AM.ADP state of dephosphorylated crossbridges (see Introduction). Based on the results shown in Fig. 2, we would favour a model in which latch bridges are generated by attachment of dephosphorylated crossbridges (Somlyo et al., 1988; Khromov et al., 1995; Butler and Siegman, 1998).

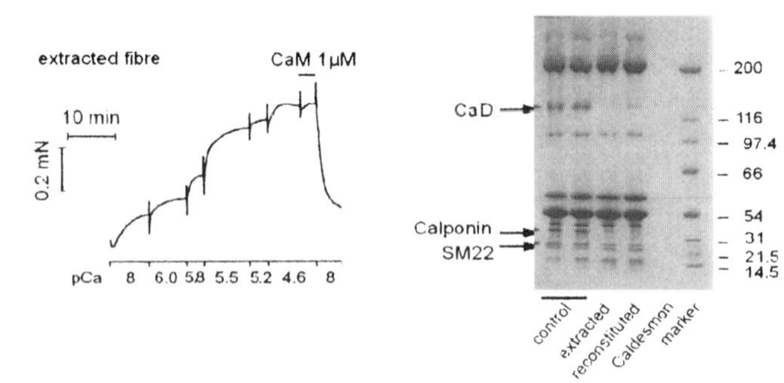

Figure 3. Effect of extraction of caldesmon (CaD) from triton skinned chicken gizzard fibers and reconstitution of extracted fibers with exogenous CaD on force. Control: fiber before extraction; note that a contraction is induced only in the presence of Ca^{2+} (pCa<6). In the extracted fiber a contraction is already seen at resting pCa (pCa >8). This is reversed after reconstitution with CaD. Lower right panel: Coomassie stained gel of chicken gizzard fibers. Typical examples of 4 independent experiments are shown.

As shown in Fig. 1 the rate of relaxation is increased from 0.26 $\pm 0.02 s^{-1}$ to 1 $\pm 0.03 s^{-1}$ when EFS is turned off at apparently basal levels of RLC phosphorylation. Furthermore, we (Katoch et al., 1997) and others (Tansey et al., 1990; Barany & Barany, 1993) have reported that smooth muscle can relax at elevated levels of RLC phosphorylation a phenomenon that is still poorly understood. These findings indicate that there must be one or more regulatory systems that increase the net detachment of crossbridges even at elevated RLC phosphorylation.

Caldesmon, which inhibits actomyosin ATPase activity, is a potential regulatory protein that could inhibit crossbridge attachment (Pfitzer et al., 1993). In line with this we found that extraction of about 80% of endogenous caldesmon from triton skinned gizzard fibers induced a Ca^{2+}-independent contraction which amounted to about 20% of maximal Ca^{2+}-activated force (Fig. 3). Reconstitution of the fibers with exogenous caldesmon reversed the Ca^{2+}-independent contraction (Fig. 3). A similar result was shown by Malmqvist et al., (1996) using a similar extraction protocol. These authors also reported that extraction of the actin binding protein, calponin, resulted in a slowly developing, RLC phosphorylation independent contraction (Malmqvist et al., 1997). We also see some extraction of calponin. However, since reconstitution with exogenous caldesmon restores the relaxed state we believe that the increase in basal force in the extracted fibers is due to loss of caldesmon. The results shown here are consistent with our earlier suggestion that caldesmon inhibits cooperative attachment of dephosphorylated crossbridges (Albrecht et al., 1997).

On the other hand, increasing the endogenous caldesmon content by incubation of skinned fibers with exogenous caldesmon decreased force at a given submaximal but not at maximal level of thiophosphorylation thereby shifting the relation between force and thiophosphorylation of RLC to the right (Fig. 4)

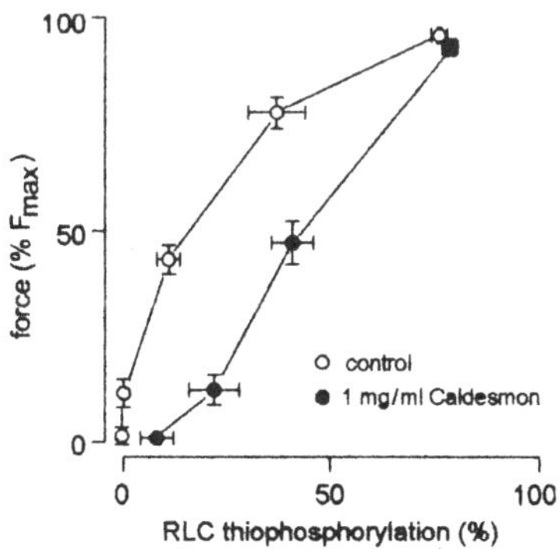

Figure 4. Caldesmon (CaD) shifts the relation between force and RLC thiophosphorylation to the right in triton skinned chicken gizzard fibers. Different levels of thiophosphorylation were achieved by incubation for 10 min with ATPγS in rigor solution with different [Ca^{2+}]. This was followed by loading with CaD or buffer (control) for 30 min in rigor solution. Tension was induced by switching to ATP containing relaxing solution

We have previously shown that the actin binding domain of caldesmon is sufficient for inhibition of contraction in skinned fibers (Pfitzer et al., 1993). The function of the myosin binding domain is less clear. Therefore we generated a mouse model lacking exon 2 of the caldesmon gene. By homology to chicken gizzard caldesmon (Li et al., 2000) this deletion removes the strong myosin binding domain.

The homozygous mice expressing the truncated form of caldesmon are viable and fertile. Preliminary experiments in triton skinned fibers showed that the half time of relaxation in ileal smooth muscle fibers from exon-2 deficient mice was 127 ± 35 sec compared to 49 ± 7.4 sec in the wild type ($n=4$, $p<0.05$). A similar result was obtained with triton skinned muscle strips from gastric fundus.

It is known that in skinned fibers dephosphorylation of RLC is much faster than the decline in tension (Kühn et al., 1990). The time course of mechanical relaxation was suggested to reflect the cooperative reattachment of crossbridges but also by the low rate of ADP dissociation from crossbridges as ADP prolongs the lifetime of dephosphorylated crossbridges (Khromov et al., 1995). In this context our results suggest that the myosin binding domain of caldesmon is involved in inhibiting the cooperative reattachment because in its absence, relaxation is slower. This is surprising because inhibition of actomyosin interaction was suggested to be mediated by the actin binding domain. The results may be explained that binding to myosin is required for proper positioning of caldesmon so that the actin binding domain can exert its inhibitory function (Lee at al., 2000).

In conclusion, there is no simple stoichiometric dependence of force on RLC phosphorylation in smooth muscle. While sustained force at low levels of RLC phosphorylation, the so called latch state, does not require additional regulatory mechanism but may be accounted for by the properties of smooth muscle myosin II, additional regulatory systems are required for regulating relaxation.

3. 1 Acknowledgments

We grateful to Dr. J.M. Chalovich, East Carolina University School of Medicine, Greenville, N.C., USA, for the generous gift of purified caldesmon. We also thank Dr. U. Schmidt and Dr. W. Fischer, then Univ. of Heidelberg, for the help with the experiments shown in fig. 1 and 4. The expert technical assistance of D. Metzler and S. Zittrich are gratefully acknowledged. This work was supported by grants from the DFG and the Medical Faculty of Cologne (Köln Fortune) to G.P.. Present address of M. Schroeter is Dept. of Biochemistry and Molecular Biology, East Carolina University School of Medicine, Greenville, N.C. USA.

4. REFERENCES

Adelstein, R.S., Conti, M.A, Hathaway, D.R., Klee, C.B., 1978, Phosphorylation of smooth muscle myosin light chain kinase by the catalytic subunit of adenosine 3′:5′-monophosphate-dependent protein kinase. *J. biol. Chem.* **253:**8347-8350.

Albrecht, K., Schneider, A., Liebetrau, C., Rüegg, J.C., and Pfitzer, G., 1997, Caldesmon promotes relaxation of skinned guinea pig taenia coli smooth muscles: inhibition of cooperative reattachment of latch bridges? *Pflügers Arch.*, **434**: 534-542.

Arner, A., and Pfitzer, G., 1999, Regulation of crossbridge cycling by Ca^{2+} in smooth muscle. *Rev. Physiol. Biochem. Pharmacol.* **134**:63-146

Barany, M., and Barany, K. 1993, Dissociation of relaxation and myosin light chain dephosphorylation in porcine uterine muscle, *Arch Biochem Biophys,* **305**:202-4.

Bialojan, C., Rüegg, J.C., DiSalvo, J., 1987, A myosin phosphatase modulates contractility in skinned smooth muscle. *Pflügers Arch.* **410**:304-12.

Butler, T. M., and Siegman, M. J., 1998, Control of crossbridge cycling by myosin light chain phosphorylation in mammalian smooth muscle. *Acta Physiol. Scand.* **164:**389-400.

Cremo, C.R., and Geeves, M.A., 1998, Interaction of actin and ADP with the head domain of smooth muscle myosin: implications for strain-dependent ADP release in smooth muscle, *Biochemistry* **37**: 1969-1978.

Dabrowska, R., Sherry, J.M.F., Aromatorio, D.K., and Hartshorne, D.J., 1978, Modulator protein as a component of myosin light chain kinase from chicken gizzard, *Biochemistry N:Y.* **17**:253-258.

Dillon, P. F., Aksoy, M. O., Driska, S. P., and Murphy, R. A., 1981, Myosin phosphorylation and the cross-bridge cycle in arterial smooth muscle, *Science* **211**:495-497.

Ebashi, S., Mikawa, T., Hirata, M., Toyo-oka T., and Nonmura Y. 1977, Regulatory proteins of smooth muscle. In: Excitation contraction coupling in smooth muscle, Eds. Casteels R., Godfraind R., Rüegg J.C. pp. 325-324.

Filo, R.S., Bohr, D.F., and Rüegg, J.C., 1965, Glycerinated skeletal and smooth muscle. Calcium and magnesium dependence. *Science* **147**:1581-1583

Fischer, W., and Pfitzer, G. 1989, Rapid myosin phosphorylation transients in phasic contractions in chicken gizzard smooth muscle. *FEBS Letters* **258**: 59-62.

Gollub, J., Cremo, C. R., and Cooke, R., 1999, Phosphorylation regulates the ADP-induced rotation of the light chain domain of smooth muscle myosin, *Biochemistry* **38**:10107-10118.

Hai, C.M., and Murphy, R.A., 1989, Ca^{2+}, crossbridge phosphorylation, and contraction. *Annu. Rev. Physiol.* **51**:285-298.

Hartshorne, D.J., and Mrwa U. 1982, Regulation of smooth muscle actomyosin, *Blood Vessels* **19**:1-18

Hartshorne, D. J., 1998, Myosin phosphatase: subunits and interactions. *Acta Physiol. Scand.* **164**:483-494.

Hemric, M.E., and Chalovich J.M. 1988, Effect of caldesmon on the ATPase activity and the binding of smooth and skeletal myosin subfragments to actin. *J.Biol. Chem.* **263**:1878-1885.

Kamm, K.E., and Stull, J.T. 1985, The function of myosin and myosin light chain kinase phosphorylation in smooth muscle. *Ann Rev Pharmacol Toxicol* **25**:593-620

Katoch, S.S., Rüegg, J.C., and Pfitzer, G., 1997, Differential effects of a K channel agonist and calcium antagonists on myosin light chain phosphorylation in relaxation of endothelin-contracted tracheal smooth muscle. *Pflügers Arch* **433**: 472-477.

Khromov, A., Somlyo, A. V., Trentham, D.R., Zimmermann, B., and Somlyo, A.P., 1995, The role of MgADP in force maintenance by dephosphorylated cross-bridges in smooth muscle: a flash photolysis study. *Biophys. J.* **69**:2611-22.

Kühn, H., Thewes, A., Gagelmann, M., Güth, K., Arner, A., and Rüegg, J. C., 1990, Temporal relationship between force, ATPase activity, and myosin phosphorylation during a contraction/relaxation cycle in a skinned smooth muscle, *Pflügers Arch. Eur. J. Physiol.* **416**:512-518.

Lee, Y.H., Gallant, C., Guo, H., Li, Y., Wang, C.A., and Morgan, K,G., 2000, Regulation of vascular smooth muscle tone by N-terminal region of caldesmon. Possible role of tethering actin to myosin. *J. Biol. Chem.* **275**:3213-20.

Li, Y., Zhuang, S., Guo, H., Mabuchi, K., Lu, R., C., and Wang, C.-L. A., 2000, The major myosin binding site of caldesmon resides near its N-terminal extreme. *J Biol. Chem.* **275**:10989-10994.

Malmqvist, U, and Arner A., 1996, Regulation of force and shortening velocity by calcium and myosin phosphorylation in chemically skinned smooth muscle, *Pflügers Arch.* **433**:42-48

Malmqvist U, Arner A, Makuch R, Dabrowska R (1996) The effects of Caldesmon extraction on mechanical properties of skinned smooth muscle fibre preparations. Pflügers Arch **432**:241-247

Malmqvist, U., Trybus, K.M., Yagi, S., Carmichael, J., and Fay, F.S., 1997, Slow cycling of unphosphorylated myosin is inhibited by calponin, thus keeping smooth muscle relaxed. *Proc.Natl. Acad. Sci. USA* **94**:7655-7660.

Marston, S.B., 1982, The regulation of smooth muscle contractile proteins. *Prog. Biophys. molec. Biol.* **41**:1-41.

Pfitzer G., 2001, Invited Review: Regulation of myosin phosphorylation in smooth muscle, *J. Appl. Physiol.* **91**:497-503.

Pfitzer, G., Rüegg, JC., Zimmer, M., Hofmann, F., 1985, Relaxation of skinned coronary arteries depends on the relative concentrations of Ca^{2+}, calmodulin, and active cAMP-dependent protein kinase. *Pflügers Arch.* **405:** 70-76.

Pfitzer, G., Zeugner, C., Troschka, M., and Chalovich, J.M., 1993, Caldesmon and a 20 kDa actin binding fragment of caldesmon inhibit tension development in skinned gizzard muscle fiber bundles. *Proc. Natl. Acad. Sci. (USA)* **90**: 5904-5908

Schirmer, R.H., 1965, The particularities of contractile proteins of the arteries (German), *Biochem Z.* **343**:269-282

Sobieszek, A, 1977, Ca-linked phosphorylation of a light chain of vertebrate smooth-muscle myosin. *Eur. J. Biochem.* **55**:49-60

Somlyo, A.V., Goldman, Y.E., Fujimori, T., Bond, M., Trentham, D.R., and Somlyo, A.P., 1988, Cross-bridge kinetics cooperativity, and negatively strained cross-bridges in vertebrate smooth muscle. A laser flash study. *J. Gen. Physiol.* **91**:165-192.

Somlyo, A.P., and Somlyo, A.V., 2003, Ca^{2+}sensitivity of smooth muscle and nonmuscle myosin II: modulated by G proteins, kinases, and myosin phosphatase. *Physiol. Rev.* **83**:1325-1358.

Tansey, M.G., Hori, M., Karaki, K., Kamm, K.E., and Stull, J.T., 1990, Okadaic acid uncouples myosin light chain phosphorylation and tension in smooth muscle. *FEBS Lett* **270**:219-221

Veigel, C., Molloy, J. E., Schmitz, S., and Kendrick-Jones, J., 2003, Load-dependent kinetics of force production by smooth muscle myosin measured with optical tweezers, *Nature Cell Biology* **5**:980-986

Wagner J., and Rüegg, J.C., 1986, Skinned smooth muscle: calcium-calmodulin activation independent of myosin phosphorylation. *Pflügers Arch.* **407**:569-71.

Wirth, A., Schroeter, M., Manser, E., Chalovich, J. M., de Lanerolle, P., and Pfitzer, G., 2003, Inhibition of contraction and myosin light chain phosphorylation in guinea pig smooth muscle by p21-activated kinase. *J. Physiol.* **549**: 489-500.

DISCUSSION

Takuwa: I would like to know the phenotype of the mice expressing caldesmon with deletion of exon 2, particularly in the vascular and non-vascular smooth muscle organ function.

Pfitzer: We have not analyzed that yet. Anyway, the animals are fertile and viable and do not show any major phenotype.

VI. OTHER ASPECTS

COMPARATIVE ASPECTS OF CROSSBRIDGE FUNCTION – SKINNED FIBRE STUDIES

J. Caspar Rüegg*

1. INTRODUCTION

As is well known the sliding filament contractile machinery is driven by the asynchronous working strokes of crossbridges, comprising the "heads" of myosin molecules (Huxley, 1969). These myosin-crossbridges connecting the sliding actin and myosin filaments have been optimized for a variety of contractile functions, from the most rapid contractile cycles of certain insect flight muscle to the extremely slow contractions of tonic smooth muscle. It has been hypothesized that muscles specialized for rapid shortening are characterized by fast cycling bridges whereas slowly cycling bridges with prolonged attached states may allow slow muscles to sustain tension ecomomically for prolonged periods (e.g. Rüegg, 1971). Clearly a greater speed in shortening requires a greater rate of filament sliding, of attachment and detachment of crossbridges and of ATP splitting, but since ATP is split at a lower rate in slow than fast muscle, the former maintains tension more economically. In the last few decades it had become possible to analyze the molecular composition as well as the contractile and energetic properties of certain fast and slow muscle fibre types in great detail (e.g. Reiser et al., 1985, Bottinelli et al., 1991, 1994, Galler et al., 1996) and it was discovered that the different myosin II isoforms largely determine the contractile and energetic properties of muscle fibres that allow them to adapt to the requirements of muscular performance according to variable functional demands. Indeed, the diversity of muscle tissues is a challenge to the comparative physiologist, who attempts to understand how crossbridges adapted to their specific roles in fast and slow muscle. However, the enormous diversity

* Department of Physiology and Pathophysiology, University of Heidelberg, D-69120 Heidelberg, Germany.

of tissues may also offer a wealth of different specimens from which a physiologist may choose when looking for the most suitable experimental material to solve a specific problem (c.f. Krebs, 1975).

One of the still unsolved mysteries of the sliding filament mechanism is the "Fenn effect" (Fenn, 1924) i.e. the phenomenon that the amount of work done by a muscle determines the extent of a chemical reaction (i.e. ATP splitting) which takes place. To account for the increased release of total energy that was observed in working muscle (e.g. Curtin et al., 1974) most molecular-mechanical contraction models predicted a strongly increasing detachment rate, when attached crossbridges become discharged during shortening (e.g. Huxley, 1957). In their more recent model of muscular contraction Huxley and Simmons (1971) proposed that crossbridges under high load proceed the attached states of the crossbridge cycle slowly wheras those under low load proceed faster. However, the details of how load affects the kinetics of crossbridges in relation to energy metabolism could at that time not be investigated in intact fibres, specifically because it was not possible to determine the rate of ATP splitting during shortening directly (without poisoning creatine kinase). Direct measurements of the myofibrillar ATP-splitting rate in working muscle fibres is now possible, however, if muscle fibre preparations are used from which the surface membrane had been removed or made highly permeable by chemical membrane skinning, while the contractile system is left intact. The mechanical events and the rate of ATP splitting of such skinned fibres can then be measured at the same time so that the "Fenn effect" can be studied *in vitro* in a "cell free" system containing ATP as the only energy source (e.g. He et al., 2000, c.f. also Sugi et al., 1998). However, when we first attempted such experiments in the nineteen sixties, the methods available were not yet adequate to study the ATP-splitting rates in skinned fast or slow skeletal muscle fibres *during* shortening. So we used a comparative approach.

2. THE "FENN EFFECT" IN VERTEBRATE AND INVERTEBRATE MUSCLE - FROM SKINNED FIBRES TO SINGLE MOLECULES

In order to study the "Fenn effect" *in vitro,* Rüegg and Tregear (1966) chose the dorsal longitudinal muscle of the giant water bug *Lethocerus maximus* because its muscle fibres responded, even after chemical skinning, to small sinusoidal length changes with delayed tension changes thereby producing oscillatory power at a steady rate for many hours, when attached to a suitable resonant or driven mechanical system (c.f. Jewell and Rüegg, 1966). Thus the relation between the rate of ATP consumption and power output under varying conditions of mechanical extension, oscillation frequency and concentration of ionised calcium, could be studied in an unequivocal way: It could be shown, for instance, that increasing the free calcium concentration and stretching the fibres at constant (buffered) free calcium increased the ATPase activity of crossbridges (to about 10 s^{-1} at 20°C), but driven oscillation of the stretched muscle fibres by up to 3%

of its rest length increased the rate of ATP splitting even further apparently in proportion to the output of power over a wide range of mechanical conditions (Rüegg and Tregear, 1966). In these experiments power output and oscillation induced extra ATPase activity (as well as ATP/P_i exchange, c.f. Ulbrich and Rüegg, 1976) depended on oscillation frequency in a bell shaped manner. The efficiency of chemomechanical energy conversion was somewhat variable, but at the optimum oscillation frequency (20 Hz at 30°C, the physiological temperature of the flying insect) it reached as much as 0.2 - 0.5 (c.f. Steiger and Rüegg, 1969). As the rate of ATP splitting presumably corresponded to the rate of crossbridge cycling (oscillatory) length changes were assumed to alter crossbridge kinetics, and Güth et al. (1981) showed later that this was probably the case: In their experiments a quick release by 0.2- 1 % of the fibre length caused a delayed fall in tension and stiffness at a rate that greatly increased with increasing amplitude of release. Assuming then that immediate stiffness was a measure of the number of crossbridges attached (as proposed in the Huxley-Simmons 1971 model), Güth et al. (1981) concluded that the (net) detachment rate of crossbridges was enhanced if the muscle was released thereby reducing the strain on the crossbridges (as indicated by a reduction in the ratio of tension and stiffness). Güth's experiments therefore suggested that the kinetics of crossbridge detachment and presumably of ATP splitting might indeed be governed by the crossbridge strain or crossbridge load as already postulated by the Huxley 1957 model. Subsequently, Güth et al. (1987) measured the influence of (repeated) rapid releases of strained crossbridges on the rate of ATP splitting and found that it increased in proportion to the frequency of applied length changes.

As in insect muscles in skinned skeletal muscle fibres the strain on crossbridges (as determined by the force to stiffness ratio) was also decreasing during shortening (c.f. Tsuchiya et al., 1982) thus causing an increase in the crossbridge detachment rate. Accordingly, the rate of ATP consumption was found to be greater during shortening than in an isometric condition (c.f. Sugi et al., 1998) and it increased in proportion to the speed of shortening (He et al., 2000). Furthermore, these results could be simulated with a kinetic scheme of the biochemical crossbridge cycle in which the rate constant controlling ADP release was sensitive to the load (or for that matter to crossbridge strain) thus accounting for the "Fenn effect". Still, it remained to be shown in a more direct manner that the mechanical load on crossbridges does in fact affect the rate of ADP release and crossbridge detachment. Recently, Veigel et al. (2003) found that this was so. A single crossbridge, or for that matter a single myosin subfragment-1 molecule, attached to an actin filament, could be shown to prolong the lifetime of the attached state in a load-dependent manner when it was strained. In these single molecule experiments the kinetics of force production and attachment time were determined by ensemble averaging the responses of smooth muscle S1 rather than of its skeletal muscle counterpart, because it was at least 10 times slower, thus allowing a much better time resolution of the mechanical events. Chicken gizzard S1 was deposited on microbeads on a solid support, and it was attaching in a cyclic manner to actin filaments held by optical tweezers while being immersed in a solution containing ATP. (To improve time resolution to about 1 ms a small amplitude 1-kHz oscillatory perturbation was applied to one of the optical tweezers thus allowing to detect the mechanical attachment of S1 to actin by the decrease in the amplitude of the 1-kHz sinusoidal signal indicating an increase in stiffness.) The attachment time depended on load as well as on ATP concentration. In the presence of 20 micromolar ATP the life time of attached crossbridges was on average 86 ± 6 ms when they carried no additional load, but 138 ± 7 ms when a load of 1.6 pN was applied.

Like myosin I, but unlike skeletal muscle myosin II (c.f. Veigel et al., 1999), chicken gizzard myosin II produced its working stroke in two steps (Veigel et al., 2003). After the first step of 4 nm occurring within a few milliseconds of myosin binding to actin there was a delay of several milliseconds until a further movement of about 2 nm occurred that was followed by crossbridge detachment. The time interval between the two substeps, the so called intermediate attached state, depended on crossbridge strain or "load", but was independent of the ATP concentration. On the other hand, the time interval between the second substep and crossbridge detachment was (almost) load independent, but shortened by increasing the ATP concentration suggesting that the phase following the second step was terminated by ATP binding. Incidentally, the size of the second substep was also found to be quite similar to the tilting of the "lever arm" (regulatory domain) of smooth muscle myosin S1 induced by the release of bound ADP as deduced from cryoelectronmicroscopy (3 nm, c.f. Whittaker et al., 1995) or EPR spectroscopy (Gollub et al., 1999). Thus, it would be tempting to speculate that the two substeps of the smooth muscle working stroke are coupled to transitions between enzymic states such as Pi release from the active site, followed by ADP release. If this were the case, the load dependent intermediate attached state, lasting from the end of the first substep to the beginning of the second substep, would probably correspond to a load dependent long-lived actomyosin-ADP state as suggested by Veigel et al. (2003). According to Cremo and Geeves (1998), opening of the ADP binding pocket to release ADP might then depend on the mechanical strain on the crossbridge. Thus, with a high load on strained crossbridges ADP release would be slower and the lifetime of attached crossbridges correspondingly longer. Clearly such a dependence of the rate constant of ADP release on crossbridge load might well account for the "Fenn effect", in particular as it was possible to simulate the dependency of the ATP consumption rate on shortening speed (or muscle load) of skinned fibres with a kinetic scheme of the biochemical crossbridge cycle in which the rate of ADP release was depending on the load (c.f. He et al., 2000).

3. CROSSBRIDGE KINETICS DURING TENSION MAINTENANCE IN VERTEBRATE AND INVERTEBRATE SMOOTH MUSCLE

During isometric tension maintenance the force on strained crossbridges is presumably much higher than during shortening, and may possibly reach as much as 3-5 pN. In this case, the life time of the intermediate attached crossbridge state (in which the actin bound myosin heads are probably loaded with ADP) would be greatly extended, and this might well provide a mechanism for economic tension maintenance in a holding function of tonic vetebrate smooth muscle. Figuratively speaking, a long-lived intermediate attached state would probably represent a "locked" state of the crossbridge lever arm, which, after performing the first phase of its working stroke, might be blocked for a moment before completing its movement, releasing ADP, and binding ATP thereby detaching from the actin filament (Rüegg et al., 2002). Such a locking mechanism may then well contribute to the so called latch state in vascular smooth muscle (c.f. Dillon et al., 1981), i.e. a tonic contractile state in which tension is maintained by "latch bridges" that are characterized by a greatly prolonged attachment and cycle time (Veigel et al., 2003). A tonic contraction of vascular smooth muscle is usually initiated by Ca^{2+} and calmodulin dependent myosin light chain kinase which phosphorylates myosin thereby initially causing relatively rapid crossbridge cycling. Subsequently, however, as tension is maintained for longer periods, (attached) crossbridges become dephosphorylated by myosin phosphatase, and this impedes crossbridge detachment according to Dillon et al. (1981) thus forming latch bridges. These are supposedly dwelling for a prolonged time in the above mentioned intermediate attached state as bound ADP is released and replaced by ATP only slowly (c.f. also Butler et al., 1990). However, ADP release can be accelerated by phosphorylation of the regulatory myosin light chain which increases the dissociation constant (and hence the ADP off-rate) of the actomyosin-ADP ternary complex (Gollub et al., 1999), thereby shortening the intermediate attached state and enhancing the rate of crossbridge cycling and ATP splitting. Conversely, ADP release may be inhibited by addition of ADP in high concentration, and this might also slow crossbridge detachment and promote a latch-like state as proposed by Khromov et al. (1998).

Further insight into the latch mechanism has been obtained by comparing latch-like states in intact and chemically skinned smooth muscle fibres in more detail. In a state of tonic contraction of vascular smooth muscle induced by high K^+ saline, the crossbridge cycle duration was about 0.75 s at 37°C (c.f. Paul et al., 1976). When the K^+ concentration of the medium was lowered again, force barely declined but was apparently maintained by very slowly cycling latch bridges characterized by a low detachment rate as shown by Rembold and Murphy (1986) and also suggested by the experiments of Siegman et al. (1985). The reason for the apparent slowing of crossbridges in the above experiments was obviously the removal of K^+ that lowered the intracellular free calcium thus causing a dephosphorylation of the myosin light chains by myosin phosphatase. Using smooth muscle strips of guinea pig *Taenia coli* Güth and Junge (1982) could reproduce a latch-like state in chemically skinned fibres.They showed that following a Ca^{2+} and Calmodulin induced contraction, the rapid removal of Ca^{2+} (and myosin dephosphorylation) does in fact impede crossbridge detachment from the actin filament as claimed by Dillon et al. (1981). Thus, force could be maintained (at low levels of myosin phosphorylation, c.f. Kühn et al., 1990) for many minutes with very little if any energy consumption as indicated by the greatly diminished rate of ATP splitting: In the absence of Ca^{2+} the rate of ATP splitting was at least ten times smaller than during the preceding Ca^{2+} induced contraction (0.25 s^{-1} at 20°C, c.f. Güth and Junge, 1982). However at high Ca^{2+} levels, the ATP-splitting rate was not stable: As shown by Kühn et al. (1990) there was a transition from an initially high ATP-splitting rate to a somewhat lower rate, and such a time dependent change in the metabolic rate was also shown in the case of chemically skinned rabbit portal vein smooth muscle where the ATP-splitting rate declined from 1.8 s^{-1} to 0.3 s^{-1} in the course of prolonged tension maintenance in Ca^{2+} activating ATP-salt solution (He et al., 1998). As ADP is then released only slowly, cycling crossbridges would probably spend a comparatively long time in the intermediate attached state mentioned above, thereby greatly reducing the tension cost (c.f. also Butler et al., 1990).

Recently Butler et al. (2001) proposed that a long-lived intermediate attached state of crossbridges (i.e. an actomyosin ADP-state, c.f. also Takahashi, 1988) may not only account for the latch state in vertebrate smooth muscle but, at least partly, also for the so called catch-phenomenon observed in tonic molluscan smooth muscles such as the slow adductor muscle of the oyster that may keep the shells closed for many hours, apparently without fatigue and barely any oxygen consumption. Another catch muscle, the anterior byssus retractor muscle (ABRM) of the edible mussel *Mytilus edulis* is capable of developing an enormous force (up to 150 N cm^{-2}) when stimulated with acetylcholine, but when it is removed the muscle does not relax. Instead, tension declines extremely slowly and it may even take more that 15 minutes until the force developed during contraction drops to half its value. Thus a remnant tension is observed known as "catch" during which force is maintained despite the fact that the cytosolic free calcium levels reach resting levels (<0.1 μM, c.f. Ishii et al., 1989, and Atsumi and Sugi, 1976). As judged from the rate of oxygen consumption (Baguet and Gillis, 1968), at least ten times less metabolic energy is then required to maintain a given tension than during the contraction proceeding the catch. The latter state could be abolished by the neurotransmitter serotonin that increases intracellular cAMP levels (Achazi et al., 1974) or, alternatively, by dibutyril cAMP which penetrates into the muscle cell. To determine the rate of ATP splitting chemically skinned rather than intact ABRM preparations were used in which a "catch-like" state could be produced as follows. First, a (isometric) contraction was induced by raising the Ca^{2+} concentration in the vicinity of the myofilaments to about 10 μM, and under these conditions tension production and ATP-splitting rate were about 50 N cm^{-2} and 0.6 - 1 s^{-1} , respectively (c.f. Schumacher, 1972, Butler et al., 1998). After lowering the Ca^{2+} concentration to below 0.1 μM, the preparation relaxed extremely slowly though the ATPase was inhibited at once by at least 90%, showing that these fibres entered a catch-like state in which a high tension could be maintained at very low tension cost (Güth et al., 1984, Butler et al., 1998). This suggested that crossbridges detached from actin very slowly, the crossbridge cycling rate being perhaps even as low as 0.2 min^{-1} (c.f. Butler et al., 2001). Addition of cAMP or of the catalytic subunit of protein kinase A rapidly abolished this catch-like state in skinned fibres (Cornelius, 1982, Pfitzer and Rüegg, 1982). Protein kinase A is known to phosphorylate two aspartic acid residues within the Ig-domains of twitchin, a kind of mini-titin located on the thick filaments of catch muscles (Funabra et al., 2003). If these residues are phosphorlylated or thiophosphorylated, skinned fibres cannot enter the catch state suggesting that twitchin phosphorylation must be implicated in the regulation of catch.

In view of the "locked actomyosin-ADP" hypothesis of the catch mechanism mentioned above it is tempting to speculate that twitchin phosphorylation might cause a decrease in the ADP-affinity of cycling crossbridges thereby accelerating the rate of ADP release, but additionally (or perhaps alternatively) it might also affect force maintaining structures different from cycling actin-myosin cross linkages (Butler et al., 1998, c.f. also Galler et al., 1999). These structures might involve non-cycling actin-myosin cross linkages (Butler et al., 2001) or thick filament interconnections of the type recently studied by Takahashi et al. (2003) that hold tension passively when twitchin is dephosphorylated. In any case, actively contracting catch muscles may enter the catch state when cycling crossbridges are trapped at low free calcium levels in a long-lived actomyosin-ADP state, stabilized perhaps by the high tension that pulls on crossbridges in the intermediate attached state described above (c.f. also Rüegg et al., 2002).

4. CONCLUSIONS AND PERSPECTIVES

Comparative approaches make it possible to find out what features of the crossbridge mechanism are common and fundamental to all muscles and which of them are variable and specializations that may have evolved because of their suitability to different functions. As we have seen, the kinetics of crossbridges differ greatly in various kinds of muscles, fast and slow, striated and smooth. While crossbridges specialized for rapid shortening are characterized by high detachment- and ATP-splitting rates, certain slow smooth muscle myosins may be capable of holding the crossbridges in a strain dependent manner for a fraction of a second or possibly even longer in an intermediate attached state (actomyosin-ADP state) before completing the lever arm movement, releasing ADP and detaching from actin (Veigel et al., 2003). Clearly a strain dependent ADP release and detachment from the actin filaments would allow crossbridges to hold tension economically when strained, but also to adapt their function to the work load thus causing the "Fenn effect". This is because the overall ATP-splitting rate would be enhanced when the elastic elements of the crossbridges are released and decreases when they are stretched under load. While it has yet to be proven in forthcoming experiments that ADP release from crossbridges does depend on crossbridge strain as postulated e.g. by Cremo and Geeves (1998), it could already be shown that AMP-PNP (β,γ imido ATP, an ADP or ATP analogue which is not split by myosin) binds to skinned fibres of the dorsolongitudinal muscle of *Lethocerus* in a crossbridge-strain dependent manner (c.f. Kuhn, 1981 for review). Indeed, chemically skinned fibres from this highly specialized cross striated muscle have been found ideal for studies on crossbridge function and in particular on the conformational changes that may be induced by ATP and its analogues (e.g. Reedy et al., 1965, Goody et al., 1976, Beinbrech et al., 1976).

Acknowledgements. I am grateful to A. Ebling for the careful preparation of the manuscript and to the "Fonds der Chemischen Industrie" for supporting the work in my laboratory as well as to Claudia Veigel for helpful discussions.

5. REFERENCES

Achazi, R. K., Dölling, B., and Haakhorst, R., 1974, 5-HT-induzierte Erschlaffung und cyklisches AMP bei einem glatten Molluskenmuskel, *Pflügers Arch. Eur. J. Physiol.* **349:**19-27.

Atsumi, S., and Sugi, H., 1976, Localisation of calcium-accumulating structures in the anterior byssal retractor muscle of *Mytilus edulis* and their role in the regulation of active and catch contractions, *J. Physiol.* **257**:549-560.

Baguet, N., and Gillis, J. M., 1968, Energy cost of tonic contraction in a lamellibranch catch muscle, *J. Physiol. (Lond.)* **198**:127-143.

Beinbrech, G., Kuhn, H. J., Herzig, J. W., and Rüegg, J. C., 1976, Evidence for two attached cross-bridge states of different potential energy, *Cytobiology* **12:**385-396.

Bottinelli, R., Schiaffino, S., and Reggiani, C., 1991, Force-velocity relations and myosin heavy chain isoform compositions of skinned fibres from rat skeletal muscle, *J. Physiol. (Lond.)* **437**:655-672.

Bottinelli, R., Canepari, M., Reggiani, C., and Stienen, G. J. M., 1994, Myofibrillar ATPase activity during isometric contraction and isomyosin composition in rat single skinned muscle fibres, *J. Physiol. (Lond.)* **481**:663-675.

Butler, T. M., Siegman, M. J., Mooers, S. U., and Narayan, S. R., 1990, Myosin-product complex in the resting state and during relaxation of smooth muscle, *Am. J. Physiol.* **258:**C1092-C1099.

Butler, T. M., Mooers, S. U., Li, C., Narayan, S., and Siegman, M. J., 1998, Regulation of catch muscle by twitchin phosphorylation: effects on force, ATPase, and shortening, *Biophys. J.* **75**:1904- 1914.

Butler, T. M., Narayan, S. R., Mooers, S. U., Hartshorne, D. J., and Siegman, M. J., 2001, The myosin cross-bridge cycle and its control by twitchin phosphorylation in *catch* muscle, *Biophys. J.* **80**:415-426.

Cornelius, F., 1982, Tonic contraction and the control of relaxation in a chemically skinned molluscan smooth muscle, *J. Gen. Physiol.* **79**:821-834.

Cremo, C. R., and Geeves, M. A., 1998, Interaction of actin and ADP with the head domain of smooth muscle myosin: implications for strain dependent ADP release in smooth muscle, *Biochemistry* **37**:1969-1978.

Curtin, N. A., Gilbert, C., Kretzschmar, K. M., and Wilkie, D. R., 1974, The effect of performance of work on total energy output and metabolism during muscular contraction, *J. Physiol. (Lond.)* **238**:455-472.

Dillon, P. F., Aksoy, M. O., Driska, S. P., and Murphy, R. A., 1981, Myosin phosphorylation and the cross-bridge cycle in arterial smooth muscle, *Science* **211**:495-497.

Fenn, W. O., 1924, The relation between the work performed and the energy liberated in muscular contraction, *J. Physiol. (Lond.)* **58**:373-395.

Funabara, D., Watabe, S., Mooers, S. U., Narayan, S., Dudas, C., Hartshorne, D. J., Siegman, M. J., and Butler, T. M., 2003, Twitchin from molluscan catch muscle: primary structure and relationship between site-specific phosphorylation and mechanical function, *J. Biol. Chem.* **278**:29308-29316.

Galler, S., Hilber, K., and Pette, D., 1996, Force responses following stepwise length changes of rat skeletal muscle fibre types, *J. Physiol. (Lond.)* **493**: 219-227.

Galler, S., Kögler, H., Ivemeyer, M., and Rüegg, J. C., 1999, Force responses of skinned molluscan catch muscle following photoliberation of ATP, *Pflügers Arch. Eur. J. Physiol.* **438**:525-530.

Gollub, J., Cremo, C. R., and Cooke, R., 1999, Phosphorylation regulates the ADP-induced rotation of the light chain domain of smooth muscle myosin, *Biochemistry* **38**:10107-10118.

Goody, R. S., Barrington, L. J., Mannherz, H. G., Tregear, R. T., and Rosenbaum, G., 1976, X-ray titration of binding β,γ–imido-ATP to myosin in insect flight muscle, *Nature* **262**:613-614.

Güth, K., and Junge, J., 1982, Low Ca^{2+} impedes cross-bridge detachment in chemically skinned *Taenia coli, Nature* **300**:775-776.

Güth, K., Kuhn, H. J., Tsuchiya, T., and Rüegg, J. C., 1981, Length dependent state of activation - Length change dependent kinetics of cross bridges in skinned insect flight muscle, *Biophys. Struct. Mech.* **7**:139-169.

Güth, K., Poole, K. J. V., Maughan, D., and Kuhn, H. J., 1987, The apparent rates of crossbridge attachment and detachment estimated from ATPase activity in insect flight muscle, *Biophys. J.* **52**:1039-1045.

Güth, K., Gagelmann, M., and Rüegg, J. C., 1984, Skinned smooth muscle: time course of force and ATPase activity during contraction cycle, *Experientia* **40**:174-176.

He, Z. H., Ferenczi, M. A., Brune, M., Trentham, D. R, Webb, M. R., Somlyo, A. P., Somlyo, A. V., 1998, Time-resolved measurements of phosphate release by cycling cross-bridges in portal vein smooth muscle, *Biophys. J.* **75**:3031-3040.

He, Z. H., Bottinelli, R., Pellegrino, M. A., Ferenczi, M.A., and Reggiani, C., 2000, ATP consumption and efficiency of human single muscle fibers with different myosin isoform composition, *Biophys. J.* **79**:945-961.

Huxley, A. F., 1957, Muscle structure and theories of contraction, *Prog. Biophys. Chem.* **7**:255-318.

Huxley, A. F., and Simmons, R. M., 1971, Proposed mechanism of force generation in striated muscle, *Nature* **233**:533-538.

Huxley, H. E., 1969, The mechanism of muscular contraction, *Science* **164**:1356-1366.

Ishii, N., Simpson, A. W. M., and Ashley, C. C., 1989, Free calcium at rest during "catch" in single smooth muscle cells, *Science* **243**:1367-1368.

Jewell, B. R., and Rüegg, J. C., 1966, Oscillatory contraction of insect fibrillar muscle after glycerol extraction, *Proc. R. Soc. (Lond.) B.* **165**:428-459.

Khromov, A., Somlyo, A. V., and Somlyo, A. P., 1998, Mg-ADP promotes a catch-like state developed through force-calcium hysteresis in tonic smooth muscle. *Biophys. J.* **75**:1926-1934.

Krebs, H., 1975, The August Krogh Principle „For many problems there is an animal on which it can be most conveniently studied", *J. Exp. Zool.* **194**:221-226.

Kuhn, H. J., 1981, The mechanochemistry of force production in muscle, *J. Muscle Res. Cell Mot.* **2**:7-44

Kühn, H., Thewes, A., Gagelmann, M., Güth, K., Arner, A., and Rüegg, J. C., 1990, Temporal relationship between force, ATPase activity, and myosin phosphorylation during a contraction/relaxation cycle in a skinned smooth muscle, *Pflügers Arch. Eur. J. Physiol.* **416**:512-518.

Paul, R. J., Glück, E., and Rüegg, J. C., 1976, Cross bridge ATP utilization in arterial smooth muscle, *Pflügers Arch. Eur. J. Physiol.* **361**:297-299.

Pfitzer, G., and Rüegg, J. C., 1982, Molluscan catch muscle: regulation and mechanics in living and skinned anterior byssus retractor muscle of *Mytilus edulis*, *J. Comp. Physiol.* **147**:137-142.

Reedy, M. K., Holmes, K. C., and Tregear, R. T., 1965, Induced changes in orientation of the cross-bridges of glycerinated insect flight muscle. *Nature* **207**:1276-1280.

Reiser, P. J., Moss, R. L., Giulian, G. G., and Greaser, M. L., 1985, Shortening velocity in single fibres from adult rabbit soleus muscles is correlated with myosin heavy chain composition, *J. Biol. Chem.* **260**:9077-9089.

Rembold, C. M., and Murphy, R. A., 1986, Myoplasmic calcium, myosin phosphorylation, and regulation of the crossbridge cycle in swine arterial smooth muscle, *Circ. Res.* **58**:803-815.

Rüegg, J. C., 1971, Smooth muscle tone, *Physiol. Rev.* **51**:201-248.

Rüegg, J. C., and Tregear, R. T., 1966, Mechanical factors affecting the ATPase activity of glycerol-extracted insect fibrillar muscle, *Proc. R. Soc. (Lond.) B.* **165**:497-512.

Rüegg, J. C., Veigel, C., Molloy, J. E., Schmitz, S., Sparrow, J.C., and Fink, R. H. A., 2002, Molecular motors: force and movement generated by single myosin II molecules, *News Physiol. Sci.* **17**:213-218.

Schumacher, T., 1972, Zum Mechanismus der ökonomischen Halteleistung eines glatten Muskels (Byssus retractor anterior, *Mytilus edulis*), *Pflügers Arch. Eur. J. Physiol.* **331**:77-89.

Siegman, M. J., Butler, T. M., and Mooers, S. U., 1985, Energetics and regulation of crossbridge states in mammalian smooth muscle, *Experientia* **41**:1020-1025.

Steiger, G. J., and Rüegg, J. C., 1969, Energetics and „efficiency" in the isolated contractile machinery of an insect fibrillar muscle at various frequencies of oscillation, *Pflügers Arch. Eur. J. Physiol.* **307**:1-21.

Sugi, H., Iwamoto, H., Akimoto, T., and Ushitani, H., 1998, Evidence for the load-dependent mechanical efficiency of individual myosin heads in skeletal muscle fibers activated by laser flash photolysis of caged calcium in the presence of a limited amount of ATP, *Proc. Natl. Acad. Sci. USA* **95**:2273-2278.

Takahashi, M., Sohma, H., and Morita, F., 1988, The steady state intermediate of scallop smooth muscle myosin ATPase and effect of light chain phosphorylation. A molecular mechanism for catch contraction, *J. Biochem.* **104**:102-107.

Takahashi, I., Shimada, M., Akimoto, T., Kishi, T., and Sugi, H., 2003, Electron microscopic evidence for the thick filament interconnections associated with the catch state in the anterior byssal retractor muscle of Mytilus edulis, *Comp. Biochem. Physiol. A. Mol. Integr. Physiol.* **134**:115-20.

Tsuchiya, T., Güth, K., Kuhn, H. J., and Rüegg, J. C., 1982, Decrease in stiffness during shortening in calcium activated skinned fibres, *Pflügers Arch. Eur. J. Physiol.* **392**:322-326.

Ulbrich, M., and Rüegg, J. C., 1976, Is the chemomechanical energy transformation reversible? *Pflügers Arch. Eur. J. Physiol.* **363**:219-222.

Veigel, C., Collucio, L. M., Jontes, J. D., Sparrow, J. C., Milligan, R. A., and Molloy, J. E., 1999, The motor protein myosin-1 produces its working stroke in two steps, *Nature* **398**:530-533.

Veigel, C., Molloy, J. E., Schmitz, S., and Kendrick-Jones, J., 2003, Load-dependent kinetics of force production by smooth muscle myosin measured with optical tweezers, *Nature Cell Biology* **5**:980-986

Whittaker, M., Wilson-Kubalek, E. M., Smith, J. E., Faust, L., Milligan, R. A., and Sweeney, H. L., 1995, A 35 - A movement of smooth muscle myosin on ADP release, *Nature* **378**:748-751.

DYNAMIC STRUCTURES OF MYOSIN, KINESIN AND TROPONIN AS DETECTED BY SDSL-ESR

Toshiaki Arata, Motoyoshi Nakamura, Shoji Ueki, Tomoki Aihara, Kazunori Sugata, Hiroko Kusuhara, and Yukio Yamamoto*

1. INTRODUCTION

In 1952 late Profs. Yuji Tonomura and Shizuo Watanabe found a rapid light scattering change (conformational change known as dissociation) of actomyosin by a stoichiometric amount of ATP[1]. This work was extended to x-ray scattering study of S1 done by us in collaboration with K. Wakabayashi[2]. Tonomura and his collaborators dicovered a rapid burst of protein-bound phosphate in the early phase of myosin ATPase reaction, formation of a key intermediate, M.ADP.P[3]. His famous hypothesis for myosin ATPase is "two pathway mechanism or non-identical two headed mechanism" which was proposed from the result that the size of phosphate burst is 0.5 mol/mol myosin head. This hypothesis was against and debated extensively with "single pathway mechanism" of Dr. Edwin Taylor[4]. This may be related to the coordination of two heads. In the other story two cDNAs were recently cloned from rabbit skeletal muscle and showed that several amino acids are different each other and positively and negatively charged in the coiled-coil neck region to make a heterodimer[5]. His hypothesis will be tested by separately expressing these two genes in muscle cells. In 1973, he and his collaborators found a direct ATP hydrolysis pathway[6] that includes weakly and strongly bound actomyosin states. During isometric contraction, strongly bound intermediate is expected to hold tension, as detected by spin-label ESR[7].

In relation to the finding of Tonomura and Watanabe[1] in 1952 the ultimate goal in biological functions of molecular motor protein actin-myosin complex[8] and Ca-regulatory protein troponin-tropomyosin-actin complex[9,10] in muscle has been to find force-generating or switching protein structural changes. The ESR spectroscopy provides dynamics, orientation, and spatial information of spin-labels specifically attached to the region of protein (SDSL-ESR). To see how motor and regulatory

* Toshiaki Arata, Motoyoshi Nakamura, Tomoki Aihara, Shoji Ueki, Kazunori Sugata, and Hiroko Kusuhara, Department of Biology, Graduate School of Science, Osaka University, Toyonaka, Osaka 560-0043, Japan. Yukio Yamamoto, Graduate School of Human and Environmental Studies, Kyoto University, Kyoto 605-8501, Japan.

proteins work in muscle, we have extensively studied on spin-labeled myosin light chain and spin-labeled troponin both exchanged into glycerinated muscle fibers. These results are presented in this paper and provide insight into the mechanism how myosin heads or troponin move their domains to function in muscle. We will extend these studies to the other molecular function systems such as microtubule (tubulin)-based motor kinesin or dynein and transporting machinery in membrane. Here we show the recent study on kinesin[11-14]. A part of the data were reported elsewhere[15,16].

2. MATERIALS AND METHODS

2.1. Spin Labeling of Myosin Light Chain

We used myosin RLC from rabbit skeletal muscle which contains two cysteines that react differentially and achieve specific spin-labeling at C-terminal domain (cys154) as will be described elsewhere and is expected to bind the heavy chain properly in a parent rabbit muscle when exchanged with native RLC. Spin-labeled RLC was prepared and exchanged into muscle fibers as described previously[17]. It was purified from rabbit skeletal myosin which had been labeled with a cysteine-specific spin probe, 4-maleimido-2,2,5,5-tetramethyl-1-pyrrolidinyloxy (MSL) by controlling an incubation time to 18 hr. Rigor buffer contained 20 mM imidazale, 5 mM $MgCl_2$, 1 mM EGTA. Relaxation and contraction buffers contained, in addition, 5 mM ATP, 30 mM creatine phosphate, and 0.25 mg/ml creatine phosphokinase; in contraction, pCa was 4.0. Ionic strength was adjusted to 200 mM with KCl, and pH was adjusted to 7.0.

2.2. Spin Labeling of Kinesin

Ser332 was substituted for a cysteine residue in a mouse kinesin K393 construct[18,19] and spin-labeling procedure was previously described in detail[15]. Immediately after the purification of expressed kinesin K393 protein, an 8-fold molar excess of MSL was reacted with purified K393 overnight at 4 °C. Excess spin label was removed by dialysis. The labeling efficiency was determined by measuring protein concentration and spin concentration estimated by double integration of ESR spectra. Tubulin was prepared from Bovine brain and polymerized to microtubules as described[20]. For ESR measurements, dimeric kinesin was incubated with either 10 mM Mg-AMPPNP, 2 mM Mg-ADP, or 5 U/ml apyrase. A sample for the sulfate-bound experiments contained 1M lithium sulfates with 0.2 mM ADP. For kinesin-microtubule samples, 10-20 μM kinesin was mixed with 40-80 μM polymerized tubulin, the complex centrifuged, and microtubule pellet measured.

2.3 Spin Labeling of Troponin

Human cardiac troponin C (TnC) was purified from *E. coli.* expressing TnC with mutational substitution to cysteine residue, and labeled with mono-(MSL) or bifunctional spin label (BSL) at excess molar ratio. BSL was synthesized as will be described elsewhere.

2.4. Preparation of Reconstituted Muscle Fibers

Skinned fiber bundles, approximately 0.5 mm in diameter, were prepared from rabbit *psoas* muscle fibers and were extracted removing the native RLC or TnC and replaced with a spin-labeled RLC or TnC as described previously[17] except that the temperature was kept at 4°C for TnC exchange.

2.5. ESR Measurements

High-resolution X-band detection from spin labeled kinesin in solution and troponin in solution and reconstituted muscle fibers was made by Bruker E-500 ESR equipment. The LC domain probe orientation was obtained from X-band ESR spectra of the fiber bundles placed at 15 - 20°C in a capillary that is set from a side hole in a specially-made TM_{110} cavity[21] with a continuous perfusion system connected to the capillary as described previously[7]. Spectral simulations and nonlinear least-squares fits for orientational distribution of spin label z-axis were performed on Igor Pro Software[17,21] (WaveMetrics, Inc., Lake Oswego, Oregon, USA). Spectral resolution into components with different rotational correlation times was analyzed as described by Freed and collaborators using stochastic Liouville approach and fully anisotropic motion in both rates and amplitudes for two spectral components[22]. Spectral broadening of the double labeled samples, compared with the two single labeled composite spectra, was analyzed using the Fourier deconvolution and the distance distribution was finally obtained[23]. Pulsed ELDOR was performed and analyzed as will be described elsewhere.

3. RESULTS AND DISCUSSION

3.1. Two Distinct Spectral (Orientational) Components of LC Domain Probe

ESR has the orientational resolution needed to detect multiple orientations of nitroxide spin labels, because each spin label orientation corresponds to a unique spectral lines. ESR spectra of skinned muscle fibers spin-labeled specifically at cys154 of RLC were measured in parallel orientation to the magnetic field (data not shown). In rigor and relaxation, the spectra showed those similar to randomly-oriented sample. However, extensive spectrum analysis showed that they are resolved into different relative intensities of two broad but distinct angular distributions. Digital subtraction of any two experimental spectra [i.e. relax – 0.81(rigor) or rigor – 0.77(relax)] unambiguously resolved the same two distinct spectral components, each of which was uniquely simulated as a single Gaussian population having central angle of 50° or 75° and full-width of ~40° with respect to the fiber axis. Therefore, the spectra in three physiological states are described by different linear combinations of the same two distinct spectral components corresponding to two distinct orientational populations.

3.2. Two Orientations of LC Probe in Relaxation, Rigor and Active Contraction

For each physiological state, the relative contributions of spectral component 1 and

2 to the experimental spectrum determine the mole fractions, p1 and p2. In relaxation, component 1 (center 48°± 4° full-width 42°± 3°) and 2 (center 75°± 3° full-width 41°± 4°) are equally populated (p1 = 0.51 ± 0.03). In the absence of ATP (in rigor), the spectrum shows a marked increase in 1 (p1 = 0.63 ± 0.03). Baker et al.[24] showed that a disordered population (~40%) was required to completely describe the experimental spectra of all three physiological states. When about 50% of randomly-oriented component, p3=0.5, was supplemented for spectrum fitting, two relative populations were determined similarly. In relaxation, component 1 (center 50°± 4° full-width 40°± 3°) and 2 (center 75°± 3° full-width 41°± 4°) are equally populated (p1 = 0.27 ± 0.02, p2=0.23 ± 0.02). In rigor, the spectrum shows a remarkable increase in 1 (p1 = 0.41 ± 0.02, p2=0.09 ± 0.02). However, it is difficult to tell which fit is better.

Calcium activation changed the spectral properties of component 2. Digital subtraction of two experimental spectra [i.e. active − 0.81(rigor)] provided a distinct spectral component, which was simulated as a single Gaussian population having central angle of 75° and full-width of ~25° with respect to the fiber axis. The full-width decreased markedly without changing population. Component 1 (center 48°± 4° full-width 36°± 4°) and 2 (center 73°± 4° full-width 24°± 4°) are equally populated (p1 = 0.50 ± 0.02). In the second model, component 1 (center 50°± 4° full-width 40°± 4°) and 2 (center 75°± 4° full- width 25°± 3°) are equally populated (p1 = 0.28 ± 0.02, p2 = 0.22 ± 0.02) with random component (p3=0.50).

Control experiments indicate that the probe nanosecond mobility does not change with respect to changes in the physiological state.

3.3. Implication from LC Probe Orientation for Mechanism of Muscle Contraction

As a result from above, ESR resolved a change in the mole fractions of two orientational populations of the LC domain (cys154) rather than a change in its actual orientation. Therefore, we have detected two broad spin-label orientations of myosin LC domain, axially separated by at least 25°, in muscle fibers (Fig. 1). In the weak-binding or relaxed state, these two LC domain orientations are equally populated and so two neck LC domains of myosin molecules are probably ordered by interactions with the core of the myosin filament. The distribution between these two myosin structures is modulated primarily by the weak (relaxed)-to-strong (rigor) actin-binding transition. A fraction (10-15%) of myosin heads undergoes a 25° rotation of the LC domain when they shift from relaxed to rigor state. If disordered population (~50%) would not be counted for, this estimation becomes larger (up to 20-30%). This rotation is consistent with other observed force-generating myosin transitions.

If the RLC domain structure of p1 is stabilized when the main head portion (catalytic domain) binds strongly to actin, myosin heads with the p2 orientation would rotate more than 25° to p1 on muscle activation, resulting in filament sliding and force generation. However, upon activation, the shift from 2 to 1 was almost undetectable. The fraction of strong-binding heads is about 20-30%7, and then the ESR-spectral shift is estimated to be only 2.0-4.5% that is undetectable. It should be noted that a large fraction (17%) of heads rotate LC spin probe upon activation of scallop muscle[24]. Interestingly, upon activation the distributions of probe orientation, especially p2 (75° orientation), become about 2-fold narrower. This suggests that the LC domain of most heads has better ordered conformation during active contraction than in relaxation. The main head portion with the LC domain orientation around 75° would undergo a weak-to-strong actin-binding transition[25,26] or translocation[27] on the actin filament during muscle activation. Alternatively, two LC orientations of a myosin molecule alternate in highly cooperative manner (see INTRODUCTION) like a flip-flop device; one head moves from 50° to 75° and simultaneously the other moves from 75° to 50°. In this model, a half of the LC domains rotate by more than 25° and the two conformations are equally populated always.

Here, we showed two orientational populations of LC domain. However, many studies showed considerably different results (Fig. 1), suggesting that the LC domain has flexibility or many conformations.

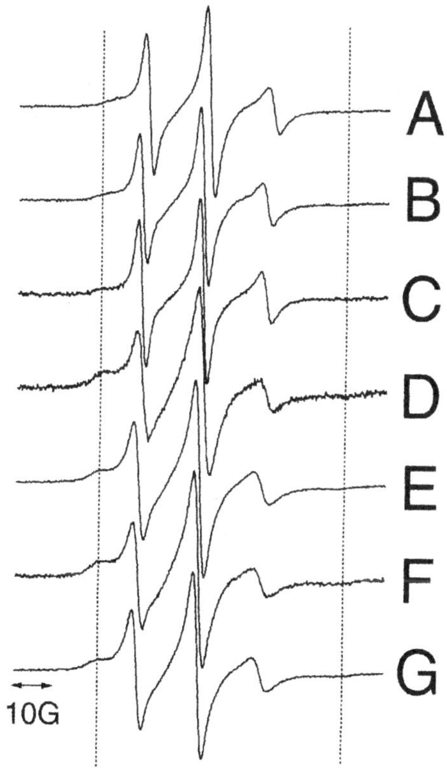

Fig. 1 ESR spectra for kinesin dimer Cys332-MSL in several nucleotide conditions, in both the absence and presence of microtubules (MTs). A: ADP, B: AMPPNP, C: No nucleotide, D: ADP + $LiSO_4$, E: ADP + MT, F: AMPPNP + MT, G: No Nucleotide + MT. Kinesin dimer bound to microtubule is drawn on the top. Two red balls indicate the position of Cys332. The vertical dashed lines in the ESR spectra show the spectral components with peaks that are indicative of restricted mobility of a spin probe attached to the neck-linker domain.

3.4. Motion of Neck Linker of Kinesin Dimer

Conventional kinesin is a highly processive motor that converts the chemical energy of ATP hydrolysis into the unidirectional motility along microtubules[11-14]. The processivity is thought to depend on the coordination between the ATPase cycles of two motor domains and their neck linkers[28-30]. Here we have used site-directed spin labeling electron spin resonance (SDSL-ESR) to determine the conformation of the neck linker (Cys332-MSL) in kinesin dimer in the presence and absence of microtubules. The spectra show that the neck linkers co-exist in both docked and disordered conformations, which is consistent with the results of monomeric kinesin[31,32]. In all nucleotide states (no nucleotide, AMPPNP, ADP), however, the spectra appear virtually the same suggesting that the neck linkers are mostly (~60%) ordered when dimeric kinesin is bound to the microtubule (Fig. 1). This result suggests that the orientation of each neck linker that is fixed rigidly controls the kinesin motion along microtubule tracks[15] or that a flip-flop device made by alternate docked-undocked conformations always works by highly cooperative manner between two motor domains.

We have measured dipolar interactions between the neck linkers of dimeric kinesin. No evidence of spectral line broadening for any of the doubly labeled samples in the absence of microtubules suggest that all of the distances are longer than 1.8-2.0 nm, the limit of the dipolar ESR method. In the presence of microtubules ESR spectra were broadened in all nucleotides, indicating that there were the spin-spin interactions. In ADP-bound state the shortest distance was 1.5 nm with broader distribution. The distance in AMPPNP or ATP□S state was 1.5 nm with the narrowest distribution. The shortest distance in the absence of nucleotides was 1.85 nm with broad distribution. These results strongly suggest that kinesin dimer really changes the distance between two neck linkers during ATPase cycle on the microtubules.

3.5. Mobility and Site-Proximity of Troponin C and I as Studied by a Mono- and Bifunctional Spin Label

Troponin is the first-discovered switch protein functioning for signal transduction in muscle, as extensively descibed by Prof. I. Ohtsuki in this issue. We have studied structures of troponin complex with SDSL-ESR (site-directed spin labeling-electron spin resonance). We have measured the distances between 35 Cys and 84 Cys on human cardiac troponin C by the method of SDSL-PELDOR (pulsed electron-electron double resonance). The distance was 2.6 and 2.3 nm in Ca2+ -saturated monomer and fiber states, respectively. The distance increased remarkably to 2.7 nm in fiber but only slightly in monomer by removal of Ca2+, suggesting that the hydrophobic patch in the N-lobe closes to much greater extent by dissociation of Ca2+ from TnC in fiber than in monomer state[33].

Spin label of native cysteine on skeletal TnI regulatory domain (TnI_{reg}) showed a large rotational mobility change from subnanosecond to nanosecond time scale by Ca^{2+} ion (data not shown), suggesting a switch movement of TnI_{reg}[10].
We are also investigating Ca^{2+} induced structural change of TnC using a new bifunctional spin label (BSL) which was synthesized for the purpose of getting more precise information from probes attached rigidly at known coordinate of TnC. ESR spectra showed a large restriction of spin label motion when the BSL replaced a monofunctional

spin label at a pair of amino acid residues: 84- and 91-residues or 84- and 98-residues. Rotational correlation time increased to about 7 ns which corresponds to that of rigid-body rotation of TnC molecule[34]. Therefore, we succeeded, for the first time, in fixing the spin label rigidly on TnC molecule in free state (either in $\pm Ca^{2+}$). This provides a promise that we can determine the precise coordinate of the spin principal axis of BSL on protein surface. This study is now in progress to know the orientation of the vector between two amino acid residues on TnC
relative to the fiber axis and to determine the coordinates of the TnC molecule on the actin filament.

ACKNOWLEDGEMENTS

We are grateful to Drs. Yuichiro Maeda and Soichi Takeda for generously giving us human TnC gene and helpful discussions. We thank Dr. Peter G Fajer for kinesin ESR measurements and analysis done in Florida State University. We also thank Dr. Hideyuki Hara for pulsed ELDOR experiments and analysis at Bruker Biospin, Tsukuba. Finally, we are supported by Special Coordination Funds for promoting Science and Technology and by Grants-in-Aid for Scientific Research, from Ministry of Education, Science, Technology, Culture and Sports of Japanese Government.

REFERENCES

1. Y. Tonomura and S. Watanabe, Effect of adenosine triphosphate on the light scattering of actomyosin solution, *Nature* 169, 112-113 (1952)
2. T. Arata, S. Kimura, Y. Sugimoto, Y. Takezawa, N. Iwasaki, and K. Wakabayashi, Structure of the monomeric actin-myosin head complex as revealed by X-ray solution scattering, *Adv Exp Med Biol.* 453, 73-77 (1998)
3. Y. Tonomura, S. Kitagawa, and J. Yoshimura, The initial phase of myosin A-adenosinetriphosphatase and the possible phosphorylation of myosin A, *J. Biol. Chem.* 237, 3660-3666 (1962)
4. Y. Tonomura and A. Inoue, There must be two pathways; E.Taylor, A single pathway for myosin ATP hydrolysis is sufficient to explain all the data, *Trends Biochem. Sci.* N32-N35 (1977).
5. A. Inoue, personal communication.
6. A. Inoue, M. Shigekawa, and Y. Tonomura, Direct evidence for the two route mechanism of the acto-H-meromyosin-ATPase reaction, *J. Biochem (Tokyo)* 74, 923-934 (1973).
7. T. Arata, and H. Shimizu, Spin-label study of actin-myosin-nucleotide interactions in contracting glycerinated muscle fibers. *J. Mol. Biol.* 151, 411-437 (1981).
8. H. E. Huxley , Fifty years of muscle and the sliding filament hypothesis, *Eur J Biochem.* 271, 1403-1415 (2004).
9. S. Ebashi, M. Endo, and I. Otsuki, Control of muscle contraction, *Q. Rev. Biophys.* 2, 351-384 (1969).
10. S. Takeda, A. Yamashita, K. Maeda, and Y. Maeda, Structure of the core domain of human cardiac troponin in the Ca^{2+}-saturated form. *Nature* 424, 35-41 (2003).
11. N. Hirokawa, Kinesin and dynein superfamily proteins and the mechanism of organelle transport. *Science* 279, 519-526 (1998).
12. K. Hirose, and L. A. Amos, Three-dimensional structure of motor molecules, *Cell Mol Life Sci* 56, 184-199 (1999).
13. E. Mandelkow, and A. Hoenger, Structures of kinesin and kinesin-microtubule interactions, *Curr. Opin. Cell. Biol.* 11, 34-44 (1999).
14. R. D. Vale, and R. A. Milligan, The way things move: looking under the hood of molecular motor proteins, Science 288, 88-95 (2004).
15. K. Sugata, M. Nakamura, S. Ueki, P. G. Fajer, and T. Arata, ESR reveals the mobility of the neck linker in dimeric kinesin, *Biochem. Biophys. Res. Commun.* 314, 447-451 (2004).

16. T. Arata, M. Nakamura, H. Akahane, T. Aihara, S. Ueki, K. Sugata, H. Kusuhara, M. Morimoto, and Y. Yamamoto, Orientation and motion of myosin light chain and troponin in reconstituted muscle fibers as detected by ESR with a new bifunctional spin label, *Adv Exp Med Biol.* 538, 279-283 (2003).
17. T. Arata, Orientation of spin-labeled light chain 2 of myosin heads in muscle fibers, *J. Mol. Biol.* 214, 471-478 (1990).
18. L. J. Brown, K. L. Sale, R. Hills, C. Rouviere, L. Song, X. Zhang, and P. G. Fajer, Structure of the inhibitory region of troponin by site directed spin labeling electron paramagnetic resonance, *Proc. Natl. Acad. Sci. U S A* 99, 12765-12770 (2002).
19. H. Shibuya, K. Kondo, N. Kimura, and S. Maruta, Formation and characterization of kinesin.ADP. fluorometal complexes, *J. Biochem (Tokyo)* 132, 573-579 (2002).
20. M. Kikkawa, T. Ishikawa, T. Nakata, T. Wakabayashi, and N. Hirokawa, Direct visualization of the microtubule lattice seam both in vitro and in vivo, *J. Cell Biol.* 127, 1965-1971 (1994).
21. T. Arata, The use of spin probes (Chapter 9), In *Current Methods in Muscle Physiology*, ed. H. Sugi, pp. 223-239, Oxford University Press, Oxford (1998).
22. D. E. Budil, S. Lee, S. Saxena, and J. H. Freed, Nonlinear-least -squares analysis of slow-motion EPR spectra in one and two dimensions using a modified levenberg-Marquardt algorithm, *J. Mag. Res.* 120, 155-189 (1996).
23. C. Altenbach, K. J. Oh, R. J. Trabanino, K. Hideg, and W. L. Hubbell, Estimation of inter-residue distances in spin labeled proteins at physiological temperatures: experimental strategies and practical limitations, *Biochemistry* 40, 15471-15482 (2001).
24. J. E. Baker, I. Brust-Mascher, S. Ramachandran, L. E. LaConte, and D. D. Thomas, A large and distinct rotation of the myosin light chain domain occurs upon muscle contraction, *Proc. Natl. Acad. Sci. U S A.* 95, 2944-2949 (1998).
25. T. Arata, Structure of the actin-myosin complex produced by crosslinking in the presence of ATP, *J Mol Biol.* 191, 107-116 (1986).
26. D. D. Thomas, S. Ramachandran, C. Roopnarine, D. W. Hayden, and E. M. Ostap, The mechanism of force generation in myosin: a disorder-to-order transition, coupled to internal structural changes, Biophys J. 68, 135S-141S (1995).
27. T. Arata, Electron microscopic observation of monomeric actin attached to a myosin head, *J. Struct. Biol.* 123, 8-16 (1998).
28. R. B. Case, D. W. Pierce, N. Hom-Booher, C. L. Hart, and R. D. Vale, The directional preference of kinesin motors is specified by an element outside of the motor catalytic domain, *Cell* 90, 959-966 (1997).¥
29. U. Henningsen, and M. Schliwa, Reversal in the direction of movement of a molecular motor, *Nature* 389, 93-96 (1997).
30. S. A. Endow, and K. W. Waligora , Determinants of kinesin motor polarity, *Science* 281, 1200-1202 (1998).
31. S. Rice, A. W. Lin, D. Safer, C. L. Hart, N. Naber, B. O. Carragher, S. M. Cain, E. Pechatnikova, E. M. Wilson-Kubalek, M. Whittaker, E. Pate, R. Cooke, E. W. Taylor, R. A. Milligan, and R. D. Vale, A structural change in the kinesin motor protein that drives motility. *Nature* 402, 778-784 (1999).
32. C. V. Sindelar, M. J. Budny, S. Rice, N. Naber, R. Fletterick, and R. Cooke, Two conformations in the human kinesin power stroke defined by X-ray crystallography and EPR spectroscopy, *Nature Struct. Biol.* 9, 844-848 (2002).
33. C. M. Slupsky, and B. D. Sykes, NMR solution structure of calcium-saturated skeletal muscle troponin C, *Biochemistry* 34, 15953-15960 (1995).
34. C. R. Cantor, and P. Schimmel, *Biophysical Chemistry Part II.*, Freeman and Company, Oxford (1980).

MUTATIONS OF TRANSCRIPTION FACTORS IN HUMAN WITH HEART DISEASE FOR UNDERSTANDING THE DEVELOPMENT AND MECHANISMS OF CONGENITAL CARDIOVASCULAR HEART DISEASE

Rumiko Matsuoka*

1. INTRODUCTION

Congenital cardiovascular heart diseases (CCVDs) are diagnosed in nearly 1 % of newborns and are the most common major birth defects. However, the Gene, Cardiovascular Disease Epidemiology Committee of the Japanese Society of Pediatric Cardiology and Cardiac Surgery reported that only 13 % of patients with CCVDs have been identified as having single mutant gene disorders with syndromes or a chromosomal abnormality, or having one of the primary environmental factors, and the rest of the patients (87%) with CCVDs still show multifactorial inheritance (Matsuoka et al., 2003-1). Familial aggregates of the most common familial CCVDs without syndromes, including 11% of atrial septal defect (ASD), 32% of ventricular septal defect (VSD), 11% of tetralogy of Fallot (TOF) and 2.8% of patent ductus arteriosus (PDA) which are seen in about 57 % of all CCVDs, were investigated (Matsuoka et al., 2003-2).

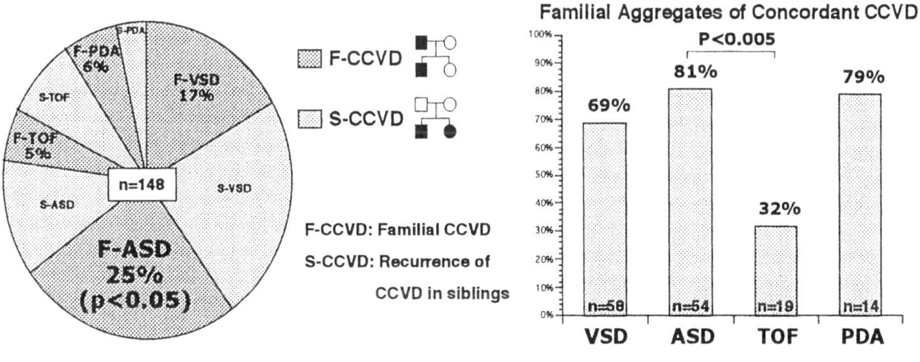

Figure 1. The results showed that familial ASD was significantly high (P<0.05) compared to all 4 familial CCVDs. Furthermore, 81 % of the families from familial aggregates of ASD had concordant-ASD, but TOF was low in these aggregates.

*Department of Pediatric Cardiology, Division of Genomic Medicine, Institute of Advanced Biomedical Engineering and Science, Graduate School of Medicine, Tokyo Women's Medical University, 8-1 Kawada-cho, Shinjuku-ku, Tokyo 162-8666, Japan

Genetic factors seem to play a predominant role in familial ASD, whereas both genetic factors and intrauterine environmental factors are more likely to be involved in the etiology of familial TOF.

Mutations in *NKX2.5* have been identified in individuals with cardiac septal defect and conduction abnormalities (Schott et al., 1998), while individuals with Holt-Oram syndrome (HOS), characterized by cardiac septal defect, conduction abnormalities, and limb anomalies, have point mutations in the transcription factor, TBX5 (Basson et al., 1997; Li et al., 1997). However, the genetic predisposition of some familial discordant CCVDs is also presumed to be single genes with different phenotypes.

Recently, *GATA4*, which encodes a zinc-finger transcription factor essential for cardiogenesis in flies, fish and mice (Molkentin et al., 1997; Kuo et al., 1997, Reiter et al., 1999, Gajewski et al., 1999) has identified non-syndromic pedigrees with cardiac septal defects such as ASD, VSD, PS, PDA or TOF/VSD+PS (Garg et al., 2003), which are seen in nearly 57 % of all CCVDs (Matsuoka et al., 2003-1). *GATA4, NKX2.5* and *TBX5* seem to interact with each other in the developmental four-chambered heart.

Therefore, to elucidate the precise molecular mechanisms of cardiac septal closure in humans, *GATA4, NKX2.5* and *TBX5* mutations in patients from familial aggregates of ASD were examined and their phenotypes were compared. For early cardiogenesis, *GATA4, NKX2.5, TBX5* and *TBX1* seem to be essential. Morphogenetic interplay and the variability were also discussed.

2. Subjects and Methods

Thirty-one patients and three of their healthy family members from 19 familial aggregates of ASD families were studied. Twenty-seven patients from 16 families had familial ASD and there was recurrence of ASD in the siblings of 4 patients from 3 families. *GATA4, NKX2.5* and *TBX5* were amplified from genomic DNA using PCR and directly sequenced. When a mutation was detected, we confirmed that it was not present in at least 100 controls. Twenty-nine of the 31 affected members had atrial septal defect (ASD), of whom 2 had additional forms of CCVDs, including ventricular septal defect (VSD) and patent ductus arteriosus (PDA). Three individuals had other forms of CCVDs, including dextrocardia, or arrhythmia. A sibling from a family in which there was recurrence of ASD, associated with VSD and PDA had several operations, including ligation of the PDA, and closure of the ASD and VSD. She had neither the cardiac conduction system nor any bone malformations as seen in Holt-Oram syndrome. However, her father had syndactyly in the second and third toes of the left foot without any type of CCVD.

3. GATA4

GATA4 belongs to a family of transcription factors that binds a consensus HGATAR DNA motif and contains two class IV zinc-finger domains (Evans et al., 1989; Tsai et al., 1989; Arceci et al., 1993). Gata1-3 are expressed predominantly in hematopoietic cells and heterozygous mutations of *GATA1* and *GATA3* cause human blood disorders and organ malformations, respectively (Nichols et al., 2000; Van Esch et al., 2000). Gata4 and its orthologues regulate numerous downstream target genes involved in cardiac development, growth and hypertrophy, typically through combinative interactions with other cardiac-specific transcription factors, including NKX2.5, EF-2 and NFATc4, which together are coexpressed uniquely in the myocardium (Molkentin, 2000). Homozygous disruption of *Gata4* in mice results in failure of midline fusion of the heart tube and early embryonic lethality (Molkentin et al., 1997; Kuo et al., 1997). Similar phenotypes occur with null mutations of *Gata5* in zebra fish or the Drosophila *Gata4* orthologue, *pannier*

(Reiter et al., 1999; Gajewski et al., 1999). In addition to its role in cardiac development, Gata4 is a critical factor for myocardial differentiation and function (Molkentin, 2000).

Two *GATA4* mutations were found in 12.5% of families with familial-ASD. Family 1 was reported in *Nature* (Garg et al., 2003). Interestingly, these family members had a frame-shift mutation of *GATA4*, which resulted in a premature stop codon that was transcriptionally inactive and truncated protein or nonsense-mediated decay of the transcript, and the type of CCVDs in these family members segregated with only ASD except in the member of the first generation, who had dextrocardia without ASD. This suggests that the *GATA4* mutation may interact with genes responsible for heterotaxia, such as *CFC1* (Bamford et al., 2000) or *ZIC3* (Gebbia et al., 1997). The second and novel *GATA4* mutation in family 2, with autosomal dominant transmission of ASD in which a mutation of *GATA4* was found in all available affected members spanning two generations, and the type of CCVDs in these family members segregated with only ASD. Similar to family 1, neither the cardiac conduction system nor other organ systems were affected in the kindred. The Ser52Phe mutation in this pedigree resulted in a missence mutation of amino acid 52, thus severely altering the encoded GATA4 protein. The segregation of *GATA4* mutations with cardiac septal defects in two unrelated families provides strong evidence that mutations in *GATA4* are the underlying cause of a subset of familial, non-syndromic cardiac septal defects, especially ASD. Consistent with this finding, deletion of the terminal end of chromosome 8p that contains *GATA4*, among many other genes, is characterized by cardiac septal defects. It was also reported that when equivalent amounts of protein were expressed, mutant Gata4 displayed 50% of the wild type transcriptional activity of *atrial natriuretic factor* (*ANF*) (Sprenkle et al., 1995) and *alpha myosin heavy chain* (*α-MyHC*) (Molkentin et al., 1994). These data suggest that the impaired DNA-binding of Gata4 may contribute to the reduced transcriptional activity of the mutant, although forced overexpression of the hypomorphic protein in tissue culture may, in part, compensate for the decreased DNA-binding affinity. The zinc-fingers of Gata4 also mediate numerous protein-protein interactions that often dictate DNA-binding specificity and, therefore, subsets of target gene activation. Therefore, we performed the mutation analysis of *GATA4* and two other genes, *NKX2.5* and *TBX5*, which encored transcriptional factors and implicated in cardiac septal formation.

3. NKX2-5

NKX2.5 is an essential gene for heart development and has been extensively investigated as a significant cause of congenital heart disease, atrial septal defect (ASD) and atrioventricular conduction disturbance (AV block). However, the relationships between its mutations and cardiac anatomical anomalies, and complication with AV block and its degrees have not been clarified. We screened mutations in familial ASD patients and examined their phenotypes. Our subjects were 16 familial ASD probands, including 4 probands with AV block and their family members. The 2 exons and flanking introns of *NKX2.5* were amplified from genomic DNA using PCR and directly sequenced. *NKX2.5* mutations were found in 18.8% of familial-ASD. Interestingly, the affected members in family 3 had a novel, frame-shift *NKX2.5* mutation, which resulted in a premature stop codon, and co-segregated with ASD and conduction abnormalities, except in the II-4 patient who only had progressive AV block. The second, previously reported *NKX2.5* mutation was found in the III-2-patient with ASD and second-degree AV block and sick sinus syndrome (SSS), and the II-2-patient with atrial fibrillation. The third *NKX2.5* and a novel mutation were confirmed using the BsaWI restriction enzyme found in a patient with ASD and third degree AV block. Our results and those of previous reports strongly suggest that arrhythmia is likely to be associated with the missense mutation in the homeodomain or the nonsense mutation of the *NKX2.5* gene. These types of mutations may cause arrhythmia but may not necessarily lead to ASD. We found 3 mutations in 3 probands with AV block and a secundum type ASD (ASD (II)) among all 16 probands. The 3 mutations were Ala88del (family 3), Arg190Cys (family 4) and Thr178Met (family 5), the former 2 mutations being novel. The Ala88del mutation was in exon1 and resulted in a frameshift, which caused the truncation of the homeodomain. These mutations were not found in 100 control subjects. Family 3-1 and all 3 siblings (family 3-2, family 3-3 and family 3-4) who had the same mutation had progressive AV block I-II. Family 3-1, family 3-2 and family 3-3 had an ASD (II) but family 3-4 did not. The Thr178Met and Arg190Cys mutations were in the homeodomain. Family 4 and family 5 who had an ASD (II) with progressive AV block underwent pacemaker implantation. Among the other 13 probands without mutations, 1 had an ASD with nonprogressive AV block I, 3 had an ASD with pulmonary stenosis and 9 had only an ASD. All 3 mutations, which we found in this study, were associated with an HD and patients with these mutations had severe AV block. These mutations may reduce DNA binding affinity and seem to cause severe AV block. Among 6 patients with AV block and mutations, there were no cardiac anatomical differences in 5 patients who had an ASD and one who did not. For atrial septation, not only *NKX2.5* but also some other genes seem to be essential. Human *NKX2.5* mutations also cause similar septal defects and NKX2.5 can physically interact with TBX5 (Hiroi et al., 2001). The reciprocal evidence that human *GATA4* and *TBX5* mutations affect interaction with one another in the setting of similar CCVDs provides further evidence that GATA4 and TBX5 may cooperate during cardiogenesis. The similar effect of TBX5 mutations on the interaction with GATA4 and *NKX2.5* raises the interesting possibility that TBX5, *NKX2.5* and GATA4 function in a common complex for regulation of a subset of genes required for cardiac septal formation.

4. TBX

Putative transcription factors of the *T-box* family genes act to control early cell-fate decisions, differentiation and morphogenesis/organogenesis. The spatial and temporal expression patterns of *T-box* genes such as *Tbx1*, *Tbx2*, *Tbx3*, *Tbx4* and *Tbx5*, are unique although they overlap in their sites of expression. This overlapping indicates that T-box genes are differentially regulated during the developmental process, particularly in areas where inductive interactions are taking place. *Tbx1* shows very little overlap with the other 2 cognate gene sets, *Tbx2/Tbx3*, *Tbx4/Tbx5*, since the divergence of *Tbx1* occurred long before the relatively recent divergence of the other 4 genes from common ancestral genes, and recent studies of mouse genetics have suggested that *TBX1* is the major candidate gene for del22q11.2 syndrome in patients (Lindsay et al., 2001; Jerome et al., 2001, Merscher et al., 2001). Although the expression patterns of *Tbx2* (Paxton et al., 2002) and *Tbx3* are similar to some temporal and spatial differences, *Tbx3* (Papaioannou et al., 1998) is a disease gene for ulnar-mammary syndrome. *Tbx5* and *Tbx4* are exclusively expressed in the vertebrate forelimb and hind limb, respectively, and *Tbx5* (Wilkinson et al., 1990; Peckam et al., 2003) is a disease gene for Holt-Oram syndrome. When mutated, these genes may produce dramatic phenotypes in a number of human congenital malformations, including cardiovascular diseases, such as the key gene of del22q11.2 syndrome, Holt-Oram syndrome, ulnar-mammary syndrome or similar syndromes. Del22q11.2 syndrome, including conotruncal anomaly face syndrome (CAFS)/velo-cardio-facial syndrome (VCFS) and DiGeorge syndrome (DGS) with conotruncal anomaly face (CAF), is the most frequent known chromosomal microdeletion syndrome, with an incidence of 1 in 3000-5000 live births. This syndrome is characterized by a 3-Mb deletion on chromosome 22q11.2, cardiac abnormalities, anomaly face, T-cell deficits, cleft palate and hypocalcemia. At least 30 genes have been mapped to the deleted region. However, the association of these genes with the cause of this syndrome is not clearly understood. To find the*TBX1* mutation, we focused on obtaining a precise diagnosis of CAFS or DGS with CAF based on a clear view of the phenotypes. In 96% (225/235) of the patients, there was a 1.5~3-Mb deletion at 22q11.2 which was not found in the remaining 10 patients (7 sporadic CAFS patients, 1 sporadic DGS with CAF patient and 2 patients from 2 CAFS families). We have identified three mutations of *TBX1* in two unrelated patients without the 22q11.2 deletion, one with sporadic CAFS/VCFS and one with sporadic DGS, and in three patients from a CAFS/VCFS family, *TBX1* is a major genetic determinant of human del22q11.2 syndrome (Yagi et al., 2003). Our results imply that the *TBX1* mutation is responsible for 5 major phenotypes in del22q11.2 syndrome, i.e., CAF, cardiovascular defects, thymic hypoplasia, velopharyngeal insufficiency with cleft palate, and parathyroid dysfunction with hypocalcemia. We conclude that *TBX1* is a major genetic determinant of the del22q11.2 syndrome in humans. In addition, we found five mutations of *TBX5* in 50% (5/10) of unrelated patients with Holt-Oram syndrome, but we did not find the mutations of *TBX3* in a patient with ulnar-mammary syndrome. It is important to emphasize that the correct diagnosis of a syndrome should be based on a clear view of the phenotypes to avoid confusion with other similar syndromes. Although little is known about specific target genes for T-box proteins in general, a 20-nt, partially palindromic sequence has been identified as the DNA target for the brachyury protein and a similar sequence has been described for several other T-box genes (Tada et al., 2001). Members of the FGF and TGF_ families are among the few known target genes containing the sequences. A recent investigation of *Fgf8* null mice indicated cardiovascular, glandular and craniofacial phenotypes seen in human del22q11.2 syndrome, and suggested that *Fgf8* is downstream of *Tbx1* (Frank et al., 2002; Vitelli et al., 2002; Abu-Issa et al., 2002; Yamagishi et al., 2003). The variability of the phenotype found in patients with the *TBX1* mutation suggests that not only environmental factors but also an altered interaction with downstream genes, which are regulated by T-box transcription factors, may play an essential role in embryogenesis.

5. Gata4-or NKX2.5-TBX5 complex

Direct sequencing of *NKX2.5* and *GATA4*, in family members from 11familial ASD families failed to reveal mutations, suggesting an alternative genetic etiology. Although we did not find *GATA4* or *NKX2.5* mutations in patients from family 6 in this study, a new *TBX5* mutation was found. The tow year-old girl had undergone several operations including ligation of PDA, closure of ASD and VSD, and removal of a hepato-hamartome. She did not have any bone malformations as seen in Holt-Oram syndrome. However, her father had syndactyly in the second and third toes of the left foot without any type of CCVD. Direct sequencing of *TBX5* in an affected family member identified a G⇒A substitution of nucleotide 1209 that predicted the substitution of glycine to alanine at codon 403 (S403A). Although TBX5 was not previously known to interact with Gata4, the two wild type proteins could immunoprecipitate with one another (Garg et al., 2003). To determine whether Gata4 cooperates with TBX5 to activate transcription, co-transfection of the *ANF* reporter with TBX5 and Gata4 in HeLa cells and primary cardiomyocytes showed that similar to Tbx5-NKX2.5 interactions (Hiroi et al., 2001), co-expression of TBX5 and Gata4 resulted in cooperative activation of the reporter (Garg et al., 2003). The similarity between cardiac septal defects observed in the setting of human *GATA4* mutations and human *TBX5* mutations is consistent with the Gata4-TBX5 complex. The combination of human genetics and biochemical analyses reveal a previously unrecognized genetic etiology for CCVDs and potential mechanisms through which certain cardiac defects may occur. The mutations in *GATA4* suggest that haploinsufficiency of *GATA4* in humans can result in the most common types of cardiac malformations and that GATA4 is essential for functional separation of the four cardiac chambers. While haploinsufficiency of *TBX5* has previously been linked with cardiac septal malformation in HOS, the specific loss of GATA4-TBX5 interaction by some *TBX5* point mutations provides a potential mechanistic understanding of how disruption of combinatorial interactions of transcription factors can lead to organ-specific birth defects.

In this study, mutations were found in 31.3% of the families with familial-ASD. *GATA4* mutations amounted to 12.5% and *NKX2.5* mutations to 18.8%. Table 1 shows reported disease genes for CCVDs and Figure. 1 shows the physical interaction between Gata4, *NKX*2.5 and TBX5 and the *ANP* gene. This may explain why the *GATA4* gene is a genetic cause of human cardiac septal defects, through its interaction with TBX5 as well as NKX2.5.

Table 1 Reported disease genes for congenital cardiovascular diseases

	Gene	Phenotype	Chromosome
ASD	***GATA4* (GATA binding protein 4)**	**ASD, VSD, PS, AVSD,**	**8p23.1-p22**
	***NKX2.5* (cardiac specific homeobox)**	**ASD+AV block, TOF**	**5q35**
	***TBX5* (T-box 5)**	**Holt-Oram syndrome**	**12q24.1**
AVSD	*GATA4* (GATA binding protein 4)	ASD, VSD, PS, AVSD, PDA	8p23.1-p22
	CRELD1 (cysteine-rich protein with EGF-like domains 1)	AVSD	3p25.3
VSD	*GATA4* (GATA binding protein 4)	ASD, VSD, PS, AVSD, PDA	8p23.1-p22
	NKX2.5 (cardiac specific homeobox)	ASD+AV block, TOF	5q35
	TBX5 (T-box 5)	Holt-Oram syndrome	12q24.1
TOF	*GATA4* (GATA binding protein 4)	ASD, VSD, PS, AVSD, PDA	8p23.1-p22
	NKX2.5 (cardiac specific homeobox)	ASD+AV block, TOF	5q35
PDA	*GATA4* (GATA binding protein 4)	ASD, VSD, PS, AVSD, PDA	8p23.1-p22
	TFAP2B (transcription factor AP2-beta)	Char syndrome	6p12

ASD: atrial septal defect; VSD: ventricular septal defect; TOF: tetralogy of Fallot

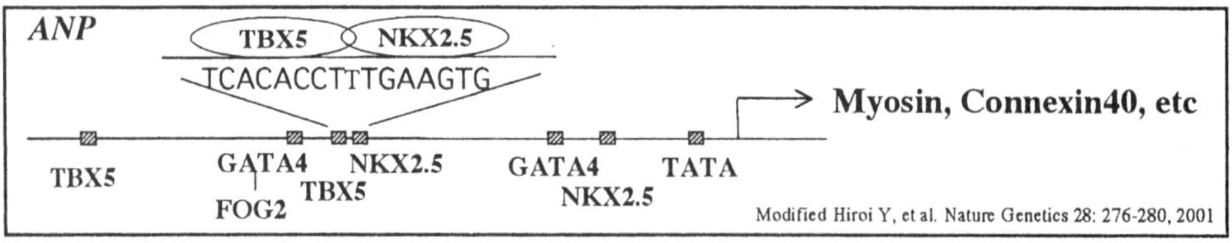

Figure 2. A physical interaction between Gata4, NKX2.5 and TBX5 on the *ANP* gene.
Equivalent amounts of protein were expressed mutant Gata4 displayed 50% of wild-type transcriptional activity of the *atrial natriuretic factor* (*ANF*) and *alpha myosin heavy chain* (α-*MyHC*). The zinc-fingers of Gata4 also mediate numerous protein-protein interactions that often dictate DNA-binding specificity and therefore subsets of target gene activation, such as two other transcriptional regulators implicated in cardiac septal formation, NKX2.5 and TBX5.

6. Genetic etiology for CCVDs

Within a single organ, it is notable that pleiotropic defects can be observed in individuals with a shared mutation, as described here. This phenomenon has also been described for *NKX2.5* mutations and suggests a role for modifying genetic, environmental or stochastic events even for monogenic causes of cardiac defects. It is interesting that in some patients, cardiac septal defects close spontaneously, presumably due to continued myocardial growth. Because GATA4 is a central factor in transcriptional regulation of cardiac growth and hypertrophy, it will be interesting to determine whether mutations of *GATA4*, as a genetic etiology of CCVDs, segregate with an increased need for surgical intervention because of failure of spontaneous closure of septal defects. Broad screening for *GATA4* mutations in humans with heart disease may provide new avenues for understanding the mechanisms and links between development of CCVDs and post-natal myocardial growth, leading to future therapeutic or preventive interventions. Therefore, the genetic predisposition of some familial-discordant-CCVDs is also presumed to be due to single genes with different phenotypes. For example, *GATA4* causes different cardiac septal defects seen in about 57% of all CCVDs. Why does this happen? Our theory is as follows. For familial ASD, genetic factors seem to play a predominant role, whereas genetic factors and intrauterine environmental factors are more likely to involve the etiology of familial TOF. The predominantly higher incidence of maternal cigarette smoking, maternal alcohol intake and low-birth-weight babies was associated with mothers from familial aggregates of TOF patients compared to other pregnant mothers according to data from welfare facilities. Furthermore, a significantly higher incidence of maternal anemia was found in mothers from familial aggregates of TOF patients compared to other CCVDs (Matsuoka et al., 2003-2).

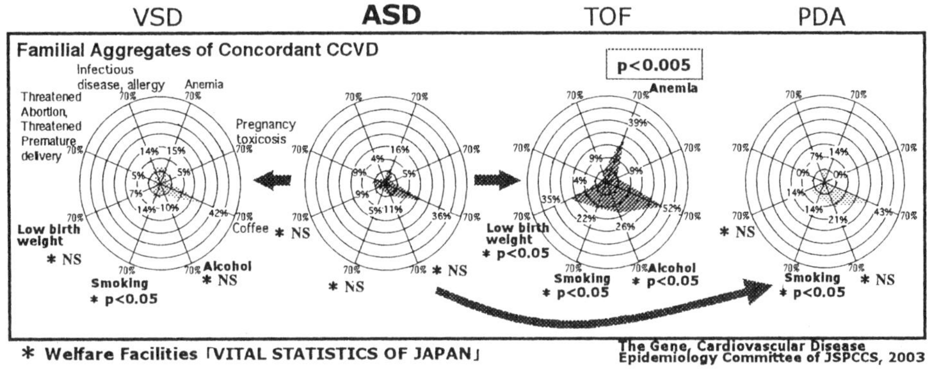

Figure 3. ASD: atrial septal defect; VSD: ventricular septal defect; TOF: tetralogy of Fallot; NS: not significant

A predominantly higher incidence of maternal cigarette smoking, alcohol intake and low-birth weight was found in familial TOF in this study compared with data from to welfare facilities [Vital Statistics of Japan] ($P<0.05$). Furthermore, a higher incidence of maternal anemia was found in familial TOF ($P<0.05$) compared to VSD, ASD and PDA. Dichotomous variables were compared using Fisher's exact probability test. The Mann-Whitney U test was used to compare two mean values of variables.

These data suggest that in familial ASD, genetic factors seem to play a predominant role, whereas genetic factors and intrauterine environmental factors are more likely to involve the etiology of familial TOF, and improvement of the intrauterine environment may reduce the frequency and complexity of CCVDs.

For early cardiogenesis, *GATA4*, *NKX2.5* and *TBX5* genes seem to be essential. The morphogenetic interplay and variability of CCVDs should be investigated.

Acknowledgments

I wish to thank Dr. Kayoko Hirayama-Yamada, Mr. Hisato Yagi, Dr. Yoshiyuki Furutani and Ms. Michiko Furutani for investigating or support of these series and technical support. I also thank Ms. B. Levene for English correction of the manuscript.

REFERENCES

Abu-Issa R, Smyth G, Smock I, Yamamura K and Meyers EN (2002). Fgf8 is required for pharyngeal arch and cardiovascular development. Development 12, 4613-4625

Arceci RJ, King AA, Simon MC, Orkin, SH and Wilson DB (1993). Mouse GATA-4: a retinoic acid-inducible GATA-binding transcription factor expressed in endodermally derived tissues and heart. Mol Cell Biol 13, 2235-2246

Bamford RN, Roessler E, Burdine RD, Saplakoglu U, dela Cruz J, Splitt M, Goodship JA, Towbin J, Bowers P, Ferrero GB, Marino B, Schier AF, Shen MM, Muenke M and Casey B (2000). Loss-of –function mutations in the EGF-CFC gene CFC1 are associated with human left-right laterality defects. Nat. Genet 26, 365-369

Basson CT, Bachinsky DR, Lin RC, Levi T, Elkins JA, Soults J, Grayzel D, Kroumpouzou E, Traill TA, Leblanc- Straceski J, Renault B, Kucherlapati R, Seidman JG and Seidman CE (1997). Mutations in human TBX5 cause limb and cardiac malformation in Holt-Oram syndrome. Nat Genet 15, 30-35

Evans T and Felsenfeld G (1989). The erythroid-specific transcription factor Eryf1: a new finger protein. Cell 58, 877-885

Frank DU, Fotheringham LK, Brewer JA, Muglia LJ, Tristani-Firouzi M, Capecchi MR and Moon AM (2002). An Fgf8 mouse mutant phenocopies human 22q11 deletion syndrome. Development 129, 4591-4603

Gajewski K, Fossett N, Molkentin JD and Schulz RA (1999). The zinc finger proteins Pannier and GATA4 function as cardiogenic factors in Drosophila. Development. 126, 5679-5688

Garg V, Kathiriya IS, Barnes R, Schluterman MK, King IN, Butler CA, Rothrock CR, Eapen RS, Hirayama-Yamada K, Joo K, Matsuoka R, Hobbs HH, Cohen J, and Srivastava D (2003). GATA4 Mutations Cause Human Congenital Heart Defects and Reveal an Interaction with TBX5. Nature 424, 443-447

Gebbia M, Ferrero GB, Pilia G, Bassi MT, Asylsworth A, Penman-Splitt M, Bird LM, Bamforth JS, Burn J, Schlessinger D, Nelson DL and Casey B (1997). X-linked situs abnormalities result from mutations in ZIC3. Nat Genet 17, 305-308

Hiroi Y, Kudoh S, Monzen K, Ikeda Y, Yazaki Y, Nagai R and Komuro I (2001). Tbx5 associates with Nkx2-5 and synergistically promotes cardiomyocyte differentiation. Nat Genet 28, 276-280

Jerome LA and Papaioannou VE (2001). DiGeorge syndrome phenotype in mice mutant for the T-box gene, Tbx1. Nature Genet 27, 286-291

Kuo CT, Morrisey EE, Anandappa R, Sigrist K, Lu MM, Parmacek MS, Soudais C and Leiden JM (1997). GATA4 transcription factor is required for ventral morphogenesis and heart tube formation. Genes Dev 11, 1048-1060

Li QY, Newbury-Ecob RA, Terrett JA, Wilson DI, Curtis AR, Yi CH, Gebuhr T, Bullen PJ, Robson SC, Strachan T, Bonnet D, Lyonnet S, Young ID, Raeburn JA, Buckler AJ, Law DJ and Brook JD (1997). Holt-Oram syndrome is caused by mutations in TBX5, a member of the Brachyury (T) gene family. Nat Genet 15, 21-29

Lindsay EA, Vitelli F, Su H, Morishima M, Huynh T, Pramparo T, Jurecic V, Ogunrinu G, Sutherland HF, Scambler PJ, Bradley A and Baldini A (2001). Tbx1 haploinsufficiency in the DiGeorge syndrome region causes aortic arch defects in mice. Nature 410, 97-101

Matsuoka R, Mori K and Ando M (2003). Working paper of the Epidemiology Committee of JSPCCS. Predisposition of congenital cardiovascular diseases in Japan. Incidence and Inheritance, 1990–1999 (2654 Families) Ped Cardiol Card Surg 19, 606-621

Matsuoka R, Mimamisawa S, Akimoto K, Ichida F, Ohta Y, Ogawa S, Ono Y, Oyama K, Kuroe K, Kosaka K, Satomi G, Joo K, Seguchi M, Takahashi E, Nakagawa M, Haneda Y, Baba K, Fukushige J, Maeda J, Murai T, Mori K, Mori K, Yoshinaga M and Ando M (2003). Working paper of the Gene, Cardiovascular Disease Epidemiology Committee of JSPCCS. Epidemiological reports for familial

congenital cardiovascular diseases. Ped Cardiol Card Surg 19, 622-628
Merscher S, Funke B, Epstein JA, Heyer J, Puech A, Lu MM, Xavier RJ, Demay MB, Russell RG, Factor S, Tokooya K, Jore BS, Lopez M, Pandita RK, Lia M, Carrion D, Xu H, Schorle H, Kobler JB, Scambler P, Wynshaw-Boris A, Skoultchi AI, Morrow BE and Kucherlapati R (2001). TBX1 is responsible for cardiovascular defects in velo-cardio-facial/DiGeorge syndrome. Cell 104, 619-629
Molkentin JD, Kalvakolanu DV and Markham BE (1994). Transcription factor GATA-4 regulates cardiac muscle-specific expression of the alpha-myosin heavy-chain gene. Mol Cell Biol 14, 4947-4957
Molkentin JD, Lin Q, Duncan SA and Olson EN (1997). Requirement of the transcription factor GATA4 for heart tube formation and ventral morphogenesis. Genes Dev 11, 1061-1072
Molkentin JD (2000). The zinc finger-containing transcription factors GATA-4, -5, and -6. Ubiquitously expressed regulators of tissue-specific gene expression. J Biol Chem 275, 38949-38952
Nichols KE, Crispino JD, Poncz M, White JG, Orkin SH, Maris JM and Weiss MJ (2000). Familial dyserythropoietic anaemia and thrombocytopenia due to an inherited mutation in GATA1. Nat Genet 24, 266-270
Papaioannou VE and Silver LM (1998). The T-box gene family. BioEssays 20, 9-19
Paxton C, Zhao H, Chin Y, Langner K and Reecy J (2002). Murine Tbx2 contains domains that activate and repress gene transcription. Gene 283, 117-124
Peckam Ea and Brook JD (2003). T-box genes in human disorders. Hum Mol Genet 12 (Suppl 1), R37-44
Reiter JF, Alexander J, Rodaway A, Yelon D, Patient R, Holder N and Stainier DY (1999). Gata5 is required for the development of the heart and endoderm in zebrafish. Genes Dev 13, 2983-2995
Schott JJ, Benson DW, Basson CT, Pease W, Silberbach GM, Moak JP, Maron BJ, Seidman CE and Seidman JG (1998). Congenital heart disease caused by mutations in the transcription factor NKX2-5. Science 281, 108-111
Sprenkle AB, Murray SF and Glembotski CC (1995). Involvement of multiple cis elements in basal- and alpha-adrenergic agonist-inducible atrial natriuretic factor transcription. Roles for serum response elements and an SP-1-like element. Circ Res 77, 1060-1069
Tada M and Smith JC (2001). T-targets: clues to understanding the functions of T-box proteins. Dev Growth Differ 43, 1-11
Tsai SF, Martin DI, Zon LI, D'Andrea AD, Wong GG and Orkin SH (1989). Cloning of cDNA for the major DNA-binding protein of the erythroid lineage through expression in mammalian cells. Nature 339, 446-451
Van Esch H, Groenen P, Nesbit MA, Schuffenhauer S, Lichtner P, Vanderlinden G, Harding B, Beetz R, Bilous RW, Holdaway I, Shaw NJ, Fryns JP, Van de Ven W, Thakker RV and Devriendt K (2000). GATA3 haplo-insufficiency causes human HDR syndrome. Nature 406, 419-422
Vitelli F, Taddei I, Morishima M, Meyers EN, Lindsay EA and Baldini A (2002). A genetic link between Tbx1 and fibroblast growth factor signaling. Development 129, 4605-4611
Wilkinson DG, Bhatt S and Herrmann BG (1990). Expression pattern of the mouse T gene and its role in mesoderm formation. Nature 343, 657-659
Yagi H, Furutani Y, Hamada H, Sasaki T, Asakawa S, Minoshima S, Ichida F, Joo K, Kimura M, Imamura S, Kamatani N, Momma K, Takao A, Nakazawa M, Shimizu N and Matsuoka R (2003). TBX1 is a major genetic determinant of human del22q11.2 syndrome. Lancet 362, 1366-1373
Yamagishi H, Maeda J, Hu T, McAnally J, Conway SJ, Kume T, Meyers EN, Yamagishi C and Srivastava D (2003). Tbx1 is regulated by tissue-specific forkhead proteins through a common Sonic hedgehog-responsive enhancer. Genes Dev 17, 269-281

GENERAL DISCUSSION PART I
SKELETAL MUSCLE MECHANICS

Chaired by B. Brenner and H. Sugi

Force-Velocity Relation of Cytoplasmic Streaming and the Duty Ratio Concept

Sugi: I would like to give a talk on the mystery of extremely rapid cytoplasmic streaming caused by ATP-dependent actin–myosin sliding in giant internordal cells of green algae. As shown in Fig. I-1, the cytoplasmic streaming is caused by relative sliding between the cytoplasmic myosin molecules attached to cytoplasmic organelles and the long and straight actin filament bundles (actin cables) fixed to the inner surface of chloroplasts (Fig. I-1). The velocity of cytoplasmic streaming is >50μm/s, being many times larger than that of actin-myosin sliding in muscle.

Using a centrifuge microscope (Oiwa et al., Proc. Natl. Acad. Sci. USA 87: 7893-7897, 1990), we determined force-velocity (*P-V*) relations of ATP-dependent sliding between the cytoplasmic myosin-coated polystyrene beads and the actin cables in the internodal cell of an alga *Chara collarina*. As with actin-myosin sliding in muscle,

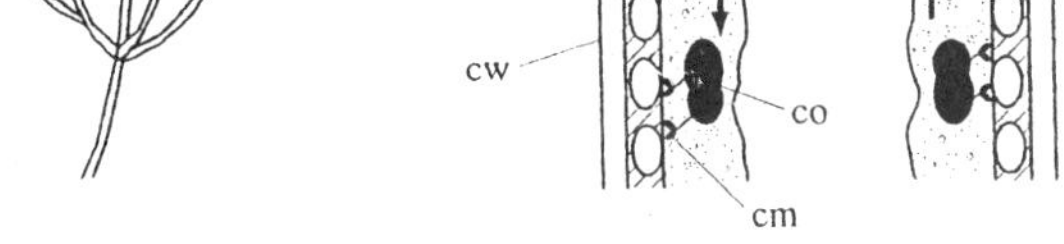

Figure I-1: Diagram of cytoplasmic streaming in the internodal cell of green algae. Ac, actin cable; ch, chloroplast; cm, cytoplasmic myosin, co, cytoplasmic organelle; cs, cell sap; cw, cell wall; cy, streaming cytoplasm; ic, internodal cell. Arrows indicate direction of cytoplasmic streaming (Sugi and Chaen, 2003).

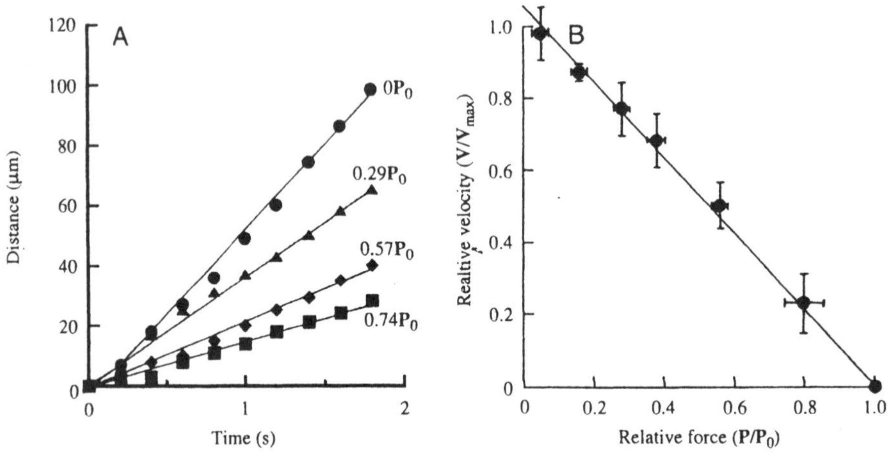

Fig. I-2 Force-velocity (*P-V*) characteristics of ATP-dependent actin-myosin sliding causing cytoplasmic streaming. (A) Constant velocity sliding of a cytoplasmic myosin-coated bead along actin cables under different positive beads. (B) *P-V* curve constructed from six different beads with large (8.6-13*p*N) maximum isometric force (P_0) (Chaen et al., 1995),

the cytoplasmic myosin-coated head moved along actin cables with a constant velocity under a constant positive load (Fig. I-2A). The *P-V* curves obtained were almost straight in shape irrespective of the amount of isometric force, i.e. the number of myosin molecules involved in the actin-myosin sliding (Fig. I-2B). In terms of the Hill equation (Hill, Proc. R. Soc. Lond. B126: 136-195, 1938), the relation between the velocity of actin-myosin sliding (V) and the load (=force, P) is expressed as $(P+a)\ V=b(P_0-p)$, where a and b are constants corresponding to two asymptotes of a rectangular hyperbola. $P=P_0$ when $V=0$, and $V=V_{max}$ when $P=0$ The value of P_0, i.e. the maximum isometric force, represents the maximum number of myosin molecules generating actin-myosin sliding, while V_{max} is thought to reflect the maximum rate of ATP-dependent actin-myosin interaction. The value of a/P_0 is a measure of curvature of the *P-V* curve, and the straight *P-V* curve is associated with a large value of a/P_0. It has been suggested that the large a/P_0 value results from a low efficiency with which chemical energy derived from ATP hydrolysis is converted into mechanical work (Woledge, J. Physiol. 197: 685-707, 1968).

On the other hand, it is generally believed that actin and myosin exhibit repeating attachment-detachment cycles, each coupled with ATP hydrolysis. The probability that all myosin heads are detached from actin (P_{off}) is written as $P_{off}=(1-p)^n$, where p is the fraction of time in which a myosin head firmly attaches to actin in one interaction cycle, and is called the duty ratio (Howard, Nature 389: 561-567, 1997) while n is the number of myosin heads interacting with actin. In the cytoplasmic streaming, only several myosin molecules are likely to be bound to each cytoplasmic organelle to produce actin-myosin sliding. This implies that the duty ratio of cytoplasmic myosin head causing the rapid cytoplasmic streaming is very large.

The duty ratio can also be expressed as $p=(d{\times}A)/V_{max}$, where d is the unitary distance of actin-myosin sliding and A is the ATPase rate per myosin head. In skeletal muscle, p is ~0.05 (Uyeda et al., J. Mol. Biol. 214: 699-710, 1990; Finer et al., Nature 368: 113-119,

1994), A is ~$20s^{-1}$ $head^{-1}$ (Toyoshima et al., Proc. Natl. Acad. Sci. USA 87: 7130-7134, 1990), and V_{max} is ~$3\mu ms^{-1}$ (Oiwa et al., Proc. Natl. Acad. Sci. USA, 87: 7893-7897, 1990). The above values of d, A and V_{max} indicate that the value of d of ~8nm, a value consistent with the rotation of myosin head producing muscle contraction (Huxley and Simmons, Nature 233, 533-538, 1971).

In the cytoplasmic streaming, p is assumed to be ~0.25, while V_{max} is ~$50\mu ms^{-1}$ (Chaen et al., J. Exp. Biol. 198: 1021-1027, 1995). Cytoplasmic myosin molecule is almost similar to muscle myosin molecule in both shape and size (Yamamoto et al., Cell Mol. Life Sci. 56: 227-232, 1999). Therefore, on the basis of myosin head rotation, the value of d for cytoplasmic myosin is constrained by the size of myosin head, being not more than 20nm. This implies that A in the cytoplasmic streaming is ~$425s^{-1}$ $head^{-1}$. According to Higashi-Fujime et al. (FEBS Lett. 375: 151-154, 1995), however, A in cytoplasmic myosin in solution is similar to that of muscle myosin.

The above puzzling situation would be solved by assuming that (1) the value of A is increased markedly without changing the value of p as a result of mechanical cooperative interaction among myosin molecules attached to the same organelle (Sugi and Chaen, J. Exp. Biol. 206: 1971-1976, 2003), and/or (2) the value of d is not constrained by the size of myosin head and is of the order of 100nm by a biased Brownian ratchet mechanism (Vale and Oosawa, Adv. Biophys. 26: 97-134, 1990).

In my feeling, the concept of duty ratio is useful to account for the results with in vitro motility assay, in which a limited number of myosin molecules are made to interest with actin filaments. In muscle contraction, however, which involves a huge number of myosin molecules interacting with actin, I am skeptical about the varidity of the duty ratio concept.

Pollack: You mentioned that the duty ratio of myosin II was so small that > 20 bridges had to be attached to an actin filament in order to sustain actomyosin sliding. Our experiments challenge that. In our experiments, with one actin and one myosin filament, there are many instances in which the number of bridges is less than 20; yet, rarely do thick and thin filaments lose contact. Once they have made contact, they start shortening.

Sugi: What is your opinion about the usefulness of the duty ratio concept?

Pollack: Well, it is a theory, and may not be correct. I therefore agree with your opinion.

Sugi: I think that the duty ratio concept is only useful in the system, in which a limited number of myosin molecules interact with actin; it assumes that the total distance of actin-myosin sliding is algebraical sum of each myosin step, and this assumption may not apply for muscle fiber shortening.

Huxley: If I understand you correctly, the ATPase measurements in solution has no structural constraint. In the case of cytoplasmic myosin versus actin cable sliding, you think that the ATPase rate is speeded up by some mechanism. Similar argument occurred in the early 1980's between Spudich and Yanagida (Toyoshima et al., Proc. Natl. Acad. Sci. USA 87: 7130-7134, 1990; Harada et al., J. Mol. Biol. 216: 49-68, 1990). Yanagida group developed a method to measure ATPase activity of a single myosin molecule interacting with an actin filament using fluorescent ATP under evanescent light illumination (Ishijima et al., Cell 92: 161-171, 1998). This is a good method to measure ATPase activity.

Sugi: Thank you for your suggestions and comments. Unfortunately, I am going to retire very soon and can not continue research work. As I gave my centrifuge microscope system to Dr. Tsuchiya in Kobe University, I hope him to continue research work of this kind.

Brenner: I also suggest you to use fluorescent ATP to measure ATPase activity within the algal cell during cytoplasmic streaming.

Sugi: I once attempted to perform such experiments. Unfortunately, in the algal cell, there are considerable ATPase activities other than actin-myosin interaction, probably arising from proton pump etc.

C-Protein Interaction with Actin as a Source of Internal Resistance against Musle Shortening

Brenner: Since this is a discussion about 'Mysteries of Muscle' I thought to bring up one such 'mystery'. If a muscle fiber is allowed to shorten under exactly constant external load, the shortening velocity does not stay constant but slows down as the shortening proceeds (e.g., Gordon AM, Huxley AF, Julian FJ, J Physiol 184:170-192, 1966; Fig. 10: over range of the tension plateau velocity decreases by almost 10%. Brenner, B J Muscle Res Cell Motil 1:409-428,1980). In in vitro motility assays this is not so. Instead the gliding velocity stays constant over time and correspondingly over long distances. This raises the question of what makes the shortening velocity in fibers to decrease.

Some years ago I had done some very careful studies and found that one can obtain constant shortening velocity if external load is reduced proportional to the distance of the shortening that takes place. So this looked like the shortening is generating some restoring force which is very closely proportional to the distance of shortening (Brenner, B Basic Res Cardiol 81:54-69, 1986). I was reminded of this observation yesterday by the presentation of Gonzales-Serratos when the question came up of what slows down movement. It again came up in the presentation by Winegrad where he showed that the myosin binding protein C can interact with the actin filament. So this raises the question whether it might be the interaction of the myosin binding protein C with actin that might act as an increase in internal load that slow down shortening velocity as movement goes on. I therefore like to ask S. Winegrad for his opinion on this point.

Winegrad: About 50 % of C-protein is closely bound around myosin filaments, and when it is phosphorylated, it interacts with actin filaments to affect actin-myosin sliding, i.e. the shortening velocity. Meanwhile, there is a striking similarity of amino acid sequence between C-protein and titin (~ 60 % homology), and this suggests that C-protein interaction with actin results in a compliant load as C-protein phosphorylation goes on during shortening (Kulikovskaya et al., J. Gen. Physiol. 122: 761-774, 2003). In cardiac muscle, partial extraction of C-protein increases shortening velocity and rate of relaxation, being consistent with the idea that C-protein-actin interaction produces compliant load.

The properties of the interaction between C-protein and actin, as modulated by phosphorylation of C-protein, are consistent with an internal load that increases in proportion to shortening. The C-protein is sufficiently compliant to remain attached during a shortening of 15 % of the sarcomeres, which is what occurs during a 50 % ejection fraction of the heart.

Gonzalez: I wonder whether your explanations on cardiac muscle can apply to skeletal muscle.

Winegrad: There are important differences between cardiac and skeletal C-protein isoforms. The skeletal isoform do not have phosphorylation sites, but it does have the myosin binding sites. Therefore, one must be careful in applying observations made on cardiac muscle to skeletal muscle.

Sugi: I would like to make a general comment. In the Huxley 1957 model (Huxley, Prog. Biophys. Biophys. Chem. 7: 255-318, 1957), the only resistance against actin-myosin sliding is negative forces exerted by cross-bridges in the negative force region. On this basis, the shortening velocity under zero external load is kept constant over a large distance. If, however, there is internal resistance against shortening other than the cross-bridges exerting negative forces, constant shortening velocity can only be maintained for a distance ~ 10 % of fiber length (i.e. the length of bare region in each sarcomeres) where the number of cross-bridges producing shortening is kept constant.

Unbinding Forces of Actin-Myosin Linkages

Brenner: In order to relate muscle mechanics to enzyme kinetics of actomyosin, we will have a talk by Dr. Ishiwata on the unbinding forces of actin-myosin linkages at the level of single molecules. I hope his talk will give you information about, for instance, the rate constant of dissociation of actin-myosin linkages.

Ishiwata: I would like to describe the unbinding force measurements on the actin-myosin (HMM or S1) single rogor bonds (Nishizaka et al., Nature 377: 251-254, 1995 and Biophys. J. 68: 75s, 1995 and Biophys. J. 79: 962-974, 2000). First of all, I would like to stress that the unbinding force is a quantity that cannot characterize the bond strength unless the loading rate (the rate of increasing the applied external load) is fixed; in practice, a molecular bond is spontaneously and stochastically broken under thermal equilibrium, implying that one does not need to apply external force to break the bond, but one only needs to wait for a while until the bond is spontaneously broken.

The lifetime of molecular bonds is also a stochastic quantity, but it has an average value, τ, which is related to the binding affinity. The number of bonds, $N(t)$, decays with time t, according to $N(t) = N(0)\exp(-t/\tau)$. The average lifetime strongly depends on the extrernal load; the larger the external load is, the shorter the lifetime is (Nishizaka et al., Nature 377: 251-254, 1995). The load dependence of the average lifetime could be experimentally formulated as follows (Fig. 1): $\tau(F) = \tau(0)\exp(-Fd/k_BT)$, where $\tau(F)$ is the average lifetime under the external load, F, and $\tau(0)$ is the average lifetime under noload, representing the binding energy at thermal equilibrium (about 1000 s for HMM and 100 s

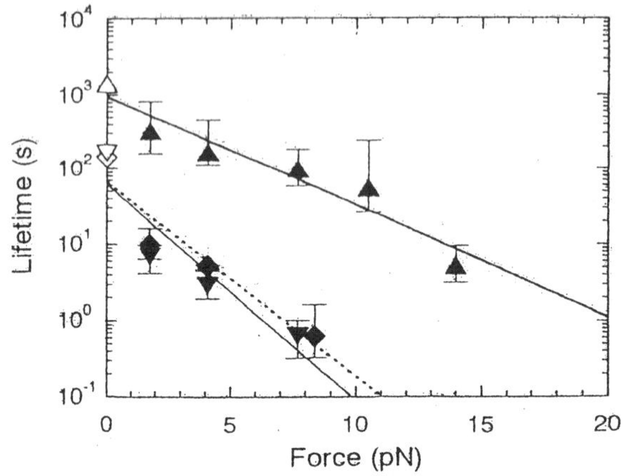

Fig. I-3: Relationship between external load and lifetime of HMM and S1 rigor bonds. Triangles and inverted triangles show the slow and fast components of HMM, respectively. Squares show the lifetime of S1. Open symbols were obtained by microscopic observation in the absence external load (Tadakuma et al., unpublished data), and filled symbols were determined by single-molecular measurements. Fitted lines show the approximation by the equation $\tau(F) = \tau(0)\exp(-Fd/k_BT)$. The thick and thin solid lines show the approximation of the slow and fast components of HMM, respectively. The dashed line shows the approximation for S1 (Nishizaka et al., Biophys. J. 79: 962-974, 2000).

for S1), d is the distance to characterize the stability of bond (1.4 nm for HMM and 2.4 nm for S1; the smaller the d value, the more stable the bond against the external load), k_B and T are, respectively, the Boltzmann constant and the absolute temperature (Nishizaka et al., Biophys. J. 79: 962-974, 2000).

Now, we define "the unbinding force" to be an external load at which the bond is broken during the increase in the external load according to $F = \alpha t$ at a constant rate α. On increasing the external load, the lifetime of the bond is exponentially declined and the bond is broken at some value of the external load. Thus, the unbinding-force distribution is obtained from the derivative of the time course of the decrease in the number of the bond. Here, note that the unbinding force thus measured is scattered but the shape of the unbinding-force distribution looks similar to the Gaussian distribution (for the relationship between the load dependence of the lifetime of the bond and the unbinding force, see Uemura et al., Pric. Natl. Acad. Sci. U.S.A. 99: 5977-5981, 2002). The average of the unbinding force thus obtained largely depends on the loading rate; the larger the loading rate is, the larger the unbinding force is (theoretically, no limit). Based on these results, I can suggest that the drug force due to the formation of non-force generating (e.g., ADP-bound) cross-bridges exponentially depends on the sliding velocity. The larger the sliding velocity is, the larger the drug force is.

Suga: I would like to clarify your unbinding force and stretch conditions. No cross-bridges were broken at a force of 10pN. You have stretched the active fiber by 100nm. Is their product equal to the energy of the bound cross-bridge?

Ishiwata: No, the 100nm is the movement of the head, but not the stretch of the unbound cross-bridge.

Suga: I see. In this case what is the amount of stretch on each cross-bridge?

Ishiwata: It is difficult to estimate, because actin filament is very compliant.

Ikebe: I wonder whether the actin binding requires HMM with two myosin heads. In other words, can single myosin head bind to actin?

Ishiwata: The actin-S1 complex has properties similar to those of actin-HMM complex. The unbinding force of actin-HMM complex is almost exactly twice the unbinding force of actin-S1 complex or that of the single-headed HMM-actin complex.

Weak and Strong Binding of Cross-Bridges to Actin during Muscle Contraction

Brenner: In terms of the actomyosin ATPase reaction scheme, cross-bridge powerstroke is normally considered to be associated with the intermediates of this reaction scheme. In this connection, I would like to have your opinions about the weak and strong cross-bridge binding to actin.

Pollack: In considering the weak/strong cross-bridge concepts based on the stretch experiments etc., how do you rule out the contribution of titin? If other people perform similar stretch experiments, is it possible to interpret the data based on titin?

Brenner: No, I don't think so. The experimental results consisted of two parts. If you stretch single skinned fibers at various sarcomere lengths and at higher ionic strength, the elastic modulus of the parallel elasticity stays constant. Titin may be a candidate for this elasticity. If, on the other hand, you lower the ionic strength, you obtain results suggesting the weakly-bound cross-bridges.

Sugi: As I talked in this meeting, Huxley and Simmons (Nature 233: 533-538, 1971) reached the idea that the T_1 and T_2 curves were scaled according to the level of isometric force, i.e. the number of force-generating cross-bridges, without data points at force levels close to zero. It was because the feedback loop did not work well if the slope of the force versus length relation was small. Thus, they overlooked the thin filament compliance.

By the way, in the early Hakone Symposia I organized, there were a great deal of discussions concerning the weak/strong binding cross-bridges. At that time, the weak/strong cross-bridge phenomena were regarded to be only obtained under rather non-physiological conditions (very low ionic strength coupled with extremely rapid ramp stretches). Nevertheless, the weak/strong cross-bridge binding concept is now widely used by almost everyone without argument. I wonder whether there are pitfalls in this concept as has been the case for the T_1-T_2 curve concept.

Brenner: I agree with you in that there may be pitfalls in this concept.

Strongly Bound Cross-Bridges and the Powerstroke

Brenner: Now, I would like to have your comments on the possible relation between the strongly bound cross-bridge and the powerstroke.

Huxley: I think the powerstroke may occur between the two strongly bound states. A cross-bridge is attached to actin, and then generates force and may release phosphate afterwards. Based on the X-ray pattern, a quick release applied to a contracting muscle produces a large mass movement. My question is whether the weakly bound cross-bridges come out during the quick release.

Brenner: It is difficult to see if something is weakly bound. I think your point is that the powerstroke may be associated with the transition from the strongly bound to the weakly bound state. One point I would like to add is that the isometric force generation may take place in the early phase of powerstroke, and I think the quick force recovery is not the same as the initial isometric force development. In your experiment, as the quick release goes on, a sort of quick force recovery occurs during the release. Do you have evidence for the wider type of states during the release?

Huxley: In the experiment using whole muscle, you do not see the force records during the release, and during the quick force recovery you see occupancy of the cross-bridges which are much closer to the rigor position. The trouble I have is that the T_2 curve suggests the total powerstroke of 12-13 nm. On the other hand, cell molecular people indicate the powerstroke of the order of 5 nm from their in vitro experiments.

Brenner: It is indeed a long-standing question. Cell molecular people show steps of roughly 5 nm steps. The question is whether there is something missing in the single molecule work.

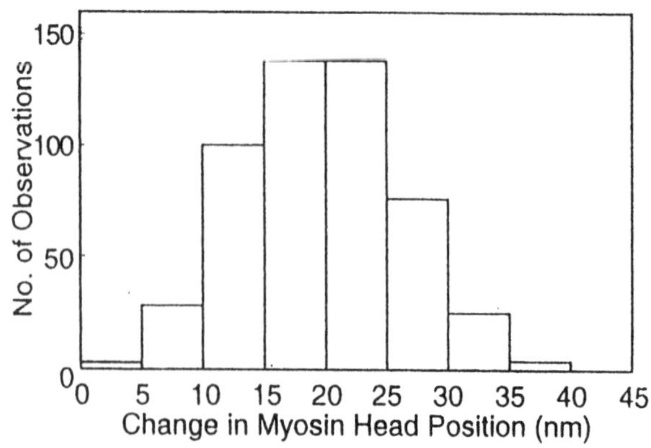

Fig. I-4: Histogram showing the distribution of distance between the two center of mass positions of the same particles in the two records, representing the amplitude of the ATP-induced myosin head movement (Sugi et al., 1997).

Sugi: I would like to call your attention to our work, in which synthetic thick filaments (rabbit skeletal muscle myosin-paramyosin complex) were kept alive in the electron microscope using a gas environmental chamber (magnification, ~100,000X, Sugi et al., Proc. Natl. Acad. Sci. U.S.A. 94: 4378-4382, 1997). The myosin heads on the filament were marked with gold particles at the 50kD-20kD junctional area. The experimental solution did not contain ATP and the thin filament. The electron microscopic image of the thick filaments with distinct gold particles, serving as myosin head position makers, were recorded on the imaging plate (spatial resolution, ~5nm).

In the absence of ATP, the position of the myosin head did not change appreciably with time, indicating that the time-averaged position of the myosin head is kept constant. When ATP was applied iontophoretically, the myosin head was found to move by ~20nm in the direction parallel to the filament long axis (Fig. I-4). On the other hand, the myosin head did not move on application of ADP.

Based on the general view that, during muscle contraction, the myosin head repeats its powerstokes when attached to the thin-filament. This implies that the myosin head also repeats its "recovery stoke" when detached from the thin filament, with an amplitude similar to that of the powerstroke. The ATP-induced myosin head movement observed in our experiments is therefore likely to be the recovery stroke (or preparatory stroke). It may be that, after performing its powerstroke, the myosin head (M) binds with an ATP molecule to perform its recovery stroke, which is associated with reactions, M+ATP→ M·ATP →M·ADP·Pi.

Thus, after perfoming the recovery stroke, the M·ADP·Pi binds to the thin filament to again perform powerstroke. The above scheme of the myosin head motion is illustrated in Fig. I-5. In this figure, the myosin head motion is assumed to largely result from length changes of the subfragment-2 (S-2) region, suggested by the late Professor Harrington and myself (Sugi et al., Proc. Natl. Acad. Sci. U.S.A. 94: 4378-4382, 1997), since the head movement of ~20nm is too large to be explained by myosin head rotaion. Expeiments are currently going on to prove this scheme.

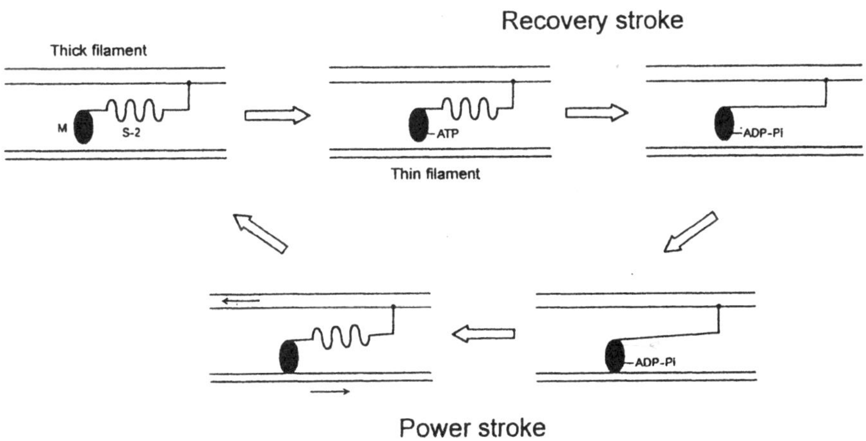

Fig. I-5: Schematic diagram showing cyclic shortening-lengthening of myosin subfragment-2 region coupled with ATP hydrolysis.

Holmes: I would like to go through the crystallography and point out what is known and what is inferred. The basic scheme we use for the inference is the Lymn-Taylor scheme (see Figure 1 in Holmes). On one half (right) of the scheme are the unattached states. These are available from X-ray crystallography since they concern the myosin cross-bridge not bound to actin. What we know about the actin bound states (left) comes either from electron microscopy or from combining electron microscopy with crystallography. The post-power stroke and pre-power stroke states are well represented by a number of independent structures, which shows that these states are well defined and are not artefacts of crystal packing. A recently published structure of nucleotide-free myosin V from Anne Houdusse's lab is a very good model for the end of the power stroke when attached to actin. The state we know least about is the start of the power-stroke. Here there is much speculation. It should be a strong binding state, a state without rapid off-rates, else it cannot function as a cross-bridge. In the movies of the power stroke that I have made I have assumed that it is similar to the crystallographic pre-power stroke state but with the actin binding cleft closed. This entails moving the upper 50K domain towards the lower 50K domain which in turn opens the nucleotide binding pocket. There needs to be a state change to signal the beginning of the power stroke. An attractive candidate would be phosphate release. The moving of switch 1 could enable phosphate release. However, crystallography is not the best way to resolve this problem. In the movies I show the converted domain and lever arm rotating around an axis during the power stroke. The position of this axis has been determined from the structures of *dictyostelium* myosin II in the pre and post power stroke states. It is a pure rotation though 60° and runs roughly parallel to the relay helix. It is also at right angles to the actin helix axis and thus is in the correct orientation to be the fulcrum for a lever arm moving in the plane containing the axis of the actin helix.

Since we are much better informed about unbound states than bound states we can speculate about the return power stroke (i.e. getting the lever arm into the correct position to start the power stroke). The two crystal structures mentioned provide crystallographic coordinates for the end of power stroke and pre-power stroke conformations. The group of Stefan Fischer in Heidelberg has taken these coordinates and used the method of conjugate peak refinement, an algorithm developed by Stefan Fischer and Martin Karplus to compute a reaction path between end states i.e. the return power stroke. One will have to ask Sefan Fischer for details of how this works. The results are shown in two movies. In the first movie we are looking (mostly) at the relay helix (shown in blue). Starting from the nucleotide-free structure firstly ATP is inserted. The effect of this is to pull in switch 2 to the CLOSED conformation. This movement can be seen in the first few frames of the movie. However, this action puts a stress on the (straight) relay helix, which responds by untwisting at one position to produce a bulge or kink. The production of the bulge rotates the distal end of the relay helix through 60°. As the converter domain (with the lever arm) is firmly anchored onto the end of the relay helix this also rotates through 60°. Note how the phenylalanine residues in the relay helix and the subsequent relay loop undergo at complex mutual rearrangement during this process. The reverse of this process (i.e. the "un-kinking" of the relay helix after binding to actin) is the quintessence of muscle contraction.

The second movie shows a view of these events looking along the rotation axis. The distal end of the relay helix forms a short two-helix bundle with the SH1 helix. The centre of the rotation axis runs about half-way between the two helices. During the transformation the two helices roll round each other but retain their local geometry. Since the SH1 group is in the middle of this helix-helix interaction it is difficult to see how one can attach bulky groups to this residue without destroying an essential element of the contractile apparatus.

Kushmerick: In the modeling, are you focusing on just some helices or on the whole molecule?

Holmes: It is not just some helices but the whole molecule is considered. But the simulation is done using the crystallographic end-states.

Maughan: How does this simulation square with the studies where people have modified lever arm length?

Holmes: They are completely consistent. Everybody has always taken the pivot point to be somewhere near the SH-2 group.

Pollack: What about consistency with the data Katayama showed yesterday on the quick freeze experiments.

Holmes: Frankly, I don't know how to interpret that. He said that what he was seeing looks like the SH1-SH2 being cross-linked. I don't see this as being part of the cross-bridge cycle. This is all what I can say.

Huxley: Ken, this simulation presumably lasted a few times 10^{-15} seconds or something like that.

Holmes: This is not actually done in real time. The way it is done is to take the end states to do a morph between them and to do mostly an energy minimization to make the various sub-states you have generated to agree with each other. So it is not a true molecular dynamics simulation. The MD is mainly used for energy minimization.

GENERAL DISCUSSION PART II
SKELETAL MUSCLE ENERGETICS

Chaired by J. A. Rall

Measurement of ATP Hydrolysis in Contracting Muscle

Rall: During the muscle energetics session we are going to discuss the evolution of techniques to measure ATP hydrolysis by muscles and motor proteins. I have asked Professor Sugi to speak first about some issues relating to muscle energetics.

Sugi: Studies on muscle energetics started in the laboratory of A.V. Hill with the measurement of heat and work during muscle contraction (Trails and Trials in Physiology, Williams and Wilkins, Baltimore, 1965). Unfortunately, heat production is a highly non-specific phenomenon that can be related to a number of events other than actin and myosin interaction. Fortunately, it is now well established that the chemical energy derived from ATP hydrolysis is the only immediate source of energy for muscle contraction and several papers have been published on the simultaneous measurement of ATP hydrolysis and muscle work production using NADH fluorescence and a sensitive Pi binding protein (see below).

Models of muscle contraction have been generated assuming that the number of cross-bridges changes according to load on a working muscle but that the force per cross-bridge is constant (e.g., Huxley, Prog. Biophys. Biophys. Chem. 7: 255-318, 1957). On this occasion, I would like to describe experiments designed to examine this issue based on a technique developed by Oiwa et. al. (J. Physiol. 437: 751-763, 1991). In these experiments a myosin-coated glass micro-needle is brought into contact with actin cables in a giant algal cell and actin-myosin sliding is induced by a fixed amount of iontophoretically applied ATP (Fig. II-1). Sliding occurs and work is done until the ATP is exhausted and then rigor ensues. With the repeated application of a constant amount of ATP, the amount of work done by actin-myosin sliding increases sharply with increasing amount of baseline force or external load, from zero to 0.4-0.6 Po, and then decreases with further increases in baseline force, reaching zero when the baseline force is Po (Fig. II-2). The increased work done with a constant amount of ATP utilized implies that the efficiency of the ATP-induced actin-myosin sliding increases sharply with external load. This bell-shaped work versus baseline force relationship

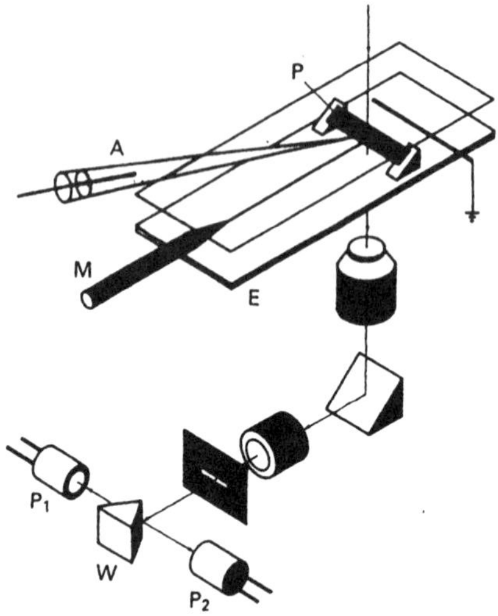

Fig. II-1: Experimental arrangement. The myosin-coated tip of a glass microneedle (M) is put in contact with the inner surface of the internodal cell strip preparation (P) at right angles to the chloroplast rows on which the actin cables are located. The experimental chamber (E) filled with ATP-free solution was mounted on an inverted light microscope. The needle was made to slide along the actin cables by applying negative current pulses to the microelectrode (A) filled with 100 mM-ATP. The needle tip movement was recorded by splitting the needle tip image with the wedge-shaped mirror (W) into two parts, each of which was projected on a photodiode (P_1 and P_2) (Oiwa et al., 1991).

is analogous to that reported by Fenn (J. Physiol. 58: 175-203, 1923) for the relationship between the load and work in tetanized frog skeletal muscle. Since the number of myosin molecules involved in the micro-needle movement is very small (< 10) and the amount of ATP utilized is constant, our results imply that the efficiency of chemo-mechanical energy conversion by individual cross-bridges changes in a load-dependent manner.

In addition, we have also presented evidence that the mechanical efficiency of dememgranated skeletal muscle fibers changes in a load-dependent manner in both auxotonic and isotonic conditions (Sugi et al., Proc. Natl. Acad. Sci. U.S.A. 95: 2273-2278, 1998; J. Exp. Biol. 206: 1201-1206, 2003).

Rall: One of the points that Professor Sugi's talk has brought up is the importance of being able to accurately measure the amount and rate of ATP hydrolysis during contraction. Measurement of the time course and amount of ATP hydrolyzed during contraction is crucial for the development of mechanistic models of muscle contraction and for the general characterization of muscle performance. Despite its importance, measurement of ATP hydrolysis during contraction has proven to be difficult and numerous techniques have been developed to achieve this goal. The first stimulus to develop these techniques can be traced back to A.V. Hill's celebrated "challenge to biochemists" (Biochim. Biophys. Acta 4: 4-11,

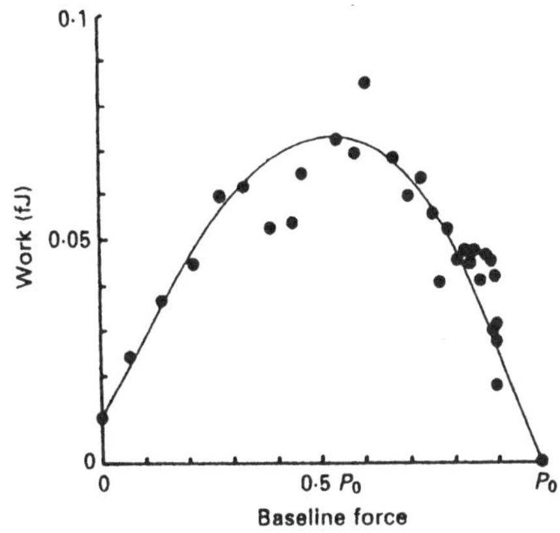

Fig. II-2: Relation between the amount of work done (in fJ) by the needle movement induced by a constant amount of ATP application and the amount of baseline force (relative to P_0). Needle movements were induced by 1 s ATP current pulses (15 nA) from various baseline forces in a random order. The curve was drawn by eye. Note a finite value of work done for zero baseline force, because the needle movement took place in the auxotonic condition (Oiwa et al., 1991).

1950) He believed that biochemists should demonstrate directly the splitting of ATP during the contraction of *living* muscle, in contrast to muscle extracts, if they wished to substantiate their claim about its preeminence as an energy donor. The challenge was taken up by several investigators but it wasn't until 1962 that Cain, Infante and Davis (Nature, Lond. 196: 214-217) were able to observe ATP hydrolysis in a contraction of a muscle that was poisoned with fluorodinitrobenzene (FDNB). FDNB blocked the creatine kinase reaction and thus blocked replenishment of hydrolyzed ATP by phosphocreatine. Martin Kushmerick was just starting in R.E. Davies' laboratory during this time and he will describe some of the events associated with this crucial development.

Kushmerick: In the 1950s there was a biochemical puzzle. The biochemists had clear evidence that purified myosin hydrolyzed ATP. In contrast the physiologically minded biochemists could not demonstrate ATP hydrolysis during muscle contraction. The only high energy phosphate reaction that was demonstrated during contraction was the breakdown of phosphocreatine by Lundsgaard in 1930 (Biochem. Z. 217: 162-177). I came to Bob Davies' laboratory as a medical student in 1959 when everyone was struggling to measure ATP hydrolysis during contraction. There was the idea at that time that there was some new biochemistry to be discovered. The thought was that some phosphorus containing compound other than ATP or phosphocreatine could be the direct energy source for contraction. Various phosphate compounds were examined and discarded as possibilities. Another possibility was that ATP was broken down during contraction but that it was re-synthesized too rapidly for the break down to be observed. The direct measurement of ATP hydrolysis during contraction was achievable through the insight that blocking the creatine kinase reaction would prevent the re-synthesis of ATP by phosphocreatine. In retrospect it is amazing how long it takes simple insights to register in one's mind. This is exactly what Cain, Infante and Davis did in 1962 and it worked. Thus the initial challenge that A.V. Hill posed evaporated and the physiologists and biochemists were very happy.

During this time great effort was spent to find ways to rapidly denature the enzymes in muscles that had been activated. One idea was to rapidly immerse a contracting muscle into liquid nitrogen but W. Mommaerts showed at a meeting that when an isolated heart is rapidly immersed in liquid nitrogen, the outside is frozen but the interior of the heart keeps beating for several seconds. M. Kretzschmar and D. Wilkie realized that to speed up the freezing of a muscle the muscle should be made as thin as possible. They did this by smashing a contracting muscle between liquid nitrogen cooled hammers (J. Physiol. 202: 66-67, 1969). They were able to achieve a tenth of a second freezing time. This technique not only made the freezing rapid but also allowed the moment of freezing to be accurately controlled.

More recently the reasons for measuring ATP fluxes has changed. First we want to make explicit the chemical reactions that underlay muscle contraction and recovery. If this can be done one can then put boundaries on the amount of work done and the amount of ATP utilized. One could then also put boundaries on the economy and efficiency of muscle contraction. This information is very important when one attempts to formulate mechano-chemical models of motors. In general the efficiency that has been measured, for example in M. Ferenczi's laboratory, turns out to be pretty high (about 40%) (He et. al., Biophys. J. 79: 945-961, 2000). I think that this is an important number. The other reason is that these ATP fluxes are intimately related to cross-bridge action and also intimately related to metabolism. It is for these reasons that measurement of ATP fluxes is very important. The questions over the years that I have been involved in this field have changed dramatically and the tools have changed. Nonetheless it is a bit disappointing that the time resolution of measurement of ATP fluxes has not dramatically improved.

Rall: In the early days, it was thought that the time-course and amount of energy liberation as measured by heat production might provide an assessment of ATP hydrolysis during contraction. This technique had the advantages of being rapid, nondestructive and sensitive. Unfortunately it was discovered that under many conditions the energy liberation did not reflect only ATP hydrolysis. Because of the nonspecific nature of the energy liberation technique, other heat producing or absorbing reactions also accompanied the observed energy liberation by ATP hydrolysis (Gilbert et. al., J. Physiol. 218: 169-193, 1971; Rall, this volume). Nonetheless it was possible to design experiments to measure energy liberation over a whole contraction-relaxation cycle and under these conditions the amount of energy liberated was proportional to the amount of ATP hydrolyzed. But this experimental design essentially wasted the rapid time resolution of the energy liberation technique.

Thus other techniques were developed. Marty mentioned direct chemical analysis for the measurement of ATP hydrolysis. As Marty emphasized, with direct chemical analysis all reactions have to be rapidly stopped by freezing the muscle. Thus you only collect data from one point in time. Plus that experiment has to be repeated many times because of the natural variation in resting ATP levels. This approach was not nearly as convenient as measuring the time course of energy liberation where one could get a continuous record in a single contraction. What one would like to have is a technique to continuously measure the ATP hydrolysis during contraction. One possibility was to utilize the oxygen consumption of the muscle as an indirect measure of ATP hydrolysis. This technique is very important in Professor Suga's research on heart energetics (this volume). Also techniques have been developed to measure the oxygen consumed during contraction of single skeletal muscle fibers (Elzinga et. al., J. Physiol. 399: 405-418, 1988). But this technique does not provide a direct measure of ATP hydrolysis. The phosphorous NMR technique has been utilized to measure ATP hydrolysis during muscle contraction (Dawson et. al., J. Physiol. 267: 703-735, 1977, Yamada, this volume). Each of these three techniques can provide accurate estimates of ATP hydrolysis during contraction but each suffers from a lack of temporal resolution. Possibly Marty Kushmerick and Kazuhiro Yamada could comment on the current temporal resolution of the NMR technique.

K. Yamada: In our studies with isolated frog muscles using NMR, we intended to improve the time-resolution of the technique as much as possible. We did this by using many muscles and accumulating signals from repeated contractions. The best time-resolution that we achieved was about ten seconds (this volume). Nonetheless we can still answer important questions with this temporal restriction.

Kushmerick: The NMR technique is inherently insensitive which means that large amounts of ATP, PC, etc are needed in order to obtain signals. However there is a nice trade off that can be applied. In mouse muscle experiments the current time resolution is no better than thirty seconds. In this case one is studying on the order of fifty milligrams of muscle. But in human muscle we are studying about fifty grams of muscle. In these studies the time resolution approaches one or two seconds. It is worth commenting on the fundamental limit of the time resolution in the NMR technique. Even though the electro-magnetic pulse that perturbs the state of the nuclei that one wants to measure is only microseconds in duration, what one actually records is the relaxation time of the perturbed magnetic field. This relaxation time takes a minimum of fifty to one hundred milliseconds. This time sets the upper limit on the time resolution of the NMR technique. While this is a disadvantage from the point of view of measuring chemical changes associated with work production, the advantage is that one can measure, repeatedly, changes in many compounds at the same time under physiological conditions.

Rall: Another possible approach is to measure ADP production associated with ATP hydrolysis by using a linked enzyme assay to couple ADP production to the oxidation of NADH, the concentration of which is measured by fluorescence (Stephenson et. al., J. Physiol. 410: 351-366, 1989). This is basically a steady state technique that can be utilized to assess ATP hydrolysis in single skinned muscle fibers. The most recent technique developed to measure ATP hydrolysis during contraction measures the P_i produced when ATP is hydrolyzed. In this case a fluorescence change is measured when P_i binds to a phosphate binding protein that has been introduced into a permeabilized muscle fiber (He et. al., J. Physiol. 501: 125-148, 1997). This technique has the advantage of high sensitivity and high temporal resolution.

Sugi: We have observed that in many respects the phosphate binding protein assay is superior to the NADH system.

Rall: The main disadvantage of the phosphate binding technique is that the contraction must be limited in duration so as not to exhaust the phosphate binding protein (one or two seconds, depending on temperature). Thus this technique has its advantages but also its limitations. Recently energy liberation, NADH and phosphate binding protein assays have been combined to assess ATP hydrolysis in single muscle fibers (West et. al., J. Physiol. 555: 27-43, 2004). This approach combines the strengths of each of these techniques while at the same time minimizing the weakness.

Gonzalez-Serratos: Using electron probe analysis we have measured the time course of calcium release from and re-uptake of calcium to the sarcoplasmic reticulum by quick freezing bundles of muscle fibers during force development (Somlyo et. al., J. Cell Biol. 90: 577-594, 1981). Analysis of surface fibers allowed for a ten millisecond resolution of the technique.

Kushmerick: One can certainly take this approach of analyzing individual fibers or subcellular domains of fibers but it would require massive parallel experiments- a very laborious task currently not warranted by the questions being asked.

I want to bring up a point that was not mentioned earlier. There is a basic problem in measuring the time course of chemical changes and inferring chemical fluxes. For example in the beating heart it is possible to show an increase in oxygen consumption when the heart does work but it is very difficult to show on a beat to beat basis the hydrolysis of ATP at discrete time points. This is true because the forward rate of ATP utilization and backward rate of re-synthesis are nearly equal. In contrast heat measurements and NMR measurements can be interpreted in terms of fluxes and this is what one needs to know.

Maughan: I want to point out that it is not only just a temporal problem but also spatial problem because of compartmentalization of these components. This has been a tremendous challenge.

Suga: With regard to oxygen consumption, we cardiac physiologists can take advantage of oxygen consumption to obtain the amount of ATP consumption in a steady state because the heart works under exclusively aerobic conditions. Unfortunately oxygen consumption measurements do not reflect ATP consumption in non-steady conditions. Thus this technique is very useful in the heart but less useful in skeletal muscle research.

Rall: Despite many years and considerable effort, there still is no flexible and sensitive technique with high temporal resolution to measure ATP hydrolysis during muscle contraction. This remains an important challenge in the muscle field.

GENERAL DISCUSSION – SKELETAL MUSCLE III

STRUCTURAL STUDIES

Chaired by John M. Squire

Squire: My own interest in muscle really started because I have always been fascinated by the way machines work. In my youth I spent many happy hours building exotic structures with Meccano and things like that and finding out how to make machines that carry out particular tasks. With this background, my approach to muscle is analogous to a Marsian coming to Earth, seeing a motor car and wondering how this strange thing works. Of course he would soon find out what it does, it goes along, he would soon find out that you have to put something into it (petrol) to make it work and when it is working there are lots of emissions (the exhaust) which he could look at and test, but in the end he would_ absolutely need to get inside the car and actually look at its inner workings. The engine, of course, is an important part of that, but it is not the whole story; there is a gearbox and the transmission, there are brakes and an accelerator, there is a starter motor and intricate electrical circuits, so he would really need to treat the car as a whole, rather than just as an engine, to see how the whole assembly works together. The engine is vital, of course, so what I thought we could do here in this Discussion Session is to think first about the whole muscle sarcomere and then, about halfway through, we can get on to talking about the engine, the crossbridge cycle itself. I think you will see that what comes earlier about the sarcomere as a whole is very relevant to analysis of the crossbridge cycle.

What I plan to do is to present short summaries of a number of aspects of sarcomere structure and then to pose some questions which I hope some of you will respond to. I will give very brief outlines of what I think we know and then highlight things which are yet to be determined.

Actin Filaments

Squire: As you know, Ken Holmes and his team have done a great job in determining the structure of actin and the actin filament (Kabsch *et al.*, Nature 347: 37-44,1990; Holmes *et al.*, Nature 347: 44-49,1990). Tropomyosin lies along the long-pitched strands of actin and the structure of troponin is coming, but not yet fully determined (see for example Takeda *et al.*, Nature 424: 35-41,2003). It is still not clear to me how troponin fits on to tropomyosin and actin and how it might move in different states (see Squire and Morris, FASEB J. 12: 761-771,1998). I think it is fairly clear from lots of different studies that tropomyosin does shift across from actin subdomain 1 to a position closer to actin subdomain 3 when calcium binds to troponin, but there remains a problem in that the conventional kind of helical reconstruction from electron micrographs that people do, or the analogous X-ray modelling with helical structures, ignore the fact that there is troponin along the thin filaments which does not lie on the actin 13/6 helix. The helical reconstruction studies assuming a 13/6 helix are such that movements of troponin would be averaged into the actin+tropomyosin 27.5 Å axial repeat and the troponin movement will affect what appears as a tropomyosin shift. However, Ken Holmes and his team have done the calculations and so have we ourselves to show that, even with troponin properly included, the troponin would not explain the whole of what is seen; there is still a tropomyosin shift although it could well be smaller than is currently talked about. When the structure of troponin is known in detail then this analysis can be done properly. For people doing electron microscopy, the reconstructions need to be done taking account of the fact that the real repeat along the actin filaments is after 385 Å not 27.5 Å.

We know a little about nebulin. It has a 55 Å axial repeat (35 residues in an α-helix), it seems to run the whole length of skeletal muscle actin filaments and it may define the thin filament length (Wang *et al.*, J. Biol. Chem. 271: 4304-4314, 1996). It does not appear to be known exactly how nebulin fits onto the actin filaments. I also tried to find out from all the experts how many nebulins there are per thin filament. But the stoichiometry of nebulin to actin, tropomyosin and troponin is not yet known. There ought perhaps to be two per thin filament, one on each long-period strand of actin, but this remains to be proved. Nebulin starts at the edge of the Z-line on the actin filament, but neither the Z-line nor the actin filament has exact 2-fold symmetry so the number of nebulins need not necessarily be two (or a multiple). At the filament tip there are capping proteins which also do not have 2-fold symmetry.

Questions About Actin Filaments

[1] Where is nebulin?
[2] How many nebulins are there?
[3] Where is troponin?
[4] How does troponin change and/or move on activation?

Pollack: In some of the experiments where there is a postulated tropomyosin shift, if one doesn't know where the nebulin's location is unknown, is it certain that what is shifting is really tropomyosin and not nebulin?

Squire: I think the answer to that is no, not yet. However, there is only half as much nebulin, assuming there are two single strands of it per thin filament, as there is tropomyosin (i.e. nebulin is a single α-helix rather than the 2-stranded α-helical coiled-coil of tropomyosin). But nebulin will contribute something to the diffraction pattern – the whole assembly needs to be modelled properly.

The Z-line and I-band

Squire: Different fibre types are very different not only in their physiology but in their ultrastructure. For example, the M-bands and Z-bands are different in different fibre types. We have a pretty good idea of the way the Z-band builds up. There are zig-zag cross-connections of α-actinin between antiparallel actin filaments, but the number of axial levels of these connections varies between muscles and fibre types. We know quite a lot about the structure of α-actinin (Blanchard *et al.*, J. Muscle Res. Cell Motil. 10: 280-289, 1989; Liu *et al.*, J. Mol. Biol. 338: 115-125, 2004). It is one of the spectrin family of proteins which have a central rod-like 3-strand coiled-coil region on one end of which is an actin binding domain. Two such molecules form the anti-parallel dimer known as α-actinin – a rod-shaped molecule with actin-binding domains at each end. The structure of the actin-binding domain is not known, but we ourselves and others are working on that. We already have crystals that diffract to better than 2 Å, but the patterns are not yet good enough to solve the structure (Govada, Chayen & Squire - unpublished data).

Another intriguing feature of the Z-line is the N-terminal end of titin which has the so-called Z-repeats in it (Gautel *et al.*, J. Cell. Sci. 109: 2747-2754, 1996; Ohtsuka *et al.*, Biochem. Biophys. Res. Commun. 235: 1-3, 1997; Atkinson et al., Biochem. 39: 5255-5264, 2000). These are differentially expressed so that different muscle and fibre types have different numbers of Z-repeats in their titin. The number of Z-repeats appears to correlate with the number of zig-zag levels of α-actinin in the Z-line (Luther *et al.*, J. Mol. Biol. 315: 9-20, 2002), but we did some careful measurements of the axial spacing between the zig-zags (Luther & Squire, J. Mol. Biol. 319: 1157-1164, 2002) and this spacing seems to be much too long to be spanned by one Z-repeat. In fact there may be two Z-repeats between successive zig-zag levels of α-actinin. The Z-repeat region of titin is therefore not long enough to span the whole Z-line, but may only cross half of the Z-line.

In summary, different fibre types have different numbers of zig-zag levels (Luther *et al.*, 2002) controlled in some way by the titin Z-repeats. Then something goes wrong and one ends up with the Z-crystals or nemaline rods found in nemaline myopathy. We have already published a 3D reconstruction of nemaline rods to about 100 Å (Morris *et al.*, 1990) and this has now been extended to about 50 Å resolution (work by P.K. Luther, E. Morris and J. Squire – not yet published).

Through the I-band there is a gradual change of the actin filament array from being on a square lattice in the Z-line to being on a hexagonal lattice in the A-band. It was Pringle (Aspects of Cell Motility 22: 67-86, 1968) who showed how this transition could be achieved by equal displacements of all the actin filaments. We were intrigued to think about how titin could be accommodated into this structure. John Trinick's group showed that there are 6 titin chains per half myosin filament (Liversage *et al.*, J. Mol. Biol. 305: 401-409, 2001). We suggested (Knupp et al., J. Mol. Biol. 322: 731-739, 2002) that the titins divide such that two sets of pairs of titin move out from opposite sides of the ends of a myosin filament and interact with actins of the same polarity in the Z-line and the remaining two titin strands move out in a plane at right angles to the first four titin strands and interact with actin filaments of the opposite polarity to those interacting with the first titin strands. The six strands of titin can therefore be distributed through the I-band in a fairly systematic way. One of the things that might be explained by this is the observation by Leepo Yu and others (J. Mol. Biol. 115: 455-464, 1977) that the lateral spacing of the Z-line changes in step with the lateral spacing of the A-band when the sarcomere length is changed. It could be the lateral forces imposed by titin,, along withthose of the displaced actin filaments, which help to keep the A-band to Z-line spacing ratio constant.

Structures that so far seem rather ill-defined are the N-lines. There are two of them. One (N1) appears to keep a constant spacing relative to the Z-line whereas the other (N2) appears to move as the sarcomere length changes as though it was placed on an elastic structure (such as titin) between the myosin filament tip and the Z-line. To my knowledge the precise roles of the N-lines, which are presumably there for a very good reason, are not yet known (see Lopate *et al.*, Muscle and Nerve 21: 1216-1219, 1998, for recent observations and other references).

Note finally that in cardiac muscle there are non-uniform thin filament lengths. This was thought to be due to the absence of nebulin in this muscle type. Only a short nebulin-like molecule called nebulette was thought to be present, representing the Z-line end of nebulin. However, it has now been reported that there may be differently expressed full length nebulin isoforms in cardiac muscle (Kazsmierski *et al.*, J. Mol. Biol. 328: 835-846, 2003).

Questions About the Z-line:

[1] How do titin and nebulin fit into the Z-band?
[2] What determines the Z-band thickness?
[3] What are the N-lines and what is their role?
[4] In cardiac muscle are there different actin filament lengths?
[5] Is this controlled by different co-expressed nebulin isoforms?

Gonzales-Serratos: I remember reading recently that long-term exercise can cause the Z-line to change in width. So are there dynamic changes in the Z-line?

Squire: Yes, over time this can certainly happen. However, on a much shorter timescale it is probably axial disordering of the Z-line which occurs.

Gonzales-Serratos: What determines the changes in the Z-line in the long term?

Squire: The changes that occur over time are caused by changes in the expression of different Z-line proteins (e.g. different titins with different numbers of Z-repeats).

Myosin Filaments

Squire: We have a pretty good idea now of how the myosin heads are organised in vertebrate muscle and insect flight muscle in the resting state (Hudson *et al.*, J. Mol. Biol. 273: 440-455, 1997; AL-Khayat *et al.*, Biophys. J. 85: 1063-1079, 2003). As far as how the myosin rods are packed into the filament backbone, there is a physically plausible model that I proposed in 1973 (Squire, 1973). We have shown that, of all the various backbone models that have been published, this 1973 model fits the high-angle X-ray diffraction data better than the other models (Chew and Squire, J. Struct. Biol. 115: 233-249, 1995; Squire *et al.*, J. Struct. Biol. 122: 128-138, 1998). However, I would not claim that we have proved that this model is correct. This is an enormously important structure to determine, for its own sake and also to see how titin and C-protein fit onto it. I do feel duty bound, having proposed it so long ago, to try to determine the backbone structure. It is possible that the new single particle analysis reconstructions of vertebrate myosin filaments which we are currently undertaking may reveal this structure if we can get the resolution high enough (AL-Khayat *et al.*, Biophys. J. 85: 1063-1079, 2004).

Another aspect of myosin filaments is what C-protein (MyBP-C) does. We now have good evidence (Squire *et al.*, J. Mol. Biol. 331: 713-724, 2003) that C-protein binds to myosin with a 429 Å repeat, but that it binds at its N-terminal end to actin, thus giving C-protein on average a slightly longer spacing than 429 Å (i.e. about 432 to 435 Å). This structure gives rise to meridional peaks at around 435 Å in X-ray diffraction patterns from vertebrate striated muscles which were shown by Rome *et al.* (Nature New Biol. 244: 152-154, 1973) using antibody labelling to be due to C-protein. These reflections get much weaker and almost disappear in patterns from active muscle. I had always thought that this must be due to the N-terminal parts of C-protein coming off actin due to activation and becoming disordered. However, Saul Winegrad has suggested that maybe there is axial movement between the myosin and actin ends of C-protein in active muscle which would reduce the intensity of the C-protein reflections even if C-protein remains bound to actin. This idea remains to be tested.

With regard to the C5 to C10 part of C-protein, we prefer to put this running axially along the myosin filament backbone (Squire *et al.*, J. Mol. Biol. 331: 713-724, 2003) rather forming a collar as in the model of Moolman-Smook *et al.* (Circ. Res. 91: 704-711, 2002). The reason for this is that the collar would itself be expected to give rise to a substantial 429 Å meridional reflection in the X-ray pattern and this would presumably remain even if the C0 to C4 part of C-protein was highly mobile or tilted to the long axis. In fact the meridional reflection that is seen in patterns from resting muscle seems to be dominated by the actin end of C-protein at about 435 Å spacing. In addition a strong 429 - 434 Å reflection is not seen in patterns from active muscle (the spacing would presumably be closer to 434 Å in active muscle because of the length change of the whole thick filament). Another important point is that domains 7 to 10 of C-protein are thought to bind to titin which itself must presumably be running more or less parallel to the myosin filament long axis.

An intriguing feature of the C-protein arrangement with three molecules radiating out from the 3-fold myosin filament and interacting with three of the six surrounding actin filaments is that the three remaining actin filaments do not interact with C-protein. One feature of the simple lattice found in bony fish muscle is that all the myosin filaments have the same orientation. This means that half of the actin filaments will bind three C-protein molecules at each axial level, whereas the other three will not bind C-protein at all. A feature of the superlattice found in higher vertebrate muscles is that there are two myosin filament orientations 180° apart in a statistical superlattice arrangement and this will redistribute the C-proteins so that each actin filament has either a double or a single interaction with C-protein at each level. This more even distribution of C-protein attachments may be an advantage of the superlattice, especially if C-protein has an important modulatory role on muscle activity.

Questions About Myosin Filaments

[1] How are the myosin rods packed into the backbone?
[2] How are the heads organised in resting muscle?
[3] Where is titin relative to the heads?
[4] Where is C-protein (MyBP-C) relative to titin and the heads?

Winegrad: I just wanted to report a preliminary observation and I want to ask you John if you can help us to interpret it. If we remove approximately a third of the C-protein from heart muscle the remaining C-protein does not behave the same way as the third that we removed. So either all C-proteins are not exactly the same or removal of some C-proteins changes the function of the remaining molecules. That is all we have observed. We don't know if it is structural of functional or what but since you have thought a great deal about C-protein perhaps you have some ideas.

Squire: Going on from that, you also made the point yourself that when you do binding of C0 and C1 to actin you get a stoichiometry of 1 to 7. I had immediately thought that this was to do with the tropomyosin/troponin system, but then you said that this was with pure F-actin. I find that result astonishing – the implication is that the binding information is in some way being propagated along the F-actin filaments. If this is the case then it is equally plausible that taking some C-protein away will have an effect on the F-actin structure that is sensed by the remaining actin-bound C-protein molecules.

Winegrad: What would interest me would be to extract some of the C-protein and then do further X-ray diffraction studies to see if there is a change in structure.

Rall: This question is for David Maughan. What functional differences are observed in muscles of the C-protein knockout mice?

Maughan: We have studied the C-protein mutant from the Seidman lab with a C-terminal dletion that includes a myosin and titin binding site (Palmer et al., Mol. Cell. Biochem. In press; Circ. Res. In press). In these mutants lacking the C-protein, dynamic stiffness is about half that of wild type and velocity is about twice that of wild type. With no C-protein to anchor the myosin rods and titin,,
The structural rigidity of the thick filament is probably compromised, resulting in a reduced dynamic stiffness of the sarcomeres. The absence of C-protein also appears to reduce viscous drag, resulting in a reduced viscous modulus and greater shortening velocity.

Pollack: I have a comment which refers to how the myosin rods might be packed in the backbone. Several years ago we reported on the ABRM thick filaments using the nanolever apparatus that I referred to earlier and carrying out stretches and releases of the filaments (Neumann, Biophys. J. 75: 938-947) If you impose tensions up to the isometric tension sustained in the muscle, the filaments will lengthen reversibly by 50% or so. At the higher resolution, we stretched these filaments and they actually stretched, as one might expect, in steps of about 2.7nm or integer multiples thereof. Knowing the step size and the charge distribution along each rod one should be able to infer much about rod packing.

Squire: We do know quite a lot about how the myosin rods are packed from X-ray diffraction. We know the lattice spacing quite well (about 20 Å), we know the pitch length, we know the charge distribution, we know how the lattice spacing changes as a result of dehydration (from 20 Å down to about 17 Å). However, it seems that there is no evidence at all, as far as I know, from X-ray diffraction that there is any significant change of filament length apart from the small compliances that Hugh Huxley and others have measured.

Pollack: There is no evidence that the 143 Å or 429 Å repeats change appreciably, which does rule out the possibility of uniform length changes along the filaments. Our point in mentioning these steps is that the filament length change occurs as a series of local length changes (each equal to an integer multiple of 2.7 nm). If the filament changes by localised length changes rather than uniform stretch, this may not be picked up by X-ray diffraction. Therefore X-ray evidence does not necessarily rule out filament length changes.

Squire: Michael Chew and I once tried some pioneering experiments in studying the 5.1 Å meridional X-ray reflection, which comes from the turn of the backbone α-helices, in patterns from resting and active fish muscle (Chew *et al.*, In synchrotron Radiation, Daresbury Annual Report pp203-204, 1996). This was an extremely difficult experiment, carried out at the Daresbury Synchrotron Radiation Source, but we had the impression that this spacing changed by nearly 0.5%.

Huxley: Would this explain some of the change of spacing from 143 Å to 144 Å or so from resting to active muscle?

Squire: It might do, but it need not necessarily contribute to this 1%. If the whole rod changes in the same way, then the 143 Å filament repeat would show the same 0.5% change due to stretching of the α-helical pitch. However, if for example the major change in the α-helix pitch is say 1% or so but only in the exposed S2 part of the myosin rod, with the tightly packed LMM part of the rod changing much less, then an average 0.5% change in the 5.1Å reflection could still be observed, but the basic myosin axial periodicity would not change from 143 Å by this mechanism alone. The rods would be longer, but still on a 143 Å repeat. In any case these experiments need to be repeated.

Huxley: Could I just spend a moment before we leave it altogether on thick filament lengths. This is just two little historical notes. One was that about 10 years ago Hernando Sosa in my laboratory at Brandeis spent his PhD time doing a study of rapidly frozen glycerinated fibres. He had them contracting under various loads and at various lengths and then rapidly froze them by plunging them into a liquid ethane mush so that the freezing of the outer fibrils was of the order of a millisecond or less. He then looked very carefully at the A-filament lengths. In that kind of preparation he was never able to see any significant change in the A-filament length. A number of people in Clara Franzini-Armstrong's laboratory looked at intact fibres frozen by a slamming method and they got quite a lot of very interesting results about the onset of crossbridge movement, but I don't recall them ever finding any changes in A-filament length. There really are very careful recent measurements about this – they are not just assumptions from 50 years ago.

Another point is a mystery and not many people have picked this up. We had a poster at a cell biology meeting about 14 years ago showing that if you take a live muscle from a frog and stimulate it electrically and then fix it chemically at a shortened length and look at the A-band lengths, as long as the muscle hasn't shortened to less than 1.6 microns sarcomere length you see A-band lengths that are essentially constant. However, if the muscles shorten below 1.6 microns you get contraction bands so obviously you can't really measure the A-band length then, but you can get A-bands that have apparently shortened. However, what is quite surprising is that, if you allow a muscle to overshorten in that way so that you know the A-bands must have been compressed and you then re-stretch it again and fix it after it has been re-stretched, you do then find A-bands that can be quite uniformly as short as a micron as though the filament ends have become depolymerised during the time that the sarcomeres were shortened. This is a very strange phenomenon that might account for some of the artefacts that have been reported. They look like normal A-bands except that they are shorter, so there is some mysterious process going on with A-band lengths in certain circumstances.

Winegrad. If you look in cardiac muscle under abnormal ischaemic conditions where you start to get proteolysis, even with the normal Z-Z sarcomere spacing you see decreases in A-band lengths as if the filaments are proteolysed (there are fragments that can be demonstrated) down to approximately 40 to 50% of the normal A-band length with the Z-line spacings unchanged. So it is much the same as you are seeing. I don't know if what you are seeing is conceivably a function of some sort of cleavage or fracture or proteolysis of some part of the filament which is more vulnerable than the half that is closer to the M-line.

Huxley: I got the impression that there was some recovery after about an hour or so, but it really needs to be done again as a little project that is relatively easy to do.

Pollack: All of that is possible and interesting. We (Periasamy et al., Biophys. J. 57: 815-828, 1990) reported in the Biophysical Journal that we took frog fibres in the same way as Huxley and Niedergerke and we found that during ordinary isometric contraction you can see reversible A-band shortening of the order of 15% and these were fibres that produced normal, not pathological, kinds of contractions. We also found the same thing with electron microscopy, with and the numbers that were roughly similar. The thick filaments were shorter when fixed in the activated state compared with the relaxed state If you look at the original Huxley and Niedergerke experiments they also reported that under certain conditions the A-bands shortened during contraction, but they interpreted it as an artefact of limited resolution. However, we repeated the experiments with increased resolution and could find no appreciable A-band length changes if we simply stretched or released the muscle, but during activated contraction there was a clear diminution of A-band length. Hugh Huxley and Saul Winegrad are talking about conditions that may be abnormal, but we could see that same thing under normal conditions.

Maughan: While we're on the subject of thick filament structure, I would like to draw attention to the close proximity and concentration of light meromyosin in the thick filament. The high concentration of light meromyosin imparts a particularly intense pressure for water to enter. Thick filament proteins must collectively have a large osmotic coefficient, and therefore a large outward pressure that forces the thick filament to expand its diameter in order to accommodate the inward flux of water driven by its large concentration gradient. It is likely that accessory proteins evolved to restrict filament expansion to minimize slippage and to thereby help keep thick filaments rigid and stiff by reducing their viscoelasticity.

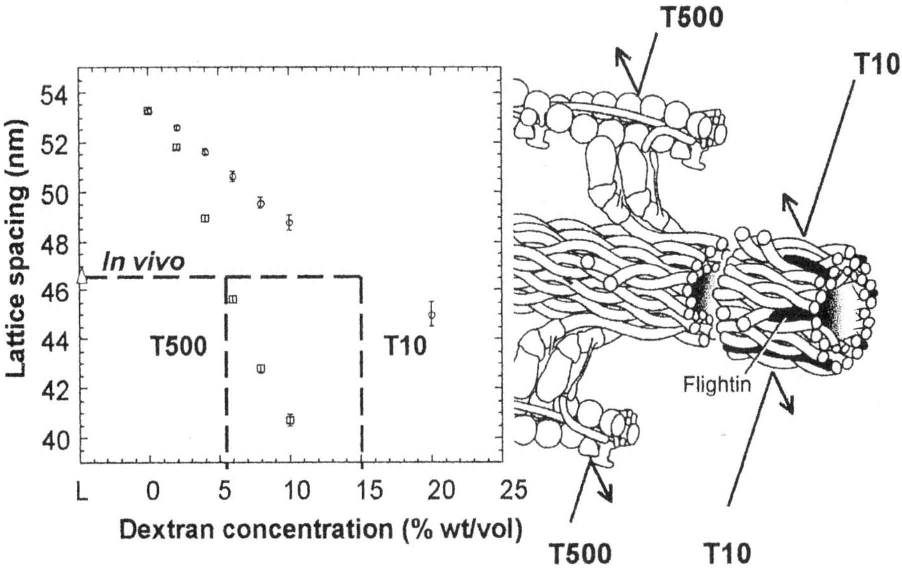

Fig. III-1. Expansion of the filament lattice upon skinning *Drosophila* indirect flight muscles, and the differential effect of compression with dextrans of different sizes. Dextran T500 and T10 are both excluded from the compact thick filament. At a given pressure, T500 compresses the lattice further than T10, consistent with its exclusion from the interfilament space due to its comparatively large size. Aspects of the lattice shown (including the location of the flightin molecules) are highly speculative.

It is tempting to speculate that in the flight muscle of the fly, the thick filament-associated protein flightin, like myosin binding protein-C (its probable counterpart in vertebrate muscle), collars or cross-links the thick filament, helping to restrict filament expansion in the face of high, outwardly-directed osmotic pressures. This hypothetical feature is shown in Fig. 1. In skinned flight muscle, one would expect both the myofilament lattice and the thick filaments to swell due to the loss of osmotically active cytosolic particles to the bathing solution. In fact, preliminary X-ray diffraction studies conducted in collaboration with Dr. Tom Irving at the Advanced Photon Source, Argonne IL, show that the skinned indirect flight muscle lattice swells by ~4 nm, i.e., to a lattice spacing 8% greater than that *in vivo* (compare living and 0% dextran spacing in plot of Fig. III-1).

That thick filaments are swollen and can be compressed is demonstrated by adding the inert polymer Dextran T10 (10 kDa) to the bathing solution (Farman *et al.*, Biophys. J. 86: 214a, 2004). Although T10 readily penetrates the filament lattice, it is likely excluded from the thick filament, including the myosin head. In contrast, T500, a much larger (500 kDa) polymer, is almost entirely excluded from the filament lattice. 5.6% w/v T500 reduces the centre-to-centre distance of the filaments by ~4 nm (i.e., back to *in vivo* spacing), whereas at roughly the same pressure T10 (~6.7 kPa; 2.5% w/v) reduces the centre-to-centre distance by ~1 nm, suggesting that thick filament expansion accounts for ~25% of the observed swelling of the myofilament lattice in skinned fibres. Preliminary X-ray data (T. Irving, G. Farman, D. Maughan) hints that the lattice swells to a significantly greater extent in the flightin null fln^0/fln^0, consistent with the notion that flightin restricts filament expansion.

M-band Structure

Squire: The M-band is very different in different kinds of muscles and in the same muscles from different animals. The M-band consists of 5 major transverse stripes of density spaced axially at 220 Å intervals. The central line M1 is flanked by two lines on each side which are related by 32 Point Group symmetry (Luther and Squire, J. Mol. Biol. 125: 313-324, 1978; Luther *et al.*, J. Mol. Biol. 151: 703-730, 1981). From one side to the other the main lines are M6, M4, M1, M4' and M6' (Sjostrom & Squire, J. Mol. Biol. 129: 49-68, 1977). What is seen in electron micrographs of different fibres and muscle types is that the relative densities of the of the different lines can vary. So-called 4-line M-bands have M1 virtually absent, 3-line M-bands have M6 and M6' virtually absent, and there are variations between these two extremes. However, M4 and M4', which probably correspond to the location of part of the M-band protein myomesin found in all muscle types (Obermann *et al.*, J. Cell Biol. 134: 1441-1453, 1996), remain a constant feature. We have never seen an M-band with M4 missing. We have shown that M-band appearance, whatever it may mean, is closely correlated with fibre type and in the case of cardiac muscles it correlates with the heart beat frequency in different animals (Pask *et al.*, J. Muscle Res. Cell Motil. 15: 633-645, 1994). Referring now to the simple lattice in bony fish muscle and the superlattice in higher vertebrate muscles, one can rationalise these structures in terms of the interactions that are made at the M-band. Since the myosin filament has 3-fold rotational symmetry, but the myosin filaments make cross-connections to six neighbouring myosin filaments, one can postulate that at M4 there are three 120°-spaced interaction sites of one kind (call them A sites) and another set of sites (call them B sites) rotated about 60° or 180° from the first. We have shown (Luther *et al.*, J. Mol. Biol. 151: 703-730, 1981; Pask *et al.*, J. Muscle Res. Cell Motil. 15: 633-645, 1994) that systematic interactions from A on one filament to B on the next will automatically produce a simple lattice structure, whereas interactions from A to A or from B to B, if preferred, cannot generate a perfect lattice, but their maximisation does generate the observed statistical superlattice in higher vertebrates.

Note that myomesin although forming much of the M4 lines is also thought to extend across the M-band from M4 to M4' (Obermann *et al.*, J. Cell Biol. 134: 1441-1453, 1996), possibly forming, together wirth the M-band part of titin, the axially aligned M-filaments seen by Luther and Squire (J. Mol. Biol. 125: 313-324, 1978). Other M-band proteins apart from myosin, titin and myomesin are M-protein and MM-CK (the muscle-specific form of creatine kinase). M-protein is thought to be located at the M1 position and is only found in fast fibres and cardiac fibres, thus explaining why M1 is weak or absent in slow fibres. MM-CK appears to be located around the M1 to M4 region of the M-band, where it interacts with both myomesin and M-protein (Hornemann *et al.*, J. Mol. Biol. 332: 877-887, 2003).

The Crossbridge Cycle

Yagi: I am introducing this paper in honour of Professor Sugi because one of his great achievements has been that his laboratory has produced not only a number of excellent papers but also a number of excellent muscle scientists. Without him the number of muscle physiologists in Japan would be halved. Producing a lot of scientists has been a major achievement. One such product is Hiro Iwamoto who is my colleague now and also Kazuhiro Oiwa and together they have recently carried out sophisticated work at Spring 8 (Iwamoto *et al.*, 2001). They used a glycerinated rabbit fibre, stretched it to non-overlap and then labelled it with myosin S1. The myosin S1 was then cross-linked to áctin using EDC and ATP was added to the fibre. In such a situation, since myosin S1 does not dissociate from actin, the ATPase rate is close to Vmax so S1 is hydrolysing ATP very quickly. They took X-ray diffraction patterns from such a fibre in the presence or absence of ATP. The result was very clear. In the absence of ATP the diffraction pattern very beautifully showed labelling of the actin layer-lines by the myosin heads. However, in the presence of ATP it looks as though no S1 was added to the muscle, the actin layer-lines were no different from those obtained without added S1. Clearly the S1-labelled structure in the presence of ATP does not allow dissociation of the heads because they are cross-linked as evidenced by the fact that on depletion of ATP the S1-labelled pattern returns. They concluded that stereospecific interactions of the heads with actin in the presence of ATP were not evident even though the ATPase rate was high. In other words, any stereospecific interactions must represent a very small part of the crossbridge cycle. If the duty ratio is very low then stereospecific binding might not be evident. A similar situation might occur in a muscle fibre shortening at maximum velocity because there is no load. In isometric conditions the situation would be different. The ATPase rate is lower, and more stereospecific interactions should be evident.

Squire: Were these fully saturated?

Yagi: Yes they were stoichiometric.

Brenner: I think that this is basically a solution condition so there is nothing constraining the movement of S1 apart from the cross-link at the actin end. There is nothing tethering the neck end of the head as there would be in the A-band. Biochemists have shown that as soon as the phosphate is off, the ADP off rate is a few thousand per second, so the strongly bound states are being rapidly depopulated. This is not even the same as doing high speed shortening where the myosin head can only be unstrained as the movement goes on.

Yagi: What can I say – the results surprised me.

Cecchi: We have shown previously that sarcomere length oscillations at frequencies between 1 and 3 kHz, and amplitude up to 4 nm/hs, applied to an activated muscle fibre produced changes in the intensity of M3 meridional X-ray reflection (I_{M3}) that were approximately sinusoidal and in opposite phase to the length and force changes. However, as the frequency was lowered below 1 kHz, intensity changes became progressively distorted and a double peak and a new minima were clearly visible during the release phase of the sinusoids (Bagni *et al.*, Biophys. J. 80: 2809-2822, 2001). We suggested that this distortion was produced when the tilting of myosin lever arm was great enough to go beyond the position at which I_{M3} was maximum (I_{M3max}). The distortion occurred exclusively at low frequency because the relatively long oscillation period allowed the active movement (the powers stroke) to contribute greatly to the lever arm movement, in addition to the elastic movement imposed by the external length oscillations. The total tilting was therefore great enough to introduce the distortion. At higher frequency, the powerstroke movement is strongly reduced and the elastic lever arm tilting alone is too small to go beyond I_{M3max} and the distortion is absent. If this hypothesis is correct, it should also be possible to obtain the double peak distortion at high oscillation frequency, providing that the elastic movement is increased so as to compensate the loss of active movement. We therefore measured I_{M3} during sinusoidal oscillations at 2.8 kHz frequency in a range of oscillation amplitudes up to 7 nm/hs, an amplitude greater that those used previously. The experiments were made on small fibre bundles isolated from the dorsal interossei muscle of the frog (*Rana temporaria*) at 2 °C. The experiments were made at SAXS beam line at the Elettra Synchrotron (Trieste, Italy). A train of sinusoidal length oscillations lasting 0.35 s was applied to the bundle at the tetanus plateau. For a given preparation, the data collected during each individual cycle (20 data points) were averaged for all the cycles and for all the tetani evoked. The sarcomere length was measured using a laser light diffractometer.

Fig. III-2 shows a typical result obtained from these experiments. It can be seen that at low oscillation amplitude, (A) IM3 has an approximate sinusoidal shape which is exactly out of phase with force. However, when the oscillation amplitude is increased (B), a clear double peak distortion occurs during the release phase of the sinusoids, as with the previous data at low frequency. This means that distortion can be obtained also when the active lever arm tilting is very small.. Intensity data were fitted with a model in which the intensity changes were calculated from the atomic model of the myosin molecule. The model assumed that filament compliance was 50% of the sarcomeres compliance. The only adjustable parameter was the mean lever arm orientation which was varied until a good fit of the experimental data was obtained. The fitting gave a mean lever arm orientation corresponding to a shift of about 0.8 nm away from the orientayion giving the maximum intensity, very similar to the value found previously at low oscillation frequency. These results are in agreement with the lever arm tilting hupothesis and show that the previous absence of distortion at high oscillation frequencies was not due to the loss of active lever arm tilting, nor to a shift of the mean lever arm position away from the position of maximum intensity, but to the relatively low oscillation amplitudes.

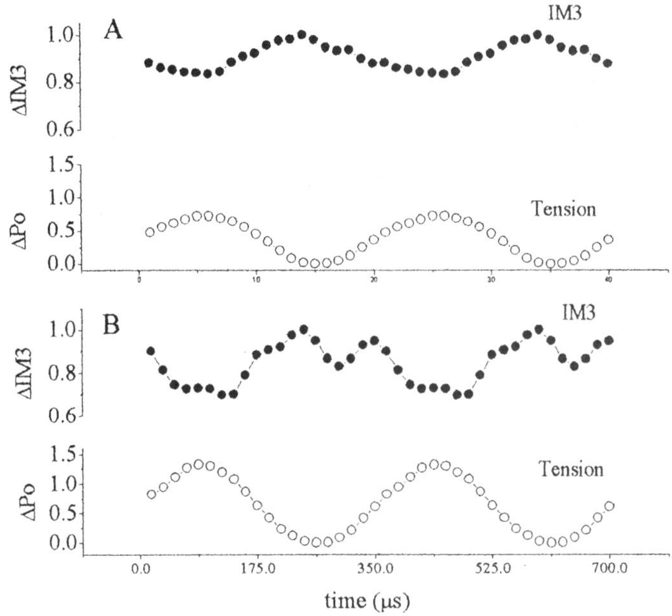

Fig . III-2. Effects of oscillation amplitude on force and IM3 changes. Oscillation amplitude was adjusted to obtain a force oscillation amplitude (p-p) of about 0.72Po in A and 1.33Po in B

Squire: What worries me a little about this kind of thing is the possibility that there are two populations of heads, say weak-binding and strong-binding

Cecchi: We have very recently elaborated our initial model to include the contribution of the detached myosin heads which are necessary to explain recent data with interference experiments (Piazzesi et al., Nature 415: 659-662, 2002). In addition, attached heads were assumed to be disordered as necessary due to the mismatch of actin and myosin helices. The results reported above are not significantly affected by these changes.

Squire: There is one thing that I would like to address which concerns the idea of actin target areas and whether they exist or not. My feeling is that the azimuthal position of the actin monomer is very important in deciding which particular actins are going to be labelled by myosin heads (Squire, J. Mol. Biol. 77: 291-323, 1972). The program MusLABEL, which is available on the CCP13 website (www.ccp13.ac.uk), can be used to investigate the pattern of labelling of actin filaments in a striated muscle A-band assuming that the actin azimuth is important. With this program you can set search parameters and see what labelling pattern is obtained. Fig. III-3 shows simulated X-ray diffraction patterns from different arrangements of heads on the six actin filaments surrounding a myosin filament in vertebrate striated muscle obtained by systematically varying the head search parameters. Each pattern has been set so that there are exactly the same numbers of heads attached, in this case close to 100% as in rigor vertebrate muscle. In order to

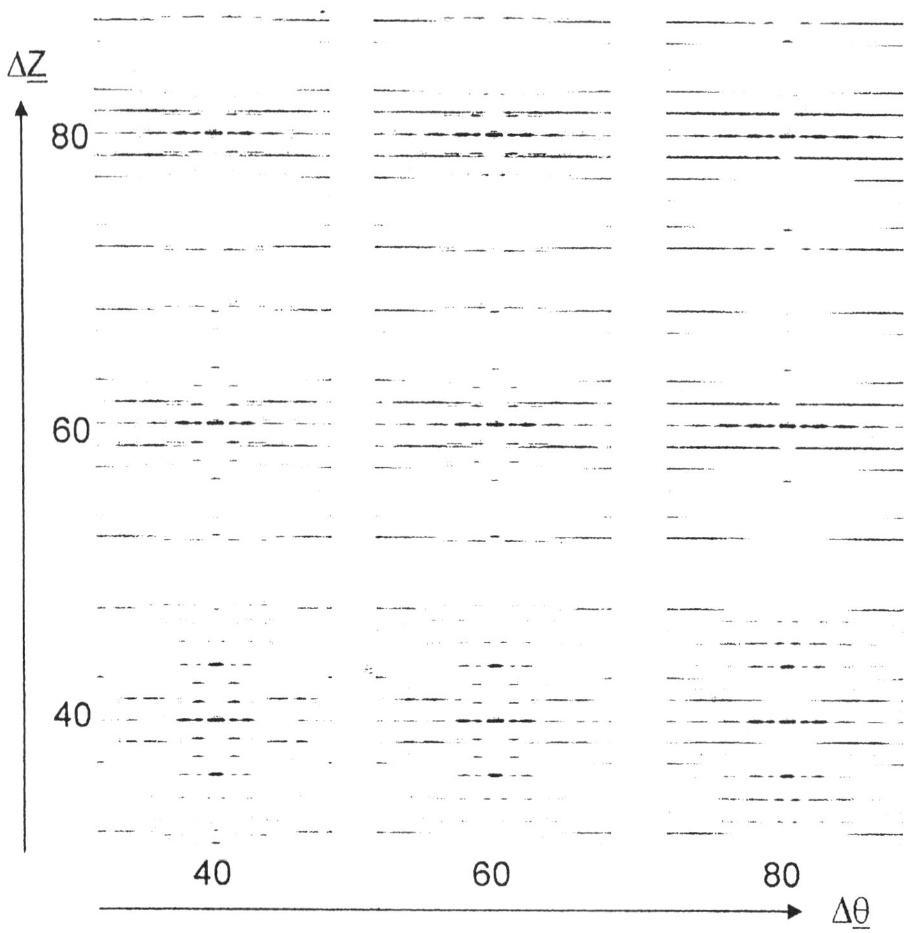

Fig. III-3. Simulated X-ray diffraction patterns from head labelling patterns on actin as calculated using the program MusLABEL. Head search parameters were ±Δθ° for the azimuthal reach of the heads and ±ΔZ Å for the axial search of the head. In each case the size of target area on the actin filaments was chasen to be just big enough to give close to 100% labelling of actin as occurs in rigor vertebrate muscle. M3 is the 143 Å meridional reflection (Squire , Harford and Knupp, in preparation).

achieve 100% labelling the actin target area needs to be large if the head search parameters are small and much smaller if the heads are allowed a much wider range of movement in finding an actin site.

The important point from this survey is that, even with the same number of heads attached throughout and even though it is assumed that the heads have exactly the same shape, the diffraction pattern can change radically. In one region of head search space, bottom right in Fig. III-3, the first actin layer-line at about 370 Å is very strong and the myosin M3 reflection is totally absent. In a second region, top left in Figure 3, there is a completely different pattern where the strongest peaks are in fact the M3 and there are layer-lines that look like the myosin layer-lines at 429 Å and 215 Å. In a third region, towards the bottom left of Figure 3, there is a third kind of pattern, actually more like the observed rigor diffraction pattern from vertebrate muscle, where the 370 Å layer-line is strong, the M3 reflection is also strong and there are layer-lines evident at around 238 Å and 179 Å which are characteristic of diffraction patterns from rigor and active muscle (see Squire and Harford, J. Muscle Res. Cell Motil. 9: 344-358, 1988; Yagi, Acta Cryst. D52, 1169-1173, 1996 ; Squire et al.., unpublished results).

Brenner: When your labelling pattern gives what you call a myosin-like pattern are the 429 and 215 layer-lines present? If so is the distribution of intensity like relaxed muscle.

Squire: Yes these layer-lines are present, but this is not modelling relaxed muscle. It is just saying that if you have heads on actin what could they do. In fact they could produce myosin-like layer-lines (see Squire and Harford, J. Muscle Res. Cell Motil. 9: 344-358, 1988). The point of showing these patterns is really to put in a word of caution to those using such things as the actin first layer-line intensity or the M3 intensity as a measure of the number of attached heads. Figure 3 shows that the intensity of this layer-line is very sensitive to the labelling pattern on actin even when the total number of attached heads remains unchanged. In my view it is essential to model the whole A-band structure fully if one is to use the observed layer-lines to determine either the number of heads attached, or the head conformation or the pattern of labelling on actin.

Squire: We come now to analysis of the crossbridge cycle itself. We have found that for our time-resolved X-ray diffraction work on bony fish muscle the cycle shown in Figure 4 is needed. In properly relaxed muscle the heads are back in an ordered configuration on the myosin filaments (Hudson *et al.*, J. Mol. Biol. 273: 440-455, 1997). However, every now and then a head will pop out of the array, as in any equilibrium structure, and then, if it has nowhere to bind to (i.e. the thin filament is switched off), the probing head will pop back again. On activation some of these probing heads will stay out because they become actin-attached and the cumulative effect of this is that as more probbing heads are captured by actin the ordered 'relaxed' head array will be lost. We also know that heads can attach to actin in at least two different structural states. The first of these is one shown both by stiffness measurements and by structural evidence to be a low-force state. For example, in the rising phase of a tetanus head attachment and stiffness both increase well before tension is produced (Harford & Squire, Biophys. J. 63: 387-396, 1992; Rep. Prog. Phys. 60: 1723-1787, 1997; Bagni *et al.*, J. Physiol. 481: 273-278, 1994; Martin-Fernandez *et al.*, J. Muscle Res. Cell Motil. 15: 319-348, 1994). The heads then move to a second attached structural state. In our analysis it is this second structural state that gives rise to almost all of the tension. The heads then come off actin, reset themselves presumably to restore the M.ADP.Pi state, and then can reattach and go through the cycle again.

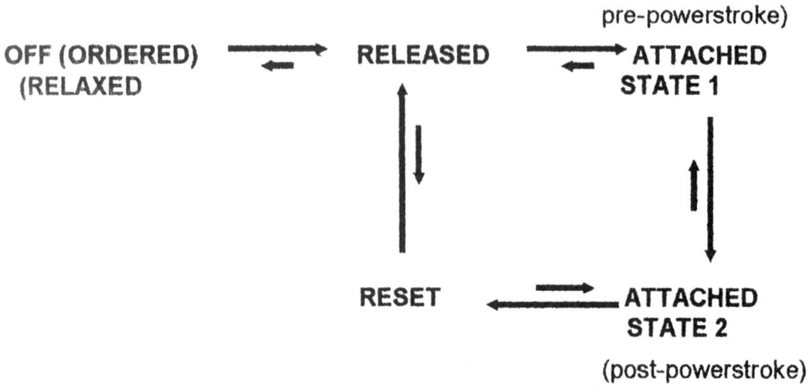

Figure 4: The <u>structural</u> crossbridge cycle from rest that we need to explain the time-resolved X-ray diffraction data from bony fish muscle. From X-ray diffraction we do not know the biochemical states that these are associated with. (Knupp, Harford and Squire, in preparation)

The question then arises as to what the two actin-attached head states are like. In our view the second attached structural state in Fig. III-4 is rather like the rigor state in that it has similar effects on the equator of the diffraction pattern to what is seen in patterns from rigor muscle. We would identify it as a stereospecific attached state. The first attached state, on the other hand, has many of the diffraction characteristics of the weak-binding state observed by Brenner *et al.* (Biophys. J. 46: 299-306, 1984) in rabbit muscle fibres at low ionic strength. Hugh Huxley mentioned in an earlier talk that he thinks it is necessary for the head to have two strong-binding stereospecifically-attached states in order for the head to carry out a working stroke. I am not sure if this is so. It seems to me that you could have heads in one biochemical state, perhaps the weakly bound state, that snap into a position where an isomeration or product release occurs, but they can't do anything like moving the neck because they are tethered at each end. Structurally they still look like weakly bound heads although they are now force-producing but strained. However, they are in the same biochemical state as Attached State 2 because if they were allowed to move they would end up in this second configuration when the strain is released.

Of course there may be a number of structural states between Attached State 1 and Attached State 2, but so far we only need two such states to model our observations. The resetting step is one which Professor Sugi discussed earlier based on his published results from the electron microscopy of filaments in an environmental cell (Sugi *et al.*, Proc. Natl. Acad. Sci. USA 94: 4378-4382, 1998).

Questions About the Crossbridge Cycle

Finally, the discussion of the crossbridge cycle leads to a number of questions to which I think answers are still needed:

[1] **What is the structure of the first attached state?** We have some ideas that it is like the weak-binding state, but this has to be proved.
[2] **Does force production require two strong states?**
[3] **Does the motor domain move on actin?** As far as I know there is very little evidence for this one way or the other. It could happen, although I know that Ken Holmes feels that such a change is not needed to explain the crossbridge cycle.
[4] **Why do force-producing heads all appear to be in the same conformation?**
[5] **Why does the evidence from crystallography suggest that the neck can swing by about 100 Å and yet those doing optical trap studies of the step size all seem to observe a step of about 50 Å?**

Holmes: Addressing point 3, there is very little evidence about what happens at the top of the powerstroke, but myosin V does give a chance of looking at this and as you know the work of Peter Knight looking at electron micrographs of myosin V bound to actin has provided the best evidence that we have at the moment. It shows that, at least to the resolution of their experiments, perhaps about 20-30 Å, the motor structures at the top and bottom of the powerstroke seem to look very alike.

Brenner: Ken, this refers to the top and bottom of the powerstroke, but it doesn't mean that there might not be a change in the docking before the powerstroke.

Holmes: Yes, but we have enough from the myosin V results to explain all that we need to explain without needing to add anything else.

GENERAL DISCUSSION PART IV
CARDIAC MUSCLE

Chaired by S. Winegrad

Winegrad: When Professor Sugi asked me to organize a discussion on heart muscle, I started out by making a list of all the things that we really should know and don't know about heart muscle. There were so many items that no time for discussion would remain, so we will focus on three or four subjects. Professor Cecchi will introduce the first subject, the relation between sarcomere length and tension, which is particularly important in cardiac muscle. Dr. Morano will follow with a discussion of the effects of some phosphorylations of contractile proteins in the heart. Then Professor Suga is going to address the question of regulation of contractility of the whole heart.

Sarcomere Length-Tension Relation

Cecchi: The only thing that I can present is some results on skeletal muscle that we got years ago, which look similar to what you can expect for cardiac muscle. The classical sarcomere length-tension curve shows a flat region between 2 and 2.2 μm and a more or less linear region going down to 3.65 μm. This is what you find in frog skeletal muscle during a titanic contraction when the fiber is fully activated. We looked at this relationship using twitch rather than tetanic contractions. In the twitch, the muscle fiber is less than fully activated. We found something interesting and comparable to what has been seen in cardiac muscle.

The twitch tension as a fraction of tetanic tension varies from 20 to 60 % of tetanic tension. If you carry out the experiments measuring the maximum peak twitch tension against the sarcomere length, you find that there is an interesting change, especially with fibers that have a very low twitch-tetanus ratio. The peak of the sarcomere length-tension relation is clearly shifted towards the region of 2.6 to 2.8 μm. If you look at a fiber with a higher twitch-tetanus ratio, indicating greater degree of activation, you see that there is a very clear shift of the sarcomere length-tension relation towards what you can get with full activation (during a tetanus). There is a clear shift toward a higher sarcomere length for the peak twitch tension with lower twitch-tetanus ratio, and the shift increases proportionally with decreasing twitch-tetanus ratio.

Winegrad: These data introduce the notion of a variable sarcomere length-tension curve.

This variability is very important in cardiac muscle. In cardiac muscle, one of the things which is very striking is that the length tension relation can deviate substantially from what you would expect from a simple overlap of thick and thin filaments. The plateau region of the sarcomere length-tension relation is much longer in cardiac muscle and extends out clearly at least to 2.4 or 2.5 μm. The broad plateau is really essential for the heart, because if the contracting heart gets onto the descending limb, cardiac function is compromised. The first idea about the mechanism involved focused on activation and EC coupling. EC coupling is probably playing important role in the phenomenon that Dr. Cecchi has just described. However, you can see a similar relation in skinned cardiac muscle even when you clamp the calcium concentration, so that it cannot vary. The next explanation, proposed by Rick Moss, had to do with the distance between thick and thin filaments and with the notion of constant volume of the filament lattice. It was expected that the distance between thick and thin filaments would decrease at increasing sarcomere length, and that would increase the likelihood of cross-bridge attachment, making the system more calcium sensitive. This has been investigated much more rigorously. The apparatus in Chicago has been used by Tom Irving and his various collaborators to actually measure the interfilament distance under these conditions and also to change the interfilament distance using Dextran as Dave Maughan has described for us. They find an interesting thing. The correlation doesn't hold. All of the change in Ca sensitivity that is associated with the addition of Dextran appears when you go from 0 to 1 % Dextran. With further increase in the concentration of Dextran and further decrease in the interfilament space, there is no change in calcium sensitivity. Since the correlation really doesn't hold, we must now look for other explanations. Dave Maughan began to allude to some possibilities in his presentation in this volume. I ask him now to continue in light of the particular problem.

Maughan: I can only comment to the extent that Tom Irving and his colleagues have reported this phenomena at the last Biophysical Society Meeting (Farman et al., Biophys. J. 86: 214a, 2004). Going to 1 or 2 % (I am not sure), Dextran produced all of the effect, and thereafter the space between the centers of filaments decreased, but not further shift in calcium sensitivity occurred. There is a shift in the intensity ratio 1:1:1:0 of the two equatorial X-ray reflections (1,1/1,0), which suggests that the myosin heads are more tightly associated with a thin filament as you compress the lattice. But that poster was largely a discussion of experiments performed last November as I showed earlier. They think that phenomenon below 1 or 2 % Dextran is a function of thick filament shrinkage and the reduction of thick filament diameter. The apparent shift of the density towards the thin filament is due to that phenomenon. They now speculate that the surface to surface spacing doesn't change a great deal in further compression, but the thick filament backbone does.

Pollack: We spent quite a few years studying length tension relations, and I think the ultimate has just recently been published in the Proceedings of the Royal Society (Rossier et al., Proc. Roy. Soc. 270: 1735-1740, 2003). We use a single myofibril, where the tension is exactly the same in every sarcomere. When you activate them you find that, during the plateau of tension, there are many different sarcomere lengths along the preparation. This means that one sarcomere at 1.9 microns is supporting the same tension as another sarcomere in series at 2.8 μm. Now the question is: Are these sarcomeres undergoing complicated dynamics? The answer is no, because we can measure every sarcomere with a resolution of the order of one or two nanometers. Within that resolution over a long period time during steady tension, those sarcomeres are locked in a constant sarcomere length. So the only reasonable interpretation we can give is that it really doesn't matter what sarcomere length you have. The same tension is sustained or generated by each of those sarcomeres in series. So getting back to your conundrum that was raised about the more-less flat length tension relation, it is possible that the flat length-tension relation is not an anomaly, but a basic fact.

Gonzales: Years ago, Rudel and Taylor showed that all myofibrils were probably not activated unless caffeine was applied.

Winegrad: However, this cannot be the explanation because the phenomenon still exists in skinned fibers, where you can clamp the calcium ion concentration.

Squire: I always assumed that this plateau in the length tension relation was due to the fact that the actin filaments have different lengths in cardiac muscle. Is that not so?

Winegrad: Tom Robinson and I (Robinson and Windegrad, J. Physiol. 286: 607-619, 1979) specifically looked at the difference in lengths of thin filaments as an explanation for this phenomenon. Tom, who did the work, found in serial transverse sections that there was a difference. The difference was moderately large in amphibian muscles, but the range was much shorter in rodents. You could explain the phenomenon in frog heart, but not in rodents. The range of difference was insufficient.

Sugi: I would like to point out that strain-dependent actin filament activation associated with low levels of Ca activation is involved in the phenomenon shown by Dr. Cecchi (twitch force in skeletal muscle fibers shows a peak at longer sarcomere lengths, say, 2.8 μm) if twitch /tetanus ratio is low (about 0.2) i.e. under small twitch force. The actin filament activation is also dependent on the past history (for example, shortening deactivation reported by Edman, J. Physiol. 246: 255-275, 1975), that twitch isometric force is very much reduced if proceeded by shortening. The shortening deactivation is largely removed in the presence of caffeine, indicating that attachment of Ca to, and its detachment from, an actin filament is associated with such past history-dependent changes in degree of activation.

Winegrad: Does anybody have any thoughts or evidence about the effect of strain on the thick filament and possible changes in function of the thick filament as a result of this. I have discussed with Neal Epstein, who is interested in stretch activation and feels quite confident that stretch activation occurs in the heart muscle, and if it does exist, it plays some role in sensitivity of the contraction to the strain on the filament. Let us move on to the second subject, phosphorylation.

Effect of Myosin Light Chain Phosphorylation

Morano: In myofibrils, almost all proteins associated with the contractile apparatus are at least partially phosphorylated. I would like to concentrate on the phosphorylation of the myosin light chains in striated muscle. Perrie, Smillie, and Perry (Biochem. J. 135: 151-164, 1973) were the first to demonstrate that the 18 kD DTNB light chain of rabbit skeletal muscle could be phosphorylated by a myosin light chain kinase (MLCK). This kinase turned out to be calcium-calmodulin dependent, and transferred the gamma-phosphorylation from ATP to a serine residue of the 18kD light chain. There is also a specific myosin light chain phosphatase which dephosphorylates the light chain. Because phosphorylation introduces some negative charges, there is a tremendous effect on the structure and hence function of the protein. What is the physiological function of phosphorylation of the light chain?

I would like to separate the phenomenon into skeletal and cardiac myosin light chain phosphorylation.

Manning and Stull (Am. J. Physiol. 242: (Cell Physiol. 11) C234-C241, 1982) and Westwood, Hudlicka and Perry (Biochem. J. 218: 841-847, 1984) demonstrated that the light chain phosphorylation should have a role in the phenomenon of post tetanic potentiation. The amplitude of twitch tension immediately after a tetanus increases markedly compared with the twitch tension prior the tetanic stimulation of a skeletal muscle. The authors demonstrated in rabbit skeletal muscle that the level of light chain phosphorylation was increased from around 10 % at rest to almost 100 % during the tetanus. This enhancement of phosphorylation slowly declines after the tetanus and correlated with the potentiated post-tetanic twitches.

There should also be some correlation between tension potentiation and myosin light chain phosphorylation in cardiac muscle. Paul England (J. Mol. Cell. Cardiol. 16: 591-595, 1984) investigated light chain phosphorylation levels during the beat to beat regulation in cardiac muscle. He found an in vivo phosphorylation level of ~45 %, and so roughly half of the light chains in the heart are phosphorylated. There was no change of the light chain phosphorylation level during beat to beat regulation of the heart: during systolic contraction, the level of phosphorylation remained at around 45 % and did not change during diastole. There was also no change in phosphorylation during either alpha or beta adrenergic stimulation. What is the physiological function of light chain phosphorylation in the heart if it is so stable and cannot be changed by pharmacological stimulation? In Caspar Ruegg's lab, I incubated skinned cardiac fibers with cardiac myosin light chain kinase and found that there was an increase in calcium sensitivity of the mechanical response: myosin light chain phosphorylation in skinned cardiac fibers caused a significant increase in tension at submaximal rather than at maximal calcium concentrations (Morano et al.; FEBS Lett. 189: 221-224, 1985). A year later, Persechini et al. (J. Biol. Chem. 260: 7951-7954, 1985) demonstrated the same phenomenon by using skinned skeletal fibers. Sweeney et al. (Proc. Natl. Acad. Sci. USA 87: 414-418, 1990) found that myosin light chain phosphorylation in skinned skeletal muscle fibers increased the rate constant for the development of tension after a brief period of unloaded shortening. The ratio between ATPase and tension as well as maximal shortening velocity did not change. Using another approach, we could demonstrate the same effect of myosin light chain phosphorylation on cross-bridge kinetics of skinned cardiac fibers (Morano et al., Acta Physiol. Scand. 154: 343-353, 1995). Rhea Levine (Levine et al., J. Struct. Biol 122: 149-161, 1998) observed that phosphorylation of the light chains in isolated thick filaments converted the orderly arrangement of myosin heads to disorder with a swinging out of myosin heads. Interestingly, chronic running training caused an increase in the in vivo level of light chain phosphorylation in the rat heart, suggesting improved cross-bridge function. Old hypertensive rats with hypertrophied hearts and attenuated contractile function revealed a decreased in vivo phosphorylation level. This suggests a pathophysiological role of decreased myosin light chain phosphorylation.

Winegrad: There are several other myofibrillar proteins that can be phosphorylated in heart muscle including TNT and TNI. TNI can be phosphorylated by PKA and on different residues by PKC, and the effect on the contraction is the opposite. At the present time, I am not aware that these reactions have been built into models of contractile regulation. These reactions need to be integrated into models of how the entire organ functions. Dr. Suga, who has worked with intact hearts throughout his career, will present some thoughts about the relationship of cellular and subcellular data to the intact heart.

Regulation of Contractility in the Whole Heart

Suga: Through the present symposium, I have understood that not only cardiac muscles but also skeletal and smooth muscles still have many mysterious aspects in their contractile mechanisms and performances. In my presentation, I have emphasized the mysterious beauty of a beating heart. This beauty of the heart at the natural, integrative level seems to me most conspicuous in the load- and contractility-independent high constant contractile or mechanical efficiency of ventricular contraction. Although I was first afraid of this simplistic mechanoenergetic relation, I recognized this integrative beauty after comparing the advanced findings of the cross-bridge behaviors at both the global and elemental levels, as I explained in my presentation.

Back to the late 1960s when I started my research on the contractility of a beating heart, Frank's ventricular pressure-volume (P-V) diagram and loops, Starling's cardiac output curve and law of the heart, and Sonnenblick's myocardial force-velocity relation and V_{max} were the conventional framework of cardiac function contractility. Although it was possible to account qualitatively for such cardiac performances by the 1957 Huxley sliding filament theory, it was still too early to do so quantitatively at that time. Since those classic characteristics had their own problems in their applications to beating hearts, I challenged first to characterize the canine beating heart performance in a better way. The challenge was fortunately crystallized in 1970-90 as my novel proposals of the E_{max} and PVA concept (Suga et al., Circ. Res. 32: 314-322, 1973; Suga, Am. J. Physiol. 236: H498-505, 1979; Suga, Physiol. Rev. 70: 247-277, 1990), as I have explained in my paper in this volume (Session V).

The E_{max} and PVA fortunately have not only beauties but also mysteries in their integrative characteristics constituted by various unitary performances of their elemental components. The greatest one of the integrative beauties seems to me that cardiac O_2 consumption (Vo_2) in excess of the basal metabolism and Vo_2 component for the excitation-contraction coupling is a load- and contractility-independent linear function of PVA over a wide physiologically working range of the beating heart. The contractile efficiency given by the reciprocal slope of the PVA-dependent Vo_2 vs. PVA relation is

~35 % and that from ATP to PVA ~60 % under the assumption of P:O ratio of 3, as I have explained in detail in my presentation. Moreover, as long as PVA is the same, the

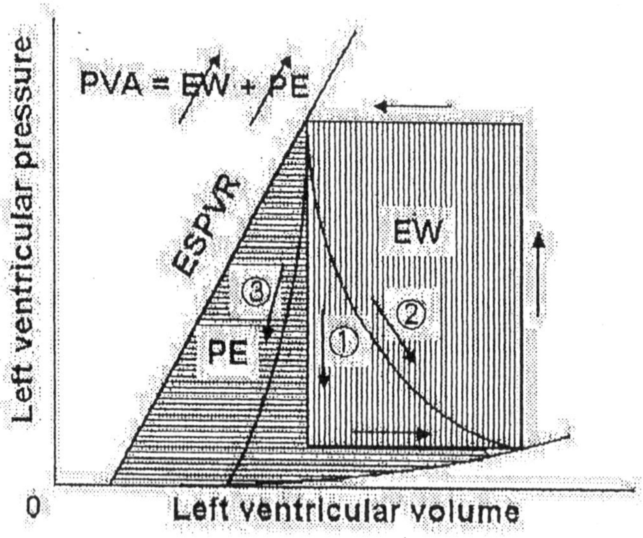

Fig. IV-1 The same PVA (pressure-volume area = total mechanical energy generated by ventricular contraction) consisting of EW (external work) and PE (potential energy) of different magnitudes under different pressure-volume trajectories (①②③) during ventricular relaxation.

relative magnitudes of its two components, external work (EW) and potential energy (PE), do not affect the PVA-dependent V_{O_2}, as shown in Fig. IV-1.

Although I neither showed this figure nor explained this aspect in detail in my presentation, this aspect is consistent with the cardiac Fenn effect, although it is obviously inconsistent with the skeletal Fenn effect. This cardiac characteristic requires multiple numbers of unitary cross-bridge steps per ATP in one cardiac cycle, as I have already explained in my presentation. When I used the unitary force of ~1.5 pN and unitary sliding step of ~10 nm referring to Sugiura's report (Sugiura et al., Circ. Res. 82: 1029-1034, 1998), the constancy of the contractile efficiency of ~60 % requires each cross-bridge to repeat the unitary sliding step multiple times per ATP in each cardiac contraction. Only a few times greater unitary step or force or both than ~10 nm and ~1.5 pN would not qualitatively affect my conclusion of the multiple unitary steps per ATP in cardiac contraction, though affecting it quantitatively. However, if they were several times more than 10 nm and ~1.5 pN, their product or even its half could be more than or close to the free energy of ATP, negating such a possibility in the beating heart. I am not yet sure whether or not the unitary force of ~10 pN of a skeletal muscle cross-bridge, which Prof. Hugh Huxley referred to as Y.E. Goldman's finding after my presentation, had better be used as a realistic value for cardiac contraction.

Taken together, we still need to obtain more evidential knowledge to bridge between the heart and its elements so that both integration and reduction become possible more easily and reliably. However, the available methodologies and technologies at present are not yet enough to solve such remaining problems in cardiac contraction.

Sugiura: Dr. Huxley pointed out that the unitary force value of cardiac muscle myosin (1.6 pN) is too small compared to the corresponding value of skeletal muscle myosin. However the paper from University of Vermont (Palmitier et al., J. Physiol. 519: 669-678, 1999) reported an even smaller value (0.8 pN). Although skeletal and cardiac muscles show common heavy chain, their light chains are different. Considering the functional importance of light chain shown by Dr. Morano, cardiac muscle myosin can be a weaker force generator.

GENERAL DISCUSSION PART V
SMOOTH MUSCLE

Chaired by J.C. Rüegg

Mechanism of the Catch State in Molluscan Smooth Muscle

Rüegg: The last part of General Discussion is concerned with problems of smooth muscle contraction mechanisms. Our object of discussion includes not only vertebrate visceral and vascular smooth muscle, but also molluscan smooth muscle with special reference to the catch state. First, I would like to ask Prof. Sugi to give a talk on his recent work on the catch mechanism in a molluscan smooth muscle.

Sugi: Somatic smooth muscles of bivalve mollusks exhibit catch state, i.e. a prolonged tonic contraction maintained with very little energy expenditure (Naus and Davies, Biochem. Z. 345: 173-187, 1966). The catch mechanism has been studied on the anterior byssus retractor muscle (ABRM) of *Mytilus edulis*. Two different hypotheses have been put forward to explain the catch state; one assumes actin-myosin linkages that dissociate extremely slowly (linkage hypotheses, Lowy and Millmaw, Phil. Trans. R. Soc. Lond. B246: 105-148, 1963), while the other postulates a load bearing structure other than actin-myosin linkages (parallel hypotheses, Heumann and Zebe, Z. Zellforsch. 85: 534-551, 1968). The parallel hypothesis is based on the electron microscopic observation that a marked aggregation of the thick filament occurs during the catch state. It has been pointed out however, that the thick filament aggregation may be an artifact due to cross linking of the filaments by glutaraldehyde fixation.

Using quick freezing and freeze substitution techniques, we have recently found that, although no thick filament aggregation is observed in the ABRM fibers frozen at rest, during active contraction and during the catch state, the thick filaments are occasionally interconnected with each other either directly or by distinct projections. The proportion of the interconnected thick filaments relative to the total thick filaments in a given cross-sectional area is much larger during the catch state than in the relaxed and actively contracting states, as shown in Fig. V-1. This suggests that the thick filament

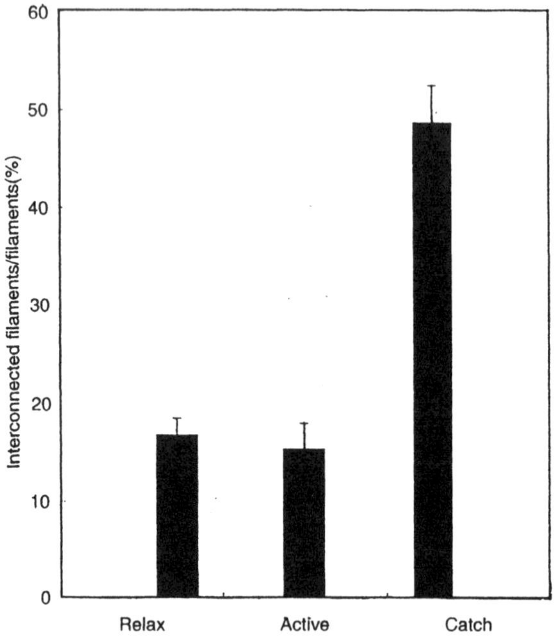

Fig. V-1. Proportion of the interconnected thick filaments as measured in cross-sections of the ABRM fibers frozen in the relaxed state, in the actively contracting state, and in the catch state (Takahashi et al., 2003)

connection constitutes the force-bearing structure responsible for the catch state (Takahashi et al., Comp. Biochem. Physiol. Part A134: 115-120, 2003).

More recently, we have shown that, when the ABRM fibers are subjected to quick increases in load after development of the maximum isometric force P_0, they could bear a load up to 10-15 times P_0, while the ABRM fibers during the early active isometric force development are lengthened rapidly under loads of only 1.5-2 P_0, as with skeletal muscle fibers (Sugi and Tsuchiya, J. Physiol. 319: 239-252, 1981)(Fig. V-2). This indicates that actin-myosin linkages are responsible for the active force development, while the catch state is maintained by the force-bearing structure other than actin-myosin linkages (Mukou et al., J. Exp. Biol. 207: 1675-1681, 2004).

A high molecular weight myosin-binding protein, twitchin, is known to occur in the ABRM. Although the arrangement of twitchin in the thick filament is unknown, it seems possible that it is twitchin that interconnects the thick filament to form the force-bearing structure. Thus, our work favours the parallel hypothesis, but not the linkage hypothesis.

Rüegg: Thank you very much Prof. Sugi for your very interesting work, providing evidence for the load-bearing system in the ABRM. It should, however, be kept in mind that twitchin phosphorylation has been shown to control the catch state in the ABRM (Siegman et al., J. Muscle Res. Cell Motil. 18: 655-670, 1997).

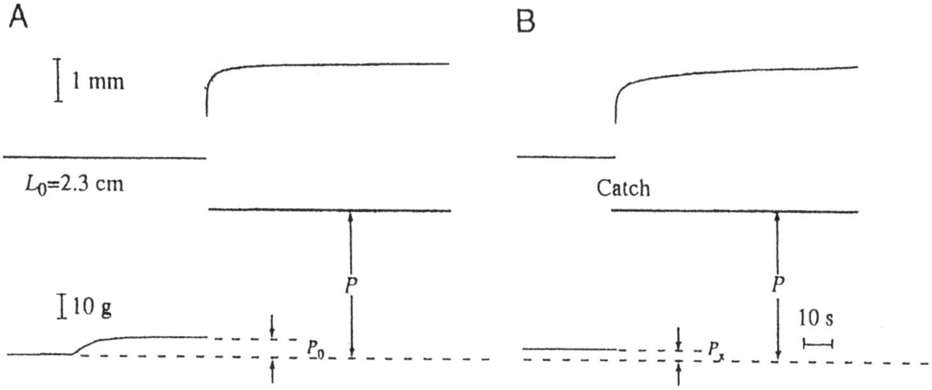

Fig. V-2. Marked load-bearing ability of the ABRM fibers during the maximum active force generation (A) and during the catch state (B). The load was increased from P_0 to P (~8 P_0) in A, and from P_x (~0.5P_0) to P (~16 P_x) in B. Records A and B were obtained from the same preparation (Mukou et al., 2004)

Gonzalez-Gerratos: What is the minimum amount of calcium concentration that start catch contraction?

Sugi: We have demonstrated electron microscopically that, in *Mytilus* catch muscle, active contraction shifts to the catch state, when myoplasmic free Ca^{2+} concentration is reduced to the resting level (Atsumi & Sugi, J. Physiol. 243: 1367-1368, 1989).

Pollack: Some years ago, we reported interconnections between adjacent thick filaments in vertebrate skeletal muscle (Suzuki and Pollack, J. Cell Biol. 102: 1093-1098, 1986). Is this similar to that you observed between the thick filaments in the ABRM?

Sugi: I think the two kinds of thick filament interconnections are entirely different from each other.

Rüegg: I would now like to discuss about problems of vertebrate smooth muscle. In vertebrate smooth muscle, we have the problem of "latch state" which seems to be analogous to the catch state in molluscan smooth muscle. The latch state may be caused by cross-bridges in an intermediate attached state. In this connection, I would like to ask Dr. Pfitzer to give a talk on this problem.

Reguratory Mechanism of Smooth Muscle Contraction with special Reference to the latch State

Pfitzer: The phosphorylation theory of smooth muscle contraction holds that, following stimulation, the intracellular Ca^{2+} concentration increases, and the Ca^{2+}-calmodulin complex is formed, which binds to myosin light chain kinase, MLCK, thereby activating it and allowing the subsequent phosphorylation of the regulatory light chains of myosin (RLC). This then allows the activation of MgATPase of smooth muscle myosin by actin and cross-bridge cycling. Relaxation is initiated when the intracellular [Ca^{2+}] decreases and the MLCK holoenzyme complex dissociates while the RLC are dephosphorylated by a specific type 1 phosphatase (MLCP). MLCP is a permanently active enzyme, the activity of which can be further increased or decreased in response to a number of intracellular signaling cascades (Pfitzer, J. Appl. Physiol. 91: 497-503, 2001; Somlyo and Somlyo, Physiol. Rev. 83: 1325-1358, 2003). It is the ratio of the activities of MLCK and MLCP at a given intracellular Ca^{2+} concentration that determines the degree of RLC phosphorylation.

The concept that phosphorylation regulates contraction of smooth muscle by enabling cross-bridge attachment predicts that force would be proportional to RLC phosphorylation. This hypothesis is consistent with results obtained in many experiments with isolated proteins and skinned fibers but it is often not consistent with results obtained in intact smooth muscle (Arner and Pfitzer, Rev. Physiol. Biochem. Pharmacol. 134: 64-146, 1999). Thus, during the sustained phase in particular of a tonic contraction, both $[Ca^{2+}]_i$ and RLC phosphorylation typically decline to lower but still suprabasal levels while tension is maintained causing an increase in the apparent Ca^{2+}-sensitivity of contraction as well as a temporal dissociation between force and RLC phosphorylation (for review cf. Kamm and Stull, Ann. Rev. Pharmacol. Toxicol. 25: 593-620, 1985). Plotting force versus phosphorylation yields a hyperbolic rather than a linear relation (Strauss and Murphy, In "Biochemistry of Smooth Muscle Contraction", Barany, pp341-353, 1996; Pfitzer and Arner, Acta Physiol. Scand. 164: 449-456, 1998). In extreme cases a high force output at only about 20% of RLC phosphorylation is observed (Fig. 2, Schmidt el al., Pflüg. Ach. 429: 708-715, 1995, Kenney et al., J. Biol. Chem. 265: 8642-8649, 1990) which suggests that unphosphorylated myosin can interact with actin and participate in force maintenance.

Often the rate of energy usage and the maximum shortening velocity follow the temporal changes in RLC phosphorylation, i.e. both are high during the rising phase of the contraction and decline towards lower values during tension maintenance, suggesting a time depending slowing of the cross-bridge cycling rate in the course of the contraction (reviewed in Hai and Murphy, Ann. Rev. Physiol. 51: 285-298, 1989). Thus, during the sustained phase, force is maintained with high economy. In analogy with the catch state in invertebrate smooth muscle, the existence of a "latch state" was proposed by Murphy and coworkers (Dillon et al., Science 211: 495-497, 1981) in which unphosphorylated cross-bridges maintain force but with a very slow ATPase cycle.

It has been difficult to induce a latch state in skinned smooth muscle. Yet, when relaxation is induced in skinned smooth muscle by rapidly switching from Ca^{2+} containing activating solution to relaxing solution (pCa >8, Kühn et al., Pflüg. Arch. 416: 512-518, 1990) or by chelating Ca^{2+} by flash photolysis of diazo-2 (Khromov et al., Biophys. J.69: 2611-2622, 1995), mechanical relaxation is much slower than the dephosphorylation of RLC. It was suggested that this slow relaxation reflects the slow detachment of unphosphorylated cross-bridges which is rate limited by the slow release of ADP from unphosphorylated cross-bridges. The slow dissociation of ADP from unphosphorylated cross-bridges explains also the low unloaded shortening velocity, because ADP release limits unloaded shortening velocity (Siemankowski et al., Proc. Natl.

Acad. Sci. USA 82: 658-662, 1985). The rate of ADP release is increased by thiophosphorylation of RLC which would then allow a rapid detachment of phosphorylated cross-bridges (Khromov et al., Boiphys. J. 86: 2318-2328, 2004). In this way, the phosphorylation dependent ADP affinity of cross-bridges would determine whether cross-bridges cycle fast or slow (Gollub et al., Biochemistry 38: 10107-10118, 1999). The results in skinned fibers, which suggest that the latch state is a strongly bound AM·ADP state, contributing to force maintenance at low levels of RLC phosphorylation, are supported by biochemical, biophysical and structural experiments (Cremo and Geeves, Biochemistry 37: 1969-1978, 1998; Gollub et al., Biochemistry 38: 10107-10118, 1999; Veigel et al., Nature Cell Biol. 5: 980-986, 2003; for further discussion cf. G. Pfitzer et al., and J.C. Rüegg in this volume). It is of interest that the intracellular concentrations of ADP are high enough to support this conclusion.

This raises the important question, how the strongly bound unphosphorylated cross-bridges are generated. In principle two models have been suggested; (i) the Hai and Murphy model (1988) which assumes that dephosphorylated attached cross-bridges are generated by the dephosphorylation of attached cross-bridges (Fig. V-3), and (ii) the cooperative model which assumes that few attached phosphorylated cross-bridges allow cooperative attachment of unphosphorylated cross-bridges (Somlyo et al., J. Gen. Physiol. 91: 165-192, 1988; Butler and Siegman, Acta Physiol. Scand. 164: 389-400, 1998). Both models predict a non-linear relation between force and RLC phosphorylation which was also found experimentally (Fig. V-4).

The basic assumptions of the Hai and Murphy model is that unphosphorylated attached cross-bridges arise from dephosphorylation of attached phosphorylated cross-bridges and not from direct attachment, which requires that the activity of MLCP is high in order to allow for a significant flux from attached phosphorylated to attached

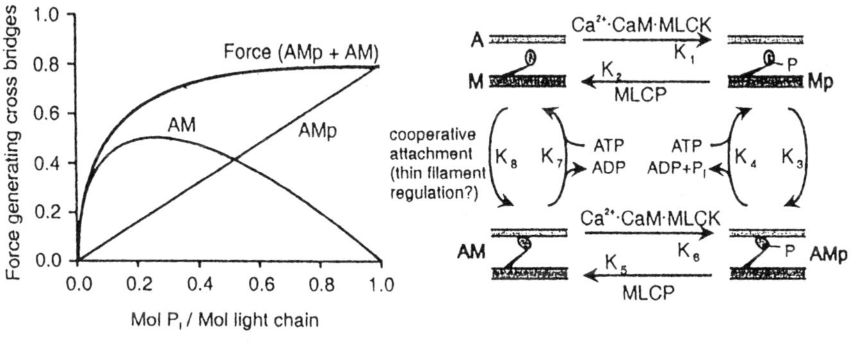

Fig. V-3. Schematic representation of a four state cross-bridge model of smooth muscle contraction (cf. Hai and Murphy, 1989). Formation of contractile linkages is by attachment of phosphorylated cross-bridges (M_p) to actin. Latch bridges (AM) are formed either by dephosphorylation of phosphorylated attached cross-bridges (AM_p, reaction 5) or by cooperative attachment of unphosphorylated cross-bridges (reaction 8). The cooperative attachment may be regulated by thin filament linked proteins. Dephosphorylated cross-bridges detach slowly compared to phosphorylated ones and, hence cycle at a slow rate. Both the Hai and Murphy model and the cooperative model would result in a hyperbolic relation between force and RCL phosphorylation whereby both attached phosphorylated cross-bridges (AM_p) and attached unphosphorylated cross-bridges (AM) contribute to force.

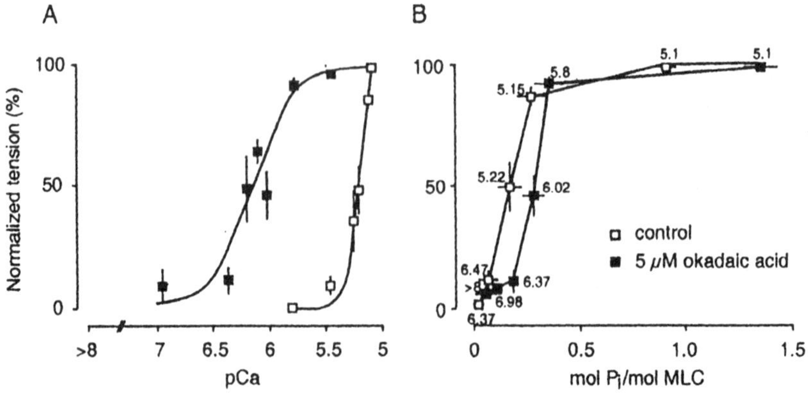

Fig. V-4. Dependence of force on pCa (A) and RLC phosphorylation (B) in triton skinned smooth muscle. The force-calcium relation was obtained by cumulatively increasing [Ca^{2+}] with and without the phosphatase inhibitor, okadaic acid (5 μM). To determine RLC phosphorylation by 2D-PAGE fibers were fixed at the indicated pCa valcues at the plateau of the contractions (B). (Modified from Schmidt et al., 1995)

dephosphorylated cross-bridges. Thus, the dependence of force on RLC phosphorylation would shift towards higher levels of phosphorylation when the activity of MLCP is low.

We tested this hypothesis in triton skinned chicken gizzard fibers in which a latch like contraction can be induced (Wagner and Rüegg, Pflüg. Arch. 407: 569-571, 1986; Pfitzer et al., this volume). The Ca^{2+}-sensitivity of contraction is low but can be increased by about one order of magnitude in the presence of the phosphatase inhibitor, okadaic acid (Fig. V-4A). The shift of the relation between force and RLC phosphorylation is only small (Fig. V-4B). A similar observation was made in portal vein (Siegman et al., Biochem. Biophys. Res. Commun. 161: 838-842, 1989).

In this context, it is of interest that during the sustained phase of contraction in vascular and intestinal smooth muscle, MLCP activity is low because of inhibition of MCLP by the Rho/Rho-kinase pathway (Pfitzer et al., J. Appl. Physiol. 91: 497-503, 2001 for a review; Somlyo and Somlyo, Physiol. Rev. 83: 1325-1358, 2003). Therefore we propose that force bearing unphosphorylated cross-bridges during the sustained phase of contraction are generated by the cooperative activation of unphosphorylated myosin rather than by the dephosphorylation of attached phosphorylated cross-bridges (Fig. V-3).

Since cross-bridges are usually phosphorylated to a certain extent under resting conditions and with certain experimental protocols, RLC phosphorylation may increase while force remains depressed (Gerthoffer et al., J. Pharmacol. Exp. Ther. 240: 8-15, 1987), the question arises what keeps unphosphorylated cross-bridges from attaching. In fact, it is not known how much phosphorylation is required to allow cooperative attachment of unphosphorylated cross-bridges. Attachment of unphosphorylated cross-bridges may be inhibited by the thin filament linked proteins, caldesmon (Albrecht et al., Pflüg. Arch. 434: 534-542, 1997) or calponin (Malmqvist et al., Proc. Natl. Acad. Sci. USA 94: 7655-7660, 1997). Both proteins inhibit actomyosin ATPase activity and are regulated by Ca^{2+} and phosphorylation (for a review, Arner and Pfitzer, Rev. Physiol. Biochem. Pharmacol. 134. 63-146, 1999). Additional physiological investigations will be required to fully establish how the attachment of unphosphorylated cross-bridges is regulated and the role for thin filament linked proteins.

Rüegg: I think the latch state may be produced by a very slowly-cycling cross-bridges, probably due to prolonged attached state and presumably to prolonged intermediate attached states. It has been known that a strain prolongs the attached states. If we consider that a cross-bridge can generate a force of 10pN, the attached state may be quite long. The central question may be whether phosphorylation-dephosphorylation story is valid in the latch mechanism or not. The history of muscle research shows that things are always changing. Then, I would like to ask Dr. Takuwa to talk on other aspects of smooth muscle.

Signaling Mechanism for Smooth Muscle Contractionn

Takuwa: I would like to present our data showing that, in smooth muscle, Ca^{2+} plays an essential role for the regulation of Rho, which is a key regulatory factor of myosin phosphatase and thereby Ca^{2+}-sensitivity. Many of excitatory receptor agonists act via heptahelical G protein-coupled receptors (GPCRs) on smooth muscle to stimulate phosphipase C and gating of plasma membrane Ca^{2+} channels, resulting in a rise in the intracellular free Ca^{2+} concentration ($[Ca^{2+}]_i$). The increase in the $[Ca^{2+}]_i$ causes the activation of the calmodulin-dependent enzyme, myosin light chain kinase (MLCK). MLCK phosphorylates the 20-kDa myosin light chain (MLC), leading to the initiation of contraction. On the other hand, smooth muscle-specific myosin phosphatase exerts negative regulation on myosin by de-phosphorylating MLC. Accumulated evidence demonstrates that excitatory receptor agonists induce inhibition of myosin phosphatase activity by a G protein-dependent mechanism in smooth muscle, which serves as a mechanism for potentiating Ca^{2+}-dependent MLC phosphorylation (Ca^{2+} sensitization). We demonstrated for the first time in vascular smooth muscle cells that Rho mediated myosin phosphatase inhibition (Noda et al., FEBS Lett. 367: 246-250, 1995). Subsequently, it was shown that Rho kinase, a Rho effector, directly phosphorylated myosin phosphatase at the 110-kDa myosin targeting regulatory subunit MYPT1/MBS to inhibit phosphatase activity *in vitro* (Kimura et al., Science 273:245-248, 1996). Rho kinase was found to phosphorylate the MYPT1 of myosin phosphatase at Thr^{695} (numbering of chicken M133 isoform) to inhibit myosin phosphatase activity. We and others also demonstrated in intact vascular smooth muscle tissues that excitatory receptor agonists stimulated phosphorylation of the MYPT1 and inhibited myosin phosphatase in a Rho kinase inhibitor-sensitive manner (Ito et al., J. Physiol. 5456: 823-836, 2003), suggesting that excitatory agonists induced downregulation of myosin phosphatase through Rho kinase-dependent phosphorylation of myosin phosphatase. More recently, another mechanism for regulating smooth muscle myosin phosphatase has attracted much attention; it was shown that the smooth muscle-specific endogenous myosin phosphatase inhibitor protein, CPI-17, is phosphorylated by a Rho kinase-and /or protein kinase C (PKC)-dependent mechanism to be activated in excitatory agonist-stimulated smooth muscle (Kitazawa et al., J. Biol. Chem. 275: 9877-9900, 2000). Thus, Rho kinase is suggested to mediate myosin phosphatase inhibition through both the inhibitory phosphorylation of MYPT1 and the activating phosphorylation of CPI-17 in smooth muscle.

Rho cycles between the GTP-bound active and GDP-bound inactive states, which is under tight regulation by the two major classes of proteins, guanine nucleotide exchange factors (GEFs) and GTPase-activating proteins (GAPs). Previous studies largely in non-muscle cells demonstrated that stimulation of GPCRs with receptor agonists including lysophosphatide acid, endothelin-1 and thrombin induced Rho activation through receptor coupling to the $G_{12/13}$ family of the heterotrimeric G proteins. Direct physical and functional interaction of either G_{12} or G_{13} with a group of Rho-GEFs was demonstrated. It was demonstrated in cultured vascular smooth muscle that $G_{12/13}$, but not G_q, mediates Rho- and Rho kinase-dependent, receptor agonist-induced contraction. We recently demonstrated in rabbit aortic smooth muscle that various excitatory receptor agonists induced an increase in the amount of GTP-bound, active Rho. For example, the thromboxane receptor agonist U46619, which induced a fully sustained contractile response, also induced sustained Rho activation with a dose-response relationship similar to that for contraction (Sakurada et al., Am J. Physiol. 281: C571-C578, 2001). We also observed that the sensitivity of receptor-induced Rho stimulation to the tyrosine kinase inhibitor was different among these agonists. These observations together suggest that receptor agonist-induced Rho stimulation may be caused by more than a single mechanism in smooth muscle.

While we were investigating Rho regulation by various stimuli, we found that membrane depolarization by KCl induced Rho stimulation in vascular smooth muscle (Sakurada et al., Cir. Res. 93: 548-556, 2003). KCl, which induced a rapid and sustained contractile response in rabbit aortic smooth muscle, induced a sustained increase in the amount of GTP-bound Rho in this smooth muscle. The time course and amplitude of 60 mM KCl-induced Rho stimulation were similar to that of 3 μM noradrenaline, which gave a similar amplitude of sustained contraction to that by 60 mM KCl. Like KCl-induced contraction, KCl stimulation of Rho was concentration-dependent with a roughly similar dose-response relation to that for contraction. Consistent with the fact that KCl stimulates Rho, we observed that either of the two structurally unrelated Rho kinase inhibitors Y27632 and HA1077 reduced not only noradrenaline-, but also KCl-induced contraction with the similar dose-effect relations. The two Rho kinase inhibitors also suppressed KCl- and noradrenaline-induced MLC phosphorylation in dose-dependent manners. Membrane depolarization with KCl activates the L-type of voltage-dependent Ca^{2+} channels (VDCC) and stimulates entry of extracellular Ca^{2+} into the cell interior. Removal of extracellular Ca^{2+}, which abolished KCl-induced contraction, completely inhibited KCl-induced Rho stimulation. The L-type VDCC blocker nitrendipine also inhibited KCl-induced force generation and Rho stimulation. On the other hand, the Ca^{2+} ionophore ionomycin, which stimulates Ca^{2+} entry into the cell interior, induced a contractile response and Rho activation. These observations strongly suggest that stimulated Ca^{2+} influx across the plasm membrane mediates Rho activation. Calmodulin is a ubiquitous Ca^{2+}-binding protein that mediates many Ca^{2+}-dependent biological processes. The calmodulin antagonist W-7 substantially inhibited

KCl-induced contraction and Rho stimulation. However, the structurally related but inactive analogue W5 did not affect either contraction or Rho stimulation. Ionomycin-induced contraction and Rho activation, like the KCl responses, were inhibited by W7, but not W5. Thus, these observations suggest the involvement of calmodulin in Ca^{2+}-dependent Rho activation in KCl-and ionomycin-contracted muscle. KCl stimulation of rabbit aortic smooth muscle and ionomycin stimulation of cultured rabbit aortic VSM cells induced an increase in the extent of Thr^{695}-phosphorylation of MYPT1. Moreover, adenovirus-mediated expression of a dominant negative RhoA mutant, N^{19}RhoA, inhibited ionomycin-induced Thr^{695}-phosphorylation of MYPT1 and MLC phosphorylation in VSM cells. These data suggest that Ca^{2+}-induced Rho activation stimulates phosphorylation at the inhibitory site of MYPT1 and contributes to MLC phosphorylation.

Excitatory receptor agonists mobilizes Ca^{2+} from both extracellular pool and intracellular pool. It is a likely possibility that excitatory receptor agonist-induced Rho activation could be Ca^{2+}-dependent. To examine this possibility, we tested the dependence on Ca^{2+} and calmodulin of receptor agonist-induced Rho activation. The combination of depletion of the intracellular Ca^{2+} stores with caffeine and removal of extracellular Ca^{2+} is known to effectively abolish agonist-induced Ca^{2+} mobilization.

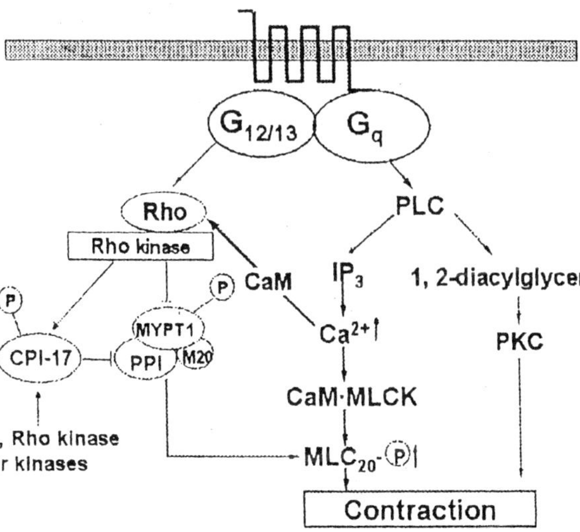

Fig. V-5. Signaling mechanisms for smooth muscle contraction. Excitatory receptor agonists act on G protein-coupled receptors (GPCRs) to stimulate phospholipase C(PLC) via G_q and Rho via both $G_{12/13}$ and G_q. The G_q-mediated Rho activation involves calmodulin (CaM) and a tyrosine kinase. PLC stimulation leads to Ca^{2+}-mobilization, resulting in the activation of the calmodulin-dependent enzyme myosin light chain kinase (MLCK). Rho inhibits myosin phosphatase via Rho kinase, resulting in potentiation of Ca^{2+}-dependent MLC phosphorylation. 1, 2-diacylglycerol (1, 2-DAG) produced via the action of PLC activates protein kinase C, which contributes to contraction independently of or synergistically with Ca^{2+}. The myosin phosphatase inhibitor protein CPI-17 is involved in inhibition of myosin phosphates activity.

This procedure inhibits contraction induced by excitatory receptor agonists including noradrenaline and U46619, a thromboxane A2 mimetic. The same Ca^{2+} depletion procedure strongly inhibited noradrenaline- and U46619-induced Rho stimulation. Noradrenaline-induced contraction and Rho activation were reduced by W7, but not by W5. Thus, receptor agonist-induced Rho activation was Ca^{2+}-dependent (Fig. V-5).

Ca^{2+} ion plays essential roles in eliciting smooth muscle contraction. The best-defined molecular target of Ca^{2+} has been the calmodulin-dependent enzyme, MLCK. Our data now demonstrate that the increase in the $[Ca^{2+}]_i$ also activates Rho, which leads to phosphorylation of myosin phosphatase via Rho kinase and resultant inhibition of myosin phosphatase. Thus, Ca^{2+} ion triggers the two events, i.e. stimulation of the MLC-phosphorylating activity and inhibition of the MLC-dephosphorylating activity. This coordinate activation of the two branches results in effective phosphorylation of MLC; Ca^{2+}-induced negative regulation of myosin phosphatase should potentiate MLC phosphorylation at a given level of $[Ca^{2+}]_i$, compared to a hypothetical situation in which no inhibition of myosin phosphatase would not occur. Thus, Ca^{2+} stimulates the Ca^{2+}-sensitivity of MLC phosphorylation, contributing to contraction. Consistent with this notion, previous studies demonstrated that relatively small increases in the $[Ca^{2+}]_i$ resulted in even large changes in the level of phosphorylated MLC in KCl- or ionomycin-stimulated vascular and nonvascular smooth muscle. However, previous investigations also showed that excitatory receptor activation induces larger increases in the level of phosphorylated MLC than KCl-induced depolarization at a given level of the $[Ca^{2+}]_i$. These observations suggest that excitatory receptor stimulation more effectively induces Ca^{2+}-sensitization, i.e. inhibition of myosin phosphatase, compared to KCl. This may be brought about by more effective Rho activation by receptor agonists compared to KCl in terms of the $[Ca^{2+}]_i$, and/or activation by receptor agonists of a second Rho-independent Ca^{2+}-sensitizing mechanism. The former could be explained by a difference in the Rho activation mechanism between excitatory receptor agonists and KCl; most likely, in addition to the Ca^{2+}-dependent Rho activation, receptor agonists employ the $G_{12/13}$-mediated and also probably the non- Ca^{2+}-dependent G_q-mediated pathways for stimulating Rho-guanine nucleotide exchange factors (see above), which are both demonstrated to operate in non-muscle cell types. The latter Rho-independent Ca^{2+}-sensitization mechanism might include stimulation of the endogenous myosin phosphoatase inhibitor phosphoprotein CPI-17.

In conclusion, we discovered the novel Ca^{2+}-dependent mechanism for activating Rho in smooth muscle. This mechanism probably involves calmodulin. This Ca^{2+}-dependent mechanism for Rho activation appears to operate on stimulation of G_q-coupled vasoconstrictor receptors, in cooperation with the $G_{12/13}$-mediated mechanism. Thus, multitudes of heterotrimeric G protein-mediated signaling pathways regulate Rho activity in smooth muscle, directing the Ca^{2+}-sensitivity regulation of MLC phosphorylation and contraction.

Morano: Could the Rho-kinase mechanism provide explanation for Ca^{2+}-independent activation of non-muscle myosin during prolonged activation of smooth muscle?

Takuwa: In our system, at least myosin light chain phosphorylation observed is Ca^{2+}-dependent. More correctly, if you use cultural cell system for permeabilization, we see a little bit of phosphorylation (~5%) in the absence of Ca^{2+}. Such phosphorylation is actually abolished by Rho-kinase inhibitor. So, it is possible that, in cultural smooth muscle cells, Rho-kinase may account for such Ca^{2+}-independent phosphorylation. But, I think that in vascular smooth muscle, the situation is a little bit different. If we incubate muscle cells for a sufficient period, the basal phosphorylation is almost absent. So, there is difference between cultured smooth muscle cells and smooth muscle tissues.

Pfitzer: We examined GTPγS-sensitization in skinned fibers in the absence of Ca^{2+}, and observed very little increase in force, being much less than that we see in the presence of Ca^{2+}. This suggests that indeed Ca^{2+} is necessary. How is Rho activated by Ca^{2+}?

Takuwa: We have no evidence for Ca^{2+}-dependent Rho activation. But I think there are several possibilities. Rho is a member of Ras superfamily. In the Ras superfamily, there are known examples of GET (quinine nucleotide exchange factor) which can be activatied by calmodulin binding, because some of them have IQ motif and calmodulin binding directly activates GEF activity. Another possibility is that some GEF is subject to phosphorylation by calmodulin-dependent protein kinase, so the calmodulin-dependent phosphorylation may be the second mechanism, but I have no actual data. In smooth muscle tissue, there could be a tissue-specific GFF.

Rüegg: Thank you. This is a very important point.

Morano: I would like to point out that different regulatory mechanisms by the second messenger pathway exist in intact smooth muscle, but they are completely destroyed in skinned fibers, and such mechanisms may not be solved by skinned fiber preparations.

Rüegg: Smooth muscle has been a "headache" muscle as Prof. Sugi mentioned in this meeting. It is indeed a very difficult tissue to study, and its behavior can not be fully explained by the phospholrylation-dephosphorylation mechanism. Probably, we should focus attention on some mechanism regulating the ADP-affinity of cross-bridges, though it may also be regulated by MLC phosphorylation. Anyway, there are still a lot of things to discover and a lot of things to be done. In this connection, I hope Prof. Sugi to organize another meeting in a couple of years. We will know much more about smooth muscle in his next meeting.

CONCLUDING REMARKS

H.E. Huxley[1]

The title of this Symposium reflects the fact that the mechanism of muscle contraction is still a very interesting topic, despite the fact that the basic underlying process – sliding of actin filaments past myosin filaments – was correctly described fifty years ago. This is because it has taken a very great deal of experimental work, and the development of many remarkable new technologies, to elucidate the underlying changes in molecular structure of the myosin crossbridges, which produce the sliding force. And even now we are still some distance from fully understanding how the chemical energy of ATP is used to produce the changes of certain groups of atoms in the myosin polypeptide chain structure which in turn produce lever arm movement and tilting of the myosin crossbridges, which powers the sliding process. Also, the detailed structural kinetics of these changes, and how they vary in different types of muscle, still present many interesting and important mysteries to solve.

For many years, the identification of significant structural changes in the crossbridges was the most intractable problem. When the sliding-filament-mechanism was first described (Huxley & Hanson, 1954; Huxley & Niedgerke, 1954), we were all quite reticent about proposing any mechanism by which the sliding force was developed, save that it was produced at interaction sites or crossbridges between myosin and actin filaments in the overlap region of the A-bands. Later than year (1954), at a Society for Experimental Biology Symposium (published 1955), Jean Hanson and I (Hanson & Huxley, 1955) suggested two possible mechanisms; one involved small changes in the actin axial periodicity coupled to a zipper-like progressive binding to myosin, followed by unbinding in such a way as to shift the actin along in a worm-like manner. The second mechanism involved a relatively large change – we suggested about 130Å – in the axial position of the myosin crossbridges, coupled in the now conventional manner to cycles of attachment and detachment and actin-activated hydrolysis of ATP. We gradually abandoned the first mechanism, because it was difficult to see how it would work at small extents of overlap. This left the second mechanism, and the need to demonstrate a substantial axial shift of the myosin heads during their force-generating interaction with actin.

When we were able to get good electron micrographs of the crossbridges in longitudinal very thin section (Huxley, 1957), we could see that they appeared tilted over a range of angles, but we could not be sure that this was not some artifact of fixation or distortion during sectioning. However, I was emboldened enough to draw a diagram of

[1] H.E. Huxley, Rosenstiel Center (MS-029), Brandeis University, 415 South St., Waltham, MA 02454-9110 USA

such a tilting action in a article published the next year in the *Scientific American* (Huxley 1958).

During the 1960's, Mike Reedy, working in my laboratory, obtained beautiful electron micrographs of insect flight muscle in rigor which clearly showed crossbridges with a systematically tilted appearance. Together with Holmes and Tregear (Reedy, Holmes, and Tregear, 1965), he showed that micrographs of the same muscle in the relaxed state showed crossbridges which were predominantly perpendicular to the filament axis. They also observed a large decrease in intensity of the 14.3 nm meridional X-ray reflection from the crossbridge repeat, as between relaxed insect flight muscle and the same specimen in the rigor state, which was consistent with the change of tilt seen in the micrographs. They therefore suggested that similar changes in tilt (of actin-attached crossbridges) might be responsible for the oscillatory action of these muscles.

Similar polarized 'arrowhead' structures were seen previously in electron micrographs of negatively stained actin filaments 'decorated' with heavy meromyosin (Huxley, 1963) but at that time the complex appearance of these images prevented unambiguous identification of tilted structures. This could only be done when the methods of 3D reconstruction were applied (Moore, Huxley & DeRosier, 1970), and the tilted and skewed attachment of the attached myosin heads was clearly demonstrated.

X-ray studies on contracting frog muscle (Huxley, Brown & Holmes, 1965; Huxley & Brown, 1967) showed that the myosin crossbridges underwent substantial movement when a muscle went from the relaxed to the contracting state. This finding was embodied in the 'tilting, swinging' crossbridge model of 1969 (Huxley, 1969), in which an elongated myosin crossbridge initially attached to actin in an approximately perpendicular orientation to the filament axis, and then either tilted on actin or underwent an equivalent change of shape, so that its distal end was displaced axially relative to the binding site on actin. If that distal end was attached to a fixed axial position on the thick filament – it was envisaged that S2 provided such a link - then the actin filament would be forced to slide past the myosin filament by a distance later characterized as the 'working stroke'.

For a long time it was not possible to obtain experimental evidence that this was what actually happened. A partial breakthrough was achieved in the early 1980's (Huxley *et al.*, 1981, 1983), showing that a large decrease in the intensity of the meridional 14.5 nm reflection from the crossbridges took place during the synchronized movement produced by a step length decrease, consistent with a tilting motion of crossbridges during the working stroke.

The overall concept of actin filament sliding, however, received very vivid and undisputable confirmation when Spudich and his colleagues (Kron & Spudich, 1986; Toyashima *et al*, 1987) first demonstrated that fluorescently labeled individual actin filaments could be seen, in the light microscope, sliding unidirectionally at the expected speeds over surfaces coated with myosin molecules or even isolated myosin subfragment one. This showed decisively that the myosin filament backbone structure played no active role in contraction, save to provide a fixed support.

When the high resolution X-ray crystallographic structure of myosin S1 was finally solved in 1993 (Rayment *et al.*), it could be seen immediately that the molecule possessed a long extension, about 10 nm in length, attached the globular domain, on approximately the opposite side of which was located the actin binding site. Clearly, such an extended structure might function as a lever arm, producing the kind of tilting movement which had been envisaged. Subsequent X-ray analysis of other myosins in the presence of nucleotide analogues mimicking different stages of the ATP hydrolysis cycle have shown that this is indeed the case (Domingues *et al*, 1988; Houdusse *et al.*, 2000.)

But this does not mean that all the problems are now solved. Besides the matter of understanding all the detailed internal molecular mechanico-chemistry to which I have already referred, there is the requirement that we must account for all the detailed properties that we can now measure in an intact muscle in terms of what we can determine about actin and myosin at the molecular level. Good progress is being made here, but there is still much to do. And there is still the lingering suspicion that some additional process, as well as lever arm movement, may be playing a role.

I think this has been an excellent meeting, to take stock of the situation, after fifty years of the sliding model. I very much hope that Dr. Sugi will, despite his official retirement, be able to organize further meetings to discuss progress and problems on this same topic, and to make further important contributions to muscle research. I know that we all want to express our very best wishes to him.

REFERENCES

Dominguez, R., Freyzon, Y., Trybus, K. M., and Cohen, C. (1998). Crystal structure of vertebrate smooth muscle myosin motor domain: visualization of the pre-power stroke state. Cell 94, 559-571.

Hanson, J., and Huxley, H. E. (1955). Structural basis of contraction in striated muscle. Symp. Soc. Exp. Biol. 9, 228-264.

Houdusse, A., Szent-Gyorgyi, A.G., and Cohen C. (2000). Three conformational states of scallop myosin S1. Proc. Natl. Acad. Sci. 97, 11238-11243.

Huxley, H. E. (1957a). The double array of filaments in cross-striated muscle. J. Biophys. & Biochem. Ctyol. 3, 631-648.

Huxley, H.E. (1958). The Contraction of Muscle. Scientific American 199, 66-73.

Huxley, H. E. (1963). Electron microscope studies on the structure of natural and synthetic protein filaments from striated muscle. J. Mol. Biol. 7, 281-308.

Huxley, H. E., Brown, W., and Holmes, K. C. (1965). Constancy of axial spacings in frog sartorius muscle during contraction. Nature 206, 1358.

Huxley, H. E., and Brown, W. (1967). The low angle x-ray diagram of vertebrate striated muscle and its behaviour during contraction and rigor. J. Mol. Biol. 30, 383-434.

Huxley, H. E., Simmons, R. M., Faruqi, A. R., Kress, M., Bordas, J., and Koch, M. H. J. (1983). Changes in the x-ray reflections from contracting muscle during rapid mechanical transients and their structural implications. J. Mol. Biol. 169, 469-506.

Huxley, H. E., Simmons, R. M., Faruqi, A. R., Kress, M., Bordas, J., and Koch, M. H. J. (1981). Millisecond time-resolved changes in x-ray reflections from contracting muscle during rapid mechanical transients, recorded using synchrotron radiation. Proc. Nat. Acad. Sci. 78, 2297-2301.

Huxley, H. E. (1969). The mechanism of muscle contraction. Science 164, 1356-1366.

Huxley, H. E., and Hanson, J. (1954). Changes in the cross-striations of muscle during contraction and stretch and their structural interpretation. Nature, Lond. 173 (4412) 973-6.

Huxley, A. F., and Niedergerke, R. (1954). Structural changes in muscle during contraction. Interference microscopy of living muscle fibres. Nature (London) 173, 971-973.

Kron, S. J., and Spudich, J. A. (1986). Fluorescent actin filaments move on myosin fixed to a glass surface. Proc. Nat. Acad. Sci. 83, 6272-6276.

Moore, P. B., Huxley, H. E., and DeRosier, D. J. (1970). Three-dimensional reconstruction of F-actin, thin filaments and decorated thin filaments. J. Mol. Biol. 50, 279-295.

Rayment, I., Rypniewski, W., Schmidt-Base, K., Smith, R., Tomchick, D., Benning, M., Winkelmann, D., Wesenberg, G., and Holden, H. (1993a). Three-dimensional structure of myosin subfragment-1: a molecular motor. Science 162, 50-58.

Rayment, I., Holden, H. M., Whittaker, M., Yohn, C. B., Lorenz, M., Holmes, K. C., and R.A, M. (1993b). Structure of the actin-myosin complex and its implications for muscle contraction. Science 261, 58-65.

Reedy, M.K., Holmes, K.C., and Tregear, R.T. Induced Changes in Orientation of the Crossbridges of Glycerinated Insect Flight Muscle. Nature 207, 1276-1280.

Toyashima, Y. Y., Kron, S. J., McNally, E. M., Niebling, K. R., Toyashima, C., and Spudich, J. A. (1987). Myosin subfragment-1 is sufficient to move actin filaments in vitro. Nature 328, 536-539.

Photo 1. (First row) J.M. Squire, J.A. Rall, J.C. Rüegg, M.J. Kushmerick, H.E. Huxley, H. Sugi, H. Gonzalez-Serratos, K.C. Holmes, M. Ikebe, G. Pfitzer, S. Winegrad. (Second row) Y. Suzuki, T. Yamada, K. Yamada, M. Endo, Y. Saeki, H. Suga, B. Brenner, I. Ohtsuki, Y. Okamoto, S. Chaen, K. Nakayama. (Third row) H. Onishi, Y. Takuwa, M.A. Bagni, T. Kobayashi, I. Morano, G. Cecchi, H. Okuyama, H. Miyata, M. Ogata, I. Shirakawa. (Fourth row) K. Kinosita, T. Tameyasu, N. Ishii, T. Arata, N. Yagi, D.W. Maughan, K. Katoh, S. Sugiura, S. Ishiwata, Y. Ohnuki.

Photo 2. (First row) S. Winegrad, M.J. Kushmerick, D.W. Maughan, Mrs. Maughan, M. Konishi/ H.E. Huxley/ During scientific session. (Second row) Y. Suzuki, H. Gonzalez-Serratos, M.A. Bagni, G. Cecchi, S. Winegrad/ B. Brenner, G. Pfitzer, I. Morano/ H. Suga, S. Winegrad.

Photo 3. (First row) J.W. Squire/ H. Sugi/ G.H. Pollack/ I. Ohtsuki, H. Sugi, I. Shirakawa, Mrs. Huxley, H.E. Huxley, H. Gonzalez-Serratos, H. Onishi, K.C. Holmes. (Second row) M.J. Kushmerick, M.A. Bagni, N. Yagi, S. Ishiwata, J.A. Rall, G. Cecchi, G.H. Pollack/ N. Ishii, J.C. Rüegg/ Mrs. Maughan, D.W. Maughan, K.C. Holmes, Mrs. Holmes, S. Winegrad, Mrs. Huxley, H.E. Huxley, H. Sugi, Mrs. Sugi.

Participants

Arata, T.
Department of biology
Graduate School of Science
Osaka University
Toyonaka, Osaka 560-0043, Japan

Brenner, B.
Department of Molecular and Cell Physiology
Medical School Hanover
D-30625 Hanover, Germany

Bagni, M.A.
Dipartimento di Scienze Fisiologiche
Universita di Firenze
I-50134 Firenze, Italy

Cecchi, G.
Dipartimento di Scienze Fisiologiche
Università di Firenze
I-50134 Firenze, Italy

Chaen, S.
Department of Applied Physics
College of Humanities and Sciences
Nihon University
Setagaya-ku, Tokyo 156-8550, Japan

Endo, M.
Department of Pharmacology
Saitama Medical School
Saitama-ken 350-0495, Japan

Gergely, J.
Boston Biomedical Research Institute
64 Grove Street,
Watertown, MA 02472, U.S.A.

Gonzalez-Serratos, H.
Department of Physiology
School of Medicine
University of Maryland
Baltimore, MD 21201-1596, U.S.A.

Holmes, K.C.
Max Planck Institute for Medical Research
Abteilung Biophysik
D-69120 Heidelberg, Germany

Huxley, H.E.
Rosenstiel Center
Brandeis University
Waltham, MA 02454, U.S.A.

Ikebe, M.
Department of Physiology
University of Massachusetts
Medical School
Worcester, MA 01655 U.S.A.

Ishii, N.
Department of Life Sciences
Graduate School of Arts and Sciences
University of Tokyo
Meguro-ku, Tokyo 153-0041, Japan

Ishiwata, S.
Advanced Research Institute for
Science and Engineering
Waseda University
Shinjuku-ku, Tokyo 169-8555, Japan

Katayama, E.
Division of Biomolecular Imaging
Institute of Medical Science
University of Tokyo
Minato-ku, Tokyo 108-8639, Japan

Katoh, K.
Neuroscience Research Institute
National Institute of Advanced Industrial
Sciences and Technology
Tukuba, Ibaraki-ken 305-8568, Japan

Kinosita, K.
Center for Integrative Bioscience
Okazaki National Research Institutes
Okazaki, Aichi-ken 444-8585, Japan

Kobayashi, T.
Department of Electronic Science
Shibaura Institute of Technology
Minato-ku, Tokyo 108-8548, Japan

Konishi, M.
Department of Physiology
Tokyo Medical University
Shinjuku-ku, Tokyo 160-8402, Japan

Kushmerick, M.J.
Department of Radiology
University of Washington
Seattle, WA 98195-7115, U.S.A.

Maughan, D.W.
Department of Molecular Physiology and Biophysics
University of Vermont
Burlington, VT 05405, U.S.A.

Miyata, H.
Physics Department
Graduate School of Science
Tohoku University
Sendai, Miyagi-ken 980-8578, Japan

Morano, I.
Max-Delbrück-Center for Molecular Medicine
D-13125 Berlin, Germany

Nakayama, K.
Department of Pharmacology
Faculty of Pharmaceutical Sciences
Shizuoka Prefectural University
Shiuoka, Shizuoka-ken 422-8526, Japan

Ogata, M.
Institute of Health Science
Kyushu University
Fukuoka, Fukuoka-ken 810-0022, Japan

Ohtsuki, I.
1-14-12 Gakuen-cho, Higashikurume-shi
Tokyo 203-0021, Japan

Okamoto, Y.
Department of Applied Chemistry
Muroran Institute of Technology
Muroran, Hokkaido 050-8585, Japan

Okuyama, H.
Department of Physiology
Kawasaki Medical School
Okayama-ken 701-0192, Japan

Onishi, H.
Department of Structural Analysis
National Cardiovascular Research Center
Suita-shi, Osaka 565-8565, Japan

Pfitzer, G.
Department of Physiology & Pathophysiology
University of Cologne
D-50931 Koeln, Germany

Pollack, G.H.
Department of Bioengineering
University of Washington
Seattle, WA 98195, U.S.A.

Rall, J.A.
Department of Physiology & Cell Biology
Ohio State University
Columbus, OH 43210, U.S.A.

Rüegg, J.C.
Physiologishes Institut
University of Heidelberg
D-69120 Heidelberg, Germany

Saeki, Y.
Department of Physiology,
School of Dental Medicine
Tsurumi University
Yokohama-shi, Kanagawa-ken 230-8501, Japan

Shirakawa, I.
Department of Physiology
School of Medicine
Teikyo University

Itabashi-ku, Tokyo 173-8605, Japan

Squire, J.M.
Biological Structure & Function Section
Biomedical Sciences Division
Imperial College
London SW7 2AZ, U.K.

Suga, H.
National Cardiovascular Center
Research Institute
Suita-shi, Osaka 565-8565, Japan

Sugi, H.
Department of Physiology
School of Medicine
Teikyo University
Itabashi-ku, Tokyo 173-8605, Japan

Sugiura, S.
Institute of Environmental Studies
Graduate School of Frontier Sciences
University of Tokyo
Bunkyo-ku, Tokyo 113-0033, Japan

Takuwa, Y.
Department of Molecular and Cell Physiology
Graduate School of Medicine
Kanazawa University
Kanazawa-shi, Ishikawa-ken 920-8640, Japan

Tameyasu, T.
Department of Physiology
School of Medicine
St. Marianna University
Kawasaki-shi, Kanagawa-ken 213-8511, Japan

Tanokura, M.
Department of Applied Biological Chemistry
Graduate School of Agricultural and Life Sciences
University of Tokyo
Bunkyo-ku, Tokyo 113-8657, Japan

Tsuchiya, T.
Department of Biology
Faculty of Science
Kobe University
Kobe-shi, Hyogo-ken 657-8501, Japan

Watabe, S.
Graduate School of Agricultural and Life Sciences
University of Tokyo
Bunkyo-ku, Tokyo 113-8657, Japan

Winegrad, S.
Department of Physiology
University of Pennsylvania
School of Medicine
Philadelphia, PA 19104-6085 U.S.A.

Yagi, N.
Japan Synchrotron Radiation Research Institute (Spring-8)
Sayo-gun, Hyogo-ken 679-5198, Japan

Yamada, K.
Department of Physiology
Oita Medical University
Oita-ken 879-5593, Japan

Yamada, T.
Department of Physics
Faculty of Science
Tokyo University of Science
Shinjuku-ku, Tokyo 162-8601, Japan

Yamashita, H.
Department of Cardiovascular Medicine
Graduate School of Medicine
University of Tokyo
Bunkyo-ku, Tokyo 113-8655, Japan

SUBJECT INDEX

ntries indicate pages on which descriptions on the subjects are readily accessible.